MICROBIOLOGY

QUESTIONS AND ANSWERS

FIFTH EDITION

ABOUT THE AUTHORS

Dr. Purshotam Kaushik is Former Professor and Head, Department of Botany & Microbiology and Dean, Faculty of Life Sciences at Gurukul Kangri University, Haridwar. He has had teaching and research stints at Panjab University, Chandigarh and Guru Nanak Dev University, Amritsar. He has also been a Visiting Professor at University of Delhi.

Dr. Purshotam Kaushik has been the Principal Investigator of the project *Environmental Biology of the Himalayan Orchids* conducted by Ministry of Environment and Forests, Government of India, and also for a project on *Lectins* conducted by U.G.C. He has visited Universities of London, Cambridge & Kew (United Kingdom) and Institute Pasteur, Paris (France).

Dr. Purshotam Kaushik has supervised 30 Ph.D. students and has published over hundred research papers. He has authored five original research monographs and two editions of the edited book *Indigenous Medicinal Plants, Including Microbes and Fungi.* He has also written a textbook on *Introductory Microbiology*. He is the President of Academy of Plant Sciences, India and has also been an elected Syndic of Gurukul Kangri University and a member of Board of Management and that of Academic Council. He has also been a member of Board of Studies of many Universities. He has been associated with designing and starting M.Sc. Microbiology courses at Gurukul Kangri University from the very inception of the courses.

Dr. Kirti Kaushik is an active researcher, currently working at La Jolla Institute for Allergy and Immunology, La Jolla, California, USA. She has taught microbiology at Rasmussen College and has worked as a Postdoctoral Research Associate at University of Minnesota, USA. She holds a doctorate from Postgraduate Institute of Medical Education and Research, Chandigarh and is a Life Member of Association of Microbiologists of India; Indian Science Congress Association; and has close Association with many academic societies.

Dr. Kirti Kaushik is recipient of multiple academic awards and has delivered invited talks at national and international conferences/ Universities. She is a prolific writer and has published many research papers in prestigious national and international journals.

MICROBIOLOGY

QUESTIONS AND ANSWERS

For Graduate & Postgraduate courses and various Competitive Examinations

FIFTH EDITION

PURSHOTAM KAUSHIK
MSc, PhD
Certificate Course in German, FAPS, FBS
Formerly Professor & Head
Department of Botany and Microbiology
& Dean, Faculty of Life Sciences
Gurukul Kangri University, Haridwar
Uttarakhand, India

KIRTI KAUSHIK
PhD
La Jolla Institute for Allergy and Immunology
La Jolla, California, U.S.A.

S. CHAND
PUBLISHING

S Chand And Company Limited
(ISO 9001 Certified Company)

S Chand And Company Limited
(ISO 9001 Certified Company)
Head Office: D-92, Sector–2, Noida – 201301, U.P. (India), Ph. 91-120-4682700
Registered Office: A-27, 2nd Floor, Mohan Co-operative Industrial Estate, New Delhi – 110 044, Phone: 011-49731800
www.**schandpublishing.com**; e-mail: **info@schandpublishing.com**

Marketing Offices:

Chennai : Ph: 23632120; chennai@schandpublishing.com
Guwahati : Ph: 2738811, 2735640; guwahati@schandpublishing.com
Hyderabad : Ph: 40186018; hyderabad@schandpublishing.com
Jalandhar : Ph: 4645630; jalandhar@schandpublishing.com
Kolkata : Ph: 23357458, 23353914; kolkata@schandpublishing.com
Lucknow : Ph: 4003633; lucknow@schandpublishing.com
Mumbai : Ph: 25000297; mumbai@schandpublishing.com
Patna : Ph: 2260011; patna@schandpublishing.com

First Edition 2000

Subsequent Editions and Reprints Second Edition 2004, Third Edition 2008, Fourth Edition 2012, Fifth Edition 2018, Reprint 2019, 2021

Reprint 2022

ISBN: 978-93-528-3122-7 **Product Code:** H6MBQ68BIOL20ENAE18O

PRINTED IN INDIA

By Vikas Publishing House Private Limited, Plot 20/4, Site-IV, Industrial Area Sahibabad, Ghaziabad – 201 010 and Published by S Chand And Company Limited, A-27, 2nd Floor, Mohan Co-operative Industrial Estate, New Delhi – 110 044.

Dedicated to the memory of

'Padmashree' Dr. P.N. Mehra (1907 – 1994)

DSc, FNA, FNA Sc

Professor Emeritus

Panjab University, Chandigarh

FOREWORD

I am happy to write a foreword to 5th edition of "Microbiology: Questions and Answers" by Dr. Purshotam Kaushik, Formerly Professor & Head, Department of Botany & Microbiology and Dean, Faculty of Life Sciences, Gurukul Kangri University, Haridwar; and Dr. Kirti Kaushik, an active researcher at La Jolla Institute for Allergy and Immunology, La Jolla, California, U.S.A.

The microbiology is considered to be one of the most applied branches of biology. It provides not merely information regarding cellular and non-cellular microorganisms, rather more important aspects of life including microbial genetics and molecular biology, which are comparatively much difficult to study in higher plants and animals. Some of the strange characteristics, viz. survival of bacterial spores, concept of enzyme energetics and metabolism make it more fascinating. Today, we not only understand viruses but viroids and prions also. The production of several important chemicals besides antibiotics, vitamins, and amino acids, and brewing find application in large scale as well as small scale in the industry.

This book provides chapters on history of microbiology, microscope, microbial diversity, the bacteria, the Bergey's Manual, the viruses, ribosomal RNA in microbial taxonomy, cultivation and growth of microorganisms, immunology, microbial genetics, microbial biotechnology, industrial microbiology, food and dairy microbiology, soil microbiology, water and waste water microbiology, biological control, mycorrhizae, biosensors, and applications of microbiology.

Prof. Purshotam Kaushik who has already authored an excellent book "Introductory Microbiology" which was first published by Emkay Publications, Delhi in 1996, has several research monographs and books to his credit and is an established author.

Prof. Kaushik received his early training under a distinguished biologist Prof. P.N. Mehra, and has professed teaching and research at Panjab University, Chandigarh, and Guru Nanak Dev University, Amritsar. He has also visited the Universities of London, Cambridge and Kew (United Kingdom) and Institute Pasteur, Paris (France), and is highly experienced in this field.

The book provides latest information in the field and will be informative and useful for graduate as well as postgraduate students besides, those interested in objective learning of microbiology for competitive examinations.

7 July 2017

Prof. P.K. Khosla
Vice-Chancellor
Shoolini University of Biotechnology
and Management, SOLAN, (HP)
& Formerly Vice-Chancellor
HP Krishi Vishwavidyalaya
Palampur, HP 176062

PREFACE TO THE FIFTH REVISED EDITION

One of us (Purshotam Kaushik) remained associated with designing and starting postgraduation courses in Microbiology at Gurukul Kangri University, Haridwar, from very inception of the courses, in early eighties of the last century and gave rigorously prepared lessons to the students with clarity of the subject that ultimately led to the 5th edition of "Microbiology: Questions and Answers". This probably was the first book of its type in style, splendid in quality and presentation of the subject perhaps the world over. The book won appreciation of the readers, within the country and abroad which, of course, prompted the authors to bring out the 5th edition.

The effort has been to follow the U.G.C. recommended syllabus of the courses. The book has been made more informative and the following new chapters have been added on *Animal Cloning* (Chapter 27); *Prokaryotes* (Chapter 8); *Viruses* (Chapter 12); *Normal Microflora, Symbiotic Relationship and Pathogenicity* (Chapter 16); *Bacterial Pathogens and Diseases* (Chapter 21); *Algae, Protozoa and Multicellular Parasites* (Chapter 19); *Infectious Diseases and Diagnostic Immunological Techniques* (Chapter 20); *Fungal Diseases or Mycoses* (Chapter 22) *and Metabolism: Catabolism and Anabolism* (Chapter 49).

This textbook will be useful for the students undergoing graduation and postgraduation courses besides those studying microbiology in the courses of pharmaceutical Sciences, BLT, MLT, Nursing and other Paramedical courses.

21 July 2017 **Purshotam Kaushik**

Webpage: purshotam.kaushik.googlepages.com **Kirti Kaushik**

PREFACE

GENESIS

In my teaching experience of more than four decades, no doubt I came across several good books of microbiology, but what I missed dearly was a study material of some sort of questions and answers which I could give to my students so that they could do self-assesment. Therefore, an idea came to my mind to prepare such a text myself which would contain in ink, my live and interactive experiences with students for over four decades. It was conceived for a text which would help students to study and test themselves, and revise on their own. Thus was born "Microbiology: Questions and Answers."

Microbiology is one of the recent facets of science. Its emergence and major developments took place in the latter half of the 19th century and the beginning of the 20th century, a period of almost 50 years, and is referred to as golden age of microbiology. Further development of microbiology is expected to usher in a "New golden age" in the 21st century. The teaching courses of microbiology have been started by different colleges and universities at both graduation and postgraduation levels. The microbiology being one of the applied branches of science has a vast scope for interdisciplinary developments.

The present text has been prepared to create quest for knowledge among the students undergoing introductory courses of microbiology both at graduation and postgraduation levels. The outstanding features of the subject have been presented in the form of objective type questions and answers. This has made the subject interesting and precise for better understanding. Almost all aspects of microbiology have been covered and the latest information in the field has been provided. The special features include development of microbiology in India, applications of microbiology, revealing microbial diversity, the concept of microbiological species, distribution of microorganisms in nature, classification of microorganisms, DNA hybridization, 16S rRNA sequencing, bacterial nomenclature, modern trends in bacterial taxonomy, Bergey's system of bacterial classification, general features of classification of bacteria, cyanobacteria, rickettsia, chlamydia, mycoplasmas, algae, archaeobacteria, actinomycetes, fungi and protozoa.

The nutritional types of microorganisms give information about autotrophs, heterotrophs, psychrophiles, mesophiles, thermophiles, hyperthermophiles, and nutritional requirement of microorganisms.

The microbial growth includes the growth curve of bacteria, measurement of growth-mass and volume measurement, cell membrane and indirect methods in addition to parameters of growth and growth in batch and continuous culture.

Purshotam Kaushik
Professor & Head
Department of Botany and Microbiology
Gurukul Kangri University
Haridwar 249404

ACKNOWLEDGEMENT

Dr. Amit Sharma, M.S., Ph.D. (Michigan State University), San Diego, California, USA.

Prof. S.C. Verma, Prof. C.L. Mandhar, Prof. M.P. Sharma and Prof. J.K. Gupta, Panjab Univeristy, Chandigrah.

Dr. A.K. Puniya, Dean, College of Dairy Science & Technology, Guru Angad Dev Veterinary & Animal Sciences University, Ludhiana - 141004.

Dr. R.S. Sandhu, Emeritus Professor, V.P. Chest Institute, Delhi University Campus.

Prof. A.K. Thukral, Prof. G.S. Virk, and Prof. Harvinder Singh Saini, Guru Nanak Dev University, Amritsar.

Prof. R.D. Gaur, H.N.B. University, Srinagar-Garhwal.

Prof. L.K. Pareek, University of Rajasthan, Jaipur.

Prof. A. Bohra and Prof. Anil Vyas, J.N.V. University, Jodhpur.

Dr. Ashish Bhatnagar and Dr. Monica Bhatnagar, M.D.S. University, Ajmer.

Prof. S.M. Singh, Prof. Karuna Verma, Prof. P.K. Singhal and Prof. Y.K. Bansal, R.D. University, Jabalpur.

Dr. Karuna Singh, B.H.U., Varanasi.

Prof. Dinesh Kumar, University of Lucknow, Lucknow.

Prof. A.K. Srivastava, Prof. Vimla Y., and Prof. M.U. Charaya C.C.S. University, Meerut.

Prof. Ravi Prakash and Prof. J.P. Yadav, M.D. University, Rohtak.

Prof. P. Gyananath, S.R.T. Marathawada. University, Nanded-431606.

Prof. K.M. Vyas and Prof. P.C. Jain, H.S.G. University, Sagar.

Prof. F.C. Garg, H.A.U., Hissar.

Late Prof. A. Mahadevan, Madras University, Chennai.

Prof. Javed Musarrat, A.M.U., Aligarh

Prof. K. Janardhan and Prof. S. Manian, Coimbatore.

Prof. Yukimasa Nishimura, Department of Applied Biological Sciences, Science University of Tokyo, 2641 Tamazaki, Noda 278, Japan.

Prof. Yogendra Singh and Dr. V.C. Kalia, Institute of Genomics & Integrative Biology, Delhi.

Prof. Subhavrat Majumdar, Bose Institute, Calcutta.

Prof. B.P. Kapadnis, University of Poona, Pune.

Prof. Sita Naik, Department of Immunology, Sanjay Gandhi Postgraduate Institute of Medical Sciences, Raebareli Road, Lucknow.

Dr. R.C. Tripathi, C.D.R.I., Lucknow.

Dr. Meena Mukherjee and Dr. Ujjala Dasgupta, Indian Institute of Chemical Biology, Calcutta.

Prof. A.K. Bhatnagar, University of Delhi, Delhi.

Dr. Madhu Bala, DRDO, Institute of Nuclear Medicine & Allied Sciences, Delhi 110054.

Dr. Anant Rai, I.V.R.I., Izatnagar.

Dr. N. Gopalkrishnan, Regional Research Laboratory (CSIR), Thiruvananthpuram (Kerala).

Prof. T.N. Lakhanpal, Prof. T.C. Bhalla, Dr. Arvind Bhatt, and Prof. Dunichand H.P. University, Shimla.

Prof. Syed Akhtar Husain, J.M.I., New Delhi.

Dr. Surbhi Kaushik, Senior Microbiologist, London, United Kingdom.

Dr. Sachin Tyagi, Yatharth Wellness Hospital, Noida.

Dr. Pankaj Goyal and Dr. Abhishek Chauhan, Amity University, Uttar Pesdesh, Noida.

CONTENTS

1
CHAPTER

HISTORY OF MICROBIOLOGY

Q.1. Which is the first book in the world to describe a fungus?

Ans. The *Rigveda* is the first book in the world to describe the fungus mushroom in Sanskrit hymns in the form of similes by comparing the head of a devil or a wicked person with the cap (**pileus**) of a mushroom which can be easily blown away even by a light current of air.

Q.2. Who discovered 'little boxes' or 'cells'?

Ans. Robert Hooke, an Englishman discovered cells in 1665.

Q.3. Who first described **'animalcules'**?

Ans. Anton van **Leeuwenhoek** was perhaps the first to describe microorganisms which he observed through magnifying lenses and wrote a series of letters on them between 1673 and 1723 to the Royal Society of London. He regarded them as **'animalcules'** which have been identified from his drawings to be bacteria and protozoa.

Q.4. Which microorganisms are isolated from the well known **'sourdough bread'** which is eight to ten times more acidic than conventional breads?

Ans. *Lactobacillus sanfrancisco* and *Saccharomyces exiguus*.

Q.5. What were the main characteristics of the microscopes made by Anton van Leeuwenhoek?

Ans. They consisted of a rectangular plate with a lens and a mounting needle which was adjustable with butterfly like knobs. The specimen was placed on the tip of the adjustable needle and viewed from the other side of a tiny and nearly spherical lens. The maximum magnification with Leeuwenhoeks, microscopes was 300 times (Fig. 1.1).

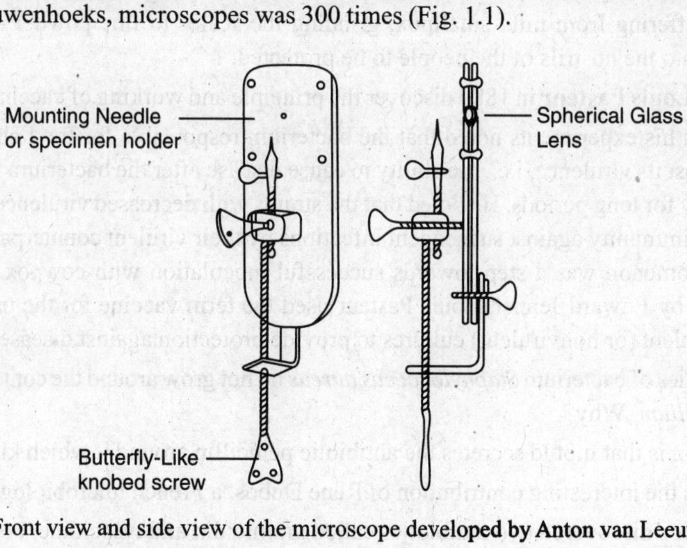

Fig. 1.1. Front view and side view of the microscope developed by Anton van Leeuwenhoek.

Leeuwenhoek (1632 – 1723) was a cloth merchant and draper, in Delft, Holland, who developed simple but very powerful microscopes of silver and copper plates. His microscope comprised of a spherical glass lens held by two metallic plates, a mounting needle or specimen holder, and screws with butterfly like knob for positioning and focussing the specimen. He sent drawings of some of his observations in the form of a series of letters to the Royal Society of London, which were published and he was honoured with fellowship of the society i.e., (FRS).

Q.6. Name the supporter of wrong **spontaneous generation** or **abiogenesis.**

Ans. John Needham in 1745.

Q.7. Who were the supporters of **biogenesis**?

Ans. Francesco Redi, an Italian, set in experiments in 1668, against spontaneous generation. Lazzaro Spallanzani also an Italian and Rudolf Virchow a German challenged spontaneous generation with the concept of biogenesis and finally it was confirmed by Louis Pasteur.

Q.8. How did Pasteur disprove the theory of spontaneous generation with his swan-necked flask?

Ans. Pasteur initially poured beef broth into a long necked flask. In the second step he heated the long neck of flask and bent it into an S-shaped curve. He found that microorganisms did not appear in the cool solution even after a long period. Some of such original vessels of Pasteur which were later on sealed did not show signs of contamination even after a period of 100 years.

Q.9. Which period is regarded as the **'Golden age of microbiology'**?

Ans. The period between 1857 and 1914.

Q.10. Who performed the first **vaccination?**

Ans. Edward Jenner, a young British physician was the first to vaccinate in 1798 with scrapings collected from cowpox blisters. He inoculated a healthy volunteer with cowpox material by scratching the person's arms with a pox contaminated needle. The scratch turned into a raised bump. The volunteer became mildly sick within a few days and recovered, but never contracted either cowpox or smallpox.

Q.11. How did the physicians in ancient China protect people against smallpox?

Ans. Ancient Chinese physicians immunized patients by collecting scales from drying pustules of people suffering from mild smallpox, grinding the scales to fine powder and inserting the powder into the nostrils of the people to be protected.

Q.12. How did **Louis Pasteur** in 1880 discover the principle and working of vaccination?

Ans. Pasteur in his experiments noted that the bacterium responsible for fowl cholera or chicken cholera lost its virulence, i.e., the ability to cause disease after the bacterium was grown in the laboratory for long periods. He noted that the strains with decreased virulence were capable of inducing immunity against subsequent infections by their virulent counterparts. Discovery of this phenomenon was a step towards successful inoculation with cowpox material against smallpox by Edward Jenner. Louis Pasteur used the term vaccine for the use of inoculation with avirulent (or nonvirulent) cultures to provide protection against disease.

Q.13. The colonies of bacterium *Staphylococcus aureus* do not grow around the contaminating colony of *Penicillium.* Why?

Ans. The reason is that mould secretes the antibiotic penicillin around it which kills the bacteria.

Q.14. What was the interesting contribution of Rene Dubos, a French microbiologist?

Ans. Rene Dubos discovered two antibiotics called gramicidin and tyrocidine, both of which were produced by the bacterium *Bacillus brevis*, cultured from soil.

Q.15. In which year did **Edward Jenner** use the first vaccine?

Ans. In the year 1798.

Q.16. How did Rebecca Lancefield classify streptococci into serotypes in 1933?

Ans. Lancefield classified various serotypes using an alphabetical system (A through O). Streptococci are responsible for a variety of diseases like sore throat (strep throat), scarlet fever and septicemia (blood poisoning).

Q.17. Who discovered in 1892 that tobacco mosaic was caused by a virus which is passed through filters fine enough to stop all known bacteria?

Ans. Dimitri Ivanovski.

Q.18. What did Wendell Stanley demonstrate in 1935?

Ans. Wendell Stanley in 1935 demonstrated that an organism called tobacco mosaic virus was so simple and homogeneous that it could be crystallized like a chemical compound.

Q.19. Which age is the **golden age of microbiology**?

Ans. The period from 1860 to 1910 tabulated below is known as the golden age of microbiology because maximum discoveries were made in that period.

Table 1.1 Golden age of microbiology

Year	*Discoverer*	*Discovery*
1857	Louis Pasteur	Microbiology of lactic acid fermentation
1861	Louis Pasteur	Disproved spontaneous generation
1864	Louis Pasteur	Pasteurization
1867	J. Lister	Aseptic surgery
1870	Abbe	Abbe condenser and oil objective.
1877	Robert Koch	Proved that anthrax is caused by bacterium
1879	A. Neisser	Discovered the cause of gonorrhoea
1880	Louis Pasteur	Developed vaccines and treatment of rabies
1880	K.J. Eberth	Causal agent of typhoid fever
1881	Robert Koch	Prepared pure cultures for the study of bacteria
1881	A. Ogston	Staphylococci cause wound infection
1882	Robert Koch	Discovered cause of tuberculosis (*Mycobacterium tuberculosis*)
1883	Robert Koch	Cause of cholera (*Vibrio cholerae*)
1883	T. Klebs	Discovered cause of diphtheria
1884	E. Metchnikoff	Phagocytosis by white blood cells
1884	Christian Gram	Gram staining technique
1884	Escherich	Discovered *Escherichia coli*
1885	A. Nicolaier	Cause of tetanus (*Clostridium tetani*)
1886	A. Fraenkel	Cause of bacterial pneumonia
1887	D. Bruce	Cause of Malta fever
1887	Petri	Petri dish or Petri plate
1887	A. Weichselbaum	Cause of meningitis
1890	E. von Bering & S. Kitasato	Discovered bacterial toxins and how to develop antitoxins.

1890	S. Winogradsky	Autotrophic growth of chemolithotrophs
1890	P. Ehrlich	Theory of immunity
1892	W. Welch and G. Nuttal	Cause of gas gangrene
1892	D. Ivanovski	First demonstration of a virus
1894	A. Yersin	Discovered the cause of plague (*Yersinia pestis*)
1896	E. van Ermengen	Cause of botulism food poisoning
1898	K. Shiga	Cause of dysentery (*Shigella dysenteriae*)
1898	E. Nocard & E.R. Roux	Pleuropneumonia of cattle by *Mycoplasma mycoides*
1900	W. Reed	Yellow fever was transmitted by mosquitoes and caused by a virus
1901	Maritnus Beijerinck	Enrichment culture method
1902	Karl Landsteiner	Human blood groups
1905	F.R. Schaudin & P.E. Hoffman	Discovered syphilis
1906	J. Bordet & O. Gengou	Discovered whooping cough caused by *Bordetella pertussis*
1908	Paul Ehrlich	Chemotherapeutic agents
1909	H.T. Ricketts	In Rocky mountains spotted fever caused by *Rickettsia rickettsii*

Q.20. List the Nobel Laureates in the field of Microbiology and their contributions.

Ans. See Table 1.2.

Table 1.2 Selected Nobel laureates in Microbiology and their contributions

Nobel Laureates	*Country of origin*	*Year of presentation*	*Contribution*
Emil A von Behring	Germany	1901	For developing a diphtheria antitoxin
Robert Koch	Germany	1905	Cultured causative agent of tuberculosis
Paul Ehrlich	Germany	1908	Stated theories on immunity
Elie Metchnikoff	Russia	1908	Described phagocytosis, i.e., intake of solid materials by cells
Alexander Fleming	Scotland	1945	Discovery of penicillin
Ernst Chain	England	(Jointly)	
Howard Florey	England		
Selman A Waksman	Ukrain	1952	Discovered streptomycin
Hans A Krebs	Germany	1953	Discovered chemical steps involved in Krebs cycle in carbohydrate metabolism
John F. Enders	USA	1954	Cultured poliovirus in cell cultures
Thomas H. Weller	USA	(Jointly)	
Frederick C Robbins	USA		
Joshua Lederberg	USA	1958	Genetic control of biochemical reactions
George Beadle	USA	(Jointly)	in *Neurospora*, a fungus

Nobel Laureates	*Country of origin*	*Year of presentation*	*Contribution*
Edward Tatum	USA		
James D. Watson	USA	1962	Identified the physical structure of DNA
Frances H.C. Cricks	England	(Jointly)	
Maurice A.F. Wilkins	New Zealand		
Francois Jacob	France	1965	Described regulation of protein synthesis
Jacques Monod	France	(Jointly)	in bacteria.
Andre Lwoff	France		
Robert Holley	USA	1968	Discovered genetic code for amino acids.
Har Gobind Khorana	India	(Jointly)	
Marshall Nirenberg	USA		
Max Delbruck	Germany	1969	Mechanisms of viral infection of bacterial
Alfred D. Hershey	USA	(Jointly)	cells.
Salvador E. Luria	Italy		
Gerald M. Edelman	USA	1972	Gave nature and structure of antibodies.
Rodney R. Porter	England	(Jointly)	
Renato Dulbeco	USA	1975	Discovered the enzyme reverse transcriptase
Howard Temin	USA	(Jointly)	and described cancer causing process of RNA
David Baltimore	USA		viruses.
Daniel Nathans	USA	1978	Described the action of restriction enzymes
Hamilton Smith	USA	(Jointly)	biological tools of genetic engineering.
Werner Arber	Switzerland		
Peter Mitchell	England	1978	Described chemiosmotic mechanisms for ATP synthesis.
Paul Berg	USA	1980	Conducted experiments in gene splicing (used in genetic engineering).
Walter Gilbert	USA	1980	Invented method of DNA sequencing.
Aaron Mug	South Africa	1982	Described the structure of TMV (Tobacco mosaic virus).
Barbara Mc Clintock	USA	1983	Discovered transposons.
Cesar Milstein	Argentina	1984	Developed technique for producing monoclonal antibodies.
George IF Kohler	Germany	(Jointly)	
Niels Kai Jerrie	Denmark		
Susumu Tonegava	Japan	1988	Discovered genetics of antibody production.
Johann Deisenhofer	Germany	1988	Described the structure of bacterial
Robert Huber	Germany	(Jointly)	photosynthetic pigments.
Hartmut Michel	Germany		
J. Michael Bishop	USA	1989	Discovered oncogenes (cancer causing genes).
Harold E. Varmus	USA	(Jointly)	

Joseph E. Murray	USA	1990	Performed first successful transplants using inummosuppressive agents.
E. Donnall Thomas	USA	(Jointly)	
Edmond H. Fisher	USA	1992	Discovered protein Kinases (the enzymes which regulate cell growth).
Edvin G. Krebs	USA	(Jointly)	
Richard J. Roberts	U.K.	1993	Revealed that genes can be separated on different segments of DNA.
Philip A. Sharp	USA	(Jointly)	
Kary B. Mulis	USA	1993	Discovery of polymerase chain reaction
Michael Smith	Canada	1993	Discovered site-directed mutagenesis.
Alfred G.,Gilman, Martin Rodbell	USA	1994	For their discovery of **G-protein** and the role of these proteins in signal transduction in cells
Edward B. Lewis	USA		
Christiane Nusslein -Volhard Eric F. Wieschaus	Germany USA	1995	For their discoveries concerning the genetic control of early **embryonic development**
Peter C.Doherty Rolf M.Zinkernagel	Australia Switzerland	1996	For their discoveries concerning the specificity of the cell mediated immune defence
Stanley B.Prusiner	USA	1997	For his discovery of **Prions** a new biological principle of infection
Robert F. Furchgott Louis J. Ignarro Ferid Murad	USA	1998	For their discoveries concerning nitric oxide as a signalling molecule in the cardiovascular system
Gunter Blobel	USA	1999	For the discovery that protein have intrinsic signals that govern their transport and localization in the cell
Arvid Carlsson Paul Greengard Eric R. Kandel	Sweden USA USA	2000	For their discoveries concerning **signal transduction** in the nervous system
Leland H. Hartwell R. Timothy (Tim) Hunt Sir Paul M. Nurse	USA UK UK	2001	For their discoveries of **key enzyme regulators** of the cell cycle
Sydney Brenner H.Robert Horvitz John E.Sulston	South Africa UK USA UK	2002	For their discoveries concerning genetic regulation of organ development and **programmed cell death**
Paul Lauterbur Sir Peter Mansfield	USA UK	2003	For their discoveries concerning **magnetic resonance** imaging

Richard Axel Linda B.Buck	USA	2004	For their discoveries of **odorant receptors** and the organization of the olfactory system
Barry J. Marshall J. Robin Warren	Australia	2005	For their discovery of the bacterium ***Helicobacter pylori*** and its role in gastritis and peptic ulcer disease
Andrew Z. Fire Craig C. Mello	USA	2006	For their discoveries of **RNA interference** gene silencing by double-stranded RNA
Mario Capecchi Sir Martin Evans Oliver Smithies	USA UK USA	2007	For their discoveries for introducing specific gene modification in mice by the use of **embryonic stem cells**
Harald zur Hausen	Germany	2008	For his discovery of human papilloma viruses causing **cervical cancer**"
Elizabeth H. Blackburn, Carol W. Greider and J ack W. Szostak	Australia USA UK	2009	For the discovery of how chromosomes are protected by telomeres and the enzyme **telomerase**
Robert G. Edwards	UK	2010	For the development of *in vitro* fertilization (IVF)
Bruce A. Beutler Jules A. Hoffman	USA Luxemborg	2011	For their discoveries Concerning activation of innate immunity
Ralph M. Steinman	Canada		For his discovery of the dendritic cells and its role in adaptive immunity
Sir John B. Gurdon Shinya Yamanaka	UK Japan	2012	For the discovery that mature cells can be reprogrammed to become pluripotent
Robert J. Lefkowitz	USA	2012	For studies of G-protein-coupled receptors
Brian K. Kobilka	USA		
James E. Rothman	USA	2013	For their discoveries of machinery
Randy W. Scheman Thomas C. Sudhoff	USA USA		regulating vesicle traffic, a major transport system in our cells
John O'Keefe May-Britt Moser Edvard Moser	USA Norway Norway	2014	For their discoveries of cells that constitute a positioning system in the brain
William C. Campbell Satoshi Moser	Ireland	2015	For their discoveries concerning a novel therapy against infections caused by roundworm parasites
Youyou Tu	China		For her discoveries concerning a novel therapy against Malaria
Tomas Lindhil Paul Modrich Aziz Sancar	Sweden USA Trurkey	2015	For mechanistic studies of DNA repair
Yoshinori Ohsumi	Japan	2016	Mechanism of AUTOPHAGY

Q.21. Fill in the blanks.

(i) India born scientist who received Nobel Prize in 1968 for discovery of genetic code for amino acids is ________ .

(ii) Transformation in bacteria was reported in 1928 by ________ .

(iii) ________ put forward a classification system for streptococci based on antigen on their cell wall, in 1934.

(iv) ________ crystallized virus in 1935.

(v) ________ received Nobel Prize in physiology or medicine for their discoveries of genetic phenomena in microorganisms.

(vi) ________ are known for their work on viral infection of bacteria (1943).

(vii) O.T. Avery, C.M. MacLeod and M. McCarty in 1944 discovered that genetic material is DNA which work they did by noting transformation of nonpathogenic pneumococci (forming small colonies) to pathogenic pneumococci (forming large glistering colonies) by DNA from heat killed ________ .

(viii) J. Lederberg and E.L. Tatum discovered that genetic material can be passed from one bacterium to another by the process of ________ .

(ix) Francois Jacob and Jacques Monod discovered in 1961 that RNA is involved in — ________ .

(x) In late 1960s Paul Berg spliced human or animal DNA to ________ .

(xi) In 1959 viral cause of cancer was reported by ________ .

(xii) In 1962 Edelman and Porter reported ________ .

(xiii) In 1981 Marqulis contributed on origin of ________ .

(xiv) Joseph Lister an English surgeon in 1867 used a dilute solution of ________ to soak surgical dressings and to spray during an operation.

(xv) To immunize against chicken cholera in 1880 Louis Pasteur used pure culture of chicken cholera bacteria ________ .

Ans. (i) Har Gobind Khorana

(ii) Fred Griffith

(iii) Rebecca C. Lancefield

(iv) Stanley, Northrup and Sumner

(v) George W. Beadle, Edward L. Tatum and Joshua Lederberg

(vi) Delbrück and Luria

(vii) pathogenic pneumococci

(viii) Conjugation

(ix) protein synthesis

(x) bacterial DNA

(xi) Stewart

(xii) Antibodies

(xiii) eucaryotic cells

(xiv) Carbolic acid also known as phenol

(xv) 8 weeks old attenuated culture.

2 CHAPTER

DEVELOPMENT OF MICROBIOLOGY IN INDIA

Q.1. Match the following.

(A)	*Saksenaea vasiformis* published in Sabouraudia: Journal of Medical and Veterinary Mycology (1985) 23: 137-140	(a)	Dr P.N. Mehra 'Padmashree' Emeritus Professor, Panjab University, Chandigarh
(B)	"Evolution of spore through the ages".	(b)	Professor (Dr) S.B. Saksena, Sagar University, Sagar
(C)	"Progress of Veterinary Sciences" a decade (1963-72) of Science in India.	(c)	Prof. A.K. Sharma of Calcutta University
(D)	"Progress of Botany" a decade of Science in India (1963-72). Veterinary College, Mumbai	(d)	Dr S.R. Rao, Retired Professor of Parasitology, Bombay
(E)	Chapter XXI "Microbiology" in A Decade (1963-72) of Science in India	(e)	P.N. Nandi, Bose Institute, Calcutta
(F)	First teaching course in Microbiology was started in 1931, upto B.Sc. level	(f)	New antifungal antibiotics by Mr M.J. Thirumalachar
(G)	Hindustan Antibiotics Ltd, Poona	(g)	Under University of Bombay at the St Xavier's College
(H)	I.D.P.L., Rishikesh and newer antibiotics	(h)	B.S. Bajaj
(I)	Gluconate and polyol metabolism in *A. niger* at Department of Microbiology, M.S. University, Baroda	(i)	K.G. Gupta
(J)	Effect of mono and divalent ions on germination of spores of *Bacillus subtilis* at Department of Microbiology, Panjab University, Chandigarh	(j)	V.V. Mody

(K)	Establishment of the then Imperial Dairy Research Institute, Bangalore.	(k)	Origin of dairy microbiology in India in 1939-40
(L)	W.L. Davies	(l)	First Director of Dairy Research appointed by Govt of India.
(M)	Dr K.K. Iya, who also remained Dy Director General I.C.A.R. and Director N.D.R.I. and F.A.O. Team Leader	(m)	First properly trained Dairy Microbiologist of India obtaining Ph.D. from the University of Wisconsin, Madison in Dairy Bacteriology and Dairy Industry in 1948.
(N)	National symposium on "Current Trends in Soil Biology', held at H.A.U. Hissar, 25-27 Febs., 1985	(n)	Was convened by Late Prof. M.M. Mishra
(O)	Vice-Chancellor of Tamil Nadu Agricultural University, Coimbatore associated with rhizosphere microflora of crop plants	(o)	Patric Tauro
(P)	Retired Professor of Eminence and Agricultural Microbiologist of H.A.U. Hissar	(p)	G. Rangaswami
(Q)	Scientist working in Indian Institute of Biochemistry and Experimental Medicine prepared a potential antitumour antibiotic from a strain of *Aspergillus niger* and named it Jawaharene	(q)	D.K. Roy
(R)	A polypeptide antibiotic mycobacillin from *Bacillus subtilis* B strain effective against dermatophytes was prepared by	(r)	S.K. Mazumdar and S.K. Bose in 1958 which was published in *Nature,* Vol. 191.
(S)	The school of research in Department of Applied Chemistry of Calcutta University was organized by	(s)	Late Prof. B.C. Guha regarding antifungal antibiotics
(T)	Initiated studies in the field of Aerobiology	(t)	H.K. Baruah
(U)	Studied microflora of Air at Guwahati in early sixties	(u)	By T. Sreerumulu and A Ramaligan working at Waltair and Mysore respectively in South India
(V)	Studied microflora of Air at Calcutta in early sixties	(v)	A.C. Das

(W)	Scientist whose group at Department of Botany of Kalyani University studied genetic transformation in *Bacillus Spp.* and *Streptomyces Spp.*	(w)	Prof. R.S. Sandhu
(X)	G.N.D.U., Amritsar, Former Professor known for his contributions in the field of medical mycology and molecular biology Professor Emeritus at V. Patel, Chest Institute, Delhi	(x)	S.P. Sen
(Y)	Known to start research on mycorrhizae in India who was a Forest Pathologist and Director, Biological Research at F.R.I. Dehra Dun.	(y)	K.S. Thind
(Z)	A Mycologist Professor of Panjab University, Chandigarh known for his contribution on forest fungi	(z)	B.K. Bakshi

Ans. [A-b], [B-a], [C-d], [D-c], [E-e], [F-g], [G-f], [H-h], [I-j], [J-i], [K-k], [L-l], [M-m], [N-n], [O-p], [P-o], [Q-q], [R-r], [S-s] [T-u], [U-t], [V-v], [W-X], [X-w], [Y-z], [Z-y].

Q.2. Which is the first book of the world to describe a mushroom?

Ans. The mushroom was first described in the *Rigveda* for the first time as *chhumb* (Sanskrit mushroom) in the form of a simile comparing the head of a devil or a wicked person with the cap or **pileus** of a mushroom which can be easily blown away even by a light current of air.

Q.3. Which was the first bacteriological laboratory established in India ?

Ans. Imperial Bacteriological Laboratory was the first of its type which was established in 1889 at Pune under leadership of Dr. Alfred Lingard. In 1893 it was shifted to Mukteshwar in Kumaon. Later on the major part of it was shifted to Izatnagar, Bareilly and in due course of Time it emerged as Indian Veterinary Research Institute (I.V.R.I.), Izatnagar, Bareilly, with some of its divisions/campus at Mukteshwar.

3

CHAPTER

MICROSCOPY

Q.1. What is a micrometer (μm)?

Ans. A micrometer, also called a micron (μ) is equal to 10^{-6} m, i.e., 1/1000th mm.

Q.2. What is a nanometer (nm)?

Ans. It was formerly called millimicron (mμ) and is equal to 10^{-9} m or 10^{-3} or 10Å.

Q.3. What is angstrom (Å)?

Ans. An angstrom is equal to 10^{-10} m.

Q.4. How many lenses were there in the microscope prepared by Leeuwenhoek in seventeenth century?

Ans. Only one lens (Fig. 3.1).

Fig. 3.1. Leeuwenhoek and his microscope.

Q.5. Who invented the first compound microscope?

Ans. Zaccharias Janssen is known to have made the first compound microscope in 1600. However, they were of poor quality.

Q.6. Who made improvements on the crude compound microscopes made by Zaccharias Janssen?

Ans. Joseph Jackson Lister, father of Joseph Lister made the improvements.

Q.7. When did Robert Hooke observe cork cells under a compound microscope?

Ans. In 1665.

Q.8. Name the important parts of a compound light microscope and also give their functions.

Ans. Illuminator or source of light, condenser (focuses light through the specimen to be seen), eye piece or ocular lens (which magnifies the image formed by objective the lens), body tube (transits the image from the objective lens to the ocular lens), objective lenses (primary lenses that magnify the object), stage (the platform that holds the slide), diaphragm (which controls the amount of light entering the condenser), coarse focussing knob (for major adjustments) fine focussing knob (for precise adjustments) (Fig. 3.2).

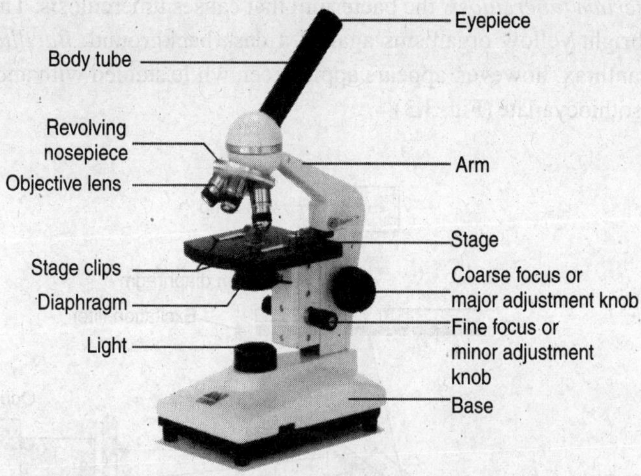

Fig. 3.2. Light microscope.

Q.9. Which lens of the objective is used for maximum magnification?

Ans. Oil immersion lens (for which a drop of immersion oil, generally cedar wood oil is placed on the slide).

Q.10. What is resolution of a microscope?

Ans. It is the resolving power of the combination of lenses to distinguish fine details.

Q.11. What is the main characteristic of a dark-field microscope?

Ans. Instead of a normal condenser a **darkfield microscope** contains an opaque disc. As there is no direct background light the specimen appears light against a black background or the dark-field. It is used to examine unstained microorganisms suspended in liquid, e.g., to see spirochetes like *Treponema pallidum* the causal organism of syphilis.

Q.12. Give the principle and main characteristics of **phase contrast microscope.**

Ans. The principle of phase contrast microscope is based on slight variations in refractive index. While the rays pass from the light source through the specimen, their velocity may be altered because of the differences in the thickness and physical properties of various portions of the specimen.

A phase contrast microscope is provided with special condensers which contain annular or ring shaped diaphragms. The diaphragm permits a ring of light to pass through the condenser focussing light on the specimen and a ring shaped diffraction plate in the objective lens. So, the diffracted and undiffracted rays are then brought into phase with each other to produce the image that meets the eye.

Q.13. What is **differential interference contrast (DIC) microscopy**?

Ans. The DIC microscopy like phase contrast microscopy uses differences in refractive indexes. However, a DIC microscope uses two beams of light instead of one. Besides it the prisms split each light beam adding contrasting colours to the specimen. In this way resolution of a DIC microscope is higher than that of a phase contrast microscope. The image is brightly coloured and is nearly three-dimensional in appearance.

Q.14. What is a **fluorescent microscope?**

Ans. A fluorescent microscope uses fluorochromes (fluorescent dye) to stain the specimen, e.g., the fluorochrome auramine O glows yellow on exposure to UV light which is strongly absorbed by *Mycobacterium tuberculosis* the bacterium that causes tuberculosis. The bacterium can be detected as bright yellow organisms against a dark background. *Bacillus anthracis* causal organism of anthrax, however, appears apple green while stained with another fluorochrome fluorescein isothiocyanate (Fig. 3.3).

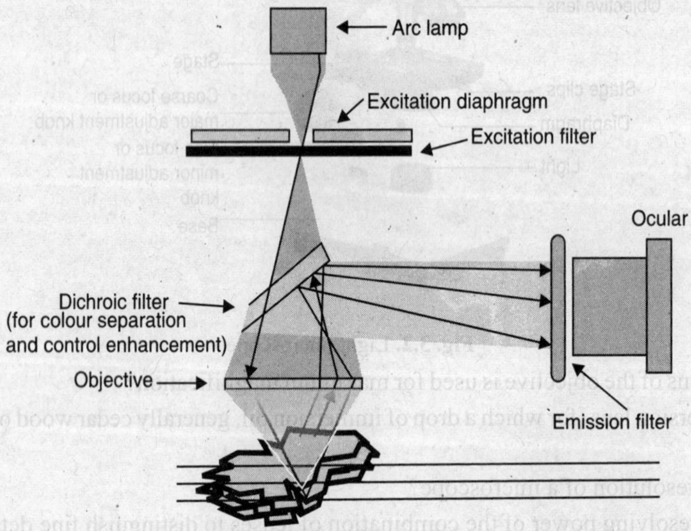

Fig. 3.3. Fluorescent microscope.

Q.15. Write the principle of **immuno fluorescence.**

Ans. *A* fluorochrome is combined with antibodies against a specific type of bacterium. On adding the preparation to bacterial cells on a glass slide the antibodies attach to the bacterial cells and the cells fluoresce on illuminating with ultra violet (UV) light.

Q.16. Give the main characteristics of an **electron microscope.**

Ans. Such a microscope uses electrons in place of light and instead of using glass lenses uses electromagnetic lenses to focus on a specimen. The beam of electrons travels through an evacuated tube. There are two types of electron microscopes. (1) Transmission electron microscope (TEM) and (2) Scanning electron microscope (SEM).

Q.17. What is a **transmission electron microscope?** What is its final image called?

Ans. In TEM a finely focussed beam of electrons from an electron gun is passed through a specially prepared ultra thin section of the specimen. The beam is focussed on a small area of the specimen by an electromagnetic condenser lens. The specimen to be observed is placed on a copper mesh grid. Finally, the electrons are focussed by an electromagnetic projector lens (instead of an ocular lens as in a light microscope) on a screen or photographic plate.

The final image in a TEM is known as transmission electron micrograph.

The salts of some heavy metals, e.g., lead, osmium, tungsten and uranium are often used for staining. These heavy metal stains are used to increase the contrast between ultra structures and the background. The metals can be fixed on to the specimen and is referred to as positive staining while if used to increase the opacity of the surrounding field it is called negative staining (Fig. 3.4 and Fig. 3.5).

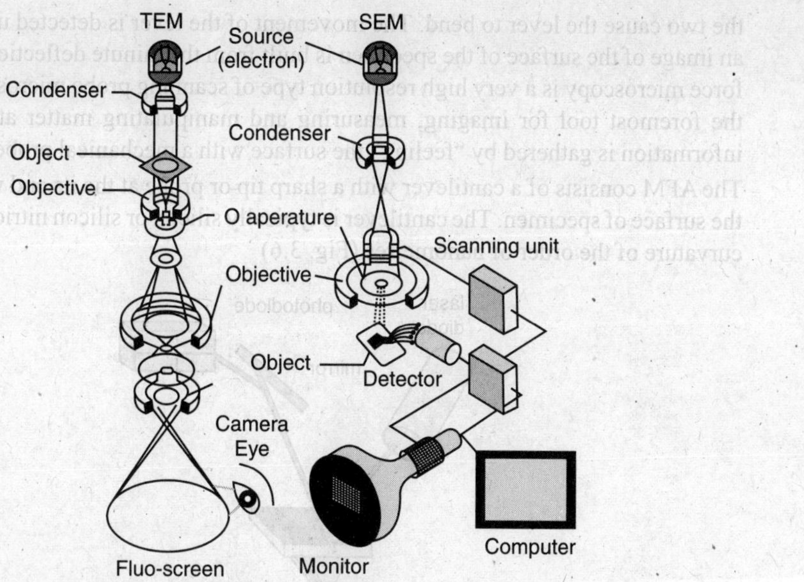

Fig. 3.4. Comparison of transmission electron microscope (TEM) and scanning electron microscope (SEM).

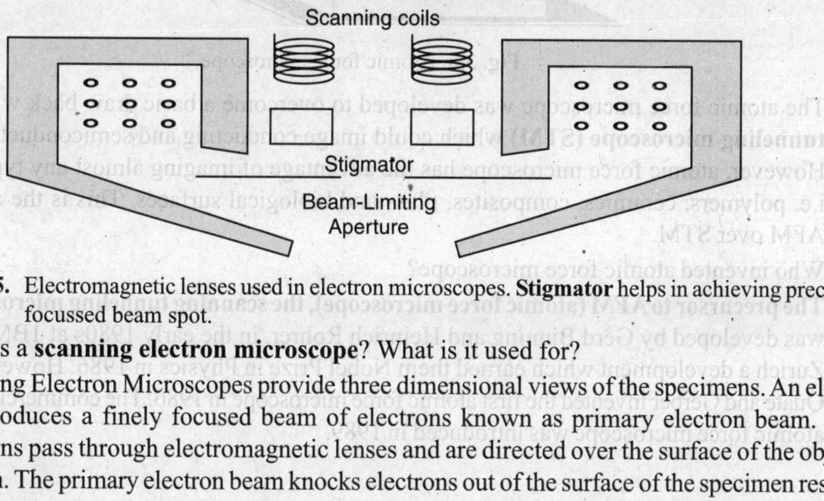

Fig. 3.5. Electromagnetic lenses used in electron microscopes. **Stigmator** helps in achieving precisely focussed beam spot.

Q.18. What is a **scanning electron microscope**? What is it used for?

Ans. Scanning Electron Microscopes provide three dimensional views of the specimens. An electron gun produces a finely focused beam of electrons known as primary electron beam. These electrons pass through electromagnetic lenses and are directed over the surface of the object to be seen. The primary electron beam knocks electrons out of the surface of the specimen resulting in the production of secondary electrons that are transmitted to an electron collector, amplified and used to give an image on the viewing screen or photographic plate.

The scanning electron microscope is used to study surface features as that of leaf-peelings, pollen grains, fungal spores and bacteria.

Q.19. Why are electron beams used instead of light in TEM and SEM?

Ans. Because electrons have shorter wavelength, and specimens smaller than 0.2 μm which can not be resolved under light microscopes can be resolved and studied with TEM and SEM.

Q.20. How does **atomic force microscope** probes the surface of a specimen?

Ans. An atomic force microscope instead of using a lens is provided with a **probe** to examine the surface of a specimen with a sharp tip which may be several micrometres in and less than 10 nm in diametre at the point near field to be examined. The tip lies at the end of the lever which may be 100 – 200 μm in length. When the tip moves across the specimen, the forces between the two cause the lever to bend. The movement of the lever is detected using a computer and an image of the surface of the specimen is built from the minute deflections of the tip. Atomic force microscopy is a very high resolution type of scanning probe microscopy. AFM is one of the foremost tool for imaging, measuring and manipulating matter at the nanoscale. The information is gathered by "feeling" the surface with a mechanical probe.

The AFM consists of a cantilever with a sharp tip or probe at the its end which is used to scan the surface of specimen. The cantilever is typically silicon or silicon nitride with a tip radius of curvature of the order of nanometres (Fig. 3.6)

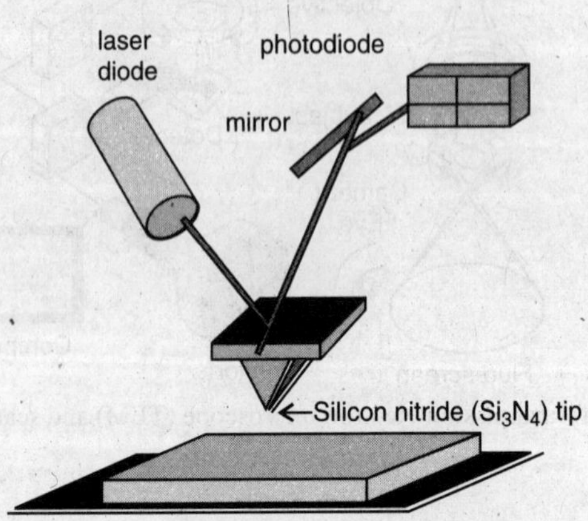

Fig. 3.6. Atomic force microscope.

The atomic force microscope was developed to overcome a basic draw back with **scanning tunneling microscope (STM)** which could image conducting and semiconducting surfaces. However, atomic force microscope has the advantage of imaging almost any type of surface i.e. polymers, ceramics, composites, glass and biological surfaces. This is the advantage of AFM over STM.

Q.21. Who invented atomic force microscope?

Ans. **The precursor to AFM (atomic force microscope), the scanning tunneling microscope (STM)** was developed by Gerd Binning and Heinrich Rohrer, in the early 1980s at 1BM Research – Zurich a development which earned them Nobel Prize in Physics in 1986. However, Binning, Quate and Gerber invented the first atomic force microscope in 1986. The commercially available atomic force microscope was introduced in 1989.

4

CHAPTER

ECOLOGICAL GROUPS OF MICROORGANISMS

Q.1. Which are the major ecological groups of microorganisms?

Ans. The microorganisms depending upon their needs to grow in their natural homes or **ecological niche** may be grouped into the following broad categories keeping in view their requirement for oxygen, source of energy (carbon) and temperature.

(*A*) On basis of their oxygen requirement they are:

(*i*) **Aerobes** or aerobic microorganisms which have capability to grow on media or substrates in contact with air, e.g. some species of *Bacillus* and *Thermoactinomyces.*

(*ii*) **Anaerobes** or anaerobic microorganisms which have capability to grow in absence of oxygen, *e.g.* species of *Bacteroides*, *Clostridium*, *Selenomonas*, *Succinivibrio* and *Veillonella.*

(*iii*) **Faculative anaerobes** or facultative anaerobic microorganisms that is they are optional in their habit and have ability to adopt an alternative life style. They are not strictly anaerobes and can grow in absence or presence of air (Oxygen).

(*B*) On the basis of their nutritional requirement or carbon source they are grouped as:

(*i*) **Autotrophs** or microorganisms which use carbon dioxide for most or all of their carbon requirement. All obligate autotrophs appear to be either chemolithotrophs or photolithotrophs. In many autotrophs CO_2 fixation occurs via the Calvin cycle or via the reductive tricarboxylic cycle. Some acetogens fix CO_2 via a different pathway sometime called the activated acetic acid pathway in which acetyl COA can be synthesized by the reduction of two molecules of CO_2. A similar pathway occurs in *Desulfovibrio* in which the methyl and carboxyl groups of acetyl CoA are derived from formate and CO_2 respectively. An analogous (similar or parallel) pathway occurs in methanogens.

(*ii*) Heterotrophs are the organisms which utilise organic compounds for most or all of its carbon requirements. The terms heterotroph is often used for **Chemoorganotrophs**. Although chemoorganotrophs and phototrophs may also be heterotrophic.

(*C*) On the basis of temperature requirement the microorganisms are grouped as:

(*i*) **Psychrophiles**, which are cold loving and grow optimally at or below 15°C and have upper limit for growth about 20°C and lower limit of growth 0°C or below.

(*ii*) **Mesophiles** are the organisms whose optimum growth temperature lies within range generally accepted as 20 – 45° C.

(*iii*) **Thermophiles** or heat loving microorganisms grow optimally above 45°C *e.g.*, *Methanococcus*, *Pyrodictium*, *Sulpholobus*, *Thermobacteroides*, *Thermophilum*, *Thermomicrobium*, *Thermoplasma*, *Thermoproteus*, *Thermothrix* and *Thermus*.

Many published ad hoc definitions of thermophile are mutually incompatible often because a given definition may be applicable is only a limited context, moreover, such definitions are frequently incompatible with the generally accepted upper limit of 'mesophile'.

(*iv*) **Hyperthermophiles** and so called **superhyperthermophiles** are the heat loving microorganisms which optimally grow above 80 to 100°C, e.g. *Thermus aquaticus* from which enzyme *Taq* polymerase have been isolated, *Pyrococcus furiousus* is an another microorganism from which the *Pfu* polymerase has been isolated. These enzymes are used in PCR technology. *Pyrolobus fumari* occur in the walls of smokers and *Methanopyrus* in deep sea chimneys and produce methane. Some such chimneys collected from, bottom of seas have been displayed in National Institute of Oceanography at Vasco in Goa in India.

(*D*) On the basis of their habitat microorganisms have been grouped as:

(*i*) Soil microorganisms which occur in soil, *e.g.*, some actinomycetes and fungi.

(*ii*) Aquatic microorganisms found in water *e.g.*, some bacterial species and aquatic fungi.

(*iii*) Aeromicroflora comprise the microorganims growing in air.

(*E*) On the basis of their nutritional habit they are grouped as:

(*i*) Saprophytes which get their nutrition from dead organic organisms.

(*ii*) Parasites, which obtain their nutrition form a live plant or animal.

(*iii*) Symbionts are the microorganisms where both host and microorganisms are benefitted.

Transformation of Metals or Metal Cycling

Q.2. What is metal cycling?

Ans. The heavy metals like mercury can exist in a variety of inorganic and organic forms. Some microorganisms can form methyl mercury and can also form methylated forms of arsenic, selenium, tin and some other metals.

Iron Cycle

Q.3. What reactions are mainly involved in biogeochemical cycling of cycling of iron?

Ans. These are the oxidation reduction reaction which reduced ferric iron to ferrous iron and oxidises ferrous iron to ferric irron (Fig. 4.1). The ferric (Fe^{3+}) and ferrous (Fe^{2+}) irons possess highly different solubility properties. The ferric form of iron precipitates in alkaline conditions as ferric hydroxide. However, the ferric iron may be reduced under anaerobic conditions into more soluble ferrous iron. If the conditions are anaerobic sufficient H_2S may be evolved to lead the precipitation of iron as ferrous sulphide. In aerobic habitats in the well drained soils the iron mainly exists in ferric form.

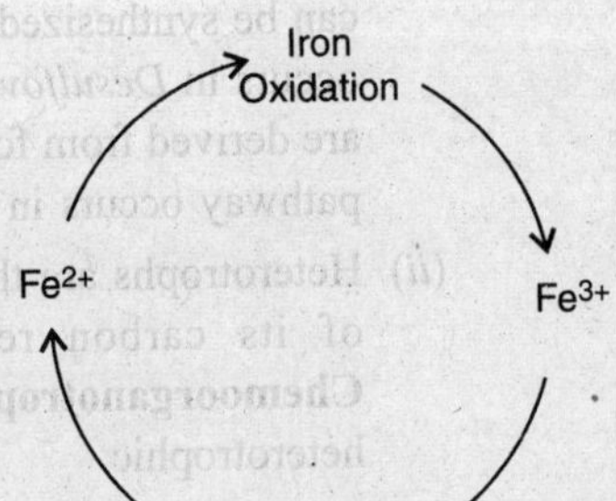

Fig. 4.1. Simplified diagram representing inter conversions of ferrous and ferric forms of iron in the iron cycle.

Q.4. How most bacteria have been evolved and gifted by nature to fulfil their iron requirements, by secreting iron chelating agents?

Ans. All microorganisms with an exception to some lactobacilli require iron employed as a cofactor by many metabolic enzymes and regulatory proteins because of its characteristic, two stable oxidation states Fe^{2+} (ferrous form) and Fe^{3+} (ferric form). Although the iron is abundant in environment yet the iron is a limiting factor for bacterial growth because it forms insoluble ferric hydroxide complexes under aerobic conditions and neutral pH.

The bacteria have developed specialised high affinity transport system to acquire sufficient amount of iron. Most bacteria bear the ability to produce and secrete molecules called **siderophores** (means iron bearers, in Greek). The siderophores are special iron chelating agents which help in solubilization and uptake of iron. Siderophores are water soluble low molecular weight (500 – 1500 Da) molecules which bind with ferric iron with great affinity.

There are two main categories of siderophores which have been studied extensively. Phenolcatechol derivatives are synthesized by enterobacteria and have been given common names enterochelin and enterobactin. These systems carry iron into the cell.

There is system to monitor the intracellular level of iron in the bacterial cell. The iron deficiency leads to excretion of siderophores. The shortage of iron in bacterial cell will bring down the growth while high concentrations of iron are toxic to bacterial cells.

Q.5. What are **manganese nodules**?

Ans. Like iron, the manganese is also oxidized and reduced. The Bacterial activity in some cases result in precipitation of manganese, and lead to formation of **manganese nodules.**

Q.6. How do microorganisms form deposits of **diatomaceous earth**?

Ans. Like other elements, silicon is also cycled between soluble and insoluble forms by microbial populations. Diatoms have silicon dioxide in their frustules and thus form deposits of **diatomaceous earth**.

Q.7. How are **coral reefs** formed?

Ans. Hard red, pink or white substances are built on the sea bed by small creatures, as a result of consumption of carbon dioxide by photosynthetic microorganisms, which shift the equilibrium away from the soluble bicarbonate to less soluble carbonate, resulting the formation of **coral reefs** in marine regions. Some times **coral islands** are also formed (Fig. 4.2).

Q.8. How can **nitrogenase activity** be measured?

Ans. The enzyme complex nitrogenase is responsible for nitrogen fixation. Nitrogenase activity can be measured by acetylene reduction assay.

Q.9. How do microorganisms lead to rock-weathering and formation of soil?

Ans. The fungi, cyanobacteria and lichens on and within the siliceous rocks in harsh environments quickly dissolve silica by releasing Carboxylic acids. In this regard 2-ketogluconic acid, citric and oxalic acids are mainly associated in dissolution of siliceous rocks. Solubilization of siliceous rocks by chelating action of the said organic acids result rock weathering and soil formation.

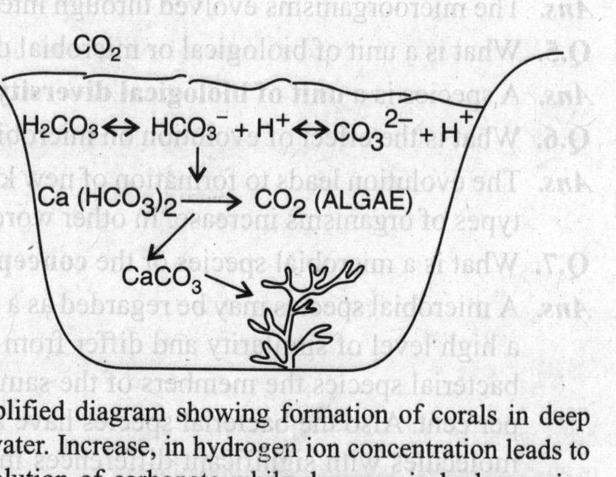

Fig. 4.2. Simplified diagram showing formation of corals in deep seawater. Increase, in hydrogen ion concentration leads to dissolution of carbonate while decrease in hydrogen ion concentration leads to precipitation of carbonate. The equilibrium between HCO^{3-} and CO_3^{2-} is influenced by CO_2 which dissolves in the water to form carbonic acid, a weak acid and its salts.

5

CHAPTER

MICROBIAL DIVERSITY

Q.1. What is microbial diversity?

Ans. The microbial diversity is the diversification (variations) among microorganisms both noncellular and cellular, and procaryotic and eucaryotic on the basis of their structure and function.

Q.2. What is **microbial mat?**

Ans. It is a complex benthic microbial community that forms a cohesive layer and is typically dominated by cyanobacteria or other photosynthetic procaryotes and occasionally micro-algae. Microbial mats are found in the regions where the environmental stress tends to exclude or reduce the number of metazoans, e.g., in intertidal and hypersaline regions, in hot springs and around hydrothermal vents. The microorganisms occurring in microbial mats include the species of *Lyngbya* and *Macrocoleus,* various purple photosynthetic bacteria *Chloroflexus* and certain aerobic and anaerobic chemotrophs as sulphur reducing bacteria.

Q.3. Name the place from which precambrian fossilized photosynthetic bacteria have been collected from fig tree formations, the fossilized stromatolite being 3.5 billion years old.

Ans. Western Australia.

Q.4. Which is older, the evolution of microorganisms or plants and animals?

Ans. Microorganisms have been evolving for over 3.8 billion years as compared to 0.6 billion years in the case of plants and animals.

Q.4A. How did the evolution of new species of microorganisms come about?

Ans. The microorganisms evolved through interactions of their genomes with the environments.

Q.5. What is a unit of biological or microbial diversity?

Ans. A species is a **unit of biological diversity.**

Q.6. What is the effect of evolution on microbial diversity?

Ans. The evolution leads to formation of new kind of microorganisms. Thereby the number of new types of organisms increase. In other words the microbial diversity increases.

Q.7. What is a microbial species or the **concept of microbial species?**

Ans. A microbial species may be regarded as a group of individual populations or strains that show a high level of similarity and differ from populations or strains in related groups. In case of bacterial species the members of the same species possess DNA homology values above 70 per cent. Also the bacterial species have a limited range of mole per cent G + C as the DNA molecules with significant differences in mole per cent G + C do not show a high level of homology. The species is a taxonomic category ranking just below a genus. The individuals of a species exhibit a high degree of mutual similarity.

In the higher organisms which reproduce by sexual reproduction, the members of a species can inbreed actually or potentially.

Q.8. Which things lead to proliferation of some kinds of organisms and extinction of others?

Ans. Genetic variations and natural selection.

Q.9. What is the phylogeny of microorganisms?

Ans. The evolutionary history of microorganisms is their phylogeny.

Q.10. What might have comprised the cytoplasmic membranes of progenitor cells?

Ans. Perhaps a mixture of glycerol ester-linked phospholipids and glycerol ethers comprised the cytoplasmic membrane of progenitor cells.

Q.11. Give an example of the bacterium which may be a descendant of deep rooted divergence and has retained the mixture of ether and ester glycerol based lipids in its cytoplasmic membrane.

Ans. *Aquifex*. The *Aquifex pyrophilus* can grow upto 95°C, however, the optimum temperature is 85°C, is an **extreme thermophile**.

Q.12. Which features of cytoplasmic membranes lead to adaptation of extremely thermophilic archaea to survive at high temperatures?

Ans. They possess cytoplasmic membranes with glycerol diethers or diglycerol tetraethers that make them adaptive for survival at high temperatures.

Q.13. How are eucaryotic organisms chimeric?

Ans. It is believed that some of the organelles of the eucaryotes are derived as a result of unions of distantly evolved microorganisms, the mitochondria, chloroplasts and perhaps some other organelles might have been acquired as a result of **endosymbiosis**.

Q.14. How could it be concluded that mitochondria originated from protobacteria or purple bacteria and chloroplasts from cyanobacteria?

Ans. Analysis of cytochromes, ferredoxins and rRNA molecules of the organelles revealed the origins.

Q.15. What is a molecular chronometer?

Ans. A measure of evolutionary time is called **molecular chronometer**.

Q.16. What is **evolutionary distance**?

Ans. Evolutionary distance is a measure of relative similarity of two microorganisms (or organisms) on the basis of similarity of nucleic acids and proteins.

Q.17. How do *Aquifex* species differ from other bacteria?

Ans. The species of *Aquifex* have glycerol diethers instead of phospholipids which are found in other bacteria.

Q.18. Why are the bacteria of *Aquifex* the first bacterial genus to be named so?

Ans. *Aquifex* means water making, they use H_2, $S_2O_3^{2-}$ (thiosulphate) and S° (sulphur) as electron donors to reduce oxygen to water and thereby are known as water makers.

Q.19. Which are the first bacterial genera to evolve?

Ans. They are *Aquifex* and *Hydrogenobacter.*

Q.20. Name an **extreme thermophile** that was isolated from a hydrothermal vent.

Ans. *Aquifex pyrophilus.*

Q.21. What is the range of temperature required by *Aquifex pyrophilus* for growth?

Ans. The optimum temperature for *Aquifex pyrophilus* is 85°C but it can grow at temperature as high as 95°C.

Q.22. Give some unique characteristics of the genus *Hydrogenobacter* of Aquificales.

Ans. (i) They are obligate autotrophs, and represent unusual lipid profile with linear C 18 : 0 and C 20: 1 as major lipids.

(ii) They possess a unique quinone, 2 methylthio -1, 4 naphthoquinone.

(iii) They have a unique carotenoid pigment.

(iv) They possess outer protein coats.

(v) They have a relatively small genome size (Ca 1.0×10^6 bp) compared with the genomes of other bacteria (Ca. 4.0×106 bp).

Q.23. What could have been the ancestral bacterial progenitor?

Ans. Perhaps some thermophilic individual which might have fixed carbon chemoautotrophically could have been the bacterial progenitor.

Q.24. What are thermotogales? Give their common characteristics.

Ans. It is a group of extremely thermophilic microorganisms which are, phylogenetically distant from bacteria. Their important characteristics are:

(i) They are Gram-negative, nonspore forming rod-shaped cells.

(ii) They are thermophilic.

(iii) They are anaerobic.

(iv) They possess fermentative metabolism.

(v) They usually possess an outer sheath-like envelope called "toga" that balloons over the ends of the cell.

(vi) They lack meso-diaminopimelic acid in their peptidoglycan.

(vii) They are sensitive to lysozyme.

(viii) They have unusually long chains of dicarboxylic fatty acids in their lipids.

Q.25. Why are the Thermotogales placed under the bacterial domain?

Ans. On the basis of comparison of bacterial DNA sequences of translation elongation factor Tu(EF—Tu), and the presence of classical Embden-Meyerhof pathway of glycolysis in the Thermotogales which is a feature of bacteria and is lacked by archaea.

Q.26. Name two **radiation resistant bacteria**.

Ans. *Deinococcus* and *Deinobacter* of the group Deinococci are highly radiation resistant bacteria. These have, therefore, been used to assess the effectiveness of radiation sterilization. This is because they have effective repair mechanism for damaged DNA. They also have a high concentration of carotenoids which absorb radiation. The carotenoids are found in all radiation resistant bacteria.

Q.27. What are **proteobacteria**? Give their main subgroups.

Ans. They are purple bacteria. The proteobacteria have been divided into five separate lineages based on ribosomal RNA sequencing which are: alpha, beta, gamma, delta and epsilon.

Q.28. Which are two major evolutionary lines of descent?

Ans. These are: (1) Gram positive bacteria with a low mole per cent G + C (clostrial lineage) and (2) Gram positive bacteria with high mole per cent G + C which is the actinomycetes lineage.

Q.29. Which pigments are found in **Cyanobacteria**?

Ans. The pigments found in Cyanobacteria are: (1) Chlorophyll-a and (2) Phycobiliproteins.

Q.30. What is common between **chlamydias** and **planktomycetes**?

Ans. Both of them lack peptidoglycan in their cell walls.

Q.31. What are the differences between chlamydia and planktomyces?

Ans. The chalmydia are exclusively intracellular in animal cells while planktomycetes are free living.

Q.32. Which are the three Kingdoms (branches) of Archaea?

Ans. They are Kingdom Crenarchaeota, Kingdom Euryarchaeota and Kingdom Korarchaeota.

Q.33. Give outstanding variations found in bacteria.

Ans. Bacteria have variable size and shape and their morphological shape varies from typical rods and cocci, grotesque forms to prosthecate bacteria. They may reproduce by binary fission, budding and forming multicellular filamentous aggregates.

Metabolically they may be chemoautotrophic depending on enormous array of organic matter or even phototrophic oxygen producing. They show anoxygenic photosynthesis to chemolithotrophy as a result of oxidation of various inorganic substances. The bacteria on the basis of their metabolism can transform almost all naturally occurring substances but they cannot break many of the man-made substances creating pollution. They occur in a wide range of habitats, e.g., from cold and dark depths of oceans to hot thermal springs, on and in roots and leaves of plants to the surface of the body of most animals including humans.

Q.34. Which group of bacteria have been called the '**undulopodia**' of eucaryotic cells and the precursors of sensory perception in animals?

Ans. The **spirochetes**.

Q.35. What are spirochetes?

Ans. The spirochetes are helically-coiled, rod-shaped Gram negative bacterial cells wound around flagella. The flagella of a spirochete cell comprise central axial filaments which are attached at the ends of the cell to flagellar motors consisting of hook, shaft and four rings. The central axial flagella are endoflagella.

Q.36. Name some of the important groups that represent bacterial diversity.

Ans. The important groups are:

(i) Spirochetes (e.g., *Treponema pallidum* and *Borrelia recurrentis*)

(ii) Aerobic/microaerophilic, motile, helical/vibroid gram negative bacteria (e.g., *Spirillum volutans, Aquaspirillum bengal, Campylobacter., Helicobacter pylori, Bdellovibrio bacteriovorus*)

(iii) Nonmotile or rarely motile Gram negative curved bacteria (e.g., *Cyclobacterium, Ancyclobacter* and *Brachyarchus*)

(iv) Gram negative aerobic/microaerophilic rods and cocci (e.g., *Pseudomonas*, acetic acid producing *Acetobacter, Gluconobacter* and *Frateuria*, nitrogen fixing *Azotobacter Chroococcum, Agrobacterium tumefaciens* and *Rhizobium*, methylotrophic bacteria *Methylomonas, Methylococcus* and *Methylobacterium*, bacteria of water bodies *Legionella* intracellular *Neisseria gonorrhoeae*)

(v) Facultative anaerobic Gram negative rods (e.g., enteric bacteria *Escherichia coli, Salmonella, Serratia marcescens*, cholera causing *Vibrio cholerae*, luminescent *Photobacterium*)

(vi) Gram negative anaerobic, straight curved and helical bacteria (e.g., *Haloanaerobium* that prefers 13% NaCl, *Halobacterioides* that prefers 8.5 to 14% NaCl, *Thermosipho* and *Thermotoga* that prefer 1-3% and 3-6% NaCl respectively)

(vii) Dissimilatory sulphate reducing and sulphur reducing bacteria (e.g. *Desulfotomaculum acetoxidans*)

(viii) Anaerobic Gram negative cocci (e.g., *Veillonella* found in saliva, tongue, cheek, mucosa and gingival crevice of human oral cavity)

(ix) Rickettsias and chlamydias

(x) Phototrophic bacteria that are of two types: Anoxygenic phototrophic bacteria (e.g., purple sulphur bacterium *Chromatium*, green sulphur bacterium *Chlorobium*) and Oxygenic phototrophic bacteria (e.g., cyanobacteria *Chroococcus, Spirulina, Lyngbya, Nostoc* and *Anabaena*)

(xi) Aerobic chemolithotrophic bacteria which are:

hydrogen oxidizing bacteria (e.g., *Alcaligenes eutrophus*), colourless sulphur-oxidizing bacteria (e.g., *Achromatium, Thiobacterium, Thiospira* and *Thiobacillus*), iron oxidized and manganese oxidizing bacteria (e.g., *Gallionella*) magnetotactic bacteria and nitrifying bacteria (e.g, *Nitrococcus* and *Nirobacter*)

(xii) Budding and appandaged bacteria (e.g., *Caulobacter crescentus, Hyphomicrobium* and *Panktomyces maris*)

(xiii) Sheathed bacteria (e.g., *Leptothrix discophora*)

(xiv) Bacteria with gliding mobility, which may be photosynthetic nonfruiting gliding bacteria (e.g., *Cytophaga* and *Simonsiella*); fruiting gliding bacteria (e.g., myxobacterium *Stigmatella*)

(xv) Gram positive cocci (e.g., *Deinococci, Staphylococcus aureus* and *Streptococcus pneumoniae*)

(xvi) Endospore forming Gram positive rods and cocci (e.g., *Bacillus sphaericus, Bacillus subtilis* and *Clostridium botulinum* and C. *perfringens*)

(xvii) Asporogenous Gram positive rods, which are regular nonspore forming Gram positive rods (e.g., *Lactobacillus, Listeria* and *Renibacterium*) irregular nonspore forming Gram positive rods (e.g., *Arthrobacter globiformis*)

(xviii) Mycobacteria (e.g., *Mycobacterium turberculosis, M. Bovis* and M. *leprae*)

(xix) Actinomycetes (e.g., *Streptomyces* and *Planomonospora*); myco-plasmas (mollicutes) or the cell wall-less bacteria (e.g., *Mycoplasma pneumonia, M. hominis* and *Spiroplasma*).

Archaeal Diversity

Q.37. What are the **"postage stamp-like sheets of cells"** of archaea found in Red Sea?

Ans. The square cells of extremely archaean *Halobacterium* species isolated from the Red Sea seem to be divided into two dimensions. They form floating gas vacuolated sheets of cells giving the appearance of "postage stamp-like sheets of cells".

Q.38. What is the most striking characteristic that differentiates archaea from other bacteria and eucaryotes?

Ans. They have 'ether' linkages in the lipids of their cytoplasmic membrane.

Q.39. Name the only individuals capable of producing methane under anaerobic and often under thermophilic conditions.

Ans. Methanogens.

Q.40. Which are the most primitive physiological and metabolic traits conserved by the archaea?

Ans. These characteristics are to survive at high temperatures and without oxygen. These are the traits that archaea might have adapted while the average temperature of the earth was much higher and there was no molecular oxygen.

Q.41. Which characteristics of the archaeal cells have enabled them to survive under most harsh conditions inhospitable to life?

Ans. These characteristics are:

(i) Ether linkages in their membrane lipids.

(ii) Diverse types of non-peptidoglycan cell wall.

(iii) Archaeal histone-like proteins which provide stability to archaeal chromosomes.

(iv) Introns within the archaeal protein and archaeal splicing mechanism.

(v) The proteins which are relatively resistant to denaturation due to the presence of amino acid sequences and hydrophobic regions.

Biodiversity of Eucaryotic Microorganisms (Fungi, Algae and Protozoa)

Q.42. What are **protoctista**?

Ans. Lynn Margulis in 1985 placed algae and protozoa together and called the group protoctista instead of calling protista.

Q.43. What is **protista**?

Ans. A kingdom of organisms that lacks true tissue differentiation which include many algae and protozoa. Ernst Haeckel who in 1886 divided organisms into plants, animals and protists, included all microorganisms—bacteria, fungi, algae and protozoa. Whittaker in 1965 who proposed 5 kingdoms, included protozoa and algae only under Protista.

Q.44. What is **Chromista**?

Ans. It is a proposed taxonomic kingdom which will include diatoms and brown algae based on occurrence of chloroplasts in the lumen of rough endoplasmic reticulum instead of being in cytosol as found in plants.

Q.45. Which are the four classical groups of protozoa based on locomotion?

Ans. The four classical groups are:

(i) Ciliophora, which include ciliates,

(ii) Mastigophora which include flagellates,

(iii) Sarcodina (pseudopod formers) and

(iv) Sporozoa (spore forming parasites).

Q.46. What are **slime moulds**?

Ans. Slime moulds are members of Gymnomycota fungi which are phagotrophic, i.e., ingest nutrients in particulate forms like most of the protozoa. They possess tubular mitochondrial cristae and form no walls in their trophic phase.

Q.47. Explain the evolution of mitochondria in protozoa.

Ans. In the process of evolution of protozoa the structure of mitochondria also evolved. The mitochondria were formed initially from endosymbiotic procaryotes -had discoid cristae. This was followed by the evolution of tubular cristae and finally flat plate like cristae evolved in protozoa. However, in some anaerobic protozoa they lost the mitochondria. The protozoa with flat platelike cristae further resulted in the evolution of fungi and animals.

Q.48. What is the recent definition of protozoa proposed by Cavalier Smith?

Ans. They are predominantly cellular eucaryotic microorganisms with often phagotrophic nutrition-without cell walls in their vegetative growth state. However, cell walls are found in spores. Some species have chloroplasts in the cytosol but have no starch. Nor do phycobilisomes, have stacked thylakoids but possess three envelope membranes instead of having two envelope membranes.

Q.49. How can you say that *Hexamita* and *Giardia* (Metamonada); and *Enterocytozoon* (Microsporodia) are the exceptional protozoa which developed before the evolution of **endosymbiotic mitochondria**?

Ans. *Hexamita* and *Giardia* (Metamonada); and *Enterocytozoon* (Microsporidia) lack mitochondria.

Q.50. Why do some protozoa and methanogenic archaea occur together in association in the rumen of some cattle?

Ans. Because some protozoa release molecular hydrogen which is utilized by the archaea.

Q.51. What are hydrogenosomes?

Ans. The **hydrogenosomes** are organelles in which enzyme hydrogenase is localized.

Q.52. What is the function of glycosomes found in some protozoa?

Ans. **Glycosomes** are specialized structures found in some protozoa in which glycolysis occurs, e.g., in trypanosomes which are flagellated protozoa and exist as blood parasites.

Q.53. What lead to the evolution of contractile vacuoles in some protozoa?

Ans. They evolved for the purpose osmoregulation, i.e., to remove water from the cell and to prevent rupture of the cell.

Q.54. What is a cytosome?

Ans. A **cytosome** is a mouth like region lined with cilia that help in passing the food particle into the groove.

Q.55. What are **rhoptries?**

Ans. These are two to several elongated electron dense bodies which extend to the cell membrane of some protozoa within the polar ring.

Q.56. What is **merogony** or **schizogony**?

Ans. They are multiple fissions which occur in some protozoa (e.g. *Plasmodium*).

Q.57. What is **plasmatomy**?

Ans. A form of sexual reproduction in which a multinucleate protozoan cell is divided to form two or more smaller but multinucleate daughter cells.

Q.58. Why are cysts formed by protozoa?

Ans. Cyst formation is a resting dormant stage for the protozoa to withstand adverse conditions.

Q.59. What is **excystation**?

Ans. The process by which the cyst returns to actively growing vegetative condition.

Q.60. What is the complex feature of malaria causing apicomplexan protozoa, **Plasmodium?**

Ans. In *Plasmodium*, the merogony and part of gametogony take place in human hosts and sporogony in mosquito hosts.

Q.61. How are the sporozoites of malarial parasite *Plasmodium* inoculated by the female mosquito?

Ans. The female mosquito injects saliva containing plasmodium sporozoites into the blood stream of vertebrates. The sporozoites move into liver parenchymal cells where they multiply by asexual reproduction after seven days.

Q.62. How did fungi evolve?

Ans. The fungi evolved from protozoa due to evolution of chitinous walls in the trophic phase.

Q.63. What is **dimorphism** in fungi? Give an example.

Ans. The nature of existing in two structural forms is called dimorphism, e.g., *Mucor rouxii* grows in a yeast like form in the atmosphere with a high percentage of CO_2 but forms filamentous mycelia at normal atmospheric concentrations of carbon dioxide.

Q.64. Name a fungus that is an opportunistic pathogen and can cause serious infections in humans.

Ans. *Mucor*.

Q.65. Name the four major classes of fungi named by classical botanists.

Ans. These are Zygomycetes, Ascomycetes, Basidiomycetes and Deuteromycetes.

Q.66. Name the fungus which has been used for expression of viral, procaryotic and eucaryotic genes by recombinant DNA technology and is useful for production of **recombinant vaccines** such as Hepatitis B.

Ans. *Saccharomyces cerevisiae*.

Q.67. What are **moulds**?

Ans. Filamentous fungi are called moulds.

Q.68. What is **mycelium**?

Ans. Interwoven hyphae or the tuft of fungal hyphae is called mycelium (plural, mycelia).

Q.69. Which is the group of fungi in which **dolipore septa** occur? What is their function?

Ans. Dolipore septa are found in Basidiomycetes. The dolipore septa are specialised connections between the cells of their hyphae. The pores in the septum allow the flow of materials from one cell to another.

Q.70. What are Zygomycetes?

Ans. Zygomycetes are filamentous fungi which typically have coenocytic mycelia. The septa in them are formed only at the time of sexual reproduction.

Q.71. How is *Neurospora* useful in genetic studies?

Ans. The spores of *Neurospora* can be isolated from the **ascus** and its genotype can be readily determined.

Q.72. Why is *Amanita phalloides* known as the death cap?

Ans. It is a deadly poison and most deaths due to mushroom poisoning have been attributed to *Amanita phalloides*.

Q.73. Which is the first fungus described in literature?

Ans. Mushroom is the fungus that was described first of all in the *Rigveda* comparing its cap with the head of a devil or a wicked person which can be easily blown away even by a light current of air.

Q.74. What is a **pyrenoid**?

Ans. The pyrenoid is a differentiated region in the chloroplast of algae where the reactions of Calvin cycle take place. The pyrenoid regions contain ribulose 1, 5-diphosphate carboxylase. The storage products are also associated with the pyrenoids.

Q.75. Give the range of variations of structure of chloroplasts in algae.

Ans. The chloroplasts in red algae possess separate thylakoids with phycobilisomes on their surfaces. The chloroplasts of green algae are in stacks and have pyrenoids with starch storage granules. The chloroplasts in diatoms and brown algae have thylakoids arranged as groups of three.

Q.76. Name the Dinoflagellate algae which causes red tide.

Ans. *Gonyaulax*.

Q.77. What are the applications of alginic acid which is one of the normal constituents of cell wall of brown algae?

Ans. The **alginic acid** is used as an additive in ice-cream and is also used as a sizer in paper and textile industry.

Q.78. What is a major characteristic of blooms of dinoflagellates at sea?

Ans. The blooms of dinoflagellates cause toxic red tides.

Q.79. What are puntae?

Ans. The holes in silica walls of diatoms are called **puntae**.

Q.80. What are **Prochlorons**?

Ans. The **prochlorons** are the individuals of the genus *Prochloron* which is one of the two genera of Prochlorohytes (order Prochlorales) which are placed under Group II of oxygenic phototrophic bacteria, according to 9th edition of Bergey Manual of Determinative Bacteriology (1994). The second genus of the Prochlorophytes (order Prochlorales) is *Prochlorothrix*. The *Prochloron* is unicellular while *Prochlorothrix* is filamentous.

The individuals of *Prochloron* are unicellular, spherical without evident mucilaginous sheath. They divide by binary fission by equatorial constriction. So far the genus is found almost exclusively associated as extracellular symbionts of colonial ascidians (chiefly didemnids) on subtropical and tropical marine shores. The only species is *Prochloron didemni*.

The Prochlorophytes resemble cyanophytes i.e., cyanobacteria from which they differ in that they form chlorophylls a, and b and lack accessory red or blue bilin pigments.

The electron micrographs of the photosynthetic bacterium *Prochloron* possesses extensive internal membranes. These membranes are known to be the sites of chemiosmotic generation of ATP by this bacterium which derives the energy for ATP formation from light energy. The **photosynthetic membranes** or **chromatophores** are cylindrically shaped vesicles in *Prochloron*.

6
CHAPTER

EXTREMOPHILES: THE ORGANISMS WHICH GROW IN EXTREME OR HARSHEST ENVIRONMENTS

Q.1. What are extremophiles?

Ans. The **extremophiles** are the organisms which grow under extreme environmental conditions like temperature, salinity, pH, anaerobic conditions (sensitive to oxygen), extreme atmospheric pressure, water stress and others under which other organisms will generally not grow or the conditions that may kill other organisms. The term 'extremophile' has generally not been used in the old literature as it came up with the discovery of a unique group of procaryotes from extreme environments. It is from Latin (*extremus* = extreme, and Greek *phila* = love or friend of).

The extremophiles include extreme **psychrophiles** (cold loving) extreme **thermophiles**, (heat loving) **alkalophiles** (bacteria that live at very high pH or alkaline conditions), **acidophiles** (microorganisms) which show preference for growth at low pH, approximately 2.0), **methanogens** (methane producing microorganisms) **osmophiles** (the organisms which grow optimally in or on media of high osmotic pressure), **barophiles** (atmospheric pressure lovers) and **sulphur metabolizers**. *Archaeoglobus profundus* a sulphur reducer bear optimum growth temperature of 82°C (Fig. 6.1)

Q.2. Which is the first extremophile to have its genome sequence?

Ans. The first extermophile to have its genome sequence was *Methanococcus jannaschii*, an archaeobacterium which lives near sea level where temperature reaches boiling point of waters and pressure enough to crush an ordinary submarine. These atmospheric pressure lovers are called **barophiles**. It exhibits high growth rate and high enzymatic activities at elevated temperatures.

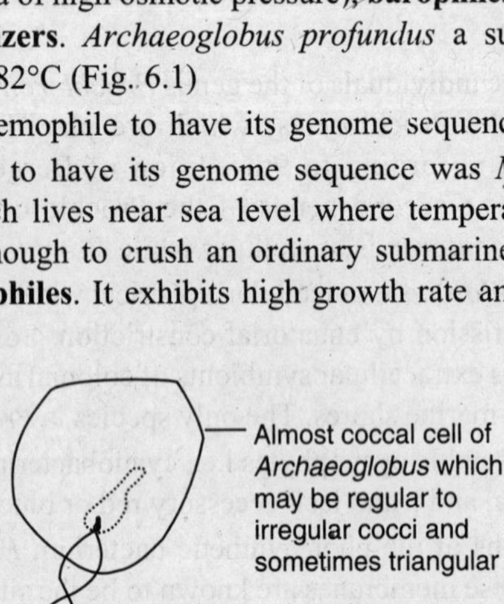

Fig. 6.1. Electron micrograph of *Archaeoglobus* the only genus of sulphate reducing archaea.

Q.3. Name the organisms other than Archaeobacteria that come under extremophiles.

Ans. The Archaeobacteria alone do not come under extremophiles rather they include bacteria like *Deinococcus radiodurans*, *Thermus aquaticus* and *Spirochaeta americane* and the animals *Pompeii* worm, psychrophilic Grydloblatlodeae (insects), antarctic krill and water bear.

Q.4. What are the unique characteristics of **cell membrane of Archaea**?

Ans. A unique characteristic of Archaea is the presence of **ether linkages** in the lipids of their cytoplasmic membranes which distinguishes archaea form eucaryotes and most bacteria. The structure of cytoplasmic membranes of many archaea is a lipid bilayer composed of **glycerol diether lipids** which is analogous (similar or parallel) to lipid layers of bacterial and eucaryotic membranes. The cytoplasmic membrane in some of the archaeobacteria or archaea are monolayers made of **glycerol tetraether lipids**. These monolayers are heat stable with hydrophilic portions (glycerol) at the cytoplasm and external interfaces and an internal hydrophobic portion (hydrocarbons). The difference lies that bacterial and eucaryotic lipids are generally based on **ester linkages**. The archaeal lipids are mainly isopranyl glycerol ethers, which are synthesized by the condensation of glycerol or other alcohols with isoprenoid hydrocarbons of 20, 25, or 40 carbon atoms. (Fig. 6.2)

A

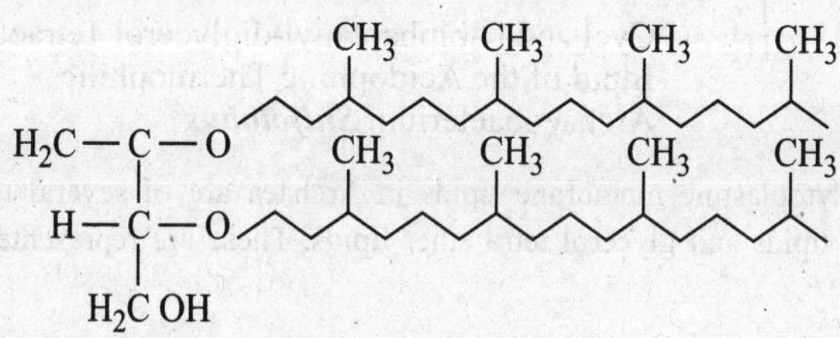

$C_{20}-C_{20}$ Diphantanyl -Glycerol Diether of the Neutrophilic halophilic Archaeo-bacterium ***Halobacterium salinarium***

B

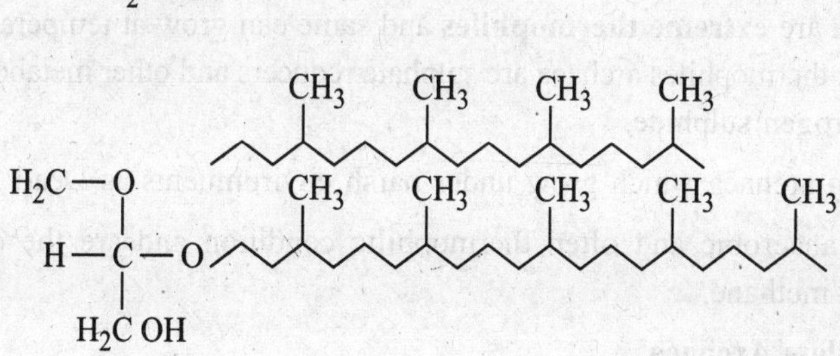

$C_{20}-C_{20}$ Diphantanyl -Glycerol Diether of the Alkaliphilic Halophilic Bacterium ***Natronobacterium gregory***

C

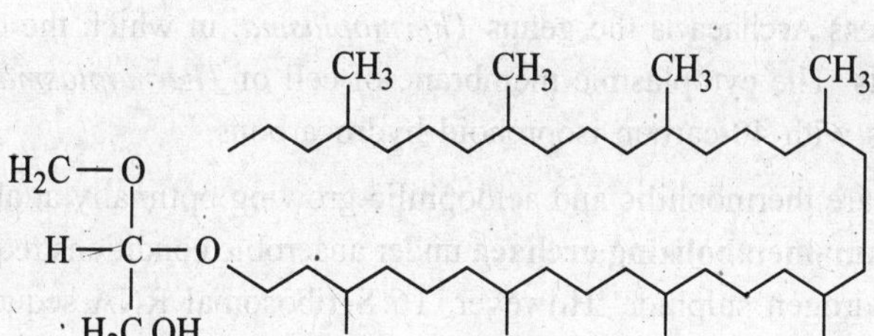

$C_{20}-C_{20}$ Diphantanyl -Glycerol Diether of the Thermophilic Methanogenic Archaeo-bacterium ***Methanococcus jannaschii***

D

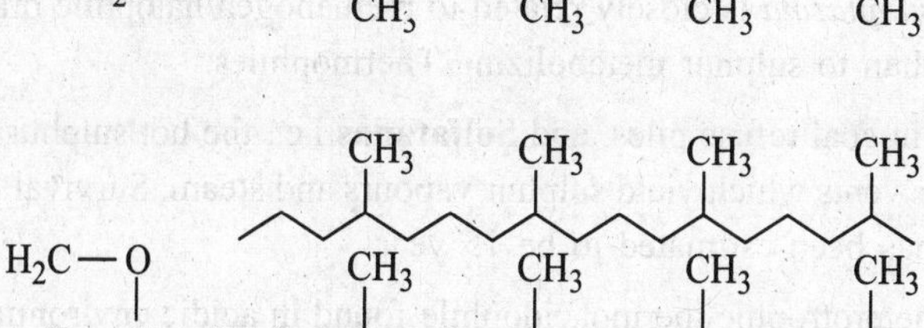

Diphantanyl Glycerol Diether of the Methanogenic Archaeobacterium ***Methanothrix soehngenii***

E

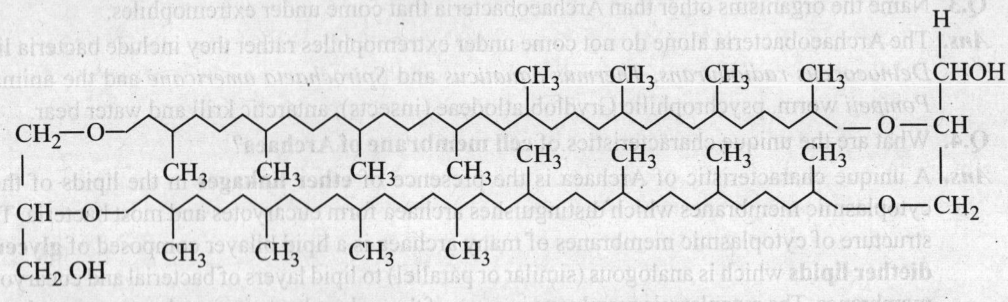

Dibiphantanyl Diglycerol Tetraether
Lipid of the Acidophilic ***Sulfolobus***

F

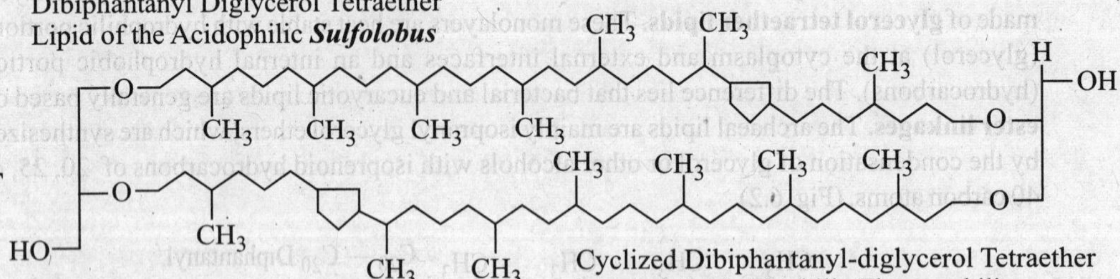

Cyclized Dibiphantanyl-diglycerol Tetraether
Lipid of the Acidophilic Thermophilic
Archaeobacterium ***Sulfolobus***

Fig. 6.2. Membrane variations. Cytoplasmic membrane lipids in Archaea are of several types including glycerol diether lipids and glycerol tetra ether lipids. These are represented in the figure parts from *A* to *F*.

Many of the archaeobacteria are **extreme thermophiles** and same can grow at temperature over 100°C. Some extremely thermophiles archaea are sulphate reducers and other metabolise elemental sulphur and hydrogen sulphide.

There are **halophiles** among archaea which grow under harsh environments as Dead Sea.

Methanogens grow under anaerobic and often thermophilic condition and are the only organisms that can produce methane.

Q.5. Name a genus of **cell wall-less Archaea.**

Ans. Typical example of cell wall-less Archaea is the genus *Thermoplasma*, in which the cells range form spheres to filaments. The cytoplasmic membrane of cell of *Thermoplasma* are made of diglycerol tetra ethers with 40-carbon isoprenoid hydrocarbons.

The species of *Thermoplasma* are thermophilic and acidophilic growing optimally at about 60°C and pH 2. They like **sulphur-metabolizing archaea** under anaerobic conditions require sulphur and reduce that to hydrogen sulphide. However, 16 S ribosomal RNA sequence homology points out that *Thermoplasma* is closely related to methanogen/halophile branch of archaeal phylogenetic tree than to sulphur metabolizing Thermophiles.

Thermoplasma grow in nature in coal refuse piles, and **Solfatanas** i.e., the hot sulphur rich environments, volcanic areas or vents which yield sulphur vapours and steam. Survival time of this cell wall less archaea has been estimated to be 15 years.

Thermoplasma acidophilus is heterotrophic thermoacidophile found in acidic environments created by chemolithotrophic sulphur oxidisers, but does not produce acid itself.

Methanogenesis

Q.6. Explain methanogenesis.

Ans. Methane producing procaryotes are a group of archaea which are capable of reducing carbon dioxide or low molecular weight fatty acids to produce methane. The methanogens are strictly anaerobic active at redox potential between – 350 and – 450 mv. They can use CO_2 as an electron acceptor. They reduce CO_2 using H_2 produced in the fermentation process. They use carbon dioxide as the only carbon source and are considered as chemolithotrophic. If carbon dioxide is considered available in carbonate form the reaction comes to be:

$$HCO_3^- + H^+ + 4H_2 \longrightarrow CH_4 + 3H_2O \ (\Delta G_0' = -32.4 \text{ kcal/mol} = -135.6 \text{ kJ/mol}) \quad ...(i)$$

In, the above equation numbered (*i*) $\Delta G_0'$ (Called Gibbs free energy) gives the indication of energy yield of the reaction is kilocalories per mole. $\Delta G_0'$ in exothermic (energy yielding) reactions is negative.

The carbon dioxide is converted to methane via a pathway involving unusual coenzymes. To begin with initially CO_2 is bound to methanofuran at the formyl reduction level and it is followed by reduction to methenyl, methylene, methyl and finally methane level, involving a number of enzymes. Transformation of CO_2 to cell material is not according to ribulose diphosphate pathway commonly found in other chemolithotrophs rather by acetyl-CoA synthase pathway.

It is significant to note that methyl group of acetate is formed by stepwise reduction of CO_2 to methyl via tetrahydro methanopterin pathway which otherwise leads to methane evolution. The methyl group gets joined to carbonyl group of acetate which is derived by the reduction of another CO_2 to CO by carbon monoxide dehydrogenase, causing catalysis of synthesis of methyl group, CO and HS – CoA to acetyl CoA that is why it has been named as acetyl CoA synthase.

The acetate in turn is processed through a reverse tricarboxylic acid cycle to give pyruvate and trioses.

There are the methanogens like *Methanosarcina barkeri* that can metabolise methanol, acetate and methylamines to methane and carbon dioxide.

The **syntrophism** also known as **cross feeding** between methanogens and other anaerobes like *Syntrophobacter* associated with methanogens broaden the range of substrates suitable for methanogenesis.

It is a matter of deep concern that part of methane escapes to the atmosphere and takes part in photochemical processes. Currently methane concentration of the atmosphere is estimated to be 1.7 ppm and it is increasing every year by one percent.

To conclude different methanogens have evolved several metabolic reaction for generating cellular energy. All are based on anaerobic respiration and all produce methane. The metabolic pathways of methanogens produce methane. The metabolic pathway of methanogens can be divided into three categories (*i*) CO_2 reducing pathway (*ii*) methylotrophic pathway and aceticlastic pathways. All the three pathways are represented in **Fig.6.3.**

A limited number of methanogens are capable to oxidise secondary alcohols for CO_2 reduction to methane. The methanogens which carry out CO_2 reducing methanogenesis use a specialized anaerobic respiration pathway (Fig. 6.4). The reduction of carbon dioxide to methane (CH_4) takes place through a series of reductive steps which generate a methyl group. This pathway employs several reducing enzymes and coenzymes F_{420} and nickel containing coenzyme F_{430}, methanofuran, mathanoprotein and coenzyme M.

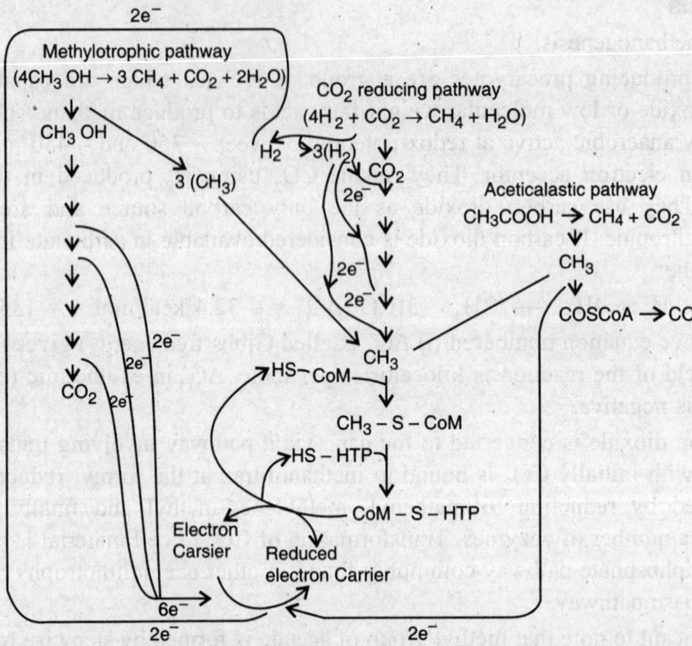

Fig. 6.3. Representation of **methanogenesis** in archaea. Out of the three probable strategies most methanogens adopt the CO_2 reducing pathway beginning with CO_2 or formate. Some use a methylotrophic pathway starting with methanol or methylamines. Very few methanogens adopt aceticlastic pathway beginning with acetate.

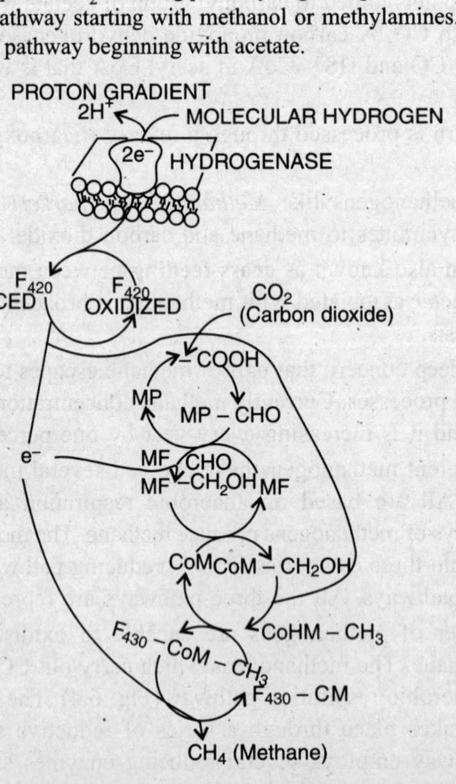

Fig. 6.4. Pathway showing conversion of carbon dioxide (CO_2) to methane (CH_4) by methanogen (methanogenic archaea). Factor $_{420}$ (F_{420}), Factor$_{430}$ (F_{430}) coenzyme M (CoM), methanoprotein (MP) and methanofuran (MF) are involved as electron carriers.

Acetogenesis

Q.7. Describe acetogenesis.

Ans. Some facultatively chemoautotrophic anaerobes are able to reduce carbon dioxide with H_2 to acetate in place of methane, *e.g. Clostridium thermoaceticum* and *Acetobacterium woodii*, the reaction for which can be represented as:

$$2CO_2 + 4H_2 \longrightarrow CH_3COOH + 2H_2O\ (\Delta G'_0 = -\ 25.6\ \text{kcal/mol} = -\ 107.5\ \text{kJ/mol})\ ...(ii)$$

In the acetogenesis shown in the above equation numbered (ii) the energy yield of the reaction is less favorable than the methanogenesis. Besides acetogenesis from carbon dioxide and hydrogen these organisms can also ferment CO, formate and methanol to acetate and have many metabolic features in common with methanogenic bacteria.

Psychrophiles

Q.8. What is a **psychrophile**?

Ans. An organism which grows optimally at or below 15°C, has an upper limit for growth at about 20°C and which has lower limit for growth of 0°C or below. This definition appeared in Bacteriological Reviews (1975), 39:144 – 167; and currently is most widely accepted definition, however, some authors still use the term very loosely to include **psychrotrophs** which is believed to be a misnomer.

Psychrophilic organism include certain algae and fungi, a number of Gram negative bacteria, *e.g.* some species of *Pseudomonas* and *Vibrio*, and a few Gram positive bacteria *Clostridium* spp.

The term psychrotroph has been used for the organism which can grow at low temperature (*e.g.* 0 – 5°C) but which has an optimum growth temperature > 15°C and an upper limit of growth > 20°C. Psychrotrophs include certain algae and fungi and various gram negative and Gram positive bacteria.

Acidophile

Q.9. What is an **acidophile**?

Ans. An organism that grows optimally under acidic condition having an optimum growth pH below 6 and some times as low as 1 or below and which typically grows poorly or not at all at or above pH 7 *e.g. Sulpholobus*, *Thermoplasma* and *Thiobacillus*.

Q.10. Can microorganisms tolerate extreme pH? Justify your statement.

Ans. The microorganisms generally cannot tolerate extreme pH values. Under highly alkaline or acidic conditions some microbial cell components may be hydrolyzed or their enzymes may get denatured. There are, however, some **acidophilic** and **alkaliphilc** bacteria which can tolerate or need extreme pH conditions for growth. Many fungi are acidotolerant but most bacteria are not. Some acidotolerant bacteria like *Lactobacillus* and **acidophiles** like *Thiobacillus* and *Sulfolobus* create their own low pH environment by producing acids.

Lactobacillus is a mixed acid fermenter and *Sulfolobus* produces sulphuric acid. *Bacillus acidocaldarius* and *Thermo acidophilus* are heterotrophic thermoacidophiles found in acidic environments created by chemolithotrophic sulphur oxidisers but do not produce acids themselves. *Thermoplasma* is an archaeon without cell wall and live in hot acid coal refuse piles. *Bacillus acidocaldarius* is the native of acid hot spring.

Applications of Acidophiles

Q.11. Give the practical importance of acidophiles that can be of applied value.

Ans. (i) *Lactobacillus* is used to prepare silage (stored cured fodder) and fermented foods.

(ii) *Thiobacillus* and *Sulfolobus* are employed in **bioleaching** of low-grade copper and uranium ores.

Alkaliphiles

Q.12. What are **alkaliphiles**?

Ans. Many bacteria and fungi are known to bear alkaline pH upto 9 but prefer pH optimum near neutrality. The examples of true alkaliphiles are *Bacillus* strains like *Bacillus alcalophilus* and *B pasteuri*. Some cyanobacteria like *Microcystis aeruginosa*, *Plectnonema nostocorum* and some species of *Spirulina* are also alkalophilic.

Halobacterium, *Natronobacterium* and *Natronococcus* is the category of alkalophiles found in saline lakes with high pH.

Application of Alkaliphiles

Q.13. Give the application of alkaliphiles.

Ans. Some alkaliphiles like *Bacillus* strains produce enzymes proteases and lipases which are stable at high temperature at alkaline pH in the presence of detergent. Therefore, these are used in some laundry detergents to clean fat and proteinaceous stains. It is interesting to note that *Clostridium paradoxum* can withstand pH greater than at 55°C and its optimal generation (doubling) time is 13 minutes at pH 9.3.

Thermophiles

Q.14. What are **thermophiles**?

Ans. The thermophiles like optimum growth temperature above 40°C. Bacterium *Bacillus stearothermophilus* grow at comparatively high temperature (55°C to 70°C). The maximum growth temperature for most thermophilic bacteria is about 99°C and many thermophilic bacteria have temperature ranging from 55° to 60°C.

Extreme Thermophiles

Q.15. What is the optimal temperature for **extreme thermophiles**?

Ans. The optimal temperature of extreme thermophiles is above 80°C. They grow in nature in hotsprings and effluents from laundromats.

Many extreme thermophiles, however, can remain live at freezing temperature and have been reported to have survived in antarctic frozen soil.

Q.16. How do thermophiles maintain their semipermeable properties at high temperature?

Ans. They have comparatively high molecular weight and branched fatty acids in their membranes which permit them to maintain their semipermeable properties at high temperatures.

Q.17. Name some unique habitats of thermophiles.

Ans. They grow in areas of volcanic activity, steam vents in such areas may have a temperature of 500°C. They also grow in hot springs, like Yellow stone National Park, and in other parts of the world.

Q.18. Write on Thermal vent communities and colonies of hyperthermophiles.

Ans. The thermal vent communities occur at depth of 800 to 1000 m where sea floor permit seawater to percolate deeply inside the crust and to react with the hot core material. The vent community of bacteria like *Thermotoga*, *Begiatoa*, *Thiomicrospira* and additional sulphide or sulphur oxidisers of various types. The archaeobacteria that may be growing are *Thermococcus litoralis, Archaeoglobus pyrodictium* and *Pyrobaculum*, (Fig. 6.5.).

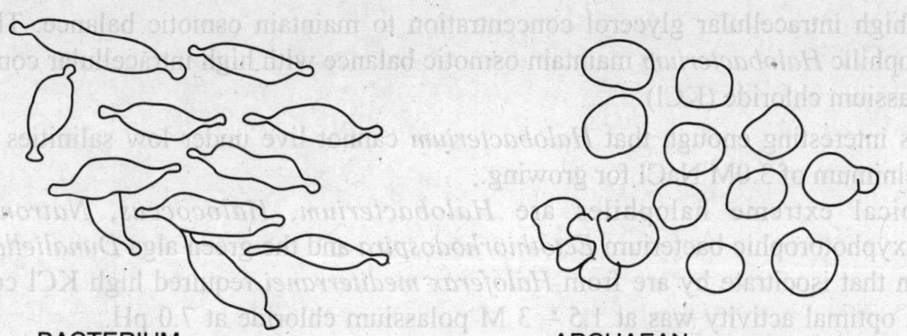

Fig. 6.5. Diagramatic scanning electron micrographs of hyperthermophilic bacterium and archaeobacterium.

Barophiles

Q.19. What are **barophiles**?

Ans. The organisms which grow best or only under conditions of high pressure in the depth of oceans are called barophiles.

Q.20. What are **barotolerants?**

Ans. The organisms which can grow under conditions of high pressure but do not show preference for growth under conditions of high pressure.

Q.21. Give the range of conditions that permited growth of barophiles.

Ans. The barophiles have been isolated from the extreme depth of 10,500 m. Some of the deep isolates were also psychrophilic and their optimum temperature and pressure influenced each other in a complex manner. An isolate from 3600 m deep waters was clearly barophilic at the *in situ* temperature of 4°C recorded for that bacterium.

Application of Barophiles

Q.22. What is the significance of slow growing barophilic marine bacteria.

Ans. They make significant contributions to biodegradation and marine food web.

Q.23. Can barophilic bacteria be cultured?

Ans. Yayanos and his coworkers succeeded in isolating deep sea **Spirillum** and grew that 15 times faster at pressure between 300 and 600 atmospheric pressure than at 1 atmospheric pressure.

Osmophiles

Q.24. What are **osmophiles**?

Ans. The microorganisms which can tolerate or prefer high concentrations of organic solutes as sugars are called osmotolerant or osmophiles. Some of the noted habitats of osmophilic micoorganisms are honey, sap flows, nectar of flowers, molasses and sugary syrups. Some yeasts like *Debaromyces hansenii* and *Zygosaccharomyces rouxii* are good examples of osmophiles. The moulds *Aspergillus* and *Penicillium* also are osmotolerant.

Halophiles

Q.25. What are **halophiles**?

Ans. Microorganisms which can tolerate or need high concentration of salt are called halophiles or halotolerants.

Q.26. How do halophiles survive under halophilic condition which in other organisms denature proteins and enzymes and dehydrate cells?

Ans. The halophiles have developed a number of peculiar modifications to adapt themselves to high salty conditions. The halophilic green alga *Dunaliella* devoid of rigid cell wall builds up high intracellular glycerol concentration to maintain osmotic balance. The obligately halophilic *Halobacterium* maintain osmotic balance with high intracellular concentration of potassium chloride (KCl).

It is interesting enough that *Halobacterium* cannot live under low salinities and requires a minimum of 3.0M NaCl for growing.

Typical extreme halophiles are *Halobacterium*, *Halococcus*, *Natronobacterium*, anoxyphotorophic bacterium *Ectothiorhodospira* and the green alga *Dunaliella*. It has been seen that isocitrate by are from *Haloferax mediterranei* required high KCl concentrations and optimal activity was at 1.5 – 3 M polassium chloride at 7.0 pH.

Some species belonging to halophilic archaea have evolved a unique mechanism of photosynthesis based on the presence of **bacteriorhodopsin** a wonderful protein pigment in their cytoplasmic membranes e.g. in *Halobacterium* (Fig. 6.6) The purple 568 nm mediates light driven proton pump which drives ATP synthesis. Bacteriorhodopsin is localised as isolated patches within the cytoplasmic membrane of *Halobacterium*.

Halophilic archaea *Halobacterium* develops chloride pumps which are based on halorho dopsin to transport chloride in the all to maintain osmotic balance and to meet out the deficiency of chloride ions which leave the all when protons are expelled, (see Fig. 6.7.)

The ribosomes of *Halobacterium* need high concentrations of potassium ions to remain stable. The enzymes of this bacterium also need high concentrations of salt to maintain their active configuration and function, and get inactivated at low salinity.

Halobacterium which is a unique heterotrophic aerobic archaeaon is devoid of murein and its all well seems to require sodium ions for stability. As mentioned earlier some strains of *Halobacterium* possess **bacteriorhodopsin bilayer membrane** components, which serve as light-driven proton pumps. They use light energy to pump protons out of the all, hence generate an electron an electrochemical potential. This all leads to ATP synthesis. Thus *Halobacterium* has adapted to live in highly saline, often saturated brine environmental conditions.

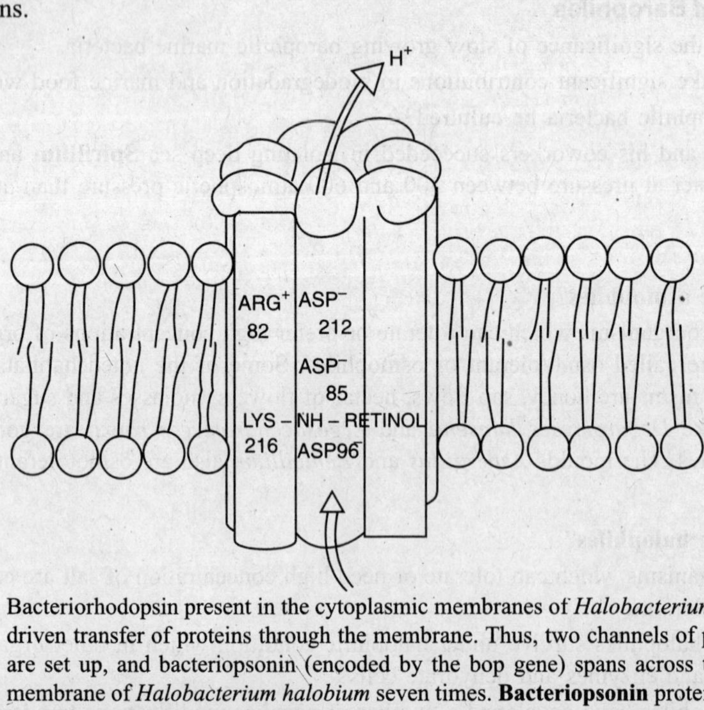

Fig. 6.6. Bacteriorhodopsin present in the cytoplasmic membranes of *Halobacterium* mediates light driven transfer of proteins through the membrane. Thus, two channels of proton transport are set up, and bacteriopsonin (encoded by the bop gene) spans across the cytoplasmic membrane of *Halobacterium halobium* seven times. **Bacteriopsonin** protein surrounds the retinol part of bacteriorhodopsin attached to the protein at a lysine residue at position 216. The Bacteriorhodopsin is a purple pigment.

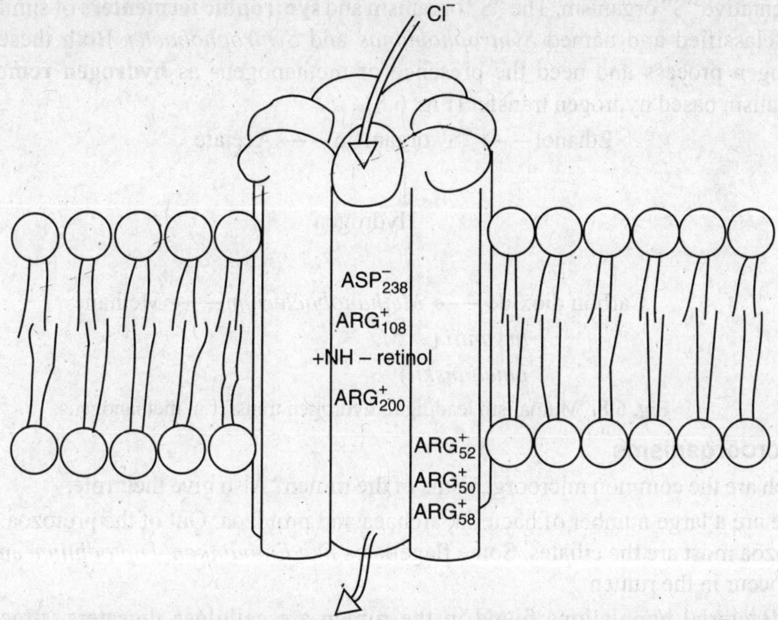

Fig. 6.7. *Halobacterium* species (Halopheles) possess halorhodopsin which works as a chloride pump. Thus, the protein haloopsin spans the cytoplasmic membrane 7 times. The arginine residue at position 108 is believed to play a role in chloride uptake while arginine residue at positions 200 take part in release of chlorine.

Biogas Production

Q.27. How do methanogens produce methane a biogas?

Ans. The methanogens are known to possess new coenzymes and pathway which use hydrogen to reduce carbon dioxide or in some cases acetate to produce methane. Decomposition of cellulosic plant residues in swamps, bogs and digestive tracts of animals is brought about in nature by methanogenic bacteria. The main metabolic reaction is the reduction of carbon dioxide (CO_2) with hydrogen (H_2) to methane (CH_4). The CO_2 and H_2 are produced in fermentation reactions. The methane and residual CO_2 are released as **biogas** also called **marsh gas**.

However, because of narrow substrate range of methanogens. They depend on anaerobic cellulose decomposers to form glucose and other fermentable carbohydrates and on mixed acid fermenters to make a range of short-chain fatty acids. Subsequently these are fermented further to give H_2 and CO_2 and other products by bacteria very closely **syntrophic** with methanogens, taking part in the phenomenon called syntrophism, in which extent of growth of an organism depends on the preparation of one or more metabolic product or nutrients given out by the organism growing in the immediate surrounding or viscinity. Archaeal populations know for methane production have fascinating synergistic relationships with bacterial and other microbial populations, *e.g.* *Syntrophomonas* species oxiside bytyrice acid caproic acid to acetate and H_2. Syntrophomonas also oxidise valeric acid and enanthic acid to acetate CO_2 and H_2 (hydrogen). On the other hand *Syntrophobacter* can oxidise propionic acid to acetate, CO_2 and H_2. The acetate and H_2 produced by these bacteria are utilized by methanogenic archaea to produce methane or the biogas.

Q.28. What is "S" organisms?

Ans. It is the organism syntrophic with methanogens. *Methanobacterium bryantii* (formerly known as *M. omelianskii*) was kept in culture collection for 26 years after its original description is 1941 before it was revealed to be syntrophic association of the methanogen proper and fermentative "S" organism. The "S" organism and **syntrophic fermenters** of similar type were later classified and named *Syntrophomonas* and *Syntrophobacter* Both these genera are hydrogen process and need the presence of methanogens as **hydrogen removers** called mutualism based hydrogen transfer (Fig. 6.8)

Ethanol ⟶ "S" organism ⟶ Acetate

↓

Hydrogen

Carbon dioxide ⟶ *Methanobacterium* ⟶ Methane
bryantii (= *M.*
omelianskii)

Fig. 6.8. Mutualism leading to hydrogen transfer in methanogens.

Rumen Microorganisms

Q.29. Which are the common microorganisms of the rumen? Also give their role.

Ans. There are a large number of bacteria, archaea and protozoa. Out of the protozoa most are the protozoa most are the ciliates. Some flagellates like *Eutodinium*, *Diplodinium* and *Sarcodina* also occur in the rumen.

The Bacterial populations found in the rumen are cellulose digesters, strach digesters, hemicellulose digesters, sugar methanogenic bacteria, proteolytic bacteria and lipolytic bacteria. Among the bacterial species are *Bacteriods*, *Ruminococcus*, *Succinimonas*, *Methanobacterium*, *Butyrivibrio*, *Selenomonas*, *Succinivibrio*, *Streptococcus*, *Eubacterium* and *Lactobacillus*.

Some of the bacterial populations produces acetate the predominant acid which is present in the rumen. Some bacteria also produce proionate the only fermentation acid which can be converted into carbohydrates by the ruminant. The wide range of bacteria community of the rumen have enzymatic capabilities needed to digest different types of plant components. Some nitrogen fixation activity bacteria but the amount of nitrogen contributed is very low to meet out the need of nutritional requirement of the animal.

The protozoan populations in the rumen also can digest cellulose and starch while other ferment the dissolved carbohydrates. Some species of protozoans are predations of bacterial populations. The proteins of the rumen protozoa are, however, digested by ruminant's enzymes. The protozoa inside the rumen store a large amount of carbohydrates. Digestion of protozoa take place in omasum and abomasum compartments of the ruminant stomach which is situated very near to rumen.

Most probably protozoa are more easily digested in comparison to bacteria because bacteria have cell walls and higher nucleic acid contents. The fungal microbes have very poor role. However, methanogenic there is sudden change of diet form dry hay to pasture grass. Production of excess of methane can distend the rumen consequently may compress lungs suffocating the animal. This conditions is called **bloat of sheep and cattle.** Once it develops only puncturing the rumen to release the excess of methane can save the animal. Bloat is also known as ruminal tympany or tympanism.

Microbial Fermentation in Rumen

Q.30. Give the outlines of fermentation occurring within the rumen.

Ans. The overall fermentation taking place is the rumen may be described by equation.

$$57.5\ (C_6H_{12}O_6) \longrightarrow 65 \text{ acetate} + 20 \text{ propionate} + 15 \text{ butyrate} + 60\ CO_2 + 35\ CH_4 + 25H_2O$$

Microorganisms present in the rumen convert cellulose, starch and other ingested nutrients to carbon dioxide, hydrogen gas, methane and low molecular weight organic acids as acetic, propionic and butyric acids. The organic acids are absorbed in the blood stream of the animal and on aerobic oxidation give out energy. Ruminants are capable to use proteins manufactured by microbial population. The CO_2 and CH_4 produced by methanogenic bacteria inside the rumen, are released outside and do not bear any nutritional benefit (Fig. 6.9).

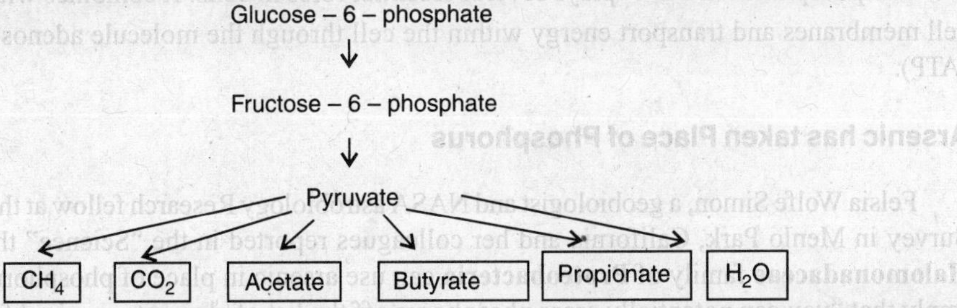

Fig. 6.9. Production of different substances as a result of microbial activity on the grasses and other plant material containing cellulose, strach, pectin and hemicellulose eaten by the animal, and transported to rumen. The fatty acids particularly acetate serve the nutritional requirement of the animal while CO_2 and CH_4 are released by the animal into the atmosphere.

Q.31. What are the **arsenic loving extremophiles** discovered recently ?

Ans. According to NASA scientist as reported in the journal Science, the first organism able to substitute one of the six chemical elements crucial to life has been found, which bacterium is found in a California Mono lake (USA) that uses arsenic, a poisonous element, in place of phosphorus.

Fig. 6.10. Mono lake, the ecological niche or natural home of salt loving extremophiles. It is fascinating that bacterium breaks the golden rule of biochemistry. So far the idea has been that life on earth is composed of at least six elements: carbon, hydrogen, oxygen, nitrogen, sulphur and phosphorus. But this bacterium is an exception to this rule and has broken the fundamental tenet of the biochemistry.

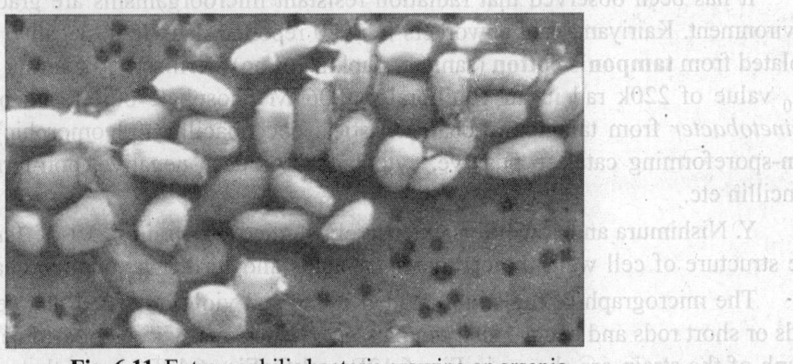

Fig. 6.11. Extremophilic bacteria growing on arsenic.

The Phosphorus in Cells of Other Organisms

The phosphate ion PO 4^{3-}plays several essential roles in cells. It combines with lipids to make cell membranes and transport energy within the cell through the molecule adenosine triphosphate (ATP).

Arsenic has taken Place of Phosphorus

Felsia Wolfe Simon, a geobiologist and NASA astrobiology Research fellow at the US Geological Survey in Menlo Park, California and her colleagues reported in the "Science" that a member of **Halomonadaceae** family of **Proteobacteria** can use arsenic in place of phosphorus. The findings imply that "you can potentially cross phosphorus off the list of elements required for life" in words of David Valentine, a geomicrobiologist at the University of California, Santa Barbara.

The arsenic is positioned just below phosphorus in the periodic table and the two elements can play a similar role in chemical reactions e.g. AsO_4^{3-} has the same tetrahedral structure and binding sites as that of phosphate. Because of this similarity of arsenate to phosphate, the arsenate can get inside cells of the arsenic eating bacterium by hijacking phosphate's transport mechanism despite the fact that arsenic is highly toxic to most organisms.

Some scientist in fun has called this arsenic user as the **oddball bacterium** because it shows a strange behavior of life and disobeys the very fundamental rule of biochemistry i.e. , it uses arsenic instead of phosphorus.

Felsia Wolfe Simon thought that the parallelism between the two elements could mean that despite its toxicity, arsenic was capable of performing jobs of phosphorus in the cell. Her search for an organism that would not just tolerate arsenic but make biological use of it took her to Mono Lake in Eastern California. The 180 square kilometer lake has extremely high arsenic concentration owing to the arsenic bearing minerals which wash down the nearby mountains.

The researchers when added radio-labelled arsenate to the solution to track its distribution they found that arsenic was present in the cellular fractions containing the bacterium's proteins, lipids and metabolites such as ATP and glucose as well as nucleic acids that made up its DNA and RNA. The amounts of arsenate detected were similar to those expected of phosphate in normal cell biochemistry suggesting that compound was being used in the same way in the cell.

Radiation Resistant Bacteria

It has been observed that radiation resistant microorganisms are gradually evolving in the environment. Kairiyama and coworkers in 1980 reported that Gram negative bacterium strain FO-1 isolated from **tampon's cotton** (sanitary napkin cotton) sterilized by **gamma radiation** denoted the D_{10} value of 220k rad in air equilibrated 0.067M phosphate buffer and belonged to the genus *Acinetobacter* from taxonomic characteristics: coccobacillus (pleomorphic), aerobic, nonmotile, non-sporeforming catalase positive, cytochrome oxidase negative, non-fermentative resistant to pencillin etc.

Y. Nishimura and coworkers extensively studied the strain F.O.1 of *Acinetobacter* described the structure of cell wall, the cellular fatty acids and further taxonomic characteristics.

The micrograph of the strain stained by crystal violet revealed that cells are cocci, coccoid rods or short rods and occur mostly in pairs and occasionally in elongated cells. The electronmicro graph of the strain are shown in Fig. 6.12B and C. Fig. 6.12B shows that cell division occurs by simple construction with formation of slight septum. The fibrils of the nuclear region have been

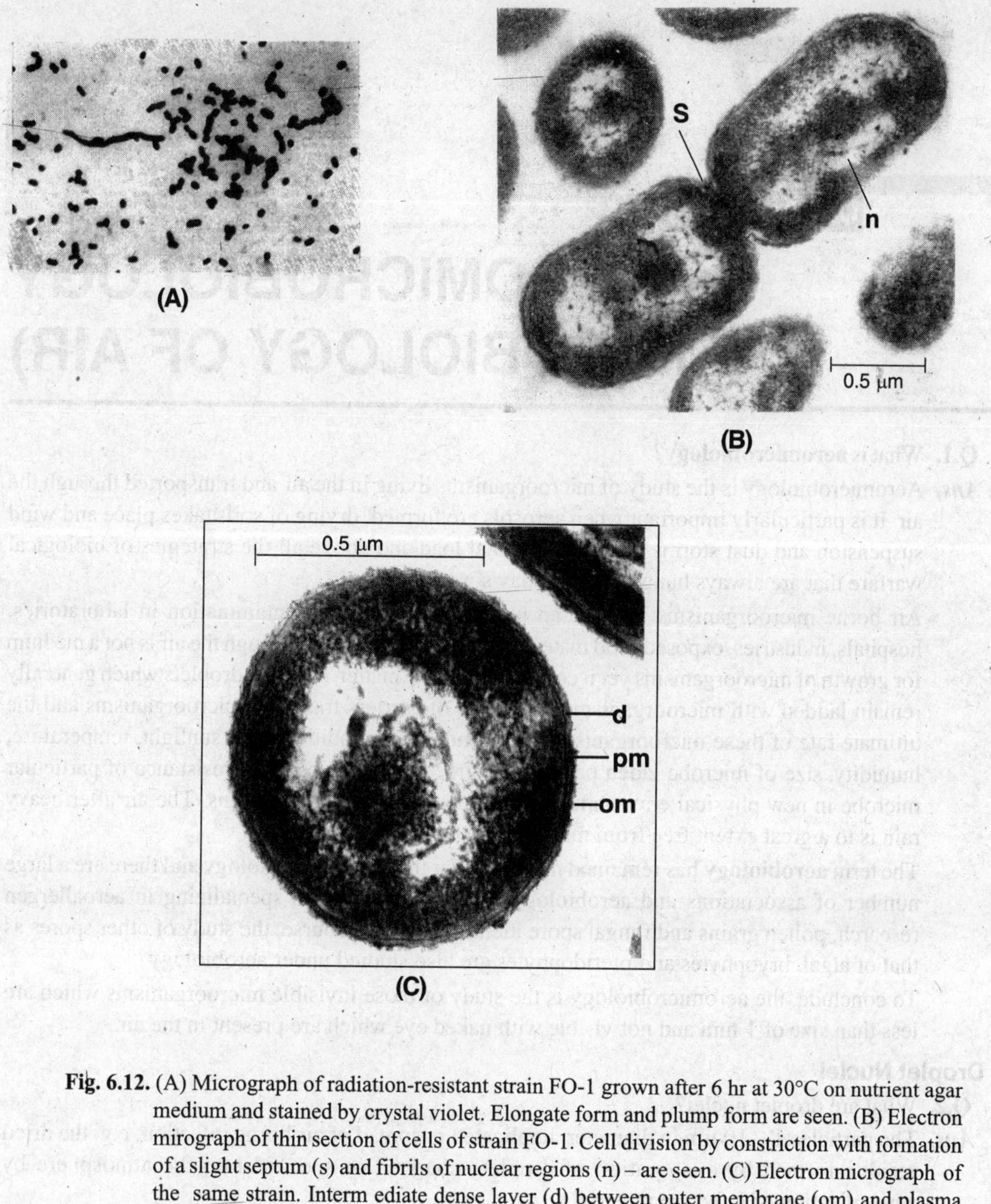

Fig. 6.12. (A) Micrograph of radiation-resistant strain FO-1 grown after 6 hr at 30°C on nutrient agar medium and stained by crystal violet. Elongate form and plump form are seen. (B) Electron mirograph of thin section of cells of strain FO-1. Cell division by constriction with formation of a slight septum (s) and fibrils of nuclear regions (n) – are seen. (C) Electron micrograph of the same strain. Interm ediate dense layer (d) between outer membrane (om) and plasma memjbrane (pm).

demonstrated. Fig. 6.12.C shows the existence of an intermediate dense layer between the outer membrane and the plasma membrane. The main components of the cellular fatty acids in strain FO-1 were oleic acid palmitic acid and palmitoleic acid. *Acinetobacter* belong to family Neisseriaceae occur in soil and water may act as opportunistic pathogen in man. *Acinetobacter* alongwith vibrios and *Micrococcus* cause spoilage of cured meats e.g. bacon.

7

CHAPTER

AEROMICROBIOLOGY (MICROBIOLOGY OF AIR)

Q.1. What is **aeromicrobiology**?

Ans. Aeromicrobiology is the study of microorganisms living in the air and transported through the air. It is particularly important when aerosols are formed, drying of soils takes place and wind suspension and dust storms pick up microbial load and above all the strategies of biological warfare that are always hanging over today's world.

Air borne microorganisms become an important sources of contamination in laboratories, hospitals, industries, exposed food materials, drinks and fodder. Although the air is not a medium for growth of microorganisms yet it carries particulate matter, dust and droplets which generally remain ladden with microorganisms. This type of carriers transport microorganisms and the ultimate fate of these microorganisms is dependent on conditions like sunlight, temperature, humidity, size of microbe laden particles, degree of susceptibility or resistance of particular microbe in new physical environment. Microbes come down with rains. The air after heavy rain is to a great extent free from microorganisms.

The term **aerobiology** has remained more popular than aeromicrobiology and there are a large number of associations and aerobiology research laboratories specializing in aeroallergen research, pollen grains and fungal spore identification. Of course, the study of other spores as that of algal, bryophytes and pteridophytes are also studied under aerobiology.

To conclude, the aeromicrobiology is the study of those invisible microorganisms which are less than size of 1 mm and not visible with naked eye which are present in the air.

Droplet Nuclei

Q.2. What are droplet nuclei?

Ans. The particles 1 – 10 μm in diameter, implicated in spread of air borne infection, e.g. the dried residue formed by evaporation of droplets coughed or sneezed into the atmosphere by aerosolization of infective material.

There is an another term droplet infection for the infection caused by inhalation of an aerosol of Saliva, mucous etc contaminated with pathogens, from an infected individual.

Aerosol

Q.3. What is an aerosol?

Ans. Technically an aerosol is a suspension of fine solid particles or liquid droplet in a gas e.g. smoke, oceanic haze, air pollution, smog, CS gas.

The term aerosol has been derived from the fact that matter floating in the air is a suspension i.e. a mixture in which solid or liquid or combined solid liquid particles are suspended in a fluid. The differentiate suspensions from the solution the term **sol** originally evolved meant to cover dispersions of tiny submicroscopic particles in a liquid. On the basis of studies of dispersion in air, the term aerosol developed and now stands for liquid droplets, solid particles and combinations of all these.

Phylloplane

Q.4. What is phylloplane?

Ans. The habitat of the microorganisms directly on the surface of leaf is called phylloplane.

The phylloplane serves as habitat of a large number of bacterial species, fungal hyphae and spores and unicellular yeast cells. The presence of viral entities on the leaf surface also cannot be ruled out. The phylloplane fungi can produce the enzyme cellular. They are also known to produce enzymes pectinase, cutinases and proteases.

There are specially pigmented populations of yeasts and bacteria on the leaf surfaces. It is believed that pigments of microbial population afford protection against otherwise harmful UV radiations; and direct sunlight on the surface of the leaf.

The populations of phylloplane of pine trees bear capability to utilise sugars and alcohol as carbon sources, in a better way than bacterial populations in the litter (leafy material and twigs left lying on the forest floor) layer. Bacterial populations in the phylloplane of rye are represented by Xanthomonads and pink chromogens in the month of may Xanthomonads and Pseudomonads in the month of July, Xanthomonads in September and listerial and staphylococci in the month of September. Populations of *Alternaria*, *Epicoccus* and *Stremphylium* have been seen as phylloplane invaders.

Phyllosphere

Q.5. What is Phyllosphere?

Ans. The phyllosphere is the habitat adjacent of the plane leaf. In other words it is the environment in immediate vicinity of plant leaf. Like phylloplane various bacterial and fungal populations occur in phyllosphere too. The principle populations in the phyllosphere region of green needles of some species of *Pinus* are *Pseudomonas,* even *Pseudomonas fluorescens.* The populations of *Sporobolomyces roseus Rhodotorula glutinis, R. mucilaginosa Cryptococcus laurentii, Torulpsis ingeniosa* and *Aureobasidium pullans* are commonly found in the phyllosphere.

Q.6. Define **antagonism.**

Ans. The antagonism is the inhibition, injury or the killing of one species of microorganisms by another microorganisms. It is an interpopulation relationship in which one population bears a deleterious or negative effect over another population of microorganisms.

8
CHAPTER

PROKARYOTES (PROCARYOTES)

Q.1. Give salient features of structure and function of prokaryotic (prokaryotic) cell

Ans. The bacteria are known for their small size and simple structure in comparison to eucaryotic cells. They exhibit a wide diversity in their shape, size, arrangement of cells and distribution of flagella on their surface. *Pseudomonas* cells bear a single flagellum for their movement.

- The bacteria can be divided into two broad types on the basis of Grams staining: (1) **Gram positive** (are stained purple to black), (2) **Gram negative** (are stained pink to red). They are also differentiated by structure of cell wall *i.e.*, Gram positive wall and Gram negative wall. The cell wall is characterised with PEPTIDOGLYCAN in procaryotes.
- Recently a better term '**cell envelope**' has been used collectively for all layers out wards to plasma membrane.
- The bacteria have capsule (to escape phagocytosis) and fimbriae (help in attachment on surface) outside their cell wall structure. They have flagella which propel them towards attractants and away from repellents Bacteria with monotrichous flagella show movement by tumble and run.
- Some bacteria produce resistant endospores to survive under adverse environmental conditions.
- The membrane bound photosynthetic tubules called THYLAKOIDS are found nearly in all cyanobacteria
- The discovery of some bacteria with **unique types of membrane systems** and electron microscopic studies revealing single layered 'so called membrane' either of protein or lipid, that enclose some inclusion bodies have created new options of demarcation among bacteria, cyanobacteria and other prokaryotic and eucaryotic cells.

Size Shape and Arrangement of Bacterial Cells

Q.2. Give size and morphology of common and unusual forms of bacteria.

Ans. The bacteria vary in size and shape. The smallest being mycoplasmas about 0.3 μm in diameter almost of the size of a large virus. Some Spirochaetes reach 500 μm in length and the cyanobacterium *Oscillatoria* in about 7 μm in diameter equal to the diameter of red blood corpuscles or cells. The smallest possible cell has been estimated to be 0.14 to 0.2 μm in diameter. However, nanobacteria or ultramicrobacteria have been reported to be smaller. However, some microbiologists believe that nanobacteria are artifacts and more research is needed to be done to ascertain. *Acanthurus nigrofuscus* a huge bacterium grows in the intestine of brown surgeonfish. *The Epulopiscium fishelsoni* attains a size of 600 × 80 μm, a little smaller than the printed hyphen and *Thiomargarita namibiensis* discovered from ocean sediments is much larger, measuring 100 to 750 μm in diameter.

Most of the bacteria are spherical cells or cocci that may be as single individual cell or in pairs called **diplococci** which are formed when cocci divide and remain together in pairs as *Neisseria gonorrhoeae*. Long chains of cells are formed when cells adhere after repeated divisions in one plane *e.g. Enterococcus, Lactococcus* and *Streptococcus*. When cocci divide in random planes irregular grape-like clusters are formed *e.g. Staphylococcus aureus*. *Micrococcus* divides into two planes to form square groups of 4-cells called **tetrads**. The cocci of *Sarcina* divide in three planes and form cubical pockets of 8-cells (Fig. 8.1)

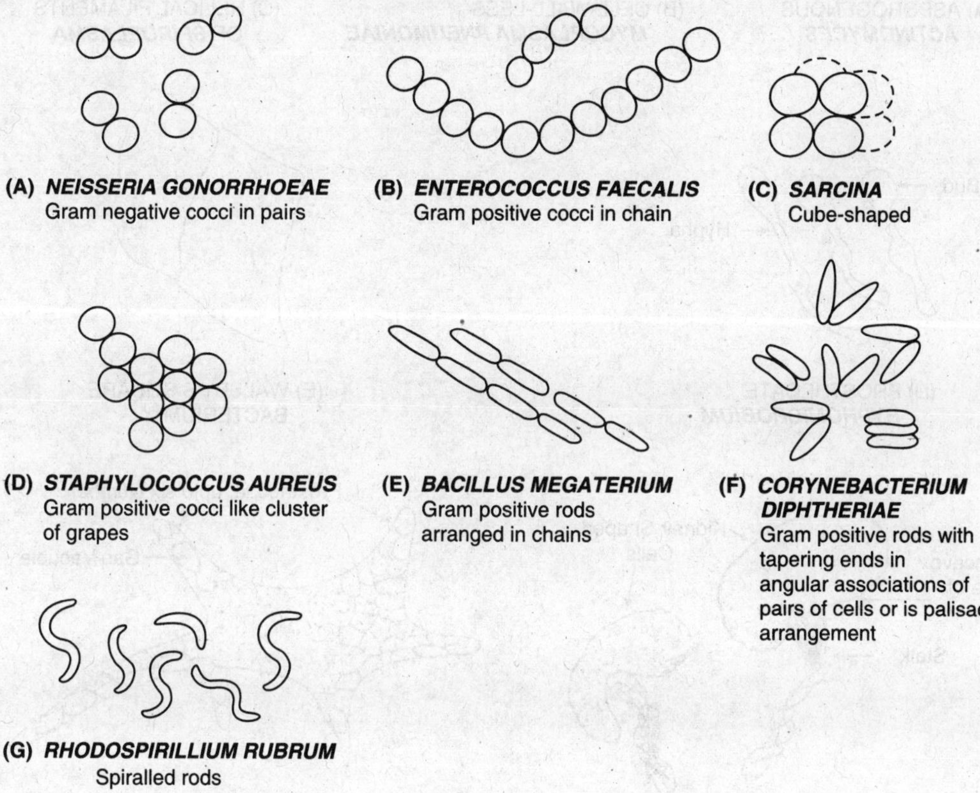

Fig. 8.1. A to G. Common representative forms of bacteria.

Another common bacterial shape is that of a rod which have been called bacilli *e.g. Bacillus megaterium*. Many rods are found singly. They may remain together to form pair called diplobacilli or in chains *e.g. Bacillus megaterium* found in long chains.

A few rod shaped bacteria are curved to form comma like structure *e.g. Vibrio cholerae*. The *Actinomycetes* called mould-like-bacteria form long filaments or hyphae which may branch to form a net work called **mycelium**. Some bacteria are shaped like long rods twisted or spiraled or in helices and are called **spiralla** if rigid and **spirochaetes** when flexible. *Hyphomicrobium* produce oval to pear-shaped bud at the end of long hypha. *Gallionella* form nonliving stalks. Some bacteria are flat, squarish to rectangular boxes living in salt pond discovered by Anthony E. Walsby, which are 2 by 2–4 μm and 0.25 μm in thickness. *Ancalochloris perfilievii*, **irregular star-like** bacteria have been isolated from water sample of hypolimnion of meromictic lake Bol'shoi, kichier, Russia (Fig. 8.2G). These are provided with **prosthecae** upto six in number, and also possess **gas vacuoles** and are nonmotile.

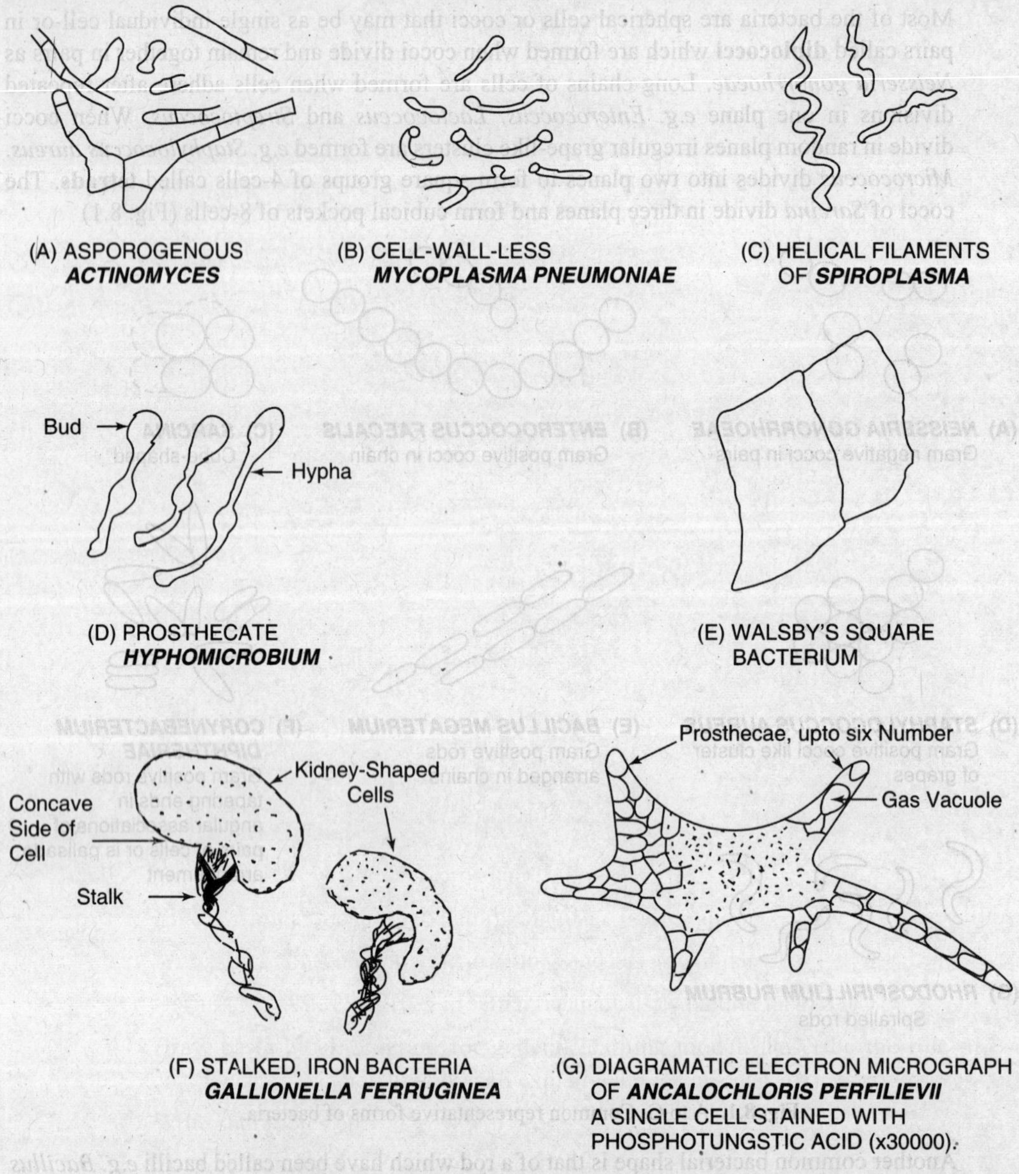

Fig. 8.2. A to F, Some Bacteria of Unusual Shapes

Q.3. What are **giant bacteria**? Explain giving suitable examples.

Ans. These are exceptionally large sized and have also been called **monstrous microbes**, as *Epulopiscium fishelsoni*, a prokaryote related to Gram positive genus *clostridium* and *Thiomargarita nambiensis*. (Fig. 8.3).

The *Epulopiscium fishelsoni* was identified by Esther Angent, Kendall Clemens and Norman Pace using rRNA sequence comparisons, which were large cigar shaped microorganisms which had already been isolated from intestinal tract of Red Sea brown surgeon fish and were wrongly regarded as protist by Fishelsón, Montgomery and Myrberg because of their large size. The generic name *Epulopiscium* was given from Latin *epulum* = a feast or banquet and *piscium* = fish and the species was named in honour of Fishelson. The cells of *E.*

fishelsoni can attain a size of 80 × 600 μm and generally range from 200 to 500 μm in length. *E. fishelsoni* is about a million times larger in volume in comparison to *Escherichia coli*. Despite of the exceptionally large size of *Epulopiscium* it bears a prokaryotic cell structure. (Fig. 8.3).

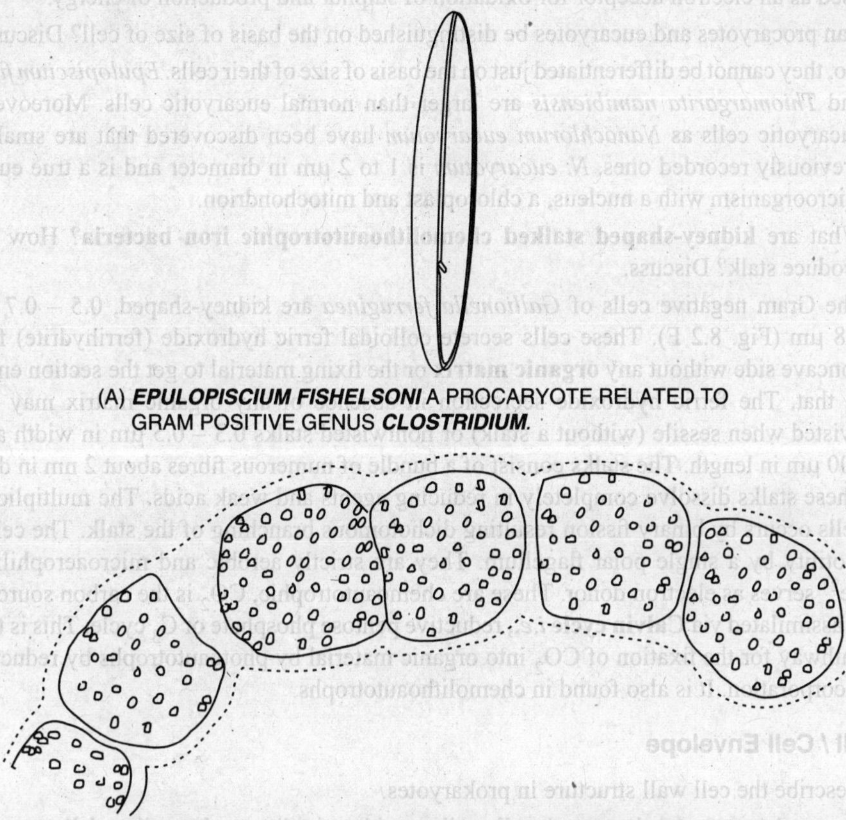

(A) ***EPULOPISCIUM FISHELSONI*** A PROCARYOTE RELATED TO GRAM POSITIVE GENUS ***CLOSTRIDIUM***.

(B) CHAINED CELLS OF ***THIOMARGARITA NAMIBIENSIS*** ISOLATED FROM OCEAN SEDIMENT OFF THE COAST OF NAMIBIA SHOWING EXTERNAL MUCOUS SHEATH AND INTERNAL SULPHUR GLOBULES.

Fig. 8.3. A to B. Wonderful huge procaryotes, the **giant bacteria**.

The *E. fishelsoni* is motile and swims about two body lengths per second *i.e.*, 2.4 cm per minute with the help of bacterial type flagella that cover the surface of the cell. The cytoplasm contains large nucleoids and many ribosomes as would be required for a large cell of this type. This organism can overcome the size limits built by diffusion by having an outer layer comprising of **convoluted plasma membrane** that increases the surface area of the cell and help in nutrient transport. It is believed that *Epulopiscium* gets transmitted between hosts through faecal contamination of fishs' food. However, the bacterium can be eliminated by starving the surgeon fish for a few days. If the juvenile surgeon fish which does not have the bacterium are placed with the infected hosts that gets reinoculated. However, it is surprising enough that this does not happen in adult surgeon fish.

Heidi Schulz in 1997 discovered the much larger prokaryote *Thiomargarita namibiensis* from the ocean sediment off the Coast of Namibia, which is a spherical bacterium between 100 and 750 μm in diameter and often form the chains of cells enclosed in slime sheath. This is more than 100 times larger in volume in comparison to *Epulopiscium fishelsoni*. A vacuole having nitrate rich fluid occupies 98 percent of cell. The cell is surrounded by 0.5 to 2.0 μm layer of cytoplasm filled with **sulphur granules**. The cytoplasm layer is almost of same thickness in all bacterial cells and thin for diffusion rates. The nitrate present in the fluid is used as an electron acceptor for oxidation of sulphur and production of energy.

Q.4. Can procaryotes and eucaryotes be distinguished on the basis of size of cell? Discuss.

Ans. No, they cannot be differentiated just on the basis of size of their cells. *Epulopiscium fishelsoni* and *Thiomargarita namibiensis* are larger than normal eucaryotic cells. Moreover, some eucaryotic cells as *Nanochlorum eucaryotum* have been discovered that are smaller than previously recorded ones. *N. eucaryotum* is 1 to 2 μm in diameter and is a true eucaryotic microorganism with a nucleus, a chloroplast and mitochondrion.

Q.5. What are **kidney-shaped stalked chemolithoautotrophic iron bacteria**? How do they produce stalk? Discuss.

Ans. The Gram negative cells of *Gallionella ferruginea* are kidney-shaped, 0.5 – 0.7 × 0.8 – 1.8 μm (Fig. 8.2 F). These cells secrete colloidal ferric hydroxide (ferrihydrite) from the concave side without any **organic matrix** or the fixing material to get the section embedded in that. The ferric hydroxide secretion in absence of any organic matrix may produce twisted when sessile (without a stalk) or nontwisted stalks 0.3 – 0.5 μm in width and upto 400 μm in length. The stalks consist of a bundle of numerous fibres about 2 nm in diameter. These stalks dissolve completely in reducing agents and weak acids. The multiplication of cells occurs by binary fission resulting dichotomous branching of the stalk. The cells show motility by a single polar flagellum. They are strictly aerobic and microaerophilic. Only Fe^{2+} serves as electron donor. These are chemoautotrophic, CO_2 is the carbon source which is assimilated via **Calvin cycle** *i.e.*, reductive pentose phosphate or C_3 cycle. This is the main pathway for the fixation of CO_2 into organic material by photoautotrophs by reduction and incorporation. It is also found in chemolithoautotrophs.

Cell Wall / Cell Envelope

Q.6. Describe the cell wall structure in prokaryotes.

Ans. The prokaryotes (prokaryotes) cell wall provides rigidity to the cell and lies outside the plasma membrane. Most bacteria except mycoplasmas and some Archaea possess cell wall. The strength and toughness provided by cell wall is mainly due to peptidoglycan.

On the basis of Gram's stain developed by Christian Gram, the cell wall has been divided into two categories. On doing Gram staining, the Gram positive bacteria are stained purple while Gram negative are stained pink to red. The Gram positive cell wall consist of 20 to 80 nm thick homogenous peptidoglycan or murein layer outer to the plasma membrane while Gram negative cell wall comparises of 2 to 7 nm peptidoglycan layer and 7 to 8 nm thick outer membrane. Peptidoglycan in Gram positive bacteria is much thicker say 3 to 4 times or more thicker than in Gram negative bacteria. The microbiologists have developed a tern **CELL ENVELOPE** for all structures outer to the plasma member or cell membrane (Fig. 8.4).

A conspicuous space is found between plasma membrane and outer membrane in Gram negative bacteria which is called **periplasmic space**. Sometimes a smaller or little space may also be seen in plasma membrane and wall (thick layer of peptidoglycan) in Gram positive bacteria. It has been revealed recently that periplasmic space may be filled with a loose network of peptidoglycan. Perhaps it is more a gell than a fluid-filled space. The material which fills up the perplasmic space has also been called **periplasm**. Gram-positive cell may have periplasm despite of the fact that perplasmic space is either absent or not conspicuous. Usually electron transport proteins are present in periplasm of denitrifying and chemolithoautotrophic bacteria. The periplasmic space also has got enzymes involved in **protein synthesis** and modification of toxic compounds which may be harmful for the cell. Gram positive wall lacks a visible periplasmic space and seems to have many periplasmic proteins, instead they secrete several enzymes which ordinarily are periplasmic in Gram negative bacteria. Such secreted enzymes are usually called **exoenzymes**. Some enzymes remain in the periplasm and are attached to the plasma membrane.

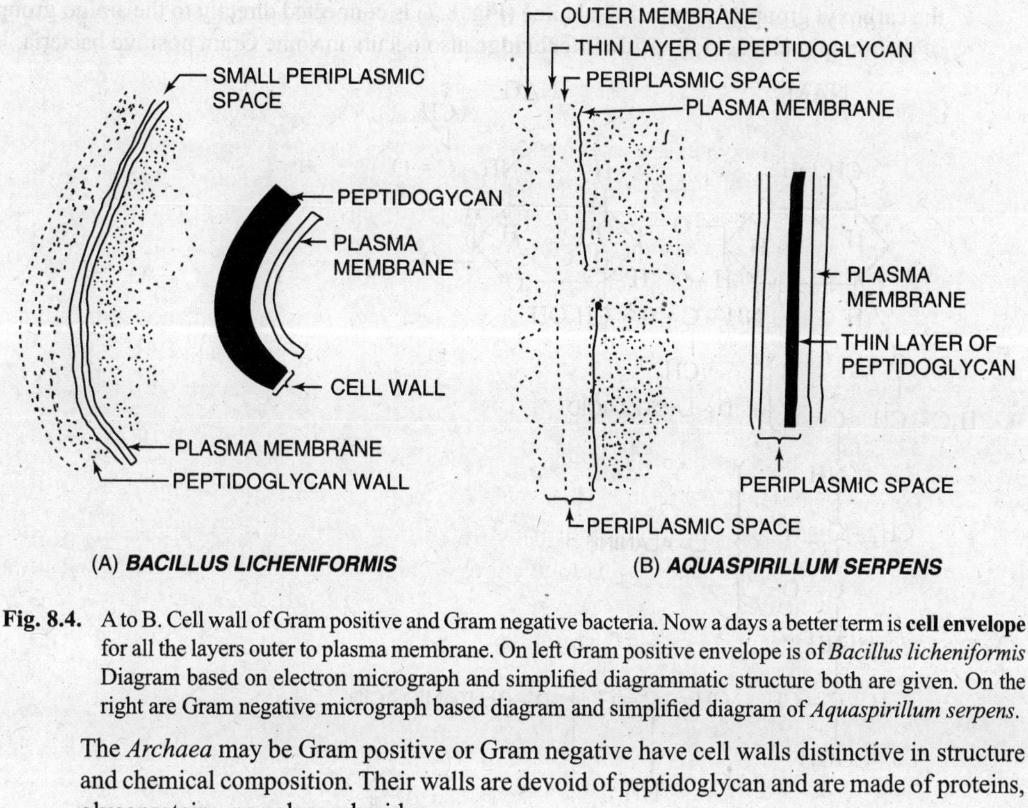

Fig. 8.4. A to B. Cell wall of Gram positive and Gram negative bacteria. Now a days a better term is **cell envelope** for all the layers outer to plasma membrane. On left Gram positive envelope is of *Bacillus licheniformis* Diagram based on electron micrograph and simplified diagrammatic structure both are given. On the right are Gram negative micrograph based diagram and simplified diagram of *Aquaspirillum serpens*.

The *Archaea* may be Gram positive or Gram negative have cell walls distinctive in structure and chemical composition. Their walls are devoid of peptidoglycan and are made of proteins, glycoproteins, or polysaccharides.

Q.7. Why the term cell envelope has been coined instead of cell wall in Gram positive and Gram negative bacterial cells? Discuss.

Ans. The concept of cell wall basically originated from plant cell wall. It is all right for Gram positive bacteria which have a thick layer of peptidoglycan outer to the plasma membrane or cytoplasmic membrane but in case of Gram negative bacteria there is an obvious periplasmic space, a thin layer of peptidoglycan, and an outer membrane. A membrane according to the existing classical concept cannot be the part of a wall. A better term **cell envelope** has been developed for plasma membrane and all layers outer to plasma membrane. This covers the plasma membrane, wall and structures like capsule or S-layer. The S-layer of eubacteria has also been regarded as **glycocalyx**.

Structure of Peptidoglycan

Q.8. Describe the structure of peptidoglycan.

Ans. The peptidoglycan also called murein is a polymer made of many identical subunits This polymer has got two sugar derivatives N-acetylglucosamine and N-acetylmuramic acid (the lactyl ether of N-acetylglucosamine) and various acids three of which (not occurring in proteins) are: D-glutamic acid, D-alanine and *meso*-diaminopimelic acid. The presence of D-aminoacids protects against attack by most of peptidases, (Fig. 8.5) The backbone of this polymer is made of alternating N-acetylglucosamine and N-acetylmuramic acid residues. A peptide chain of four alternating D-and L-amino acids is connected to the carboxyl group of N-acetyle muramic acids. Some bacteria have got L-lysine in third position for *meso*-diaminopimelic acid. Peptide interbridge of Glycine is shown in (Fig. 8.7). The chains of linked subunits of peptidoglycan are joined by cross-links between the peptides. Generally, the carboxyl group of terminal D-alanine (Fig. 8.7) is connected directly to the amino group of diaminopimelic acid. **Peptide interbridge** also occurs in some Gram positive bacteria.

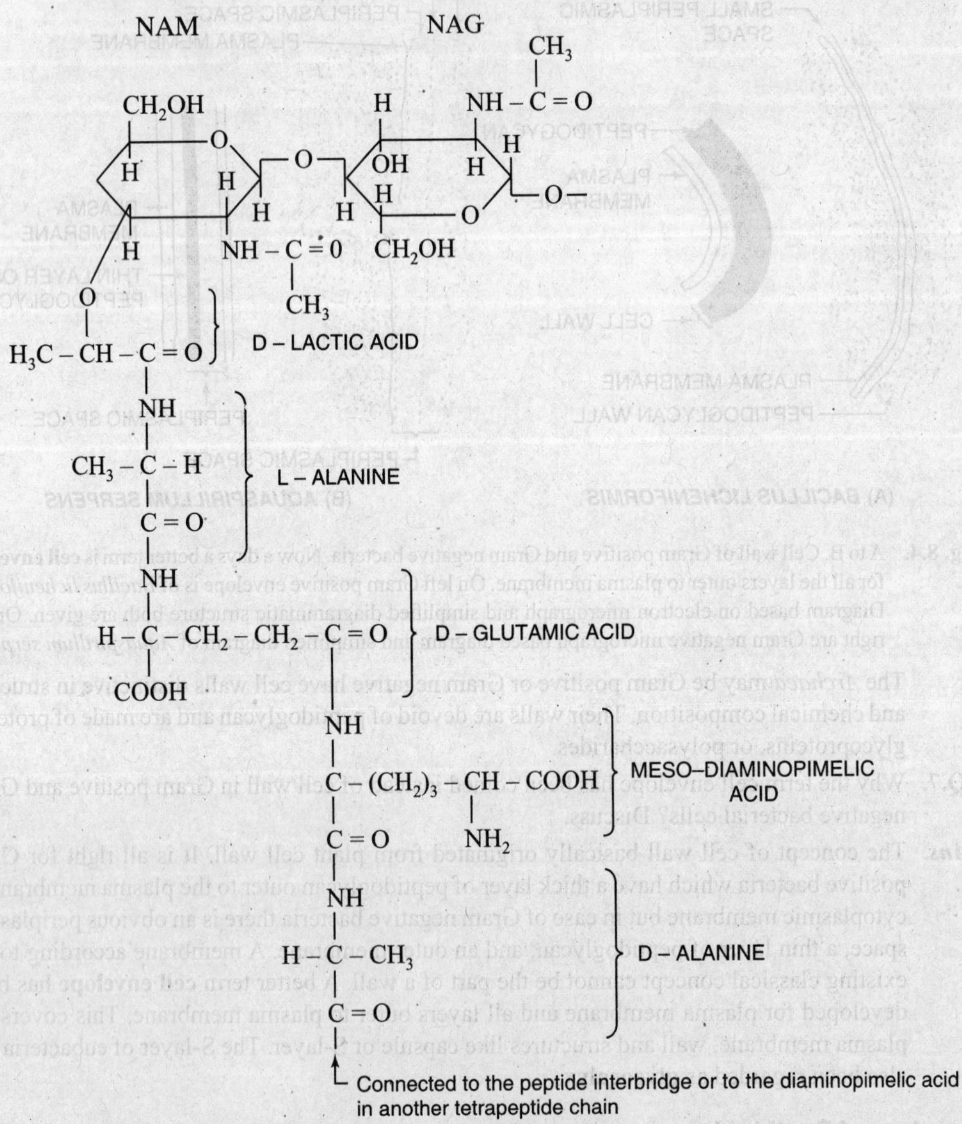

Fig. 8.5. Composition of peptidoglycan subunit of *Escherichia coli* and most other Gram negative bacteria. Same structure also happens in many Gram positive bacteria. NAG stands for N-acetylglucosamine and NAM for N-acetylmuramic acid. NAG with lactic acid is attached by an Ether linkage.

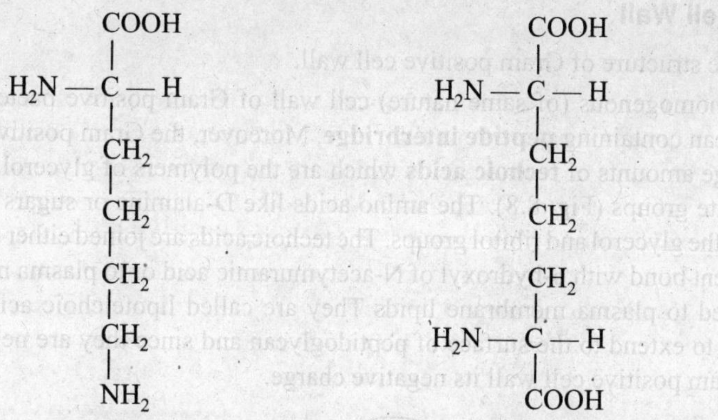

L – LYSINE *meso* – DIAMINOPIMELIC ACID

Fig. 8.6. Diamino acids found in **peptidoglycan**, a large polymer composed of long chains of alternating N-acetylglucosamine and N-acetylmuramic acid residues. The polysaccharide chains are linked to each other through connections between tetrapeptide chains attached to N-acetylmuramic acids. The rigidity and toughness shown by bacterial cell wall is due to peptidoglycan.

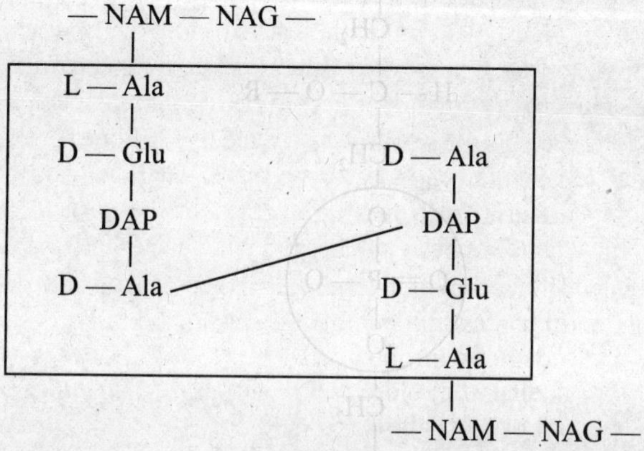

(A) ***ESCHERICHIA COLI*** PEPTIDOGLYCAN WITH DIRECT CROSS-LINKING CHARACTERISTIC OF GRAM NEGATIVE BACTERIA, IN GENERAL

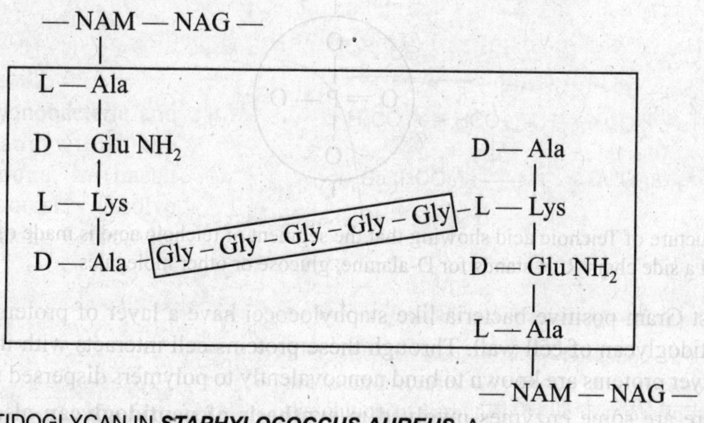

(B) PEPTIDOGLYCAN IN ***STAPHYLOCOCCUS AUREUS***, A GRAM POSITIVE BACTERIUM

Fig. 8.7. A and B, Cross-links in peptidoglycan. NAM stands for N-acetylmuramic acid while NAG stands for N-acetylglucosamine. Gly stands for Glycine.

Gram Positive Cell Wall

Q.9. Describe the structure of Gram positive cell wall.

Ans. Thick and homogenous (of same nature) cell wall of Gram positive bacteria is mainly of peptidoglycan containing **peptide interbridge**. Moreover, the Gram positive cell walls also contain large amounts of **techoic acids** which are the polymers of glycerol or ribitol joined by phosphate groups (Fig. 8.8). The amino acids like D-alamine or sugars like glucose are attached to the glycerol and ribitol groups. The techoic acids are joined either to peptidoglycan or to covalent bond with 6-hydroxyl of N-acetymuramic acid or to plasma membrane lipids. When joined to plasma membrane lipids They are called lipoteichoic acids. The teichoic acids seem to extend to the surface of peptidoglycan and since they are negatively charged provide Gram positive cell wall its negative charge.

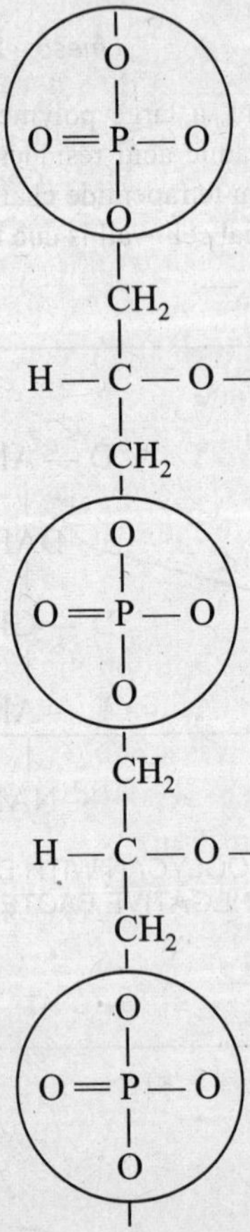

Fig. 8.8. Structure of Teichoic acid showing that the segment of teichoic acid is made of phosphate, glycerol and a side chain R. R stands for D-alanine, glucose or other molecules.

Most Gram positive bacteria like staphylococci have a layer of proteins on the surface of peptidoglycan of cell wall. Through these proteins cell interacts with its environment. The S-layer proteins are known to bind noncovalently to polymers dispersed throughout the wall.

There are some enzymes involved in **synthesis of peptidoglycan** also appear to interact noncovalently with the cell wall. However, other surface proteins are attached to peptidoglycan covalently. The enzyme **sortase** catalyses the attachment of surface proteins to the peptidoglycan of Gram positive cells of bacteria.

Gram Negative Cell Walls of Procaryotes

In Gram negative bacteria peptidoglycan is thin and constitute 5 to 10 percent of the cell wall. It is about 2 nm thick and have one or two layers or sheets of peptidoglycan in *Escherichia coli*. The thin peptidoglycan lies between plasma membrane and outer membrane (Fig. 8.4B). The major membrane protein is Braun's lipoprotein, a small lipoprotein covalently joined to the underlying peptidoglycan and embedded in the outer membrane by its hydrophobic ends. The outer membrane and peptidoglycan are so firmly linked by means of lipoprotein that they can be isolated as one unit. Twenty to 100 mm areas of contact between two membranes have been detected in plasmolyzed cells of *Escherichia coli*.

The lipopolysaccharides (LPSs) are the unique constituents of outer membrane which are made up of lipid and carbohydrate and comprise of lipid A, core polysaccharide and O-side chain. The **lipid A** part contains two glucosamine sugar derivatives each with three fatty acids and phosphate or polyphosphate attached. It is buried in the outer membrane and the remainder of the 'LPS' molecule projects from the surface. The **core polysaccharide** part is joined to lipid A and is constructed of 10 sugars in *Salmonella*. The **O-side chain** or **O-antigen** is a polysaccharide chain stretched outwards from the core. It has got several unique sugars and varies in composition in different bacterial strains. Although the host antibodies can recognize O-side chains but surprising enough that the Gram negative bacteria foil the host defenses by rapidly changing the nature of O-side chain to escape detection.

Also lipid A is toxic thereby LPS can act as endotoxin, and is responsible for the symptoms which appear in infections caused by Gram negative bacteria.

The outer membrane prevents or retards the entry of bile salts, antibiotics which may kill or harm the bacterium. Thus outer membrane in Gram negative bacteria serves the protective function. Interesting enough that outer membrane is more permeable that plasma membrane and permits the entry of small useful molecules like glucose and other monosaccharides, which is facilitated by the presence of special **porin proteins**. Three porin molecules get grouped together and span the outer membrane through which molecules smaller than 600 to 700 daltons can pass.

Flagellum

Q.10. Describe the structure of Gram negative and Gram positive flagella.

Ans. The bacterial flagellum in Gram negative as well as Gram positive bacteria comprises of three parts. 1, an elongated and conspicuous **filament** made of helical protein. 2, **Hook** is a very tightly coiled right handed helix which is curved through about 90° and is composed of protein subunits which are distinct from flagellin but are assembled in a similar fashion. 3, **Basal body** varies with species but essentially a stack of ring-shaped structures on a central hollow, rod-shaped core. In *Escherichia coli* and related Gram negative species there are four rings (Fig. 8.9) In Gram positive *Bacillus subtilis* there are only two rings. In *Caulobacter crescentus* there are five rings. In Campylobacter spp there is a **"cart wheel structure"** with eleven spokes (*i.e.* like bars or wire rods connecting the hub or central part of a wheel with the rim or edge. In *E. coli* and most Gram negative bacteria the outer most L-ring is associated (coplanar) with outer membrane the next P-ring is coplanar with the wall. Peptidoglycan. The S and inner most M rings are associated with plama membrane or cytoplasmic membrane.

Mechanism of Gram Staining

Q.11. How is dye crystal violet-iodine complex retained during short decolourization by alcohol in Gram positive bacteria but not in Gram negative bacteria? Explain.

Ans. The peptidoglycan is thick in Gram positive bacteria. It shrinks on decolourization by ethanol or alcohol consequently the pores of peptidoglycan shrink therefore crystal violet-iodine complex is retained in the Gram positive bacteria. On the other hand the peptidoglycan in Gram negative bacteria is very thin and not highly cross-linked and with larger pores, permits the decolounization. Moreover, alcohol treatment also extracts enough lipid from the Gram negative wall hence further increases its porosity. Thus alcohol takes out the pruple crystal violet-iodine complex from Gram negative bacteria and instead these are stained pink to red on staining with differential stain safanin.

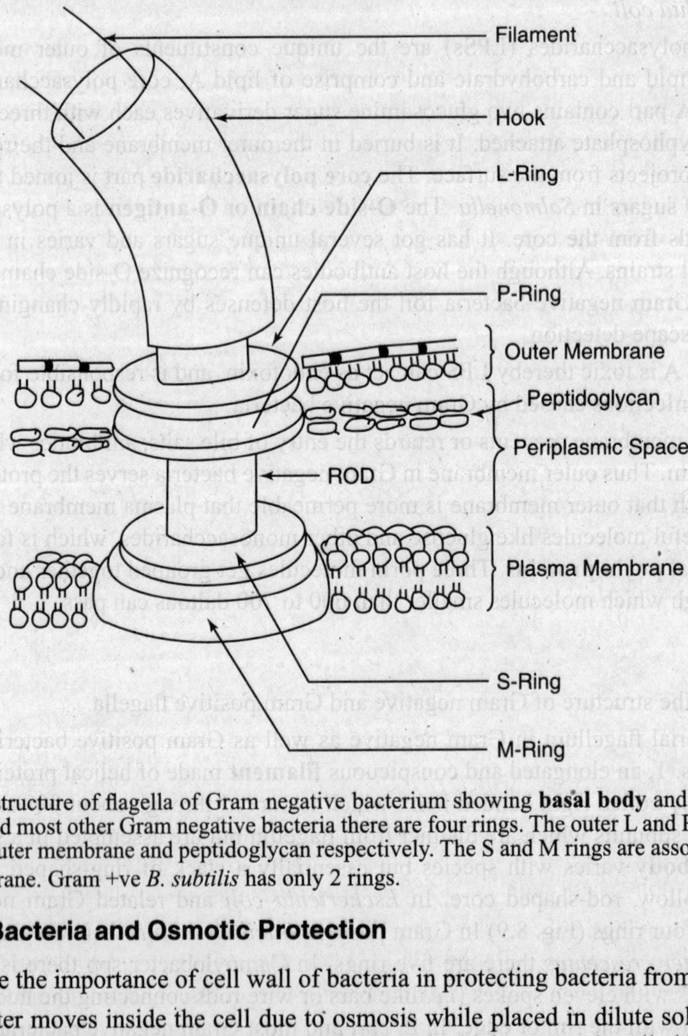

Fig. 8.9. Ultra structure of flagella of Gram negative bacterium showing **basal body** and **hook**. In *Escheria coli* and most other Gram negative bacteria there are four rings. The outer L and P rings are coplanar with outer membrane and peptidoglycan respectively. The S and M rings are associated with plasma membrane. Gram +ve *B. subtilis* has only 2 rings.

Cell Wall of Bacteria and Osmotic Protection

Q.12. Describe the importance of cell wall of bacteria in protecting bacteria from osmotic lysis.

Ans. The water moves inside the cell due to osmosis while placed in dilute solutions. Thus cell of bacteria swells and ruptures and gets destroyed which process is called **lysis**. This can be demonstrated by treating the bacteria with enzyme **lysozyme** which hydrolyzes the bond which connects the N-acetylmuramic acid with carbon four of N-acetyl glucosamine. The **penicillin** inhibits the synthesis of peptidoglycan.

However, if bacteria are incubated with penicillin in an **isotonic solution** (*i.e.* of the same concentration as that of bacterial cell) the Gram positive bacteria are converted to a **protoplast** (*i.e.* cell minus cell wall) and continue to grow when isotonicity is maintained even they lack cell wall. The Gram negative bacteria retain their outer membrane after treatment with penicillin and form a **spheroplast**.

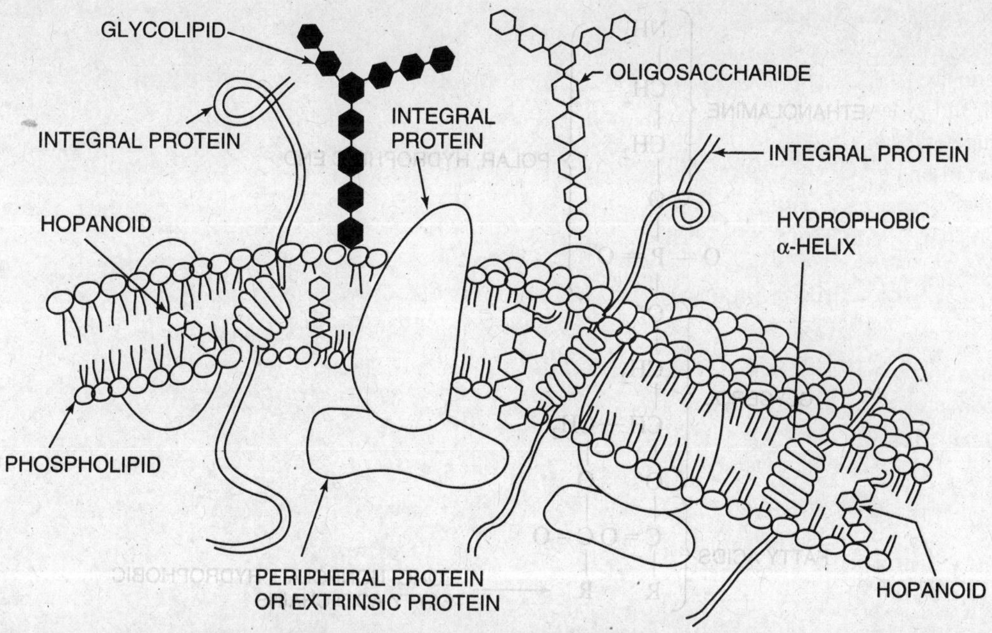

Fig. 8.10. Structure of plasma membrane based on **fluid mosaic model** for bacterial membrane proposed by S. Jonathan Singer and Garth Nicholson. Small spheres represent the **hydrophilic ends** of membrane phospholipids while the tails are for **hydrophobic fatty acid chains**.

Procaryotic Cell Membranes

Q.13. Describe structure and function of the **plasma membrane**.

Ans. The bacterial plasma membrane is considered to have high proportion of proteins than eucaryotic membranes perhaps they meet out different functions which are performed by organelle membranes in eucaryotes. Most of the membrane associated lipids in bacteria (Fig. 8.11), are structurally asymmetric with polar and nonpolar ends and are known to be amphipathic. Polar ends are hydrophilic and interact with water and nonpolar ends are hydrophobic and are insoluble in water. Thus lipids form a bilayer in membranes. Many bacteria have pentacyclic sterol like molecules called **hopanoids** in their membrane (Fig. 8.12 B). It has been estimated that total amount of hopanoids in sediments is about 10^{11-12} tons. which is equal to total mass of organic carbon in all living organisms *i.e.* 10^{12} tons. The hopanoids have contributed significantly to the **formation of petroleum**. Hopanoids are synthesized from the same precursors as steroids. Like steroids in eucaryotes hopanoids stabilize the bacterial membrane. Archaeal membranes differ from bacterial membranes in having a monolayer with lipid molecules spanning the whole membrane. **Peripheral proteins** are soluble in water and comprise 20 to 30 percent of membrane proteins, while **integral proteins** are 70 to 80 percent. Integral proteins are amphipathic. Their hydrophobic regions are buried in liquid and hydrophilic parts project from the membrane surface.

Internal Membrane Systems

The **mesosomes** in bacteria are found next to septa or cross walls, and are as invagination of plasma membrane. Recently bacteriologists believe that the mesosomes are artifacts generated during the chemical fixation for electron microscopy and are not true structures.

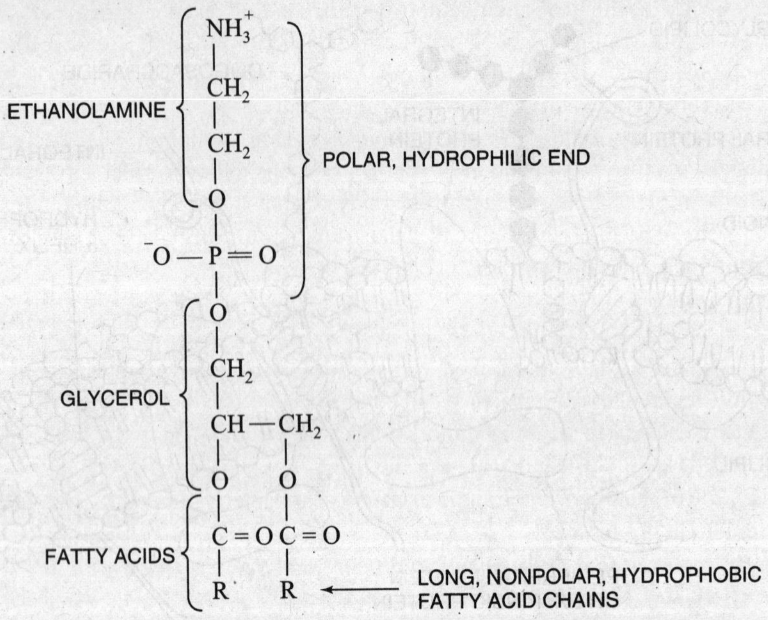

Fig. 8.11. Polar membrane lipid's structure showing phosphatidylethanolamine, an amphipathic phospholipid commonly found in bacterial membranes. **Phosphatidylethanolamine (PE)** is more common and more abundant in Gram negative species. However, **phosphatidylglycerols (PG)** appear to occur in all bacteria.

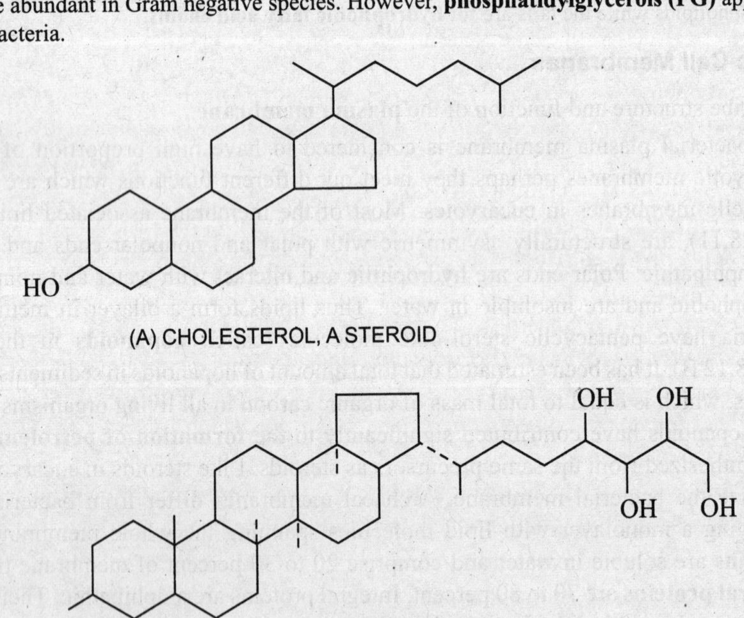

Fig. 8.12. A to B. Membrane steroids and hopanoids. **Steroids** are found in eukaryotic membranes while **hopanoids** are found in the membranes of some bacteria.

Some bacteria represent some internal membrane systems which are altogether different from mesosomes (Fig. 8.14). The internal plasma membrane infoldings are present as **lamellar stacks** in *Ectothiorhodospira* and other member of 'Anoxygenic phototrophic bacteria'. The infoldings become extensive and complex in cyanobacteria, purple bacteria or in bacteria with high respiratory activity, *e.g.* nitrifying bacteria.

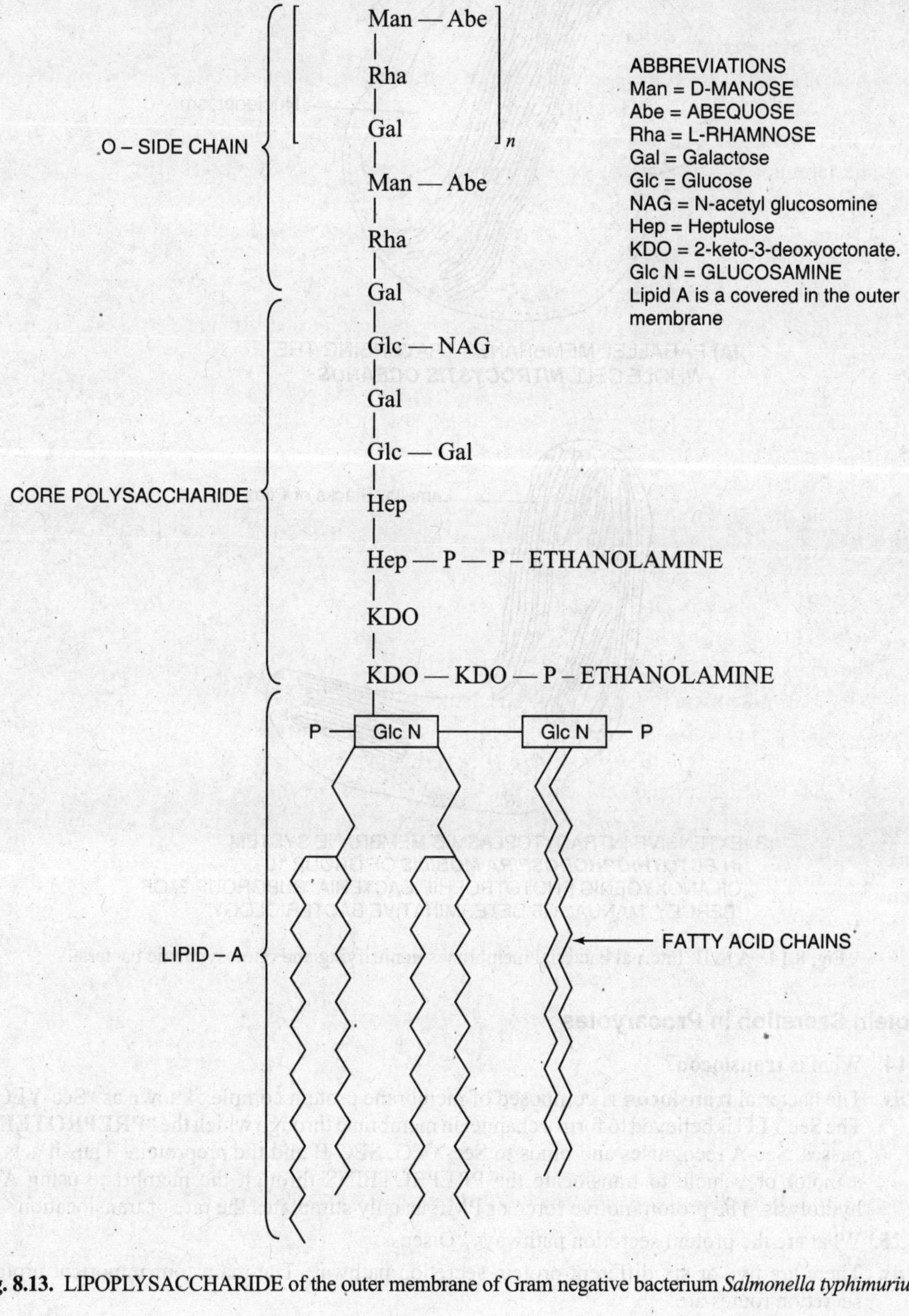

Fig. 8.13. LIPOPLYSACCHARIDE of the outer membrane of Gram negative bacterium *Salmonella typhimurium*.

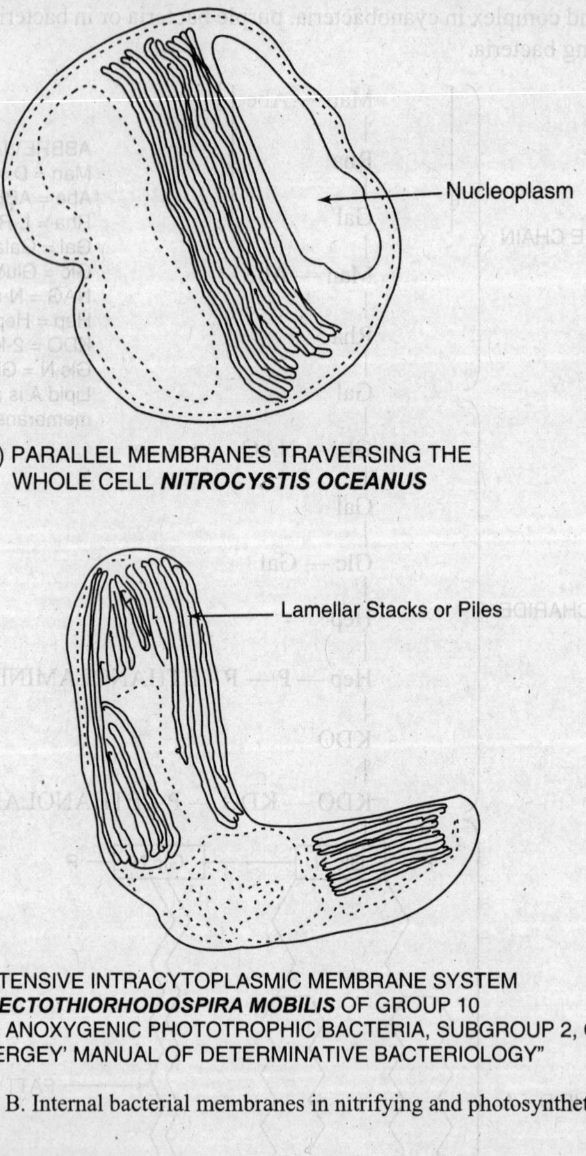

Fig. 8.14. A to B. Internal bacterial membranes in nitrifying and photosynthetic bacteria.

Protein Secretion in Procaryotes

Q.14. What is **translocon**?

Ans. The bacterial **translocon** is composed of membrane protein complex known as "Sec-YEG". The See YEG is believed to form a channel in membrane through which the "**PREPROTEIN**" passes. Sec-A recognises and binds to Sec-YEG, SEC-B and the preprotein. Thus it acts as a motor or vehicle to translocate the PREPROTEINS through the membrane using ATP hydrolysis. The proton motive force or PMF usually stimulates the rate of translocation.

Q. 15. What are the protein secretion pathways? Discuss.

Ans. There are five or six different protein secretion pathway. The main four important protein secretion routes are:

1. Sec-dependent pathway. 2. Type II protein secretion pathway. 3. Type I protein secretion pathway or the ABC protein secretion pathway. 4. Type III protein secretion pathway.

The **Sec-dependent pathway** has also been called the general secretion pathway (GSP) which either translocates proteins through the plasma membrane or integrates them into the membrane. The **PRESECRETORY PROTEIN** or preprotein bears a **signal peptide** at its amino terminal or N-terminal end.

The signal peptide delays protein folding after its synthesis on ribosome and is specifically recognized by the Sec machinery. Where these **PREPROTEINS** are synthesised by the ribosomes, the **CHAPERONES** or the **molecular chaperones** bind to them. The chaperones keep the preproteins unfolded and assist them to reach the transport machinery or **translocon**. The most of used chaperones in the *Escherichia coli* are Sec-B and the signal recognition particles (SRP). Sec-B is found in Gram negative bacteria and SRP occurs in all procaryotes.

The Sec-B bind to Sec A component of translocon and the preprotein is tranferred to Sec-A. In other procaryotes the SRP-pre-protein complex moves to SRP receptor. The pre protein is released with the support of GTP hydrolysis and the preprotein associates with the translocon, perhaps with the Sec-A as in case of in *Escherichia coli*.

To conclude, after the preprotein gets bound to translocon it is transferred through the membrane. The bacterial translocon is made of membrane protein complex known as Sec YEG, dissociable Sec-A and other accersory proteins *i.e.* Sec-D, Sec-F. The Sec-YEG is believed to form a channel in the membrane through which the preprotein passes.

The Sec-A recognizes and binds to Sec YEG, Sec-B and the preprotein. This complex works as a motor to translocate through the membrane using ATP hydrolysis. Proton motive force or the PMF generally stimulates the rate of translocation.

Type II PROTEIN SECRETION PATHWAY: Some Gram negative bacteria adopt the pathway to transport proteins across the outer membrane after they reach periplasm usually through the activity of Sec-dependent pathway.

The type II protein secretion pathway is most important of the routes discussed here and is found in pathogens like *Erwinia carotovora, Pseudomonas aeruginosa* and *Vibrio cholerae*. This pathway is responsible for secretion of proteins as cellulases, pectinases, proteases, lipases, cholera toxins and pili proteins. This is a quite complex and contain twelve to fourteen proteins most of which are found in integral membrane proteins.

TYPE I PROTEIN SECRETION PATHWAY OR ABC PROTEIN SECRETION PATHWAY: A large number of Sec independent secretion pathways are members of the ATP binding cassette (ABC) family of transport systems. The type I protein secretion pathway is associated in secretions of proteins like toxins (α-haemolysin), proteases, lipases and specific peptides. The proteins secreted often have C-terminal secrection signals. The ATP hydrolysis is used to drive translocation in this system. The protein moves through a channel developed by hydrophobic domains. A protein links the plasma membrane and outer membrane during transport in *Escherichia coli*. A channel is formed in the outer membrane by some other protein. Tol C trimer appears to have this role for at least some type I systems in *E. coli*. In this way the protein translocation takes place in one step through both membranes.

TYPE III PROTEIN SECRETION PATHWAY: Some Gram negative pathogens have type III protein secretion pathway and use this systems to secrete virulence factors and inject them into plant and animal host cells, *e.g.* by *Salmonella, Yersinia, Shigella, E. coli, Bordetella, Pseudomonas aeruginosa* and *Erwinia*. The contact between these bacterial pathogens and their host cells stimulates the process and an activation signal such as a low calcium ion level is needed for efficient secretion. Nearly twenly various proteins become components of type III systems. The type III secretion machinery is shaped like a needle. The slender needle like part of the system extends out of the surface of the cell. A cylindrical base is connected to both the outer membrane and plasma membrane appears like basal body of the flagellum. It is believed that the proteins may move through the translocation channel. Four different types of protein are secreted by type III system. These are: (1) Needle proteins or proteins needed for secretion. (2) Secretion regulators. (3) Proteins which help in secretion of proteins in target cells. (4) Effector proteins which change the host cell functions. Effector proteins are sometimes enzymes which serve many purposes i.e. they may act as cytotoxins, as inhibitors of phagocytosis, stimulators of reorganization of cytoskeleton or promoters of **apoptosis** (cell death in pathological circumstances). Small chaperones are believed to be involved in delivery of proteins required for the secretory system.

Cytoplasmic Matrix

Q.16. What is **cytoplasmic matrix**?

Ans. The substance or material between plasma membrane (cytoplasmic membrane) and the nucleoid is called cytoplasmic matrix, and the plasma membrane and every thing inside it is called a **protoplast**.

Inclusion Bodies

Q.17. What are inclusion bodies?

Ans. The **inclusion bodies** are numerous types of granules of organic and inorganic material seen in the cytoplasmic matrix (Fig. 8.16). Typical bilipid layer membranes are absent in inclusion bodies but some inclusion bodies are enclosed in a single layered so called membrane which is 2.0 to 4.0 mm thick. The membrane enclosed inclusion bodies of this type are poly-β-hydroxybutyrate granules, some glycogen, sulphur granules, carboxysomes, gas vacuoles and vesicles. The single layered so called membrane of inclusion bodies may be some protein or lipid. The quantity of inclusion bodies depend on their requirement as polyphosphate granules get depleted in fresh water habitats which are deficient in phosphate. The inclusion bodies not bound by any sort of membrane are polyphosphate granules, cyanophycin granules and some glycogen granules. Glycogen granules are in both forms. Some inclusion bodies also serve purposes other than storage *e.g.* **magnetosomes** (Fig. 8.15) which help bacteria to arrange and orient in magnetic field of the earth.

Nucleoid

Q.18. What is a **nucleoid**?

Ans. It is prokaryotic chromosome located in irregular shaped region, which is also known as nuclear body, chromatin body or nuclear region. The membrane bound nucleus is absent in procaryotes which is found in eucaryotes. Usually procaryotes have a single circle of double stranded DNA but some have a linear DNA chromosome. Currently it has been discovered that bacteria like *Vibrio cholerae* have more than one chromosome. Many bacteria have **plasmids** also.

Chlorosomes

Q.19. What are **chlorosomes**? Discuss.

Ans. The chlorosomes are the elongated pigment vesicles of bacteria attached to plasma membrane. The chlorosomes are a characteristic of subgroup 5 and 6 of Group 10 of Anoxygenic phototrophic bacteria of 9th edition of "Bergey's Manual of Determinative Bacteriology" published in 1994. The typical genera that exhibit chlorosomes are *Chlorobium* and *Chloroflexus*. The chlorosomes are elongated intracellular membranous vesicles formed in all members of CHLOROBIINEAE grown under appropriate conditions . These are vesicles which contain most of the light harvesting pigments of photosynthesis. These have also been called **antenna of photopigments**. The chlorosomes are attached to cytoplasmic membrane but differ in structure. The chlorosomes in the members of chlorobiaceae contain bacteriochlorophyll c, d and e and various carotenoids *e.g.* chlorobactene in green cells, isorenieratene in brown cells. In Chloroflexaceae they possess bacteriochlorophyll c and y and β-carotene (beta carotene). A small amount of 'antenna' Bchl – a is present in Chlorosomes of *Chloroflexus aurantiacus* and also appears to occur in chlorosomes of members of Chlorobibiaceae. The cytoplasmic membrane in Chlorobiineae contains the reaction centres Bchl-a together with small amounts of 'antenna' Bchl-a.

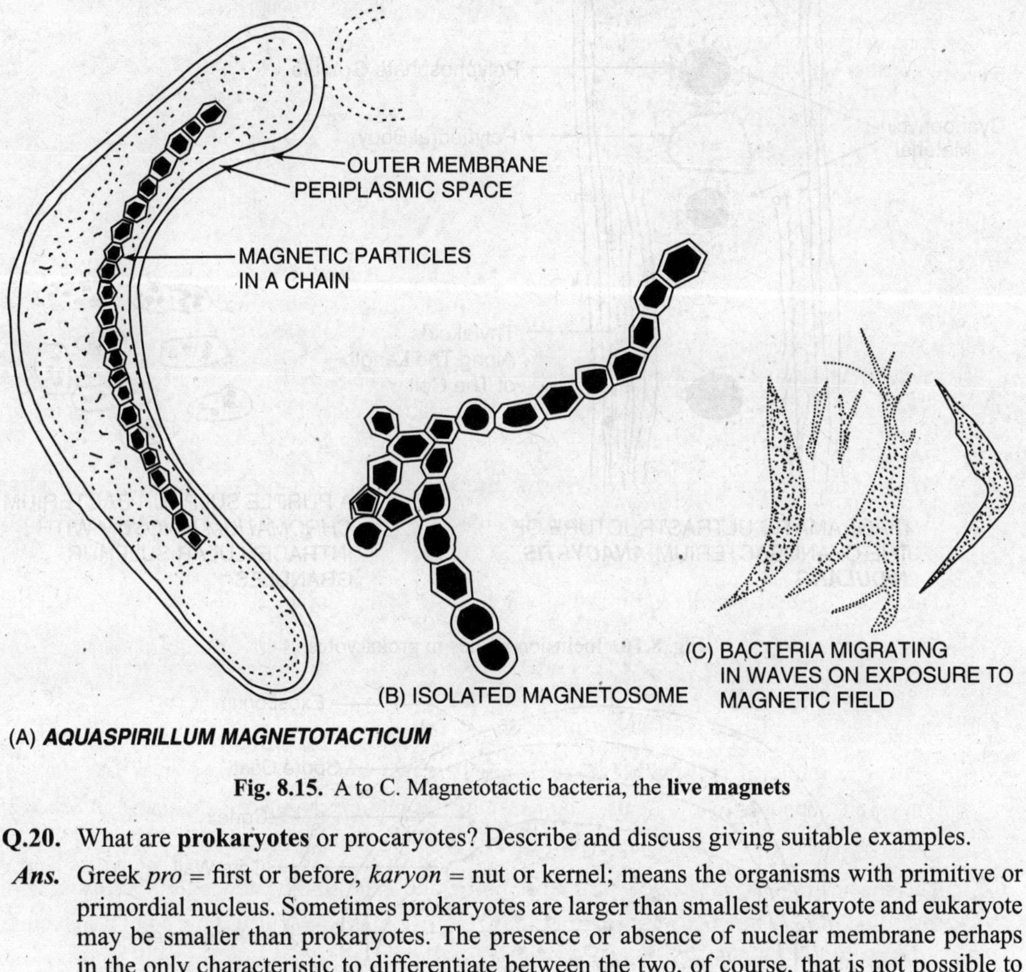

Fig. 8.15. A to C. Magnetotactic bacteria, the **live magnets**

Q.20. What are **prokaryotes** or procaryotes? Describe and discuss giving suitable examples.

Ans. Greek *pro* = first or before, *karyon* = nut or kernel; means the organisms with primitive or primordial nucleus. Sometimes prokaryotes are larger than smallest eukaryote and eukaryote may be smaller than prokaryotes. The presence or absence of nuclear membrane perhaps in the only characteristic to differentiate between the two, of course, that is not possible to differentiate between the two without help of a transmission electron microscope (TEM). Thus the **cellular microorganisms** in which the cells are devoid of true membrane bound or membrane enclosed nucleus and their genetic material is located in a **nucleoid**, *i.e.* in prokaryotes chromosomes are not separated from cytoplasm by a specialized membrane. There is only a **single chromosome** in bacteria but recently it has been reported that some bacteria like *Vibrio cholerae* possess more than one chromosome. The nucleoid can be seen under light microscope on staining with Feulgen stain which reacts with DNA specifically and stains it. In the process of cell division a cell may show more than one nucleoid as the genetic material has been duplicated. Although it has been authentically reported that the **mesosomes** are absent in bacteria and the reports of presence of mesososomes are because they are artifacts due to faulty staining techniques and not true structures, yet the careful electron microscopic studies have revealed nucleoid to be found in contact with either mesosome or the plasma membrane. Therefore, bacterial DNA is attached to membranes which may be involved in separation of DNA into daughter cells in process of division of cell. Nucleoids have been isolated free from membranes as well as intact with membranes.

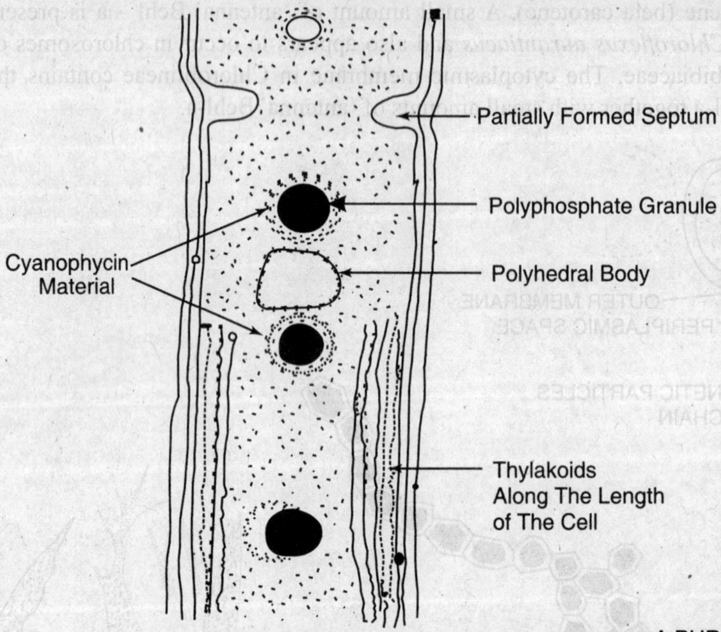

DIAGRAMATIC ULTRASTRUCTURE OF THE CYANOBACTERIUM ***ANACYSTIS NIDULANS***

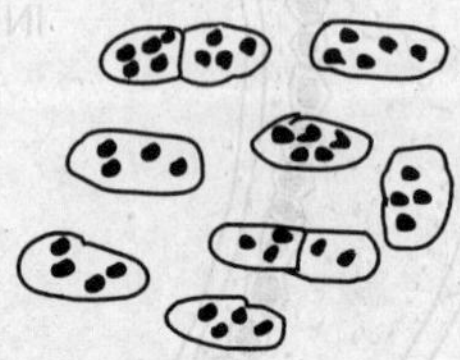

A PURPLE SULPHUR BACTERIUM *CHROMATIUM VINOSUM* WITH INTRACELLULAR SULPHUR GRANULES

Fig. 8.16. Inclusion bodies in prokaryotes.

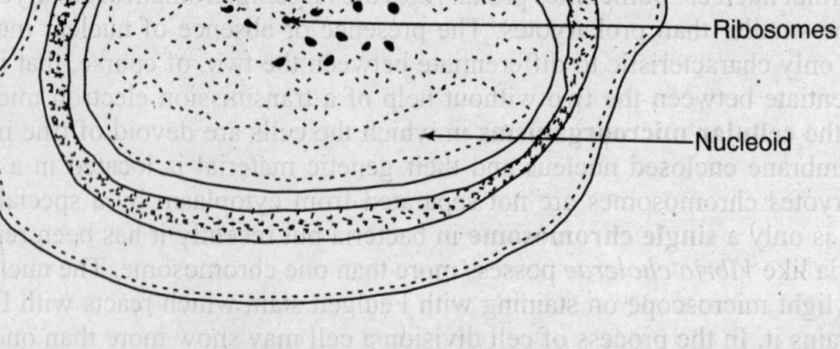

Fig. 8.17. Structure of endospore.

Chemical composition of nucleoids have been found to be 60% DNA, 30% RNA and 10% protein by weight. The rod shaped cell of *Escherichia coli*, 2 to 6 μm, possesses closed circle of DNA which is 1400 μm in circumference. DNA is looped and coiled to a great extent, perhaps with the aid of RNA and nucleoproteins. However, these proteins are different from histone proteins found in eucaryotic nuclei.

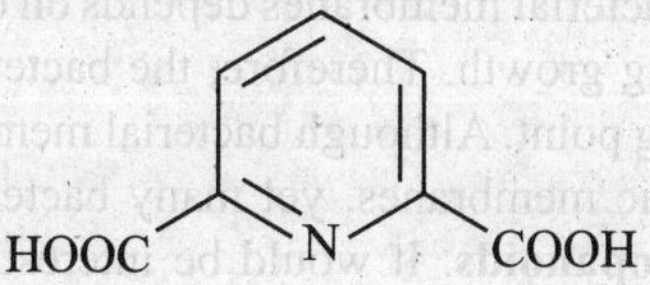

Fig. 8.18. Dipicolinic acid, which is found in high concentration in the bacterial endospore. It provides heat resistance to endospore.

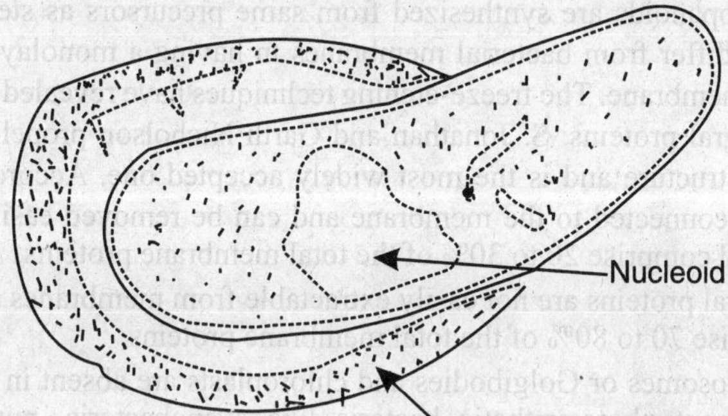

Fig. 8.19. Germination of endospore of *Clostridium pectinovorum*.

The group that claims the absence of mesosomes in bacteria has made it evident by using better techniques of staining in electron microscopy in which dehydration is not required.

Strange enough some exceptions have been reported regarding the membrane in some nucleoids. DNA containing regions are seen in two genera of **Planctomycetes**. The *Pirellula* shows a single membrane that surrounds a region called **pirellulosome** that has a fibrillar nucleoid and ribbon like particles. Also, the nuclear body of *Gemmata obscuriglobus* is enclosed by two membranes which so far remain exceptions that have not been well understood.

The **plasmids** have been reported in many bacteria in addition to chromosome, which are double stranded DNA molecules, often circular, and can exist and replicate independently of the chromosome or may be integrated with the chromosome. The plasmids in both these cases usually are inherited and passed on to the progeny. The plasmids are not attached to plasma membrane and sometimes are lost to one of the progeny cells during the process of cell division. The plasmids are not needed for the host growth and reproduction, no doubt, they carry genes which provide their bacterial host a selective advantage, as the plasmid genes may render bacteria drug resistant, provide them new metabolic activities, turn them pathogenic or may provide them with other properties.

Plasma Membrane

The bacterial cytoplasmic membrane or the cytoplasmic membrane is typically devoid of sterols. The **plasma membrane** contains both proteins and lipids. The plasma membrane of bacteria possesses greater amount of proteins in comparison to eukaryotes perhaps because the bacterial membranes perform many other function which are performed by organelle membrane in case of eukaryotes.

Generally membrane associated lipids are structurally amphipathic. The polar end interact with water, thus are **hydrophilic** and the nonpolar ends are **hydrophobic** and are insoluble in water. In this way lipids form a bilayer in membranes. The outer surface are hydrophilic while the hydrophobic ends are buried in the enterior away from surrounding water. Most of the amphipathic lipids are phospholipids.

The lipid composition of bacterial membranes depends on environmental temperature so that the membrane remains fluid during growth. Therefore, the bacteria growing at lower temperature have fatty acids with lower melting point. Although bacterial membranes do not possess sterols like cholesterol that occur in eucaryotic membranes, yet many bacterial membranes have pentacyclic sterol-like molecules known as **hopanoids**. It would be interesting to note that big quantities of hopanoids are present in our natural environment. The total amount of hopanoids in sediments has been estimated to be about 10^{11-12} tons that is as much as total amount of carbon in all living organisms, estimated to be 10^{12} tons. The hopanoids are also known to have contributed to the formation of petroleum. The hopanoids are synthesized from same precursors as steroids. Many archaeal membranes, however differ from bacterial membranes in having a monolayer with lipid molecules spanning the whole membrane. The freeze-etching techniques have revealed the presence of peripheral proteins and integral proteins. S. Jonathan and Garth Nicholson propelled the **fluid mosaic model** for membrane structure and is the most widely accepted one. According to them peripheral proteins are loosely connected to the membrane and can be removed easily which are soluble in aqueous solutions and comprise 20 to 30% of the total membrane proteins.

On the other hand integral proteins are not easily extractable from membranes are insoluble in aqueous solutions and comprise 70 to 80% of the total membrane proteins.

The mitochondria, dictyosomes or Golgibodies and chloroplasts are absent in prokaryotes. Moreover, the prokaryotes except photosynthetic bacteria like cyanobacteria, purple bacteria or bacteria with high respiratory activity like nitrifying bacteria do not have internal membrane other than plasma membrane or membrane bound tubules. In cyanobacteria which are prokaryotes membrane bound tubules called **thylakoids** are found. But all bacteria possess plasma membrane which is a requirement of all living cells.

In prokaryotes ribosomes are of 70S type. A cell wall is typically present except in **mycoplasmas** which are also called Mollicutes or cell wall-less bacteria. The peptidoglycan or pseudomurein is present in the cell wall. In some cases a S-layer, a regularly structured layer composed of protein or glycoprotein lies on the surface of many bacteria. The cell wall protects bacterium and provides rigidity. The capsule, flagella and fimbriae are appendages present outside the cell wall. Flagella when present are relatively simple. The flagella help bacteria to move towards attractants and away from repellents. The cytoplasmic matrix typically has several constituents which however, are not membrane enclosed as inclusion bodies, ribosomes or nucleoid (genetic material). The storage compounds usually include glycogen and **poly-β-hydroxybutyrate** (PHB), as inclusion bodies. The granules of organic and inorganic material can also be seen in the cytoplasmic matrix that can be seen with a light microscope.

The resistant spores are present in some bacteria which help bacteria to survive under unfavourable extreme environmental condition.

The nitrogen fixation and chemoautotrophy occur in prokaryotes only. The term prokaryote is synonymous with bacteria including Archaeobacteria and Eubacteria. They generally are placed under kingdom procaryotae, 4 divisions of which have been proposed in 'Bergey's Manual of Determinative Bacteriology' which are: Firmicutes, Gracilicutes, Mendosicutes and Tenericutes.

Now a days **Eubacteria** comprise a phylogenetic category of the rank of domains or empires according to Carl Woese, which includes all those prokaryotic organisms not classified in the Archaeobacteria. The prokaryotes include most Gram positive bacteria, all the cyanobacteria, members of Rhodospirales, all enterobacteria, pseudomonads and prochlorons.

Eubacterial cytoplasmic membrane contains ester-linked lipids. The cell wall while present typically contains peptidoglycan. The **introns** which are noncoding intervening sequences in a split or interrupted gene coding for RNA that is missing from the final product, have not been detected in eubacterial nucleic acid.

To conclude according to 9th edition of 'Bergey's Manual of Determinative Bacteriology' published in 1994 prokaryotes are all the 35 groups covering following:

(i) Archaeobacteria *i.e.* **methanogens** comprising of 18 genera all beginning with prefix methano as *Melhanobacterium, Methanobrevibacter, Methanosphaera, Methanothermus. Methanococcus*; **archaeal sulphate reducers** comprising of two species with *Archaeoglobus fulgidus* and *A. profundus*, **extremely halophilic archaeobacteria** or **halobacteria** comprising of 6 genera *e.g. Haloarcula, Holobacterium, Halococcus, Haloferax, Natronobacterium* and *Natronococcus*; **Cell wall-less archaeobacteria** *e.g. Thermoplasma acidophilum*; and **Extremely thermophilic and Hyperthermophilic S°-metabolizers comprising** of 14 genera *e.g. Acidians, Desulfurolobus, Metallosphaera, Sulpholobus, Pyrobaculum, Thermophilum, Thermoproteus, Desulfurococcus, Hyperthermus, Pyrococcus, Pyrodictium Staphylothermus, Thermococcus* and *Thermodiscus*.

(ii) Bacteria: Gram negative eubacteria, which have cell wall, Gram positive eubacteria that have cell wall, and cell wall less bacteria or mycoplasmas.

(iii) 11th Group of Oxygenic phototrophic bacteria comprising of Cyanobacteria, and Prochlorophytes, order Prochlorales comprising of two genera *Prochloron* and *Prochlorothrix*.

Q.21. Learn to define/describe the technical terms related to prokaryotes given below in alphabetical order.

Ans. **ABC protein secretion**: These are the members of ATP-binding casette (ABC). This has also been called Type I protein secretion pathway which is involved in secretion of proteins like toxins (α-haemolysin proteases lipases and peptides.

Amphitrichous: Presence of a single flagellum (or *e.g.* fimbria) at each pole of a cell.

Axial filament: Composed of periplasmic flagella occuring in (spirochaetes) moving in mucous or mud by spinning movement.

Basal body: The part of flagellum embedded in the cell.

Capsule: A layer of material external to but contiguous with the cell wall. This is well organized and not easily washed off. Made of poly-D-glutamic acid as in *Bacillus anthracis*. In some bacteria (*e.g. Beijerinckia*) a capsule may enclose more than one cell. Commonly divided into three categories as **macrocapsules** or true capsules; **microcapsules** that cannot be observed by light microscopy but by electron microscopy or serological techniques *e.g.* Quellung phenomenon or **Quellung reaction** in which the capsule swells on reacting with bacteria specific antiserum; and **slime layers**.

Carboxysomes: Are polyhedral bodies 100 nm in diameter and contain enzyme ribulose 1,5 biphosphate carboxylase.

Chemoreceptors: These are special proteins which bind chemicals and transmit signals to other components of chemosense system.

Chemotaxis: The movement towards attractants and away from repellents.

Chemotaxonomy: The taxomy based on chemical criteria such as DNA base composition, cell wall lipids 16S rRNA structure etc.

chlorosomes: These are vesicles containing light harvesting pigments also called antenna photopigments found in *Chloroflexus* and Chlorobiaceae

Coccus: A spherical bacterial cell.

Core: In bacteriology pertains to endospore sense but in virology a protein nucleic acid complex surrounded by a capsid.

Core polysaccharide: It is joined to lipid A.

Cortex: Nearly half of the spore volume which rests beneath the spore coat.

Cyanophycin granules: The inclusion bodies of cyanobacteria composed of large polypeptides having approximately equal amount of amino acids and aspartic acid.

Cytoplasmic matrix: The substance lying between plasma membrane and the nucleoid.

Deoxyribonucleic acid: It is D.N.A. The prokaryotes usually contain a single circle of double stranded D.N.A. Some have linear D.N.A. chromosome. *Vibrio cholreae* has more than one chrome.

Diplococcus: The arrangement of cocci in pairs.

Endospore: The heat and chemical resistant thick-walled dormant spores *e.g. Bacillus* and *Clostridium* which are rods while *Sporosarcina are cocci*.

Envelope: The structures outer to plasma membrane. It includes wall and structures like capsule.

Exoenzyme: These are secreted enzymes.

Exosporium: Thin delicate covering of the spore. Below exosporium lies spore coat.

Filament: The longest and most obvious part of the flagellum is filament. The flagellum comprises of filament, basal body and hook.

Fimbriae: (Singular FIMBRIA). These are thin proteinaceous, chromosome or plasmid encoded filaments which extend from cell surface of some types of microbial cells. The fimbriae which often occur in large numbers on a given cell are functionally distinct from both PILI (Singular PILUS) and flagella and are typically able to facilate cell to cell or cell to substratum adhesion. The adhesive fimbriae of certain pathogens are important virulence factors. A given cell may have more than one type of fimbriae. Fimbriae are common on Gram negative bacteria including on cyanobacteria and uncommon on Gram positive bacteria.

The fimbriae also occur on a members of fungi *e.g. Candida albicans, Saccharomyces cerevisiae* and various smut fungi. But the below given account refers specifically to the fimbriae of Gram-negative bacteria.

The fimbriae have a uniform width typically 2–10 nm and 100–5000 nm in length. In atleast some cases on individual FIMBRIA appears to be in the form of a loosely or tightly coiled helix. The smallest fimbriae *e.g.* k 88 fimbriae typically form tangled masses on the cell surface and may appear as an amorphous layer at low magnifications under electron microscope. Such fimbriae are sometimes called **fibrillae** or as capsular material. Such type of fimbri are reported to be anchored in the cell envelope in a way similar to that of bacterial flagella while others are apparently assembled at the cell surface and are presumed to be attached only on the outer membrane. The fimbriae may occur in a polar or peritrichous arrangement on the cell. Polarity fimbriae may exhibit a motality called **'TWITCHING MOTILITY**.

Each fimbria consists of linearly repeating protein called **fimbrillin** subunits. The fimbrillin often contains a high proportion of nonpolar amino acids. Thus fimbriate cells are commonly believed but not always hydrophobic than are afimbriate cells. The hydrophobicity in this context is indicated by the tendency of cells to congregate at an air-water or oil-water interface.

The occurrence of number of fimbriae on cells depends on various factors which include *e.g.* the phase of growth and growth temperature. Therefore, k 88 and k 99 fimbriae are noted below at 18°C temperature.

On the basis of their adhesive and haemagglutination properties fimbriae have been categorized into four main types. They are:

Type 1 fimbriae also called F 1 fimbriae are chromosome encoded fimbriae which are common on *Escherichia coli* and other enterobacteria. They are about 7 nm in diameter and 4000 nm in length and each comprises subunits of fimbrillin (about 17000 Molecular Weight, and is the product of pil A gene) and are arranged in a right handed helix. Type 1 fimbriae adhere apparently via lateral rather than terminal sites on the fimbria to various surface of cells including erythrocytes of guinea pigs horse resulting HAEMAGGLUTINATION and such adherence is inhibited by D-mannose or by methyl α-D-mannoside that means binding is mannose sensitive. Thus type 1 fimbriae cause mannose sensitive haemagglutination (MSHA). No specific connection appears in type 1 fimbriae and bacterial virulence in man.

Type 2 fimbriae occur in some strains of **Salmonella** and they do not show adhesive and haemagglination properties of type 1 fimbriae.

Type 3 fimbriae are seen in various members of Enterobacteriaceae they bear adhesive properties and are not inhibited by mannose, that means adhesion is mannose resistant (MR). They promote haemagglutination of oxerythrocytes only when these have been reacted with tannic acid which has been called 'tanned ox' haemagglutination.

Type 4 fimbriae have been noticed in strains of *Proteus*. They possess adhesive properties and promote mannose resistant haemagglutination abbreviated as MRHA.

Interesting enough that plamid encoded fimbriae (specific adhesions occur on enterotoxigenic strains of *Escherichia coli* that cause disease in man and domestic animals. These fimbriae include the colonization factors I and II (= F2 and F3 fimbriae respectively), on man.

Flagellin: A protein of which flagella are made of.

Flagellum: Plural flagella. These are locomotry appendages arising from plasma membrane and cell wall.

Fluid mosaic model: It is a model for membrane structure proposed by S. Jonathan Singer and Garth Nicolsan.

Gas Vascuole: A subcellular organelle found only in procaryotes which consist of a cluster of gas vesicles. The gas vacuole appear bright and refractile under the light microscope. They are found in planktonic cyanobacteria specially in those which from bloom.

Gas vesicles: These are the aggregates of numerous, small, hollow, cylindrical structures. Their walls do not contain lipids rather are made up of single protein.

General secrection pathway (GSP): It is protein secrection pathway which has also been called secrection dependent pathway. It is atleast of 6 types.

Germination: It is breaking of dormant stage of spore of prokaryotes after activation by heat etc.

Gliding motility: Type of motility exhibited by myxobacteria involving type IV fimbriae.

Glycocalyx: It is a network of polysaccharides extending from the surface of bacteria.

Glycogens: Are polymers of glucose units composed of long chains formed by α (1–4) glycosidic bonds and branching chains connected to them by α (1–6) glycosic bonds.

Hook: Short curved segment of flagellum connecting filament to its basal body.

Hydrophilic: Water loving molecules as polar ends of membrane lipids called polar membrane lipids.

Hydrophobic: Non polar end of membrane are of long fatty acid chains.

Inclusion body: Granules of organic and inorganic material present in cytoplasmic matrix.

Integral proteins: These comprise seventy to eighty percent of membrane proteins which cannot be easily extracted from membranes and are insoluble in aqueous solutions.

Lipid A: It is one of three parts of lipopolysaccharides (LPSs) buried in the outer membrane.

Lipopolysaccharides (LPSs): Are most unusual constituents of outer membrane.

Lophotrichous: The tuft of flagella at one or both ends.

Lyses: Destruction or disruption of the cells.

Lysozyme: The enzyme that attacks peptidoglycan hydrolyzing the bond which connects N-acetyl muramic acid with carbon 4 of N-acetyl glucosamine.

Magnetosomes: Are the inclusion bodies of iron in the form of magnetite. This helps bacteria to orient in Earth's magnetic field.

Metachromatic granules: The bacteria store phosphate or phosphate granules or volutin granules. The phosphate is a linear polymer of orthophosphates joined by ester bonds. The **volutin granules** function as storage reservoirs of phosphate.

Monotrichous: It refers to the presence of only flagellum (or *e.g.* pilus) per cell. If located at one end it is a polar flagellum.

Murein: It is 20–80 nm thick homogenous peptidoglycan.

Nucleoid: It is irregularly shaped region of procaryotic cell in which the prokaryotic chromosome is located. It is also called nuclear body, chromatin body or nuclear region.

O-antigen: One of the three parts of lipopolysaccharides (LPSs) called O-side chain.

O-side chain: One of the three parts of lipopolysaccharides (LPSs) also called O-antigen.

Osmosis: The process in which water moves from dilute solutions to concentrated solutions.

Outer membrane: It is found in Gram-ve bacteria which is 7 to 8 nm thick.

Penicillin: It is an antibiotic which inhibits synthesis of peptidoglycan.

Peptide interbridge: A pentaglycine interbridge which connects peptide chains.

Peptidoglycan: It is a tough substance, 20–80 nm thick murein (peptidoglycan) layer in Gram + ve and 2–7 nm in periplasmic space in Gram – ve bacteria.

Peripheral proteins: These are loosely connected to the membrane and can be removed.

Periplasm: The substance or material which occupies the periplasmic space.

Periplasmic space: This is the space between plasma membrane and outer membrane of Gram – ve bacteria. Sometimes similar but smaller gap may be seen in Gram + ve bacteria also.

Peritsichous flagella: It is a condition when flagella are distributed on whole surface of bacteria.

Plasma membrane: It is the membrane surrounding cytoplasm of both procaryotic and eucaryotic cells.

Plasmid: The extrachromosomal double stranded DNA molecules usually circular or in ring form which can exist and replicate independently of chromosome or may be integrated with chromosome.

Plasmolysis: The water flows outside the cell consequently cytoplasm shrinks and pulls away from cell wall.

Pleomorphic: The bacteria with variable shape.

Polar flagellum: One flagellum located at one end of the bacterium.

Poly-β-hydroxybutyrate (PHP): It contains β-hydroxybutyrate molecules joined by esterbonds between carboxyl and hydroxyl molecules of adjacent molecules. Poly-β-hydroxybutyrate is a linear polymer of β-hydroxybutyrate D (–)-3-hydroxybutyrate: H – [O . CH (CH_3) .

CH_2 . CO$]_n$–OH. It occurs in refractile granules of various sizes within the cells of various prokaryotes, *e.g. Azotobacter, Bacillus, beggiatoa, Beijerinckia, Paracoccus, Pseudomonas, Rhizobium, Sphaerotilus* and *Spirillum* spp, certain cyanobacteria and members of chromatiaceae and Rhodospirillaceae but apparently not in eucaryotes. The granules are usually bounded by non-unit-type memberane about 2.0 – 4.5 nm thick. The PHB granules can be stained with Nile Blue-A.

In many bacteria PHB accumulates when nutrients other carbon source become limiting and degraded when those conditions are reversed. Therefore, PHB acts as a reserve of carbon and energy. *Azotobacter beijerinckii* accumulates very large quantities of PHB say upto 80% of the cell mass. Typically, PHB is synthesized via the pathway: acetyl-CoA → Acetoacetyl-CoA → β-hydroxybutyrate-CoA → PHB. The enzymes associated are β-ketothiolase, β-hydroxybutyryl-CoA dehydrogenase and PHB synthetase. PHB synthetase occurs bound to the membrane which delimits the PHB granules. PHB degredation occurs via the pathway PHB → β-hydroxybutyrate → acetoacetate → acetoacetyl CoA → acetyle CoA. The first enzyme in this pathway PHB depolymerase is granule bound in some bacteria while soluble in others. Some bacteria e.g. certain pseudomonads can degrade exogenous PHB.

Porin proteins: These permit the passage of small molecules like glucose and other monosaccharides.

Protoplast: Plasma membrane and the contents within it *i.e.* cell minus cell wall.

Ribosome: These are complex objects made of protein and RNA, which mediates protein synthesis. Different types of ribosomes are characterised by different sediment coefficients on ultracentrifugation.

Run: Movement of bacteria in a straight or slightly curved line.

Sec-dependent pathway: Also called general secretion pathway (GSP) is the most commonly used pathway which either translocates proteins across the plasma membrane or integrates them into membrane itself.

Self-assembly: In it subunits spontaneously aggregate.

Sex pili: Are appendages 1 to 10 per cell often longer than fimbriae about 9 to 10 nm in diameter. The number of fimbriae may be upto 1000.

Signal peptide: Occurs at amino terminal or N-terminal end of presecretory protein or preprotein.

S-layer: It is a surface layer in Gram negative, Gram positive bacteria and Archaea.

Slime layer: A zone of diffuse, unorganized material which is removed easily.

Spheroplast: The Gram negative cells retain their outer membrane after treatment with penicillin.

Spirilla: Bacterial rods rigidly twisted into spirals or helices.

Spirochaete: The bacterial rods twisted or spiraled into flexible shape.

Sporangium: Mother cell of spore.

Spore cell wall: It is also called core wall. It is inside the cortex and surrounds the protoplast or core.

Spore coat: It lies below the exosporium and is composed of several protein layers and may be fairly thick and impermeable to toxic molecules and provides resistane.

Sporogenesis: It is sporulation or formation of spores.

Svedberg unit: It is a unit of sedimentation coefficient.

Techoic acid: It occurs in Gram positive bacteria but not in Gram negative. Techoic acid appears to extend to surface of peptidoglycan. Being negatively charged gives Gram positive cell wall its negative charge.

Translocon: The preproteins are kept unfolded by "CHAPERONS" and help them to reach transport machinery or the TRANSLOCON.

Tumble: After a run bacteria stop and tumble or twiddle about. The tumble is followed by a run in different direction.

Twiddle: It is a tumble.

Type-I Protein secrection pathway: It is ABC (ATP-binding Cassette) protein secrection pathway.

Type-II protein secrection pathway: It occurs in some Gram negative bacteria (pathogens *Erwinia carotovora, Pseudomonas aeruginosa* and *vibrio cholerae*) to transport proteins across outer membrane (O.M.).

Type-III protein secrection pathway: It occurs in several Gram negative pathogens which secrete out virulence factors and inject them into plant and animal host cells.

Vibrio: Curved rods forming shape of commas or incomplete spirals.

Volutin granules: These store phosphate as phosphate granules. The phosphate is a linear polymer of orthophosphates joined by ester bonds. These are also called metachromatic granules because they show metachromatic effect.

Walsby's square archaeon: Salty squares isolated from hypersaline pond near red sea in 1980 by British microbiologist Anthony Walsby. These are shaped flat square to rectangular boxes, approximately 2 μm × 2 to 4 μm and just 0.25 μm in thickness.

9 CHAPTER

BACTERIA

"O Lord, I beg upon my knees that all my various syntheses may not turn out to be inferior to those conducted by bacteria".—Anon's quotation boldly points out the importance of bacteria to human beings. The importance of bacteria was further highlighted by J.W. Foster (1964) in his statement "Never underestimate the power of bacteria."

Bacteria are the simplest of all cellular organisms even though they have undergone evolution for a period of approximately 3.5 to 4 billion years. The layer of viscous material surrounding some bacteria as a capsule or slime layer which has also been referred to as **glycocalyx** is of great significance to bacteria.

They are very small in size and are represented in different shapes. One trillion or 10^{12} bacterial cells weigh only 1 g. The **generation time** of *Escherichia coli* to divide by binary fission is about 20 minutes. In comparison to it a mammalian cell if cultured in a laboratory will take 13 to 24 hours to divide into two cells.

The flagella (singular, flagellum) and the pili (singular, pilus) are the bacterial appendages meant for movement and assist in reproduction respectively. The flagella propel bacteria to move and if compared are faster than most land animals. Bacteria with F pilus are known as donor cells. The **cell wall** and the **cell envelopes** are the terms referred to for the protective layers of the bacteria. The bacterial cells lack membrane bound organelles like mitochondria and chloroplasts found in eucrayotic cells but the cytoplasmic membranes in many of them extend to form **mesosomes.** The internal cell structures are **ribosomes** and **nucleoid.** The chemical substances which are found accumulated are sulphur granules, volutin granules and poly-beta-hydroxybutyrate (PHB). Spores and cysts are metabolically inactive resting forms of bacteria (Fig. 9.1).

There are several techniques of handling bacteria in the laboratory, the most commonly employed methods are microscopy, staining, isolation, cultivation, environmental influences, oligo-dynamic action, evaluation of disinfectants, physiological characterization and those associated with dairy, soil, water and wastewater, and industry.

After isolating bacteria in pure cultures an important step is to keep them alive by preserving the pure cultures for use in future. If these cultures are to be maintained only for a few days or months they can be stored under refrigerated conditions at temperatures ranging from 4° to 10°C. However, for long term storage bacteria like other microorganisms can be stored in tanks of liquid nitrogen at –196°C or in deep-freezers at –70° to –120°C. For safe delivery bacteria can be frozen and then dehydrated and sealed under vacuum, the whole process is referred to as **lyophilization.** The lyophilization or freeze drying assures the viability of bacterial cultures for many years.

The archaeobacteria when seen through the microscope appear similar to bacteria but the two differ in their chemical composition and activities and the environment in which they can survive.

Many archaeobacteria survive in adverse conditions under high levels of 'salt or acid or high temperature. Some of the archaeobacteria are endowed with unique chemical activities such as production of methane gas from carbon dioxide and hydrogen. The methane producing archaeobacteria live in habitats without oxygen like deep swamp mud or the intestines of cattle and sheep that chew cud.

The eubacteria on the other hand are necessary in recycling wastes and the production of antibiotics such as streptomycin. Streptococcal sore throat, tetanus, plague, cholera and tuberculosis are some of the diseases caused by the eubacteria.

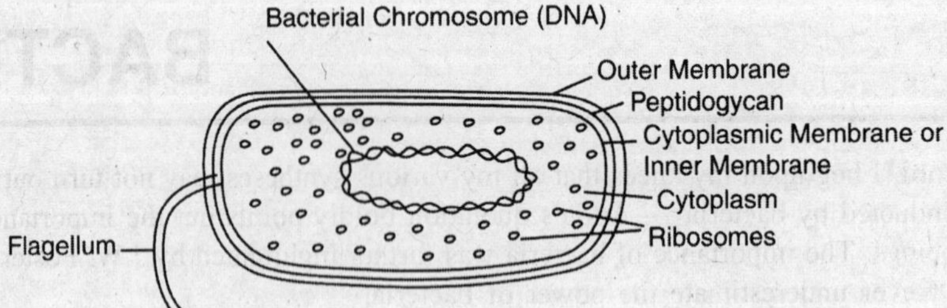

Fig. 9.1. A simplified diagram of a Gram negative bacterial cell, showing **cytoplasmic membrane, peptidoglycan,** and **outer membrane.** The space between two membranes housing the peptidoglycan layer is called **periplasmic space.** Cytoplasmic membrane also called inner membrane encloses the cytoplasm and a **bacterial chromosome** (DNA) which stores the hereditary information. There are numerous membrane bound **ribosomes** in the cytoplasm where protein synthesis takes place. It is interesting that **mesosome** has not been shown because recent studies revealed that the ribosomes are artifacts, and not true structures.

Peptidoglycan in Gram negative bacteria is thin $\left(\frac{1}{3} \text{ to } \frac{1}{4}\right)$ than Gram positive bacteria.

Bacteria

Q.1. Write technical terms for the following:

(i) Spherical bacteria.
(ii) The cocci of bacteria that remain in pairs after dividing.
(iii) The bacterial cocci that remain in chains.
(iv) Bacterial cocci arranged in cubes of eight cocci.
(v) Bacterial cocci arranged in grape-like clusters.
(vi) Bacilli arranged in pairs.
(vii) Bacilli arranged in chains.
(viii) Bacteria shaped like curved rods.
(ix) Bacteria that maintain a single shape.
(x) Bacterial cells with many shapes.

Ans. (i) Coccus (plural cocci),
(ii) Diplococci,
(iii) Streptococci,
(iv) Sarcinae,
(v) Staphylocci,
(vi) Diplobacilli,
(vii) Streptobacilli,

(viii) Vibrios,

(ix) Monomorphic and

(x) Pleomorphic.

Q.2. Match the following:

(A)	A viscous gelatinous polymer over the cell wall	(a)	Tetrads
(B)	Bacterial cocci in groups of four	(b)	Glycocalyx
(C)	Glycocalyx firmly attached to the cell wall	(c)	Slime layer
(D)	Glycocalyx loosely attached to the cell wall	(d)	Capsule
(E)	Long filamentous appendages which propel bacteria	(e)	Flagella
(F)	Single polar flagellum	(f)	Amphitrichous
(G)	Single flagellum at each end of the bacterial cell	(g)	Monotrichous
(H)	Two or more flagella at one or both poles	(h)	Lophotrichous
(I)	Flagella distributed over entire cell	(i)	Peritrichous

Ans. [A b], [B a], [C d], [D c], [E c], [F g], [G f], [H h], [I i].

Q.3. How do spirochetes move? Name two spirochetes.

Ans. Spirochetes move by axial filaments which are bundles of fibrils. *Treponema pallidum,* the causative agent of syphilis and *Borrelia burgdorferi,* the causative agent of lyme disease are two well known examples of spirochets.

Q.4. How does corkscrew motion or giggling motion help *Treponema pallidum?*

Ans. It helps to penetrate well inside the body tissue.

Q.5. What substance provides rigidity to bacterial cell walls? Give its structure.

Ans. The **peptidoglycan** which is a macromolecular network and consists of repeating disaccharide attached to polypeptides, provides rigidity to the bacteria] cell walls. The disaccharide comprises two monosaccharides N-acetyl glucosamine (NAG) and N-acetyl muramic acid (NAM). The linkage between the two monosaccharides is known as, β -1, 4 linkage.

The tetrapeptide side chains form the polypeptide links. The parallel tetrapeptide side chains are linked by peptide cross bridge.

Q.6. What are Gram positive cell walls?

Ans. Cell walls which stain positive are Gram positive. They consist of peptidoglycan (murein) and are more thick than peptidoglycan in Gram negative bacteria. The cell walls of Gram positive bacteria also possess teichoic acids which are of two types (1) lipoteichoic acid and (2) wall teichoic acid. The teichoic acid gives the cell wall antigenic specificity, the characteristic of which is also used to **identify bacteria** in serological tests.

Q.7. What are Gram negative cell walls?

Ans. The bacterial cells which give negative Gram's stain test possess Gram negative cell walls. The peptidoglycan in them is joined with lipoproteins (lipids covalently linked to proteins) on the **outer membrane.** There is a space between the outer membrane and the **plasma membrane** which is referred to as **periplasmic space.** The high concentration of degradative enzymes and transport proteins are housed in the periplasmic space. Gram negative cell walls are devoid of a substance called, **teichoic acid,** a polysaccharide found in Gram + ve cell walls.

The outer membrane of Gram –ve bacteria comprises lipoproteins, lipopolysaccharides and phospholipids. The outer membrane protects the cell from antibiotics like penicillin, digestive enzymes such as lysozyme, detergents heavy metals, bile salts and some dyes. Embedded in the outer membrane are certain **proteins** called **porins** which permit passage of molecules, e,g., nucleotides, disaccharides, peptides, amino acids, vitamin B_{12} and iron. The portion of lipopolysaccharide known as **O-polysaccharide** in lipopolysaccharide of outer membrane provides the bacterial cell with its antigenic characteristics useful for **identification and differentiation** of Gram negative bacteria. The lipid portion of lipopolysaccharide known as **Lipid A** is an endotoxin and is toxic to the host's blood stream or gastrointestinal tract. It is held responsible for fever and shock to the patient infected with pathogenic Gram negative bacteria.

Q.9. Why are some β -lactam antibiotics more effective than penicillin against Gram negative bacteria?

Ans. Because some β -lactam antibiotics are able to penetrate the outer membrane of Gram negative bacteria.

Q.9. Why is the antibiotic penicillin effective against Gram positive bacteria but not harmful to human red blood cells?

Ans. The penicillin interferes with the formation of peptidoglycan in Gram positive bacteria. Red blood corpuscles do not contain peptidoglycan and are not interfered with.

Q.10. The *Mycoplasma* is a wall less bacterium. How is its cell protected from **osmotic lysis**?

Ans. The plasma membrane of *Mycoplasma* possesses sterols, a class of lipids.

Q.11. What is a **fluid mosaic model** of plasma membrane (or cytoplasmic membrane)?

Ans. The plasma membrane comprises phospholipid bilayers and proteins which are not static but move freely within the membrane surface. This dynamic arrangement is known as fluid mosaic model.

Q.12. Recent studies reveal that **mesosomes** are artifacts and not true structures. How are they seen in electron-micrographs?

Ans. Infoldings of plasma membrane are formed during the dehydration process in electron microscopy. The mesosomes are not seen in electron micrographs requiring no dehydration in some modern techniques.

Q.13. How do the movements of materials across a membrane take place?

Ans. The movements of materials take place by means of active processes like active transport and group translocation or by passive processing as simple diffusion, facilitated diffusion and osmosis.

Q.14. What is the cytoplasm in a bacterial cell?

Ans. The bacterial cell is made of thick, aqueous, semitransparent and elastic internal matrix. The cytoplasm contains proteins (enzymes), carbohydrates, lipids, inorganic ions and several low molecular weight compounds.

Nuclear Area of Bacteria

Q.15. What is nucleoid?

Ans. The nuclear region in a bacterial cell is called **nucleoid.** The **nucleoid** comprises a single long circular molecule of double stranded DNA also referred to as **bacterial chromosome**. There is, however, no nuclear membrane like in eucaryotic cells and it also lacks histones. In addition to a chromosome bacteria also contains small circular double stranded (extrachromosomal) DNA molecules known as **plasmids**.

The plasmids may contain 5 to 100 genes and the bacteria can survive with and without them. The genes carried by plasmids are known to be responsible for antibiotic resistance, tolerance to toxic metals, production of toxic substances, and synthesis of certain enzymes. The plasmid DNA is used for genetic manipulations in biotechnological targets.

Q.16. Why do bacterial cells possess ribosomes?

Ans. Bacterial cells like eucaryotic cells possess **ribosomes** which are the sites of protein synthesis. The ribosomes consist of two subunits and each subunit further consists of protein and a type of RNA popularly called ribosomal RNA or **rRNA.**

The procaryotic ribosomes are 70 S ribosomes while in eucaryotic cells there are 80 S ribosomes. Here S stands for **Svedberg unit,** named after its discoverer and is based on sedimentation under ultra-high speed centrifugation. The 70 S ribosome comprises of small 30 S subunits containing one molecule of r RNA and large 50 S subunits containing two molecules of ribosomal RNA. The antibiotics streptomycin, neomycin and tetracycline are effective because of their characteristic to inhibit protein synthesis on the ribosomes.

Bacterial Inclusions

Q.17. Name the important inclusions of bacteria.

Ans. Important bacterial inclusions are metachromatic granules, polysaccharide granules, lipid inclusions, sulphur granules, carboxysomes and gas vacuoles.

Q.18. What are metachromatic granules? Name the bacterium of which they form a characteristic feature.

Ans. The metachromatic granules are collectively known as **volutin** that represents a reserve of inorganic phosphate (polyphosphate) that can be used in the synthesis of ATP. These large granules have been named so because they sometimes stain red with certain blue dyes like methylene blue. The metachromatic - granules are a characteristic feature of *Corynebacterium diphtheriae,* the causative agent of diphtheria.

Q.19. Match the followings.

(A)	Glycogen and starch granules while stained with iodine appear reddish brown and blue respectively	(a)	A common lipid storage substance (Lipid inclusion)
(B)	Poly-β -hydroxybutyric acid	(b)	Polysaccharide granules
(C)	Sulphur granules in the cells	(c)	Gas vacuoles/gas vesicles to maintain buoyancy
(D)	Carboxysomes	(d)	*Thiobacillus*
(E)	Cyanobacteria, anoxigone photosynthetic bacteria and halobacteria	(e)	Enzyme 1, 5-diphosphate carboxylase
(F)	*Clostridium* and *Bacillus* possess resting spores called	(f)	Germinated
(G)	7500 years old endospores of *Thermoactinomyces vulgaris* from freezing muds of Elk Lake in Minnesota when rewarmed and placed on nutrient medium	(g)	Endospores

Ans. [A b], [B a], [C d], [D e], [E c], [F g], [G f]

Q.20. Write technical terms for the following statements.

(i) Endospore like structures in Gram-negative bacteria that can withstand heat and chemicals and can be stained with endospore stain (giving positive results), are found in the bacterial species

(ii) The process of formation of endospores in bacteria

(iii) Return of a bacterial spore to its vegetative condition

(iv) Extrachromosomal rings of DNA in bacteria

(v) Energy is expended to modify chemicals to carry them across the membrane

(vi) Materials travel from low to high concentration by permeases making the cell to expand energy

(vii) The movement of substances across the membrane where the substances move from higher to lower concentration without energy expansion by the cell, is called

(viii) After the cell wall of Gram positive bacteria are damaged by the enzyme lysozyme all that is left is called

Ans. (i) *Coxiella burnetii*

(ii) Sporulation

(iii) Germination

(iv) Plasmids

(v) Group translocation

(vi) Active transport

(vii) Passive movement (or passive process)

(viii) Protoplast.

Q.21. Match the following:

(A)	Ribosomes	(a)	Attachment to surfaces
(B)	Plasma membrane	(b)	Transfer of genetic material
(C)	Pilus	(c)	Formation of cell wall.
(D)	Glycocalyx	(d)	Mobility
(E)	Cell wall	(e)	Protect from osmotic lysis
(F)	Flagellum	(f)	Protection from phagocytosis
(G)	Fimbriae	(g)	Resting stage
(H)	Endospore	(h)	Protein synthesis
(I)	Peptidoglycan	(i)	Selective permeability

Ans. [A h], [B i], [C b], [D f], [E e], [F d], [G a], [H g], [I c]

10 CHAPTER

PLASMIDS

Q.1. What is a plasmid? Discuss.

Ans. The plasmids have been reported from procaryotic as well as eucaryotic cells. A **plasmid** may be defined as extrachromosomal genetic element which is not essential for the growth of the cell and has no extra cellular form. In procaryotes a plasmid is typically a relatively short usually circular double stranded DNA molecule but in eucaryotes may be short circular or linear double stranded DNA molecule. The plasmids are distinct from chromosomal DNA may be mitochondrial DNA (mt DNA) or cytoplasmid DNA (ct DNA). The plasmids as circular, supercoiled DNA molecules are present in most species or bacteria may be absent in some strains. They are small ranging from about 0.2 to 4 per cent that of the bacterial chromosome. In size they range from 1.5 kb to 300 kb and MWt (molecular weight) 10^6 to 2×10^8. Many procaryotes contain more than one plasmids in addition to their chromosome. They make bacterial cell adaptable to the environment, e.g. R plasmid provides resistance against an antibiotic or heavy metal. They are valuable to a geneticist for two reasons: (1) To construct partial diploids and (2) to use in gene cloning in genetic engineering by using plasmids as gene vectors in recombinant DNA technology.

Q.2. Why bacterial plasmids have been regarded as **tiny chromosomes**.

Ans. DNA molecules which possess their own origins of replication and are capable of self duplication are known as **replicons**. The circular main bacterial chromosome is a replicon and so are the plasmids inspite of their much small size i.e., they are autonomously replicating circular DNA molecules inside the host cell.

Q.3. Whether the following are true or false?

(i) Plasmid DNA usually encode no essential functions and the bacteria lacking plasmids multiply normally.

(ii) F factors or F plasmids do not encode the proteins that allow bacteria to conjugate.

(iii) R 1 a naturally occurring 94 kilobase complex plasmid confers resistance to five antibiotics chloramphenicol (Cm), streptomycin (Sm), sulphonamide (Su), ampicillin (Ap), and kanamycin (Km).

(iv) Col plasmids do not synthesize proteins collectively called colicins that can kill closely related bacterial strains which lack a Col plasmid of the same type.

(v) Many species of the *Pseudomonas* can utilise several toxic organic compounds as carbon sources e.g., camphor, toluene and octane because of a set of plasmids known as **degradation plasmids**.

(vi) The linear plasmid does not occur in *Streptomyces* sp.

Ans. (i), (iii), (v) True; (ii), (iv), (vi) False.

Nomenclature of Plasmids

Q.4. How is the **nomenclature of plasmids** done?

Ans. The plasmids are designated in various ways. The Colicin plasmids are designated according to the colicins they encode, e.g. Col E1-K30 encodes colicin E1 (lead to membrane changes) Col V, I-K 94 encodes colicin V and Ia. Many antibiotic resistance plasmids are designated by the letter 'R' followed by the number, e.g., R1, R4 etc; others include R6K, RK2, RP1. The recombinant plasmids are indicated by a prefix 'p' e.g. pML31. A number of plasmids are given trivial names (e.g. Δ), phage plasmid λdv for λ genes (lambda genes) cro, cl, O, P and the names of some indicate particular functions e.g., Ti plasmid tumour inducing plasmid and TOL plasmid. Some individual plasmids are also known as COLEI plasmid, delta plasmid, PBR 322 for high copy number, pBR 345 for Col E1-type replication, PMB1, PMB31, psc 101, R1 plasmid, R100 plasmid and Rp1 plasmid.

Q.5. How are plasmids detected?

Ans. The plasmids are detected by two methods.

(i) By conducting genetic experiments. The first plasmid was discovered for F factor in *Escherichia coli*. A strain of *E. coli* with phenotype Met^- Bio^- Thr^+ Bio^+ was mixed with second strain Met^+ Bio^+ Thr^- Leu^- and the mixture was placed on minimal agar. New colonies appeared on minimal agar at a frequency of about 10^{-7} which possessed the phenotype Met^+ Bio^+ Thr^+ Leu^+ being a recombinant.

(ii) By conducting physical experiments with bacterial extracts. The plasmids are detected electrophoretically. For it a single colony is taken from a plate and is lysed by lysozyme-detergent procedure and subjected to gel electrophoresis. The bacterial chromosome being very large cannot penetrate the gel but the plasmid DNA being very short can penetrate. Plasmid DNA form a narrow band at a position in the gel characteristic of its molecular weight. The band can be visualised by staining the gel with ethidium bromide, which binds tightly to the DNA and flouresces while irradiated with UV light.

Q.6. How is the plasmid DNA purified?

Ans. The purification of plasmid DNA is done in the following stages.

(i) The **cell lysate** which is a translucent solution is made from plasmid containing bacteria by **lysozyme-detergent treatment**.

(ii) On centrifugation of the cell lysate, the **bacterial chromosome complex** containing proteins and RNA settle at the bottom of centrifuge tube while the plasmid DNA remains in the clear supernatant known as **cleared lysate**.

(iii) The chromosomal DNA that might have been left in the cleared lysate is removed by adding CsCl and ethidium bromide and centrifugation.

(iv) The plasmid DNA being covalently circular shows higher density than linear chromosomal fragments and gets separated.

Q.7. Which are the stages in **transfer process of plasmid DNA** in bacterial **conjugation**?

Ans. The four stages in **bacterial conjugation** are:

(i) Formation of specific donor (F^+) and recipient (F^-) pairs which is known as **effective contact**.

(ii) Preparation for transfer of DNA from donor to recipient cell called mobilisation.

(iii) Transfer of DNA.

(iv) Formation of replicative functional plasmid.

Q.8. Choose the correct answers (more than one option may be correct).

(i) The **conjugative plasmid** carry genes for effective contact function in bacterial conjugation.

(ii) The preparation for transfer of DNA in conjugation is made by **mobilizable plasmid**.

(iii) A self **transmissible plasmid**, e.g., F plasmid is conjugate as well as mobilizable.

(iv) The sex pilus is present on the recipient bacterial cell.

(v) The **sex pilus** is present on the donor bacterial cell.

(vi) The **F pilus** comprises of a very hydrophobic protein known as **pilin**.

(vii) The purified pili can bind to recipient bacterial cells.

(viii) Some strains of *Streptococcus faecalis* have a self-transmissible plasmid.

(ix) In some strains of *Streptococcus faecalis* plasmid free recipient produce a **mating protein** analogous to the pheromones of the female insect while such mating protein is not formed in the plasmid containing donor cells.

(x) In the bacterial strain *Streptococcus faecalis* mating protein or the pheromones lead the donor cells to make a protein known as **adhesin** which coats the donor cell enabling cells to make donor-recipient pairs.

(xi) Looped rolling-circle mechanism is not used by phage ϕX174.

(xii) Looped rolling-circle mechanism is used by phage ϕX174.

(xiii) Looped rolling-circle mechanism explains transfer of F plasmid DNA from an F^+ donor to F^- recipient.

(xiv) **Fertility inhibition** is a common feature in R plasmids.

(xv) Most R plasmids never have any active ***fin*** **repressor**.

Ans. (i), (ii), (iii), (v), (vi), (vii), (viii), (ix), (x), (xii), (xiii), (xiv), (xv) Correct; (iv) and (xi) Wrong.

Q.9. What is ***tra*** **operon**?

Ans. The genes which are essential to cause transfer are called *tra* operon or the ***tra*** **genes of F**.

Q.10. Name the **genes for pili formation**.

Ans. These are *tra A, tra B, tra C, tra E, tra F, tra G, tra H, tra J, tra K, tra L, tra U, tra V, tra W*.

Q.11. Match the following genes of F plasmid to their functions.

(A)	*tra J*	(a)	Mating aggregation
(B)	*tra G, tra N*	(b)	Initiation of transfer
(C)	*tra L, tra M*	(c)	Transfer
(D)	*tra M*	(d)	Operator for *fin O* gene
(E)	*tra O*	(e)	Subunits of endonuclease that nicks *Ori T*
(F)	*tra Y, tra Z*	(f)	Structural genes for pilin
(G)	*Ori T*	(g)	Lethal zygosis (Killing of females by excess Hfr cells)
(H)	*Ori V*	(h)	Surface exclusion i.e., inhibition of mating between F containing cells; and encode membrane proteins
(I)	*tra S, tra T*	(i)	DNA replication origin
(J)	*ilz A, ilz B*	(j)	Origin of transfer DNA synthesis
(K)	*frp*	(k)	F replication
(L)	*inc*	(l)	Incompatibility of Inc F group
(M)	*fin O, fin P*	(m)	Fertility inhibition found in Col R and F-like plasmids

Ans. [A f], [B a], [C b], [D c], [E d], [F e], [G j], [H i], [I h], [J g], [K k], [L l], [M m]

Plasmid Replication (Reproduction of Plasmids)

Q.12. How does a plasmid replicate?

Ans. A plasmid replicates inside the host cell. Therefore, all plasmids native to the same host species show same mode for replication. It is interesting that like phages there exist much variation in both the enzymology mechanics of plasmid DNA replication. The polymerase III is the major replication protein in *E. coli* plasmids. The replication generally has been reported to occur by butterfly or rabbits ear mode. In it partially replicated molecules have untwisted replicated portions like θ replication and a supercoiled unreplicated part.

Q.13. What are stringent and relaxed plasmids?

Ans. The **stringent plasmids** have **low copy number** in which number per cell may be 1 to 2 while in **relaxed plasmid** there is **high copy number** ranging from 10 to 100 per cell.

Q.14. What is **amplification of plasmid**? What is its use?

Ans. It is an increase in number of plasmids. On addition of chloramphenicol which is an inhibitor of protein synthesis initiation of replication of chromosome DNA is inhibited but replication of plasmid DNA continues and the number of plasmids per cell increases to 1000 or more even. Thus amplification can be used to increase the amount of plasmid DNA.

Q.15. Name the inhibitors of plasmid replication.

Ans. **Acridines** like proflavin acridine orange inhibit the replication of plasmids without inhibiting replication of chromosomal DNA. This sort of inhibition causes the loss of plasmids known as **acridine curing.**

Q.16. What is integrative suppression?

Ans. Integration of the F into certain bacterial mutants bearing defects in DNA gives rise to the phenomenon called integrative suppression.

Q.17. What is **RTF**?

Ans. Most of the R-plasmids consist of two contiguous segments of DNA. One of the segments is known as RTF or resistance transfer factor. It carries genes that regulate replication of DNA, copy number, transfer genes and some times gene for tetracycline resistance called *tet* and bears a molecular weight of 11×10^6.

Q.18. What are colicinogenic or **Col plasmids**?

Ans. These are found in *Escherichia coli* and produce colicins, the proteins that inhibits the growth of susceptible bacterial strains that lack Col plasmid.

Q.19. What is **lacuna**?

Ans. When a colicin producing cell is placed in a lawn of sensitive bacterial cell the colicin producing bacterium inhibits the growth of sensitive cells and form a clear zone called **lacuna**.

Q.20. Match the following:

(A)	Colicin B and Colicin 1 b	(a)	Is encoded by *mob* gene
(B)	Plasmid-encoded nuclease for Col E1	(b)	Damages cytoplasmic membrane
(C)	Specific base sequence for Col E1 nuclease which contains cutting site is called	(c)	*bom* (basis of mobility)
(D)	Degrades DNA	(d)	Cleaves 16 S rRNA
(E)	Colicin E_3	(e)	Colicin E_2
(F)	Integrated fragment of Ti plasmid	(f)	Broad host range plasmids

(G)	Inc P (R plasmid or incompatibility group of *E. coli*) and group Inc P1 of *Pseudomonas aeruginosa*	(g)	Breaks the hormonally regulated system that controls the cell division

Ans. [A b], [B a], [C c], [D e], [E d], [F g], [G f]

Q.21. Choose the appropriate words from within the words in large brackets given below and fill up the blanks in the following sentences [(a) hemolysis, (b) traveller's diarrhoea, (c) degradation plasmids, (d) θ replication, (e) a nick]

(i) Ent plasmids in *Escherichia coli* synthesize enterotoxins that cause ________.

(ii) *Escherichia coli* strains isolated from pigs possess plasmid called Hly which causes ________.

(iii) The species of *Pseudomonas* can consume toxic substances e.g. camphor, toluene and octane, as their carbon source because they have plasmids that are called ________.

(iv) The mode of replication adopted by plasmids for their reproduction is ________.

(v) Early enzymatic step required in transfer replication but not required for typical θ replication is ________ at the transfer origin to generate a 3′-OH group.

Ans. (i) b, (ii) a, (iii) c, (iv) d, (v) e.

CHAPTER 11

BERGEY'S MANUAL

Q.1. Give an account of the **History of Bergey's Manual**.

Ans. The Society of American Bacteriologists presently known as the American Society of Microbiology began compiling the first edition of Bergey's Manual by appointing an editorial board with Dr David Hendricks Bergey as its Chairman. The other members of the board were Francis C. Harrison, Robert S. Breed, Bernard W. Hammer and Frank M. Huntoon. The Society of American Bacteriologists published the Journal of Bacteriology as a service to science. The first edition of Bergey's Manual was published in 1923. The Board with some change in membership and Dr David H. Bergey continuing as chairman brought out a second edition of the manual in 1925 and a third edition in 1930.

Dr Bergey, while preparing the 4th edition in 1934, requested the Society of American Bacteriologists to make available the royalties received by the Society from the sale of the earlier editions, to meet the expenses of preparing the fourth edition for publication. Such provision was made by the Society, but the use of the Society's financial machinery proved a burden to the Society as well as to the Editorial Board.

Consequently an agreement between the Society and Dr Bergey was made that the Society would transfer to Dr Bergey all its rights, title and interests in the manual and that Dr Bergey would in turn create an educational trust to which all the rights would be transferred. Hence, Dr Bergey became the nominal owner of the manual. On January 2, 1936 he executed a Trust Indenture, designating Dr David H. Bergey, Robert S. Breed and E.G.D. Murray as the initial trustees, and transferred to the Trustees and their successors the ownership of the Manual, its copyrights and the rights to receive the income earned from its publications. The trust is a non-profit organization.

The trustees till now have published successively the 4th, 5th, 6th, 7th, 8th, and 9th editions of the Manual in 1934, 1939, 1948, 1957, 1974 and 1994 respectively. The Trust in 1977 also published an abbreviated version of the 8th edition entitled. 'The Shorter Bergey's Manual of Determinative Bacteriology'.

The other publications of the Trust include 'INDEX BERGEYANA' published in 1966 and 'SUPPLEMENT to INDEX BERGEYANA' published in 1981. The trust published the first edition of 'Bergey's Manual of Systematic Bacteriology' in four volumes. Volumes I and II were published in 1984 and 1986, respectively while volumes III and IV were both published in 1989. These volumes are considered to have much broader scope than the previous publications and were intended to act as an amplified source for revision of the Determinative Manual.

The trust promotes and supports various aspects of taxonomic research. It recognizes those individuals who have made outstanding contributions in the field of bacterial taxonomy through the **Bergey Award,** an effort jointly supported by funds from the Trust and M/S Williams & Wilkins who have been involved in the production of the Manual since its inception.

Q.2. Which points or steps one will have to keep in mind while using "Bergey's Manual of Determinative Bacteriology?"

Ans. One will have to keep the following points in mind while using the Manual.

Step 1. Whether your isolate is a Procaryotic Microorganism or Eucaryotic Microorganism.

Step 2. To which major category of bacteria ones isolate belongs:

Whether:

(I) Gram negative eubacteria that have cell walls.

(II) Gram positive eubacteria that have cell walls.

(III) Eubacteria lacking cell walls.

(IV) Archaeobacteria.

The chapter IV of Begey's Manual of Determinative Bacteriology 9th edition (1994) provides characteristics that allow differentiation of these four categories.

Step 3. To which group does one's isolate belong?

Step 4. To which group does one's genus belong?

Step 5. To which species does one's isolate belong?

Q.3. Name the two standard publications of Bergey's Manual Trust regarding Bergey's classification of Procaryotes.

Ans. These are as given below.

(i) 'Bergey's Manual of Systematic Bacteriology' published by Williams & Wilkins, Baltimore, in four volumes. Each volume has several sections and each section contains a number of related genera. The four volumes provide information as categorized below.

Volume I was published in 1984 and contains information on Gram-negative bacteria of medical and commercial importance, e.g., spirochetes, spiral and curved bacteria, Gram-negative aerobic and facultatively aerobic rods, Gram-negative obligate anaerobes, Gram negative aerobic and anaerobic cocci, sulphate and sulphur reducing bacteria, rickettsias and chlamydias, and mycoplasmas.

Volume II was published in 1986. It contains information on Gram positive bacteria of medical and commercial importance, e.g., Gram positive cocci, Gram positive endospore forming and nonsporing rods, mycobacteria and nonfilamentous actinomycetes.

Volume III was published in 1989, and contains, information on the remaining Gram-negative bacteria and the Archaea, e.g., phototrophic, gliding sheathed, budding, and appendaged bacteria, cyanobacteria, chemolithotrophic bacteria, methanogenes, extreme halophiles, hyperthermophiles, Thermoplasma and other Archaea.

Volume IV was published in 1989. It contains information on filamentous actionomycetes and related bacteria.

(ii) The second standard publication is the 9th edition of 'Bergey's Manual of Determinative Bacteriology' (1994) published by Williams & Wilkins, Baltimore.

The 9th edition is intended solely for the identification of those bacteria that have been described and cultured. The arrangement of the book is strictly phenotypic and no attempt has been made to offer a natural classification. The arrangement chosen is utilitarian and is intended to aid in identification of bacteria. The publication divides bacteria into 35 **Groups** which are comparable to the **'Parts'** in the 8th edition of 1974 and the **'Sections'** in the volumes I to IV of Bergey's Manual of Systematic Bacteriology.

The Headquarters of the trust and editorial office of the Manual is housed in the Department of Microbiology at Michigan State University. Comments on this edition are welcome and have been invited by Bergey's Manual Trust, Deptt of Microbiology, Giltner Hall Michigan State University, East Lansing, M.I., USA, 48824-1101.

Q.4. How many 'Sections' are there covering all procaryotes (bacteria) in 'Bergey's Manual of Systematic Bacteriology'?

Ans. In all there are 33 Sections of procaryotes (bacteria) in 'Bergey's Manual of Systematic Bacteriology'. The Volume I covers Sections 1 to 11, Volume II covers Sections 12 to 17, Volume III covers Sections 18 to 25, and Volume IV covers Sections 26 to 33.

Q.5. Give the classification of **Rickettsias** and **Chlamydias**.

Ans. According to 'Bergey's Manual of Systematic Bacteriology' (1984) Rickettsias and Chlamydias are placed in Volume 1, Section 9. Their classification is as under.

Section 9. RICKETTSIAS AND CHLAMYDIAS

Order I. Rickettsiales
- Family I. Rickettsiaceae
  - Tribe I. Rickettsieae
    - Genus I. *Rickettsia*
    - Genus II. *Rochalimaea*
    - Genus III. *Coxiella*
  - Tribe II. Ehrlichieae
    - Genus IV. *Ehrlichia*
    - Genus V. *Cowdria*
    - Genus VI. *Neorickettsia*
  - Tribe III. Wolbachieae
    - Genus VII. *Wolbachia*
    - Genus VIII. *Rickettsiella*
- Family II. Bartonellaceae
  - Genus I. *Bartonella*
  - Genus II. *Grahamella*
- Family III. Anaplasmataceae
  - Genus I. *Anaplasma*
  - Genus II. *Aegyptianella*
  - Genus IV. *Eperythrozoon*

Order II. Chlamydiales
- Family I. Chlamydiaceae
  - Genus I. *Chlamydia*

According to 9th edition of 'Bergey's Manual of Determinative Bacteriology' (1994), the Rickettsias and Chlamydias have been placed under Group 9.

This group includes two subgroups I. Rickettsias; II. Chlamydias

Some Rickettsias appear ring-shaped in stained preparations, some have flagellum, and some may stain Gram positive.

The chlamydia require very specialized cultivation techniques for study. Only certain laboratories are equipped for their identification. The Group 9 has been described only on one page in a Manual of 787 pages.

Q.6. Give classification/systematic position of **cyanobacteria**.

Ans. **Cyanobacteira,** popularly known as blue green algae have been placed in Volume III (1989) of 'Bergey's Manual of Systematic Bacteriology'. In it the cyanobacteria are placed in Group I of Section 19.

Section 19. Oxygenic Photosynthetic Bacteria

Group I. Cyanobacteria

Subsection I. Order Chroococcales

Genus I. *Chamaesiphon*

Genus II. *Gloeobacter*

Genus III. *Gloeothece*

Subsection II. Order Pleurocapsales

Genus I. *Dermocarpa*

Genus II. *Xenococcus*

Genus III. *Dertnocarpella*

Genus IV. *Myxosarcina*

Genus V. *Chroococcidiopsis*

Subsection III. Order Oscillatoriales

Genus I. *Spirulina*

Genus II. *Arthrospira*

Genus III. *Oscillatoria*

Genus IV. *Lyngbya*

Genus V. *Pseudanabaena*

Genus VI. *Starria*

Genus VII. *Crinalium*

Genus VIII. *Microcoleus*

Subsection IV. Order Nostocales

Family I. Nostocaceae

Genus I. *Anabaena*

Genus II. *Aphanizomenon*

Genus III. *Nodularia*

Genus IV. *Cylindrospermum*

Genus V. *Nostoc*

Family II. Scytonemataceae

Genus I. *Scytonema*

Family III. Rivulariaceae

Genus I. *Calothrix*

Subsection V. Order Stigonematales

Genus I. *Chlorogloeopsis*

Genus II. *Fischerella*

Genus III. *Stigonema*

Genus IV. *Geitleria*

In 'Bergey's Manual of Determinative Bacteriology' 9th edition (1994) the cyanobacteria have been placed under Group 11. Oxygenic Phototrophic Bacteria.

The Group 11 itself comprises two separate groups. The most thoroughly studied and understood group is the cyanobacteria. They have chlorophyll 'a' and phycobilins. It comprises 5 subgroups: Chroococcales, Pleurocapsales, Oscillatoriales, Nostocales and Stigonematales.

Q.7. Give the classification/systematic position of **Actinomycetes.**

Ans. According to 'Bergey's Manual of Systematic Bacteriology' Vol. IV (1989). The **Filamentous Actinomycetes** and Related bacteria are included in Section 26 to Section 33.

Section 26. Nocardioform Actinomycetes include genera: *Nocardia, Rhodococcus, Nocardiodes, Pseudonocardia, Oerskovia, Saccharopolyspora, Faenia (Micropolyspora), Promicromonospora, Intraporangium, Actinopolyspora, Saccharomonospora, Amycolatopsis* and *Amycolata.*

Section 27. Actinomycetes with multilocular sporangia include three genera: *Geodennatophilus, Dermatophilus* and *Frankia.*

Section 28. Actinoplanetes include five genera: *Actinoplanes, Ampulariella, Pilimelia, Dactylosporangium* and *Micromonospora.*

Section 29. Streptomycetes and related genera include four genera: *Streptomyces, Streptoverticillium, Kineosporia* and *Sporichthya.*

Section 30. Maduromycetes include genera : *Actinomadura, Microbispora, Microtetraspora, Planobispora, Planomonospora, Spirillospora* and *Streptosporangium.*

Section 31. *Thermomonospora* and related genera include *Thermomonospora, Actinosynnema, Nocardiopsis* and *Streptoalloteichus.*

Section 32. Thermoactinomycetes contain genus *Thermoactinomyces*

Section 33. Other genera contain *Glycomyces, Kibdelosporangium, Kitasatosporia* and *Saccharothrix.*

In 9th edition of 'Bergey's Manual of Determinative Bacteriology' (1994) Actinomycetes have been covered under Groups 22 to 29.

Q.8. Give classification/systematic position of the **Mycoplasmas.**

Ans. According to 'Bergery's manual of Systematic Bacteriology' Vol. I (1984), the **Mycoplasmas** fall under Section 10.

The Mycoplasmas

Order Mycoplasmatales

Family I. Mycoplasmataceae

Genus I. *Mycoplasma*

Genus II. *Ureaplasma*

Family II. Acholeplasmataceae

Genus I. *Acholeplasma*

Family III. Spiroplasmataceae

Genus I. *Spiroplasma*

Other Genera

Genus *Anaeroplasma*

Genus *Thennoplasma*

Mycoplasma-like organisms of plants and invertebrates.

According to 'Bergey's Manual of Determinative Bacteriology' 9th edition (1994), the Mycoplasmas (or Mollicutes): cell wall-less bacteria have been placed under Group 30.

Keys to the genera of mycoplasmas (mollicutes) are:

I. Facultatively anaerobic or microaerophilic

A. Sterol required

1. Cell helical in logarithmic growth

Genus *Spiroplasma*

2. Cells not helical
   (a) Urease positive
       Genus *Ureaplasma*

B. Sterol not required
       Genus *Acholeplasma*

II. Obligately anaerobic

A. Sterol required
       Genus *Anaeroplasma*

B. Sterol not required
       Genus *Asteroleplasma*

Q.9. Give classification/systematic position of **Archaea (old name, Archaeobacteria)**.

Ans. The Archaea or Archaeobacteria in 'Bergey's Manual of Systematic Bacteriology' Volume 3 (1989) have been placed under Section 25. Section 25. Archaeobacteria

Group I. Methanogenic Archaeobacteria

Order I. Methanobacteriales

Family I. Methanobacteriaceae

Genus II *Methanobacterium*

Genus II. *Methanobrevibacter*

Family II Methanothermaceae

Genus *Methanothermus*

Order II. Methanococcales

Family Methanococcaceae

Genus *Methanococcus*

Order III. Methanomicrobiales

Family I. Methanomicrobiaceae

Genus I. *Methanomicrobium*

Genus II. *Methanogenium*

Genus III. *Methanospirillum*

Family II. Methanosarcinaceae

Genus I. *Methanosarcina*

Genus II. *Methanococcoides*

Genus III. *Methanolobus*

Genus IV. *Methanothrix*

Other taxa

Family Methanoplanaceae

Genus *Methanoplanus*

Others

Genus *Methanosphaera*

Group II. Archaeobacterial Sulfate Reducers

Order Archaeoglobales

Family Archaeoglobaceae

Genus *Archaeoglobus*

Group III. Extremely Halophilic Archaeobacteria

Order Halobacteriales

Family Halobacteriaceae

Genus I. *Halococcus*

Genus II. *Halobacterium*

Genus III. *Haloferax*

Genus IV. *Haloarcula*

Genus V. *Natronococcus*

Genus VI. *Natronobacterium*

Group IV. Cell Wall less Archaeobacteria

Genus. *Thermoplasma*

Group V. Extremely Thermophilic S° Metabolizers

Order I. Thermoproteales

Family I. Thermoproteaceae

Genus I. *Thermoproteus*

Genus II. *Thermophillum*

Family II. Desulfurococcaceae

Genus I. *Desulfurococcus*

Genus II. *Thermococcus*

Genus III. *Thennodiscus*

Genus IV. *Pyrodictium*

Order II. Thermococcales

Family. Thermococcaceae

Genus I. *Thermococcus*

Genus II. *Pyrococcus*

Order III. Sulfolobales

Family Sulfolobaceae

Genus I. *Sulfolobus*

Genus II. *Acidianus*

In 'Bergey's Manual of Determinative Bacteriology' 9th edition (1994) the Archaeobacteria have been covered from Group 31 to Group 35. These groups are: Group 31. Methanogens, Group 32 Archaeal Sulfate Reducers, Group 33. Extremely Halophilic, Aerobic Archaeobacteria (Halobacteria), Group 34. Cell wall-less Archaeobacteria, and Group 35. Extremely Thermophilic and Hyperthermophilic S°-Metabolizers.

The Need for Pure Cultures (Purification of Bacteria)

Q.10. Why does one need pure cultures for identification?

Ans. Despite the fact that some bacteria may be having remarkable characteristics for identification, their isolation and preparation of pure cultures is a necessity. Modern techniques of identification using gene problems often with enhancement by a polymerase chain reaction PCR can be performed using mixed populations. However, these techniques are limited to a small number of clinically important species. As pointed out in the 'Bergey's Manual of Determinative Bacteriology' under chapter 11 'The Nature of Bacterial Identification Schemes' the single selection of a colony from a plate does not assure purity. This particularly is true while selective media are used.

Live but non-growing contaminants may often be present in or near a colony and can be subcultured along with the chosen organisms. That is why non-selective media are preferred for final isolation because they allow such contaminates to develop into visible colonies.

In non-selective media too, some well isolated colonies may not grow soon. Inoculants may be slow growing and may appear on the plate only after a longer incubation period.

In case of bacteria that form extracellular slime the filaments become firmly embedded and are difficult to separate. It is difficult to obtain pure cultures in cyanobacteria too as the contaminants get embedded in the gelatinous sheaths surrounding the cells.

In general, however, the colonies obtained as a result of streaking for pure cultures are similar to one another providing evidence of purity. Although this may be true, there may exist variations, e.g., capsular variations, pigmented or non-pigmented variants that are selected by certain media, temperature or other growth conditions.

Another criterion of purity is morphology. The organisms from a pure culture usually exhibit a high degree of morphological similarity in stains or wet mounts.

There are exceptions depending upon the age of the culture, the medium used and other growth conditions, e.g., coccoid body formation, cyst formation, spore formation and pleomorphism. A good example is that the both culture of marine spirillum after 2 or 3 days may lead one to conclude that the culture is contaminated with cocci. But experienced microbiologists are aware that such spirilla often develop into thin walled coccoid forms after a period of active growth.

Isolation of Pure Cultures

Q.11. Why is isolation of pure cultures required?

Ans. Generally the study of physiological, serological and other characteristics of bacteria are authoritative and meaningful when such studies are conducted from **pure culture.**

It means the growth of a single strain of bacteria which is free from contamination by other species of bacteria or other microorganisms. Therefore, for diagnostic examination of mixed infective material the first important step is the isolation of a particular organism in a pure culture. Pure cultures are also needed for proper identification and differentiation of bacteria.

The important techniques employed for isolation of pure cultures or creating pure cultures are as under:

(i) Planting out on a solid culture medium.

(ii) Use of selective, enrichment or indication media.

(iii) Use of selective growth conditions.

(iv) Selective treatment of the specimen before culture.

(v) Animal inoculation.

Bergey's Classification of Bacteria

Q.12. Give the classification table of bacteria.

Ans. The classification of bacteria according to the 9th edition of 'Bergey's Manual of Determinative Bacteriology' published in 1994 is as under. In it the bacteria have been classified under 35 Groups.

Group 1 The spirochetes

Genus *Borrelia*
Genus *Brachyspira*
Genus *Cristispira*
Genus *Leptonema*
Genus *Leptospira*
Genus *Serpulina*
Genus *Spirochaeta*
Genus *Treponema*
Genera of insect gut *Spirochetes*

Group 2 Aerobic/Microaerophilic, Motile, Helical/Vibrioid Gram-Negative Bacteria

Genus *Alteromonas*
Genus *Aquaspirillum*
Genus *Azospirillum*
Genus *Bdellovibrio*
Genus *Campylobacter*
Genus *Cellvibrio*
Genus *Halovibrio*
Genus *Helicobacter*
Genus *Herbaspirillum*
Genus *Marinomonas*
Genus *Micavibrio*
Genus *Oceanospirillum*
Genus *Spirillum*
"Spirillum minus"
"Spirillum pleomorphum"
"Spirillum Pulli"
Genus "Sporospirillum"
Genus *Vampirovibrio*
Genus *Wolinella*

Group 3 Nonmotile (Or Rarely Motile), Gram-Negative Curved Bacteria

Genus *Ancylobacter*
Genus "Brachyarcus"
Genus *Cyclobacterium*
Genus *Flectobacillus*
Genus *Meniscus*
Genus "Pelosigma"
Genus *Runella*
Genus *Spirosoma*

Group 4 Gram-Negative Aerobic/Microaerophilic Rods and Cocci SUBGROUP 4A

Genus *Acetobacter*
Genus *Acidiphilium*
Genus *Acidomonas*
Genus *Acidothermus*
Genus *Acidovorax*
Genus *Acinetobacter*
Genus *Afipia*
Genus *Agrobacterium*
Genus *Agromonas*
Genus *Alcaligenes*
Genus *Alteromonas*
Genus *Aminobacter*
Genus *Aquaspirillum*
Genus *Azomonas*
Genus *Azorhizobium*
Genus *Azotobacter*
Genus *Beijerinckia*
Genus *Bordetella*
Genus *Bradyrhizobium*
Genus *Brucella*
Genus *Chromohalobacter*
Genus *Chryseomonas*
Genus *Comamonas*
Genus *Cupriavidus*
Genus *Deleya*
Genus *Derxia*
Genus *Ensifer*
Genus *Erythrobacter*
Genus *Flavimonas*
Genus *Flavobacterium*
Genus *Francisella*
Genus *Fratfuria*
Genus *Gluconobacter*
Genus *Halomonas*
Genus *Hydrogenophaga*
Genus *Janthinobacterium*
Genus *Kingella*
Genus *Lampropedia*
Genus *Legionella*
Genus *Marinobacter*
Genus *Marinomonas*
Genus *Mesophilobacter*

Genus *Methylobacillus*
Genus *Methylobacterium*
Genus *Methylococcus*
Genus *Methylomonas*
Genus *Methylophaga*
Genus *Methylonphilus*
Genus *Methylovorus*
Genus *Moraxella*
Genus *Morococcus*
Genus *Neisseria*
Genus *Oceanospirillum*
Genus *Ochrobactrum*
Genus *Oligella*
Genus *Paracoccus*
Genus *Phenylobacterium*
Genus *Phyllobacterium*
Genus *Pseudomonas*
Genus *Psychrobacter*
Genus *Rhizobacter*
Genus *Rhizobium*
Genus *Rhizomonas*
Species *Rochalimaea henselae*
Genus *Roseobacter*
Genus *Rugamonas*
Genus *Serpens*
Genus *Sinorhizobium*
Genus *Sphingobacterium*
Genus *Thermoleophilum*
Genus *Thermomicrobium*
Genus *Thermus*
Genus *Variovorax*
Genus *Voleaniella*
Genus *Weeksella*
Genus *Xanthobacter*
Genus *Xanthomonas*
Genus *Xylella*
Genus *Xylophilus*
Genus *Zoogloea*

SUBGROUP 4B

Genus *Bacteroides*
Genus *Taylorella*
Genus *Wolinella*

Group 5 Facultatively Anaerobic Gram-Negative Rods

SUBGROUP 1 Family Enterbacteriaceae

Genus *Arsenophonus*
Genus *Budvicia*
Genus *Buttiauxella*
Genus *Cedecea*
Genus *Citrobacter*
Genus *Edwardsiella*
Genus *Enterobacter*
Genus *Erwinia*
Genus *Escherichia*
Genus *Ewingella*
Genus *Hafnia*
Genus *Klebsiella*
Genus *Kluyvera*
Genus *Leclercia*
Genus *Leminorella*
Genus *Morganella*
Genus *Obesumbacterium*
Genus *Pantoea*
Genus *Photobacterium*
Genus *Pragia*
Genus *Proteus*
Genus *Providencia*
Genus *Rahnella*
Genus *Salmonella*
Genus *Serratia*
Genus *Shigella*
Genus *Tatumella*
Genus *Xenorhabdus*
Genus *Yersinia*
Genus *Yokenella*

SUBGROUP 2 Family Vibrionaceae

Genus *Aeromonas*
Genus *Enhydrobacter*
Genus *Moellerella*
Genus *Plesiomonas*
Genus *Vibrio*

SUBGROUP 3

Family Pasteurellaceae

Genus *Actinobacillus*

Genus *Haemophilus*

Genus *Pasteurella*

SUBGROUP 4

Other Genera

Genus *Calymmatobacterium*

Genus *Cardiobacterium*

Genus *Chromobacterium*

Genus *Eikenella*

Genus *Gardnerella*

Genus *Streptobacillus*

Genus *Zymomonas*

Group 6 Gram-Negative, Anaerobic, Straight, Curved, and Helical Bacteria

Genus *Acetivibrio*

Genus *Acetoanaerobium*

Genus *Acctofilamentum*

Genus *Acetogenium*

Genus *Acetomicrobiurn*

Genus *Acetothermus*

Genus *Acidaminobacter*

Genus *Anaerobiospirillum*

Genus *Anaerorhabdus*

Genus *Anaerovibrio*

Genus *Bacteriodes*

Genus *Butyrivibrio*

Genus *Centipeda*

Genus *Fervidobacterium*

Genus *Fibrobacter*

Genus *Fusobacterium*

Genus *Haloanaerobium*

Genus *Halobacteroides*

Genus *Hyobacter*

Genus *Lachanospira*

Genus *Leptotrichia*

Genus *Malonomonas*

Genus *Megamonas*

Genus *Megamonas*

Genus *Mitsuokella*

Genus *Oxalobacter*

Genus *Pectinatus*

Genus *Pelobacter*

Genus *Porphyromonas*

Genus *Prevotella*

Genus *Propionigenium*

Genus *Propionispira*

Genus *Rikenella*

Genus *Roseburia*

Genus *Ruminobacter*

Genus *Sebaldella*

Genus *Selenomonas*

Genus *Sporomusa*

Genus *Succinimonas*

Genus *Succinivibrio*

Genus *Syntrophobacter*

Genus *Syntrophomonas*

Genus *Thermobacteroides*

Genus *Thermosipho*

Genus *Thermotoga*

Genus *Tissierella*

Genus *Wolinella*

Genus *Zymophilus*

Group 7 Dissimilatory Sulfate or Sulfur-Reducing Bacteria

SUBGROUP 1

Genus *Desulfotomaculum*

SUBGROUP 2

Genus *Desulfobulbus*

Genus *Desulfomicrobium*

Genus *Desufomonas*

Genus *Desulfovibrio*

Genus *Thermodesulfobacterium*

SUBGROUP 3

Genus *Desulfobacter*

Genus *Desulfomonde*

Genus *Desulfobacterium* | Genus *Desulfonema*
Genus *Desulfococcus* | Genus *Desulfosarcina*

SUBGROUP 4

Genus *Desulfurella* | Genus *Desulfuromonas*

Group 8 Anaerobic Gram-Negative Cocci

Genus *Acidaminococcus* | Genus *Syntrophococcus*
Genus *Megasphaera* | Genus *Veillonella*

Group 9 The Rickettsias And Chlamydias

I. Rickettsias

II. Chlamydias

Group 10 Anoxygenic Phototrophic Bacteria

SUBGROUP 1

Genus *Amoebobacter* | Genus *Thiocystis*
Genus *Chromatium* | Genus *Thiodictyon*
Genus *Lamprobacter* | Genus *Thiopedia*
Genus *Lamprocystis* | Genus *Thiospirillum*
Genus *Thiocapsa*

SUBGROUP 2

Genus *Ectothiorhodospira*

SUBGROUP 3

Genus *Rhodobacter* | Genus *Rhodopila*
Genus *Rhodocyclus* | Genus *Rhodopseudomonas*
Genus *Rhodomicrobium* | Genus *Rhodospirilluni*

SUBGROUP 4

Genus *Heliobacillus* | Genus *Heliobacterium*

SUBGROUP 5

Genus *Ancalochloris* | Genus *Pelodictyon*
Genus *Chlorobium* | Genus *Prosthecochloris*
Genus *Chloroherpeton* | Consortia, symbiotic aggregates

SUBGROUP 6

Genus *Chloroflexus* | Genus *Heligthrix*
Genus *Chloronema* | Genus *Oscillochloris*

SUBGROUP 7

Genus *Erythrobacter*

Group 11 Oxygenic Phototrophic Bacteria

I. THE CYANOBACTERIA

The Classification of Cyanobacteria

System of Cyanobacterial Classification

SUBGROUP 1 (= Order: Chroococcales)

Genus *Chamaesiphon* | Genus *Microcystis*
Genus *Cyanothece* | Genus *Myxobaktron*
Genus *Gloeothece* | Genus *Synechococcus*
Genus *Gloeobacter* | Genus *Synechocystis*
Genus *Gloeocapsa*

SUBGROUP 2 (= Order: Pleurocapsales)

Genus *Chroococcidiopsis* | Genus *Myxosarcina*
Genus *Dermocarpa* | Genus *Pleurocapsa*
Genus *Dermocarpella* | Genus *Xenococcus*

SUBGROUP 3 (=Order: Oscillatoriales)

Genus *Arthrospira* | Genus *Pseudanabaena*
Genus *Crinalium* | Genus *Spirulina*
Genus *Lyngbya* | Genus *Starria*
Genus *Microcoleus* | Genus *Trichodesmium*
Genus *Oscillatoria*

SUBGROUP 4 (=Order: Nostocales)

SECTION A (Family: Nostocaceae)

Genus *Anabaena* | Genus *Nodularia*
Genus *Aphanizomenon* | Genus *Nostoc*
Genus *Cylindrospermum*

SECTION B (Family: Rivulariaceae)

Genus *Calothrix*

SECTION C (Family: Rivulariaceae)

Genus *Calothrix*

SUBGROUP 5 (=Order: Stigonematales)

Genus *Chloroglocopsis* | Genus *Geitleria*
Genus *Fischerella* | Genus *Stigonema*

II. THE PROCHLOROPHYTES (ORDER PROCHLORALES)

Genus *Prochloron* | Genus *Prochlorothrix*

Group 12 Aerobic Chemolithotrophic Bacteria and Associated Organisms

SUBGROUP 1 COLOURLESS SULFUR-OXIDIZING BACTERIA

SECTION A MORPHOLOGICALLY CONSPICUOUS BACTERIA

Genus *Macromonas* | Genus *Thiospira*
Genus *Thiobacterium* | Genus *Thiovulum*
Genus *"Thiodendron"*

SECTION B LESS CONSPICUOUS SULFUR-OXIDIZING BACTERIA

Genus *Thermothrix* | Genus *Thiomicrospira*
Genus *Thiobacillus* | Genus *Thiosphaera*

SECTION C Related Bacteria

Genus *Acidiphilium*

SUBGROUP 2 Iron-And Manganese-Oxidizing And/or -Depositing Bacteria

Genus *Gallionella* | Genus *"Siderocapsa"*
Genus *"Leptospirillum"* | Genus *"Naumanniella"*
The Magnetotactic Bacteria | Genus *"Ochrobium"*
Genus *"Metallogenium"* | Genus *"Siderococcus"*
Family *"Siderocapsaceae"* | Genus *Sulfobacillus*

SUBGROUP 3 Nitrifying Bacteria

SECTION A Nitrite-Oxidizing Bacteria

Genus *Nitrobacter*
Genus *Nitrospina*
Genus *Nitrococcus*
Genus Nitrospira

SECTION B Ammonia-Oxidizing Bacteria

Genus *Nitrosomonas*
Genus *Nitrosococcus*
Genus *Nitrosospira*
Genus *Nitrosolobus*
Genus *"Nitrosovibrio"*

Group 13 Budding And/or Appendaged Bacteria

SUBGROUP 1 Prosthecate Bacteria

Genus *Anacalomicrobium*
Genus *Asticcaculis*
Genus *Caulobacter*
Genus *Dichotomicrobium*
Genus *Filomicrobium*
Genus *Hirschia*
Genus *"Hyphomicrobium"*
Genus *Hyphomonas*
Genus *Labrys*
Genus *Pedomicrobium*
Genus *Prosthecobacter*
Genus *Prosthecomicrobium*
Genus *Stella*
Genus *Verrucomicrobium*

SUBGROUP 2 Order Planctomycetales

Genus *Gemmata*
Genus *"Isophaera"*
Genus *Pirellula*
Genus *Planctomyces*

SUBGROUP 3 Other Budding And/Or Appendaged Bacteria

Genus *Angulomicrobium*
Genus *Blastobacter*
Genus *Ensifer*
Genus *Gallionella*
Genus *Gemminger*
Genus *Nevskia*
Genus *Seliberia*

Group 14 Sheathed Bacteria

Genus "Clonothrix"
Genus *Crenothrix*
Genus *Haliscomenobacter*
Genus *Leptothrix*
Genus *"Lieskeella*
Genus *"Phragmidiothrix"*
Genus *Sphaerotilus*

Group 15 Nonphotosynthetic, Nonfruiting Gliding Bacteria

SUBGROUP 1 Single Celled, Rod Shaped Gliding Bacteria

Genus *Capnocytophaga*
Genus *Chitinophaga*
Genus *Cytophaga*
Genus *Flexibacter*
Genus *Flexithrix*
Genus *Lysobacter*
Genus *Microscilla*
Genus *Sporocytophaga*
Genus *Thermonem*

SUBGROUP 2 Flattened, Filamentous Gliding Bacteria

Genus *Alysiella*
Genus *Simonsiella*

SUBGROUP 3 Sulfur-Oxidizing, Gliding Bacteria

Genus *Achromatiuin*
Genus *Beggiatoa*
Genus *Thioploca*
Genus *"Thiospirillopsis"*
Genus *Thiothrix*

SUBGROUP 4 The Pelonemas

Genus *"Achroonema"*
Genus *"Desmanthos*
Genus *"Pelonema"*
Genus *"Peloploca*

OTHER GENERA

Genus *Agitococcus*
Genus *Desulfonema*
Genus *Herpetosiphon*
Genus *"Isosphaera"*
Genus *Leucothrix*
Genus *Saprospira*
Genus *Toxothrix*
Genus *Vitreoscilla*

Group 16 The Fruiting, Gliding Bacteria

The Myxobacteria

Genus *Angiococcus*
Genus *Archangium*
Genus *Chondromyces*
Genus *"Corallococcus"*
Genus *Cystobacter*
Genus *"Haploangium"*
Genus *Melittangium*
Genus *Myxococcus*
Genus *Nannocystis*
Genus *Polyangium*
Genus *"Sorangium*
Genus *Stigmatella*

Group 17 Gram-Positive Cocci

Genus *Aerococcus*
Genus *Coprococcus*
Genus *Deinobacter*
Genus *Deinococcus*
Genus *Enterococcus*
Genus *Gemella*
Genus *Lactococcus*
Genus *Leuconostoc*
Genus *Marinococcus*
Genus *Melissococcus*
Genus *Micrococcus*
Genus *Pediococcus*
Genus *Peptococcus*
Genus *Peptostreptococcus*
Genus *Planococcus*
Genus *Ruminococcus*
Genus *Saccharococcus*
Genus *Salinicoccus*
Genus *Sarcina*
Genus *Staphylococcus*
Genus *Stomatococcus*
Genus *Streptococcus.*
Genus *Trichococcus*
Genus *Vagococcus*

Group 18 Endospore-Forming Gram-positive rods and Cocci

Genus *Amphibacillus*
Genus *Bacillus*
Genus *Clostridium*
Genus *Desulfotomaculum*
Genus *Oscillospira*
Genus *Sporohalobacter*
Genus *Sporolactobacillus*
Genus *Sporosarcina*
Genus *Suffidobacillus*
Genus *Syntrophospora*

Group 19 Regular, Nonsproing Gram-Positive Rods

Genus *Brochothrix*
Genus *Carnobacterium*
Genus *Caryophanon*
Genus *Erysipelothrix*
Genus *Kurthia*
Genus *Lactobacillus*
Genus *Listeria*
Genus *Renibacterium*

Group 20 Irregular, Nonsporing Gram-Positive Rods

Genus *Acetobacterium*
Genus *Acetogenium*
Genus *Actinomyces*
Genus *Aeromicrobium*
Genus *Agromyces*
Genus *Arachina*
Genus *Arcanobacterium*
Genus *Arthrobacter*
Genus *Aureobacterium*
Genus *Bifidobacterium*
Genus *Brachybacterium*
Genus *Brevibacterium*
Genus *Butyrivibrio*
Genus *Cascobacter*
Genus *Cellulomonas*
Genus *Clavibacter*
Genus *Coriobacterium*
Genus *Cotynebacterium*
Genus *Curtobacterium*
Genus *Dermabacter*
Genus *Eubacterium*
Genus *Exiguobacterium*
Genus *Falcivibrio*
Genus *Gardnerella*
Genus *Jonesia*
Genus *Lachnospira*
Genus *Microbacterium*
Genus *Mobiluncus*
Genus *Rubrobacter*
Genus *Pimelobacter*
Genus *Propionibacterium*
Genus *Rarobacter*
Genus *Rothia*
Genus *Sphaerobacter*
Genus *Terrabacter*
Genus *Thermoanaerobacter*

Group 21 The Mycobacteria

Genus *Mycobacterium*

Groups 22-29 The Actinomycetes

Growth and Examination of Actinomycetes. Some Guidelines
A Practical Guide to Generic Identification of Actinomycetes
A Brief Guide to Generic Groups

Group 22 Nocardioform Actinomycetes

SUBGROUP 1 Mycolic Acid-Containing Bacteria

Genus *Gordona*
Genus *Nocardia*
Genus *Rhodococcus*
Genus *Tsukamurella*

SUBGROUP 2 Pseudonocardia and Related Genera

Genus *Actinobispora*
Genus *Actinokincospora*
Genus *Actinopolyspora*
Genus *Amycolata*
Genus *Amycolatopsis*
Genus *Kibdelosporangium*
Genus *Pseudoamycolata*
Genus *Pseudonocardia*
Genus *Saccharomonospora*
Genus *Saccharopo~yspora*

SUBGROUP 3 Nocardiodies and Terrabacter

Genus *Nocardioides*
Genus *Terrabacter*

SUBGROUP 4 Promicromonospora and Related Genera

Genus *Jonesia*
Genus *Oerskovia*
Genus *Promicromonospora*

Group 23 Genera With Multilocular Sporangia

Genus *Dermatophilus*
Genus *Frankia*
Genus *Geodermatophilus*

Group 24 Actionoplanetes

Genus *Actinoplanes*
Genus *Ampullariella*
Genus *Carellatospora*
Genus *Dactylosporangium*
Genus *Micromonospora*
Genus *Pilimelia*

Group 25 Streptomycetes and Related Genera

Genus *Intrasporangium*
Genus *Kineospora*
Genus *Sporichthya*
Genus *Streptomyces*
Genus *Streptoverticillium*

Group 26 Maduromycetes

SUBGROUP 1 Streptosporangium and Related Taxa

Genus *Microbispora*
Genus *Microtetraspora*
Genus *Planobispora*
Genus *Planomonospora*
Genus *Spirillospora*
Genus *Streptosporangium*

SUBGROUP 2 Actinomadura

Genus *Actinornadura*

Group 27 Thermomonospora and Related Genera

Genus *Actinosynnema*
Genus *Nocardiopsis*
Genus *Streptoalloteichus*
Genus *Thermomonospora*

Group 28 Thermoactinomycetes

Genus *Thermoactinomyces*

Group 29 Other Genera

Genus *Glycomyces*
Genus *Kitasatosporia*
Genus *Saccharothrix*

Group 30 The Mycoplasmas (or Mollicutes): Cell Wall-Less Bacteria

Genus *Acholeplasma*
Genus *Anaeroplasma*
Genus *Asteroleplasma*
Genus *Mycoplasma*
Genus *Spiroplasma*
Genus *Ureaplasma*

Group 31 The Methanogens

SUBGROUP 1

Genus *Methanobacterium*
Genus Methanobrevibacter
Genus *Methanosphaera*
Genus *Methanothermus*

SUBGROUP 2

Genus *Methanococcus*
Genus *Methanocorpusculum*
Genus *Methanoculleus*
Genus *Methanogenium*
Genus *Methanolacinia*
Genus *Merhanomicrobium*
Genus *Methanoplanus*
Genus *Methanospirillum*

SUBGROUP 3

Genus *Methanococcoides*
Genus *Methanohalobium*
Genus *Methanohalophilus*
Genus *Methanolobus*
Genus *Methanosarcina*
Genus *Methanothrix*

Group 32 Archaeal Sulfate Reducers

Genus *Archaeoglobus*

Group 33 Extremely Halophilic, Aerobic Archaeobacteria (Halobacteria)

Genus *Haloarcula*
Genus *Halobacterium*
Genus *Halococcus*
Genus *Haloferax*
Genus *Natronobacterium*
Genus *Natronococcus*

Group 34 Cell Wall-Less Archaeobacteria

Genus *Thermoplasma*

Group 35 Extremely Thermophilic and Hyperthermophilic S°-Metabolizers

SUBGROUP 1

Genus *Acidianus*
Genus *Desulfurolobus*
Genus *Metallosphaera*
Genus *Sulfolobus*

SUBGROUP 2

Genus *Pyrobaculum*
Genus *Thermofilum*
Genus *Thermoproteus*

SUBGROUP 3

Genus *Desulfurococcus*
Genus *Hyperthermus*
Genus *Pyrococcus*
Genus *Pyrodictium*
Genus *Staphylothermus*
Genus *Thermococcus*
Genus *Thermodiscus*

12 CHAPTER

VIRUSES

Q.1. Give salient features of viruses.

Ans. The viruses are noncellular entities comprising of **core** of one or more molecules either of DNA or RNA but never both. The core is enclosed in a **capsid** (protein coat) which may be icosahedral, helical or complex structure. Sometimes lipids and carbohydrates are also found in addition to protein. They are obligate intracellular parasites and can reproduce only within a living cell.

- A complete virus particle representing the extra cellular phase in called a **VIRION**.
- The viruses are cultured by inoculating living host or cell cultures with a **VIRION PREPARATION**. The virus concentration can be determined by virion count or from number of infectious units.
- The Simian virus 40 abbreviated as SV–40 possesses pentameric capsomeres.
- SV–40 is a small double stranded polyoma virus with 72 capsomeres.
- SV–40 is the causative agent of central nervous disease in rheus monkeys and is known to produce tumours in **HAMSTER**, a group of short tailed rat-like hibernating rodents, *cricetus cricetus* and allied species. SV–40 was first isolated from cultures of monkey kidney cells while preparing POLLIO VACCINE.
- **Nucleocapsid**-a protein nucleic acid complex' is helical in TMV, the tobacco mosaic virus.
- **Ebola virions** like all filoviruses are filamentous particles which my appear in shape like a **Shepherd's crook** *i.e.* stick with rounded head at one end or in shape of "U" or "6". They may be coiled and branched.
- **Fruit bats** are believed to carry and spread the virus without being affected.
- **Ebola virus disease (EVD)** or **Ebola haemorrhagic fever (EHF)** is a disease of humans and other primates (Highest order of mammals including men, apes, monkeys and lemurs).
- The symptoms of EVD or EHF (Ebola haemorrhagic fever start from two days to three weeks after infection accompanied with a fever, sore throat, muscle pain and headaches. Typically vomitting, diarrhoea and rash follow along with decreased functioning of liver and kidneys. Around this time affected people may begin to bleed both within the body and externally.
- The male survivors of EHF or EVD may be able to transmit the disease via semen for nearly two months.
- Brahmins of India used small **pledget of cotton charged with the variolous matter** which he/they moistened with two or three drops of the Ganges water and applied that to wound to inoculate against small pox. The cotton which he preserved in a **double caligo rag** was saturated with matter from preceding year and never inoculated with fresh matter.

History of Viruses

Q.2. What do you envisage of the 'CONCEPT OF ORAL VACCINATION' in the worship of Goddess SHEETLA? Also give outlines of discovery of viruses.

Ans. The ancient Indians perhaps might have been aware of viruses which have been called the "MARUTS" microorganisms and entities travelling in the air, described in the Vedas.

They were fully cognizant of the diseases caused by the viruses as different types of poxes and neurological diseases like rabies. The previous night cooked, stale food, is offered to the Goddess Sheetla early in the morning and served to the members of family after that. The goddess is worshiped on "SHEETLA-ASHTMI" *i.e.* Ashtmi of Krishna Paksha, 8th day of the first fortnight of the month of Chaitra-the first month of Vikram Samvat also called the Hindu Calendar. Such worships continue for a week. This perhaps is a type of **ORAL INOCULATION** against poxes. There are large or small temples of SHEETLA almost in every village and town. The big temples include the temple of Sheetla at Nagori Gate, Jodhpur, Near Daksha Temple at Haridwar, and at Gurgaon (Haryana) today's information technology hub of India. The *Rigveda* oldest scripture of the world make mention of the small pox. Rigveda contains prayer to waters of river to keep humans free from disease *shipada*. The term *shipada* (reference to small pox a skin disease is due to malevolence or "ill will" of female demon *Simida* [*Rigved*: Mandal 7, Sukta 50, Mantra 4].

INDIANS KNEW HOW TO INOCULATE PEOPLE AGAINST SMALL POX: Although Edward Jenner performed vaccination in 1798 but Indians performed it much earlier. According to Y.L. Nene (2006) "A glimpse at viral diseases in ancient period" published in Indian Journal of virology, 17 (2) : 59–66, gives reference to two accounts one written by Rev Coult and sent to Dr. Oliver Coult on diseases in Bengal dated 10th February 1731 and other by Dr. J.Z. Holwell, F.R.S. sent to President and members of The College of Physicians in London in 1767 AD on **"inoculating for small pox in the East Indians".** Both these accounts provide evidence that Indians knew how to immunize people against small pox. Rev. Coult (1731) makes mention that "operation of inoculation of small pox" was performed in Bengal taken from concurring accounts of several **BHAMANS (Brahmins) and physicians** of this part of India and the operation was called *tikah* (teeka) by natives was being performed at least since 1581 AD. Rev. Coult further mentioned according to **BHAMANIAN (Brahminian) records** operation of inoculation was first performed by **Dhanwantri**.

J.Z. Holwell (1767) also makes mention of this operation by **BRAMINS OF INDOSTAN** (= Brahmins of India) who used variolous matter moistened with two or three drops of the Ganges water and applied to the wound fixing it on with a slight bandage. J.Z. Holwell further adds that the cotton which he (Brahmin) preserves in a double caligo rag is saturated with matter from the inoculated pustules of the preceding year, for they never inoculate with fresh matter nor with matter from the disease caught in the natural way. The published records reveal that how diseases have been instrumental in **Early Colonization of America**. The population of native people of North America got reduced 90% in a period of about 100 years of contact with Spanish. There might have been more than 72000 native people in AD 1600 of which only 8600 remained by 1674 AD. The population of the natives of Huron decreased from 32000 to 10000 in 10 years and those of Massachusetts decreased from 3000 to 300. It is apparent that European diseases like small pox decimated native American populations to which the disease was new and hence it paved the way for colonization of North American Continent by Europeans. Some American cities as Boston, Philadelphia and Plymouth-grew upon the sites of previous native American villages. Hernán Cortés conquered Empire of Aztec in Mexico because epidemic of small pox destroyed the city of Mexico. Perhaps the virus was first introduced to Mexico in 1520 which was brought by relief expedition sent to join Cortes. The small pox killed Aztec King Cuitlahuac, the nephew and the son-in-law of slain emperor and 1/3 of population of natives. The lady Wortley Montagu the wife of an English Ambassador had noticed that the Turkish women inoculated their children against small pox. The children gained immunity and the small pox was mild. Montagu wanted to educate the English public with her experience gained from Turkish women, but her efforts were not much successful. **Edward Jenner** and English (country) Gloucestershire physician was perhaps the first to perform vaccination of a boy named James Phipps in a scientific manner, in 1798, with the cow pox lesion material while he heared the statement of a milk maid that "she could not catch small pox since she had cow pox". The boy James Phipps who was inoculated by Edward Jenner with cow pox material, had mild fever but was protected against small pox. He sent the results of his first case for publication in 'Transactions of Royal Society'. When his paper was rejected he carried on more work and published the results of his 23 successful vaccinations in the form of a book in 1798.

Charles Chamberland developed porcelain bacterial filters known as **Chamberland's filters** which were useful in making the discovery of viruses. The term already existed for harmful agents from Latin *virus* = poison or venom. Chamberland, a man of multiple interests was also the invertor of autoclave. Dimitri lvanowski, a Russian in 1892, revealed that the leaf extract from infected tobacco mosaic plants induced tobacco mosaic disease in healthy tobacco plants despite of the fact that bacteria were removed in process of filtration. The occurrence of disease he attributed to some toxin in the filtrate. Martin W. Beijerinck who was working independently on tobacco mosaic disease demonstrated in 1898 that filtered sap of the diseased plants was still infections was some agent other than bacteria, a virus that multiplied only on living plants cell.

The contemporaries of Beijerinck who were Friedrich Loeffler and Paul Frosch of Germany observed that **"Hoof and Mouth Disease"** of Cattle (also called **"Foot and Mouth Disease"** was also caused by a filterable virus.

Walter Reed in 1900 found that "Yellow Fever" disease which was spreading in Cuba, was due to a filterable virus which was transmitted by mosquitoes.

Vilhelm Ellermann and Oluf Bang in Copenhagen, on advent of 20th century recorded that **leukemia** can be **transmitted between chickens** by cell free filtrates, perhaps the virus. Peyton Rous of Rockefeller Institute of New York demonstrated in 1911 that malignant muscle tumour of chickens was caused by a virus.

An interesting discovery was made by Frederrick W. Twort in 1915, who isolated **bacterial viruses** (now called bacteriophages) which could infect and destroy micrococci and intestinal bacilli. Unfortunately Twort could not follow his findings so Felix d' Herelle confidently confirmed the existence of bacterial viruses. Interesting enough, that d' Herelle also isolated bacterial viruses from patients of dysentery perhaps caused by *Shigella dysenteriae*, Gram negative bacteria which are intestinal pathogens of man and other primates. D' Herelle noted that viruses isolated by him can reproduce only on live bacteria and can cause lysis of cells in the bacterial lawns. He named these viruses as **BACTERIOPHAGES** which were capable of making holes in the lawns by eating bacteria and causing lysis of the cells.

Wendell M. Stanley crystallized the tobacco mosaic virus in 1935, found that TMV was made of a protein was a step towards the chemical nature of viruses. However, the complete information was provided in 1930s by Frederick C. Bawden and Norman W. Pirie, who separated particles of TMV into protein and nucleic acid.

Morphology of Viruses

Q.3. Give morphological characteristics of size and shape of viruses.

Ans. The viruses generally range from 14 nm to 300 nm in size, but some exceptions are not ruled out. The smallest virus is little larger than ribosomes. Both rhinovirus a common cold virus

14 nm and poliyovirus have diameters of 30 nm. In them proteins assemble into a stable structure known as an iscosahedron. The bacteriophage ϕ x 174 is also one of the smallest and has a diameter of 18 nm. The vaccinia virus is one of the largest viruses, the virion of which are almost as large as the smallest bacteria, and can be seen under a light microscope. The virus used in todays small pox vaccine is called **vaccinia** virus. It is probably not cow pox rather a highly attenuated form of small pox. The rod shaped tobacco mosaic virus is 300 nm long and 18 nm in diameter. Some flexuous filamentous viruses may reach a length of 1200 2000 × 12 nm as Closteroviruses-a group of plant viruses. Thus some viruses are larger than some bacteria like Chlamydia and Mycoplasma.

The viruses are of various types may be rod shaped (TMV) flexuous (Closteroviruses), spherical, criss-crossed, hexagonal to shaped like mirror with handle (T-even phage), and bullet shaped (animal rhabdoviruses *e.g.* rabies virus). The plant rhabdoviruses are usually bacilliform but sometimes bullet shaped typically about 200 to 350 × 70 to 95 nm in size. The viruses differ in shape and complexity of structure, *e.g.* HIV is relatively simple in structure, but devastating in action, and causes AIDS

Structure (Ultrastructure) of Viruses

Q.4. Describe the ultrastructure of viruses. shape of viruses.

Ans. The virions, the complete and mature particles of viruses outside the host cell comprise of a nucleic acid (either DNA or RNA but never both) surrounded by a protein coat or **capsid**. The nucleic acid which comprise the central part or **core** and the capsid (protein coat) have been also called **nucleocapsid**. In other words protein cover and nucleic acid (core) complex is called nucleocapsid. Besides nucleocapsid (capsid, plus nucleic acid) some viruses have an **envelope**, the outer membrane layer of 'lipoproteins' where the lipid components of viral envelope are derived from host cell membranes while the protein components are virus-encoded. Many of the proteins project from the surface as **spikes or peplomers**.

The enveloped viruses are almost spherical but exhibit variability in shape even though their nucleocapsid may be icosahedral or helical. Some viruses have capsid symmetry neither absolutely icosahedral nor symmetrical. They possess tail and other structures, and therefore, have been called **complex viruses**. They include structures as bacteriophages (from Greek phagein = 'to eat') or have complex multilayered walls surrounding the nucleic acid as pox virus like vaccinia (Fig. 12.1).

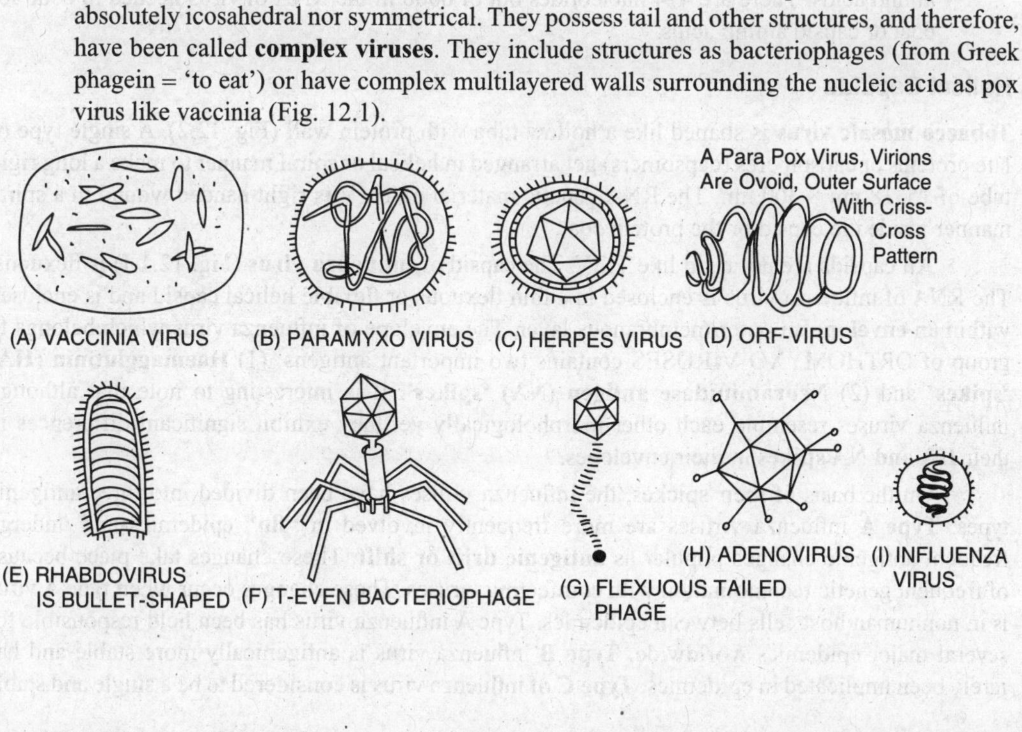

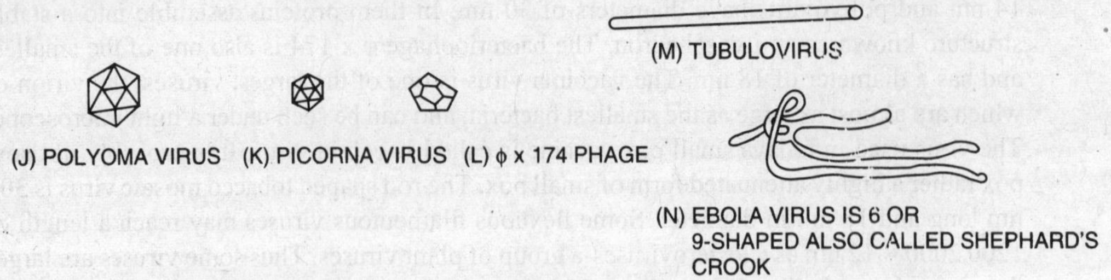

Fig. 12.1. Range of shape and structure of viruses: A. **Vaccinia** virus is a useful vector for inserting foreign DNA into animal cells, has a broad host range. B. **Paramyxovirus** is enveloped virion with helical nucleocapsid. C. **Herpes** virus is about 120–200 nm in diameter, is enveloped, contains a core of DNA surrounded by icosahedral capsid. D. Orf-virus showing criss-cross surface pattern. E. Bullet-shaped Rhabdo (Rabies) virus. F. T-even virion capable of infecting *E. coli*. G. Flexuous tailed phage. H. Adenovirus I. Influenza virus known for antigenic drift or shift. J. Polyoma virus. K. Picorna virus (*pico* = small). L. φ x 174 phage, one of the smallest bacteriophage. M. Tubulovirus, a tube like virus. N. Ebola virus.

The viruses which do not have envelope are called naked. Morphologically some capsids are **icosahedral** in shape. An icosahedron is a regular polyhedron with 20 equilateral triangular faces and 12 vertices as adenoviruses, T-even bacteriophage and φ x 174 phage (Fig. 12.1F and 3). Such capsids while observed under low magnification of electron microscope appear to be spherical.

Second category of capsids in naked virions are helical in the form of hollow protein cylinder that may be either rigid or flexible.

In **lytic cycle** the first formed proteins are called **early proteins** and their production continues until DNA replication is complete. This is followed by synthesis of **late proteins** which are called **capsomeres** or **capsomers**. These are the proteins used in virion assembly and regulation. TMV capsid comprises of single type of small subunits each possessing 158 amino acids. There are 474 nucleotides out of 6000 in the RNA of virus needed to code for coat or capsid amino acids.

Helical Capsids

Tobacco mosaic virus is shaped like a hollow tube with protein wall (Fig. 12.2). A single type of late proteins or capsomeres (capsomers) get arranged in helical or spiral manner to make a long rigid tube of 15–18 nm × 300 nm. The RNA genetic material of TMV is right handed wound in a spiral manner inside the capsid or the protein coat.

All capsids are not rigid like TMV. The capsid of **influenza virus** (Fig. 12.1-I) is flexuous. The RNA of influenza virus is enclosed in a thin flexuous or flexible helical capsid and is enclosed within an **envelope** an outer membranous layer. The envelope of influenza virus which belongs to group of ORTHOMYXO VIRUSES contains two important antigens: (1) **Haemagglutinin (HA) 'spikes'** and (2) **Neuraminidase antigen (NA) 'spikes'**. It is interesting to note that although influenza viruses resemble each other morphologically yet they exhibit significant differences in their **HA and NA spikes** in their envelopes.

On the basis of their spickes, the influenza viruses have been divided into three antigenic types. **Type A** influenza viruses are more frequently involved in **"flu"** epidemics and undergo frequent antigenic changes popular as **antigenic drift or shift**. These changes take place because of frequent genetic recombination with coinfecting viruses. These changes occur when type A virus is in nonhuman host cells between epidemics. Type A influenza virus has been held responsible for several major epidemics worldwide. **Type B** influenza virus is antigenically more stable and has rarely been implicated in epidemics. **Type C** of influenza virus is considered to be a single and stable antigenic form which occurs sporadically. Secondary bacterial pneumonia caused by streptococci and staphylococci may occur in severely debilitated (weakened) or elderly persons. Cardiovascular and renal diseases may also occur as complications. Recently **Guillian-Barre syndrome** and **Reye's syndrome** have been recorded with increasing frequency. **Reye's syndrome** is an acute disease of children and adolescents which involves non inflammatory encephalopathy with fatty infiltration and dysfunction of the liver characterized by persistant and profuse vomiting and neurological dysfunction, sometimes progressing to delirium, coma and death. Reye's syndrome can occur after various viral infections but more frequently after type B influenza infections.

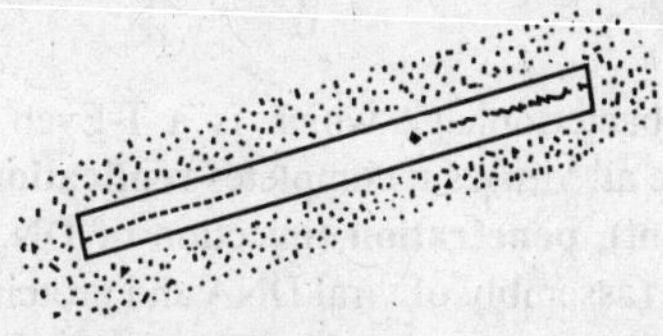

(A) ELECTRONMICROGRAPH BASED DRAWING OF A LONG APPARENTLY ROD SHAPED, NEGATIVE STAINED HELICAL CAPSID OF TMV

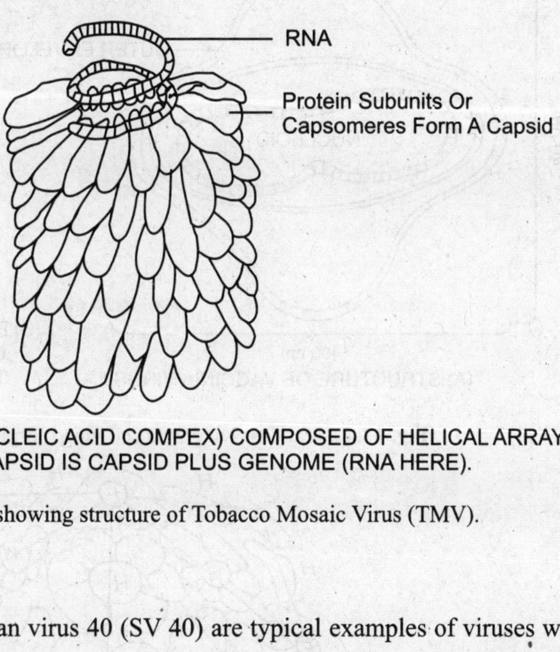

(B) NUCLEOCAPSID (A PROTEN NUCLEIC ACID COMPEX) COMPOSED OF HELICAL ARRAY OF CAPSOMERES. A NUCLEOCAPSID IS CAPSID PLUS GENOME (RNA HERE).

Fig. 12.2. A and B. Diagrams showing structure of Tobacco Mosaic Virus (TMV).

Icosahedral Capsids

Adenoviruses, papiloma virus, and simian virus 40 (SV 40) are typical examples of viruses which form icosahedral capsids .

In viruses there are few genes and sometimes only one which can encode for protein that may self assemble to form the capsid. The viruses adopt the icosahedral shape because it is the most efficient style to enclose the space. The capsids are made of ring-shaped or knob-like structures or units **late proteins** called **capsomeres** or **capsomers**. Each capsomer in turn is made of 5 or 6 **early proteins** or **protomers**. The **pentamers** also called pentons have five smaller subunits. The **hexamers** or hexons have six protomers. The pentamers are at the 12 **vertices**-the top points or higher points opposite the base of a triangle or cone like geometrical shapes. The hexamers form the edges and faces of the icosahedron. The capsid of this type has 42 capsomeres (capsomers). The adenoviruses have 252 capsomers in their capsid. The pentamers and hexamers in some plant and bacterial RNA viruses are made of only one type of subunit **early proteins** but in adenoviruses pentamers are made of different proteins than that of hexamers. The early proteins or **protomers** join to form **capsomers** by means of noncovalent bonds. The bonds between proteins in pentamers and hexamers are stronger than that in separate capsomers. Thus empty capsid can dissociate into separate capsomers.

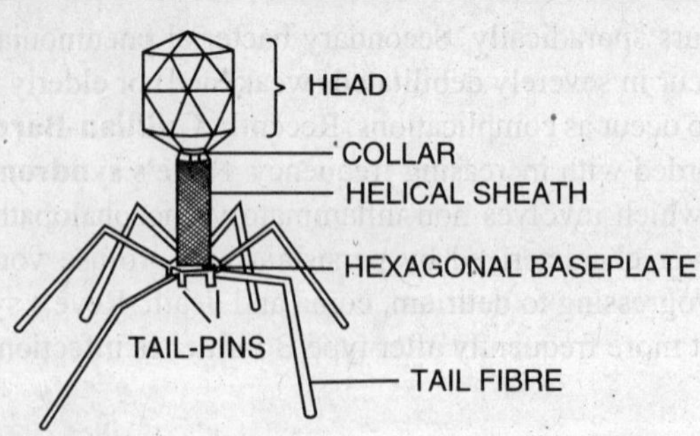

Fig. 12.3. The structure of T4 bacteriophage which is a T-Even Coliphage. It is **lytic** (virulent) dsDNA containing virion. Like all viruses it completes **replication** (multiplication cycle) through the stages **adsorption** (attachment), **penetration** (injection of DNA), **biosynthesis** (making viral DNA and proteins), **maturation** (assembly of viral DNA and proteins) and **lysis** (cell burst to release virions).

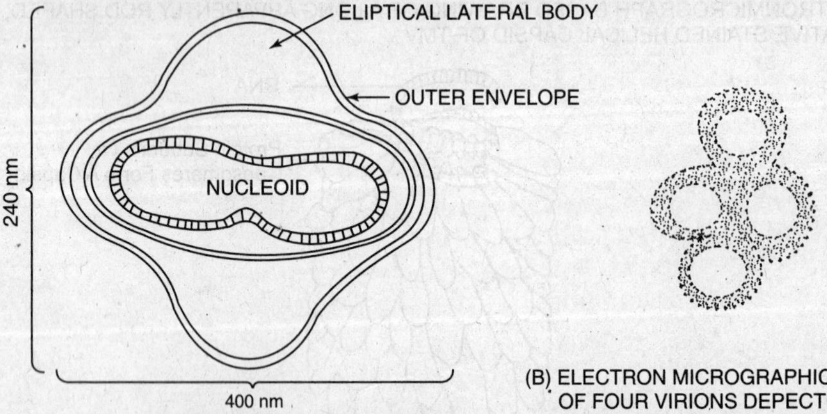

(A) STRUCTURE OF VACCINIA VIRUS

(B) ELECTRON MICROGRAPHIC DIAGRAM OF FOUR VIRIONS DEPECTING THICK COAT OF SURFACE FIBRES

Fig. 12.4. A and B. Diagrams showing morphology of vaccinia virus.

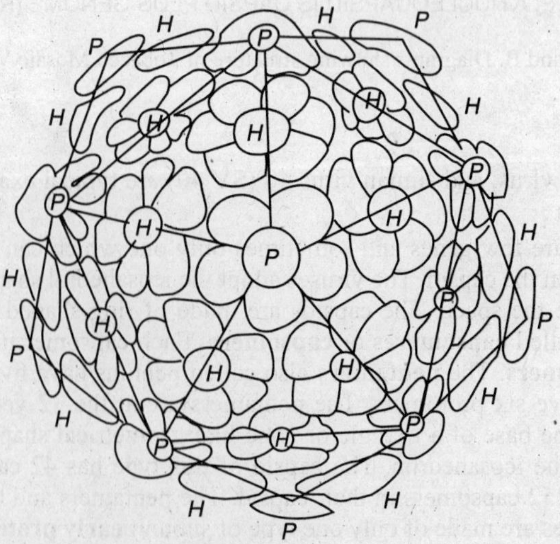

Fig. 12.5. Structure of an **Icosahedral capsid** comprised of pentamers and hexamers. The pentamers represented by P are present at the 12 vertices. (peaks). The Hexamers represented by H form the edges and faces of the icosahedron. The capsid consists of 42 **capsomeres** the late protein subunits. Each capsosomere comprises of 5 or 6 identical **protomers** (early proteins).

Most icosahedral capsids comprise of both pentamers and hexamers but recently another type has been discovered, *i.e.* in **simian virus 40 (SV–40)** which is a small double stranded DNA virus (of genus polyomavirus) has only pentamers. (Fig. 12.6) SV–40 is made of 72 cylindrical pentamers with hollow centres. The five flexible arms extend from the edge of each pentamer. Twelve pentamers are at icosahedron's vertices and associate with five neighbours, just like viruses in which hexamers are also present.

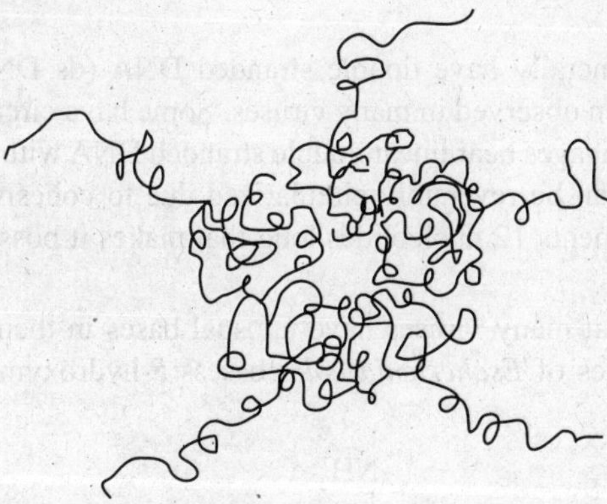

(A) DIAGRAMATIC ELECTRONMICROGRAPH OF A **PENTAMER** OF SV-40 CAPSID DEPECTING A TANGLED-WEB-LIKE STRUCTURE WITH EXTENDED ARMS

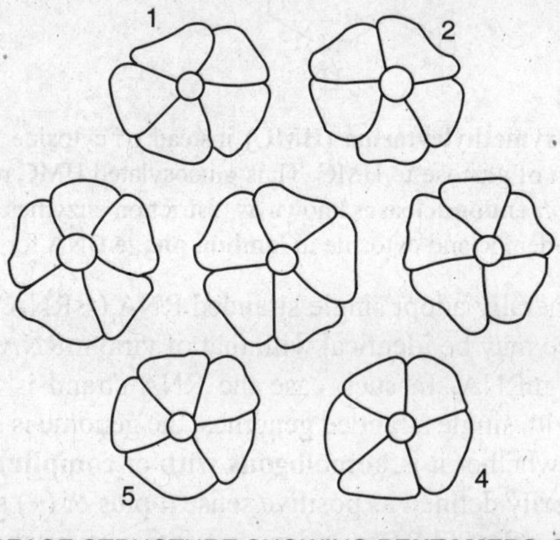

(B) A PART OF SURFACE STRUCTURE SHOWING **PENTAMERS.** EACH PENTAMER IS SHOWN BY A FIVE-PETALED FLOWER. EXTENDED ARMS HAVE NOT BEEN SHOWN TO KEEP THE DIAGRAME SIMPLE. THE OUTERMOST PROTOMERS ARE NUMBERED CLOCK-WISE DEMARCATING ONE AT THE **VERTEX** (PEAK ON APEX)

Fig. 12.6. A and B. Icosahedral capsid of Simian virus 40 which is made of **pentamers** or pentons only.

NUCLEIC ACID: VIRAL GENOME OR FULL SET OF GENES

The viruses most often have feasible four types of nucleic acids. These are: single stranded DNA; double stranded DNA; single stranded RNA; and double stranded RNA.

The plant viruses for instance most often have single stranded RNA as their genome. Some of the phages may have single stranded DNA or single stranded RNA. Bacterial viruses (abbreviated as phages usually contain double stranded DNA. Undoubtedly some viral nucleic acids show variation from this generalization.

The genome of MS 2 and Q β viruses is perhaps smallest, *i.e.* 1 × 10^6 daltons that can code from three to four proteins. On the other hand the genomes of T-even bacteriophages, herpes virus and vaccinia virus are 1.0 to 1.6 × 10^8 daltons and can synthesize more than 100 proteins.

The **bacteriophage** (also abbreviated as phage) ϕ x 174 a microvirus, M 13 an Inovirus or filamentous phage, and Parvoviruses have single stranded DNA (ss DNA) genomes. Some viruses are provided with linear pieces of DNA while the genome of others is comprised of closed circle of DNA.

The DNA viruses generally have double stranded DNA (ds DNA). Linear dsDNA with various modifications has been observed in many viruses. Some have circular dsDNA. The **lambda phage**, a group of temperate phages bear linear double stranded DNA with cohesive ends. The linear DNA of the lambda phage can be reversibly circularized due to cohesive ends. There are single stranded complimentary segments 12 nucleotides long that makes it possible to cyclize when they base pair their counter parts.

Interesting enough that many viruses have unusual bases in their DNA instead of normal nucleotides *e.g.* T-even phages of *Escherichia coli* possess 5-hydroxymethylcytosine in place of cytosine (Fig. 12.7).

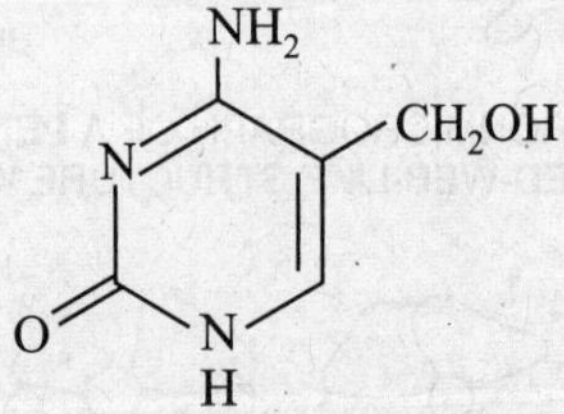

Fig. 12.7. Unusal base **5-hydroxymethylcytosine (HMC)** instead of cytosine found in **T-even phage**. It is glucosylated by addition of glucose to HMC. Thus glucosylated HMC residues protect T4 DNA from attack by *Escherichia coli* **endonucleases** known as restriction enzymes. The methyl groups are added to the amino groups of adenine and cytosine in **lambda phage** DNA for such protection.

The RNA viruses generally adopt single stranded RNA (ssRNA) for their genetic material. The viral RNA base sequence may be identical with that of viral mRNA, *i.e.* positive sense ssRNA genome can act directly as mRNA. In such case the RNA strand is said to be **plus strand** or **positive strand**. In viruses with single stranded genomes, the genome is said to be **positive-sense** or negative-sense according to whether it is **homologous with** or **complimentary to** respectively, the **viral mRNA** which is arbitrarily defined as positive sense = plus or (+) strand. The examples of the positive strand viruses are the polio, TMV, brome mosaic (a tripartite ssRNA containing plant virus), and Rous sarcoma viruses originally isolated from naturally occurring sarcoma of chicken, some strains can induce tumourigenesis (tumorigenesis) on inoculation into mice, rabbits and monkeys). The examples of negative strand viruses are rabies, mumps, measles and influenza viruses.

Numerous **RNA genomes** are **segmented** into 10–12 segments. All segments are enclosed in the same protein coat or capsid. It is considered that each fragment or segment codes for one protein. The brome mosaic virus is made of four segments distributed among three different virus particles. Thus all three of the largest fragments are needed to cause infection. Inspite of this complex and inefficient arrangement with fragments scattered to different capsids brome mosaic visions manage to infect the same host successfully. It is wonderful that RNA viruses have gained EXTRA ADVANTAGE by having their genome resemble eucaryotic mRNA. The plus strand RNAs can manage protein synthesis immediately after entering the cell of the host. It is because the plus strand viral RNA usually resembles mRNA in more than the equivalence of its nucleotide sequence. Like eucaryotic mRNA usually has a 5′ cap of 7 methylguanosine. Many plant and animal RNA genomes are capped. Besides most or all plus strand RNA animal viruses also possess a poly-A-sequence at 3′ of their genome, therefore, closely resemble eucaryotic mRNA with respect of structure of both ends.

It is fascinating that a number of single stranded plant viral RNAs have 3′ ends which resemble eucaryotic transfer RNA or tRNA and the genomes of tobacco mosaic virus will accept amino acids. However, the capping has not been observed in RNA bacteriophages.

Currently one of the largest outbreak has been the **2014 West African Ebola outbreak** Like all mononegative viruses ebola virions contain linear nonsegmented single strand RNA genome of negative polarity which possesses inverse-complementary 3′ and 5′ termini. Ebola viruses do not possess a 5′ cap, are not polyadenylated and are not covalently linked to a protein. Ebola virus genomes are approximately 19 kilobase pairs long and contain seven genes in the order *3′–UTR*–NP–VP35–VP40–GP–VP30–VP24–L-*5′–UTR*.

Types of Viral Envelopes (Peplos) and Their Composition

The **envelope** (synonym **peplos**) is a lipoprotein membrane which forms the outermost layer of the virion in certain animal viruses, plant viruses and bacteriophages. The animal virus envelopes have been studied most extensively. The lipid components of the viral envelope are often derived from host membranes (nuclear membranes or plasma membranes) while the proteins typically glycoproteins are virus encoded. Many of the proteins project from the surface of the envelope which are called **spikes** or **peplomers**. (Fig. 12.8) These spikes enable the virus particle in attachment to the surface of host cell. As they differ in different viruses are of **taxonomic significance** and can be used in identification of some viruses. Since the envelopes are membranous structure the enveloped viruses have frequently variable shape, therefore, these have been referred to be **pleomorphicviruses**. Contrary to it the envelope of the bullet-shaped rabies virus are strongly attached to the nucleocapsid lying below the envelope and provide the **rabies virus** a characteristic constant shape (Fig. 12.1 E). The enveloped of some viruses are denatured are called **ether sensitive** and lipid mediated activities are blocked and the virus become inactive. The **spikes** or **peplomers** in **influenza virus** (Fig. 12.8) project about 10 nm from the surface of the envelope at the intervals of 7 to 8 nm.

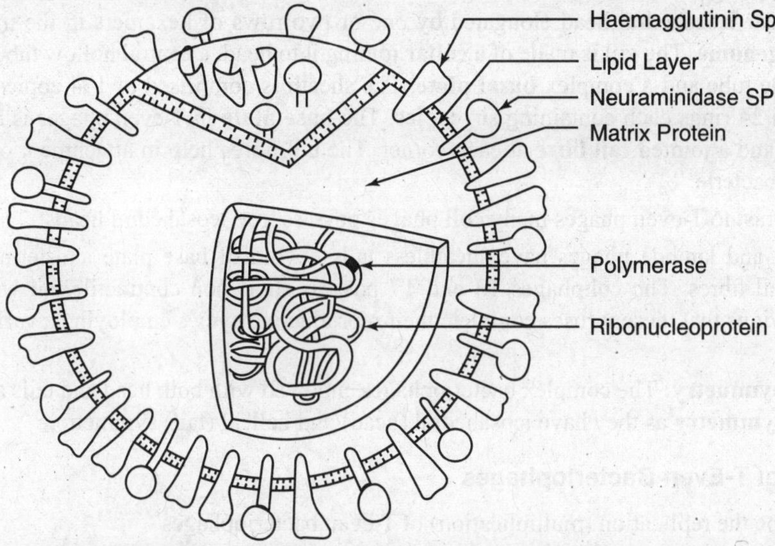

Fig. 12.8. Diagramatic structure of the **human influenza virion**. The **envelope** of the virion is flexible and the **spikes** made of glycoproteins **haemagglutinin** (HA) and **Neuraminidase** (NA) project from the surface of the virus. HA **trimers** and NA **tetramers** form spikes and mushroom shaped projection. During infection influenza virus virions attack (via the HA protein) to sialic acid containing receptors on plasma membrane of a host cell. Penetration take place by endocytosis followed by HA dependent membrane fusion.

The enzyme neuraminidase is present in some spikes which help the virus to penetrate mucous layers of the respiratory epithelium and to reach host cells. On the other hand some spikes possess haemagglutinin proteins as they bind the membranes to cell membranes of red blood cells (RBCs) and lead to haemagglutination and clustering of RBCs. Thus haemagglutinins are involved in attachment virions on the cells of the host. The **M or matrix proteins** a nonglycosylated protein also occur on the inner surface of the envelope and hence provides stability. However, the spike proteins which are exposed on the outer envelope surface are usually glycoproteins *i.e.* proteins have carbohydrate attached to them.

Most **viral enzymes** perhaps are located inside the capsid. Many of these enzymes are used in replication of nucleic acid. The **influenza virus** (Fig. 12.8) uses RNA as its genetic material and keeps an RNA dependent RNA polymerase which works as a replicase (RNA replicase the enzyme that synthesizes RNA using an RNA template used for replication by RNA viruses) and RNA transcriptase carries on synthesis of m RNA as directed by genomic RNA of influenza virus. The **polymerase**, the enzyme that directly make phosphodiester bonds and called so because of polymeric nature of nucleic acids (the polymerases that make RNA are called RNA polymerases and that make DNA are called DNA polymerases) is amalgamated with ribonucleoprotein.

Obviously, the viruses are devoid of true metabolism and cannot reproduce independently outside the host cells but the viruses keep one or more enzymes for completion of their life cycle.

Complex Viruses

Poxviruses: The poxviruses are the largest of all animal viruses 250–450×170–260 nm in size. They bear double stranded DNA associated with protein. DNA is the genome that is core or a **nucleoid** (Fig. 12.4A) in the form of a biconcave disc and is surrounded by a membrane. Two **elliptical** or **lateral bodies** lie between the nucleoid and its **outer envelope** a membrane and thick layer covered, by an array or protective costume of tubules or fibres.

T-even phages: T2, T4 and T6 phages which infect *Escherichia coli* have been studied in detail. They bear icosahedral head elongated by one or two rows of hexamers in the middle and have DNA as **genome**. The tail is made of a **collar** joining it to head, a central hollow tube, a sheath surrounding the tube and a complex **basal plate**. The sheath is comprised of 144 copies of gp 18 protein kept in 24 rings each containing six copies. The **base plate** in T-even phages is hexagonal and has a **pin** and a jointed **tail fibre** at each corner. The tail fibres help in attachment of virus on the surface of bacteria.

In contrast to T-even phages many coli phages possess true icosahedral heads.

T1, T5 and lambda phages have sheathless tails devoid of base plate and terminate into rudimentary tail fibres. The coliphages T3 and T7 possess short non contractile tail without tail fibres. It is obvious that these viruses complete their reproductive cycles employing a variety of tail structures.

Binal symmetry: The complex bacteriophages endowed with both head and tails are known to bear **binal symmetry** as they have icosahedral (head) and helical (tail) symmetry.

Lytic Cycle of T-Even Bacteriophages

Q.5. Describe the replication (multiplication) of T-even bacteriophages

Ans. The multiplication of T-even bacteriophages T2, T4 and T6 in their host *Escheria coli* is by **lytic cycle**. The virions of T-even bacteriophages are large, complex and nonenveloped with a typical head and tail structure (Fig. 12.3). The DNA of these phages is only 6% of that of the DNA found in *Escherichia coli*. The bacteriophage (also called phage) has enough DNA for 100 genes. Like all other viruses multiplication of T-even phages can be given into five distinct stages (Fig. 12.9). First stage is **Adsorption** or **attachment** in which the T-even phage gets attached to its host cell surface by tail fibres. These phages resemble one another morphologically and their genomes can undergo recombination with each other when present in the same host cell. They differ *e.g.* in their host cell receptor sites. The T4 receptor on *Escherichia coli* B strain is α-glucosyl-(1 → 3)-glucose terminus of the rough lipopolysaccharide (LPS) core.

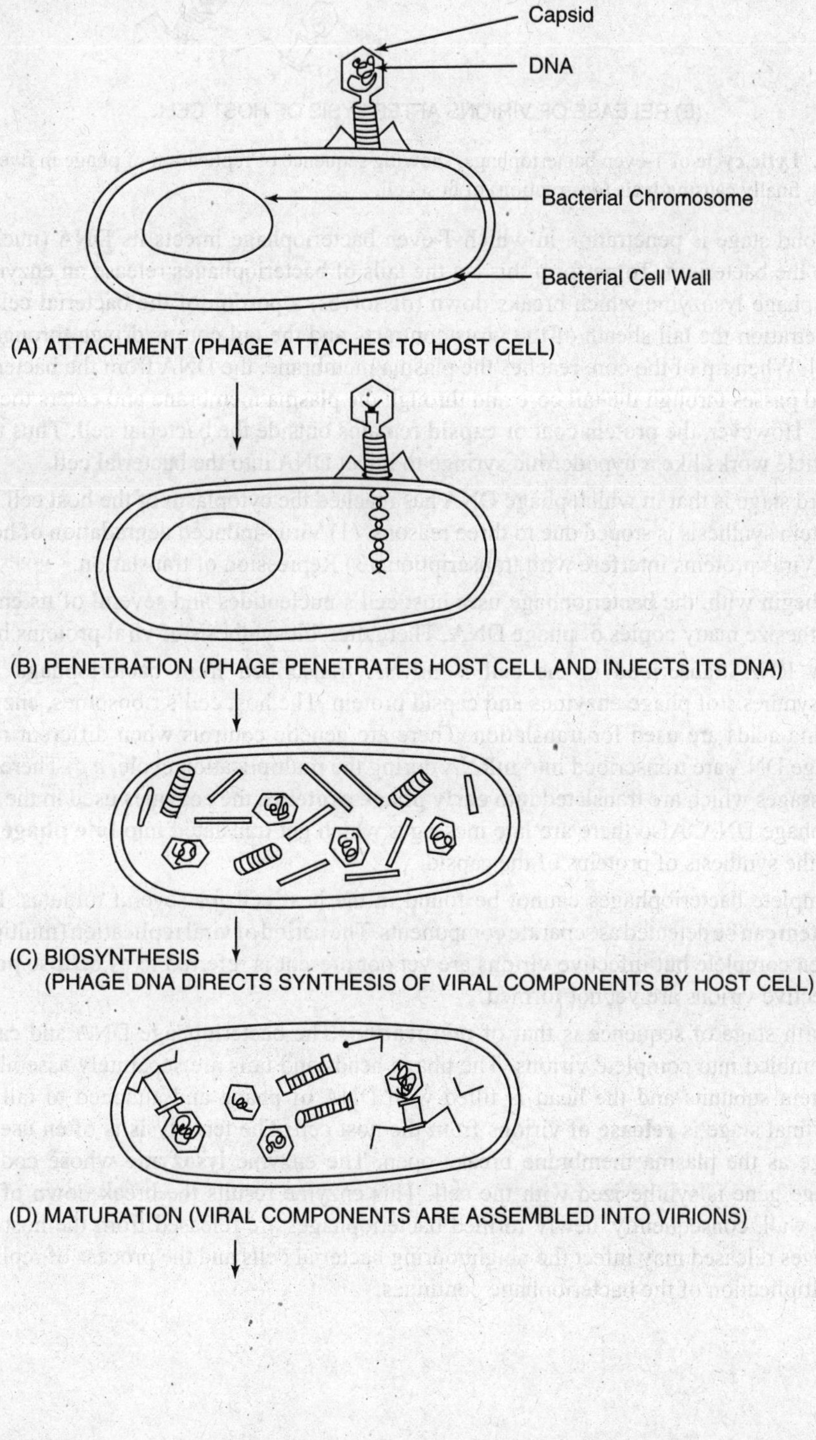

(E) RELEASE OF VIRIONS AFTER LYSIS OF HOST CELL

Fig. 12.9. Lytic cycle of T-even bacteriophage, showing sequence of replication of phage in five stages, finally causing **lysis** (destruction) of host cell.

Second stage is penetration in which T-even bacteriophage injects its DNA (nucleic acid) into the bacterium. To perform this act the tails of bacteriophages release an enzyme called the phage lysozyme which breaks down (dissolves) a portion of the bacterial cell wall. In penetration the tail sheath of the virus contracts and the tail core is driven through the cell wall. When tip of the core reaches the plasma membrane, the DNA from the bacteriophage-head passes through the tail core and through the plasma membrane and enters the bacterial cell. However, the protein coat or **capsid** remains outside the bacterial cell. Thus the phage particle works like a hypodermic syringe to inject DNA into the bacterial cell.

Third stage is that in which phage DNA has reached the cytoplasm of the host cell. The host protein synthesis is stoped due to three reasons. (1) Virus-induced degradation of host DNA. (2) Viras proteins interfere with transcription.. (3) Repression of translation.

To begin with, the bacteriophage uses host cell's nucleotides and several of its enzymes to synthesize many copies of phage DNA. Thereafter, biosynthesis of viral proteins begins.

Any RNA transcribed in the cell is mRNA transcribed from bacteriophage DNA for biosynthesis of phage enzymes and capsid protein. The host cell's ribosomes, enzymes and amino acids are used for translation. There are genetic controls when different regions of phage DNA are transcribed into mRNA during the multiplication cycle, *e.g.* There are early messages which are translated into **early phage proteins**, the enzymes used in the synthesis of phage DNA. Also there are late messages which get translated into **late phage proteins** for the synthesis of proteins of the capsid.

Complete bacteriophages cannot be found in the host cell for several minutes. DNA and protein can be detected as separate components. The period of viral replication (multiplication) when complete but infective virions are yet not present is referred to as **eclipse period**. *i.e.* infective virions are yet not formed.

Fourth stage of sequence is that of **maturation**. The bacteriophage DNA and capsids are assembled into complete virions. The phage heads and tails are separately assembled from protein subunits and the head is filled with DNA of phage and attached to tail. Fifth or the final stage is **release** of virions from the host cell. The term **lysis** is often used for this stage as the plasma membrane breaks open. The enzyme **lysozyme** whose code is by a phage gene is synthesized with the cell. This enzyme results the break down of bacterial cell wall, consequently, newly formed bacteriophages are released from the host cell. The phages released may infect the neighbouring bacterial cells and the process of replication or multiplication of the bacteriophage continues.

The time spent from the attachment of the phage to release is called **burst time** which ranges from 20 to 40 minutes. The number of bacteriophage particles released from a single cell is called **burst size** in which 50 to 200 virions are released.

The different stages that take place in the multiplication of bacteriophages can be shown graphically by **one-step growth curve**.

Lysogenic Cycle: Incorporation of Phage DNA into the Host Cell Without Lysis.

Q.6. Describe the **lysogenic cycle** of bacteriophage lambda.

Ans. Some viruses do not cause lysis and death of the host cell when they replicate (multiply) as it happens in T-even bacteriophages. Thus some phages or bacteriophages can either proceed through a lytic cycle or begin a lysogenic cycle by incorporating their DNA into DNA of host cell. In the latter said case the process is known as **lysogeny** *i.e.* bacteriophages or phages are carried passively and transmitted at the time of division of the cell of the host. The phages which can multiply in this way are called **lysogenic phages** or **temperate phages**. The bacterial host cells involved in the process are called **lysogenic cells**.

The bacteriophages λ (lambda) is one of the intensively studied temperate phages which adopt the lysogenic cycle. (Fig. 12.10)

First, on penetrating into an *Escherichia coli* cell the linear phage DNA forms a circle. This circle can multiply and be transcribed leading to production of new phage and to cell lysis *i.e.* lytic cell resulting death of the cell and releasing phage virions.

Alternatively, the circle can combine with the DNA of host cell and becomes the part of it *i.e.* adopts the **lysogenic cycle**. The inserted DNA of the bacteriophage is known as **prophage**. Most of the prophage genes are repressed or suppressed by two repressor proteins which are the products of phage genes. This means that the represses genes inhibits functioning of the phage genes. The repressors stop transcription of all other phage genes by binding to operators. Thus it is obvious that phage genes which would otherwise direct the synthesis and release the new virions are turned off in the same manner as the genes of *Escherichia coli*, *lac* operon are turned off by the *lac* represser.

In this way the host cell's machinery replicates or multiplies the bacterial chromosome automatically also replicates prophage DNA also. However, a rare spontaneous event or the action of ultraviolet light or that of certain chemicals can lead to excision or removal of phage DNA and initiation of lytic cycle.

There are three notable results of lysogeny:

(1) The lysogenic cells are immune to reinfection by the same phage. But the host cell is not immune to infection by other phage types.

(2) The host cell may show new properties, *e.g.* The bacterium *Corynebacterium diphtheriae* the causative agent of diphtheria of children, is a pathogen whose disease-producing properties are related to the synthesis of a toxin. The bacterium *C. diphtheriae* can produce toxin only when it carries a temperate phage because the prophage carries the gene coding for the toxin. There is another example that the streptococci carrying a temperate phage can produce the toxin associated with scarlet fever. The toxin produced by *Clostridium botulinum* causative agent of botulism is also because it is incoded by a prophage gene.

(3) Lysogeny makes **specialized transduction** in which only certain bacterial genes are transferred.

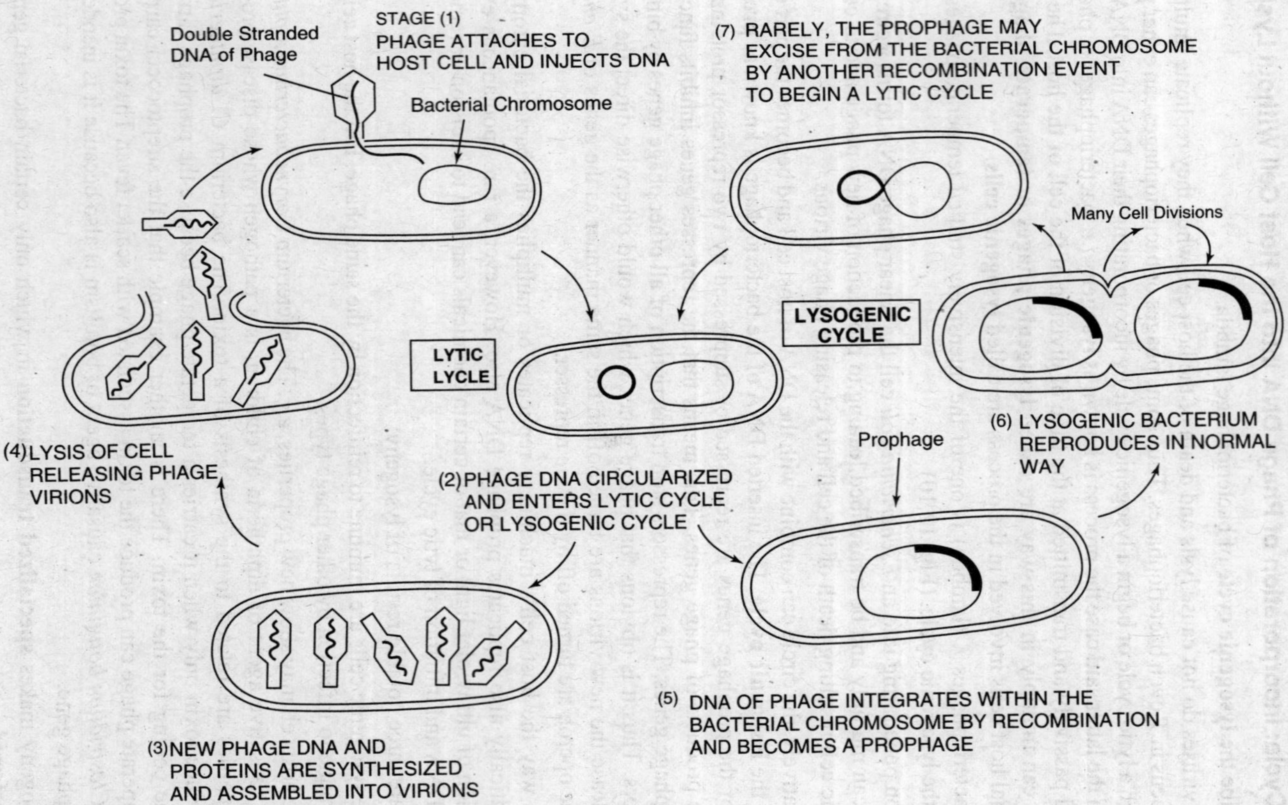

Fig. 12.10. **Lysogenic cycle** of Bacteriophage (phage) λ (lambda) in *Escherichia coli.* Both. **lytic (virulent)** cycle in which cell of host ruptures and **Lysogenic cycle** carrying viral prophage in which cell is not destroyed are shown.

The bacterial genes can be picked up in a phage coat and transferred to another bacterium in process known as **generalized transduction**. Any bacterial genes can be transferred by generalized transduction because the host chromosome is broken down into fragments any of which can be packaged into a phage coat.

On the other hand in **specialized transduction** (Fig. 12.11) only certain bacterial genes can be transferred. The specialized transduction takes place through a lysogenic bacteriophage that packages bacterial DNA along with its own DNA, in the same capsid. When a prophage is excised from the host chromosome adjacent genes from either side may remain attached to the DNA of phage, *e.g.* the genes which may remain attached to the bacteriophage λ are genes for galactose utilization and synthesis of biotin. Thus the phage particle newly synthesized from this DNA may thus contain one or more of the bacterial genes along with phage DNA. The phage may provide or grant new characteristics to the cell on lysogenization of a new host cell.

A cell infected by bacteriophage lambda can gain the ability to metabolize galactose or synthesis of biotin.

There is an another example that bacteriophage lambda (λ) can pick up the gal gene for galactose fermentation from its galactose-positive host. The phage carries this gene to a galactose negative cell which consequently becomes galactose positive. (Fig. 12.11). Some animal viruses can undergo process very similar to lysogeny. Animal viruses can remain latent in cells for long periods without multiplying or causing disease may become inserted in a host chromosome or remain separate from the host DNA in a repressed state like lysogenic phages. To conclude, the cancer causing viruses may also be latent.

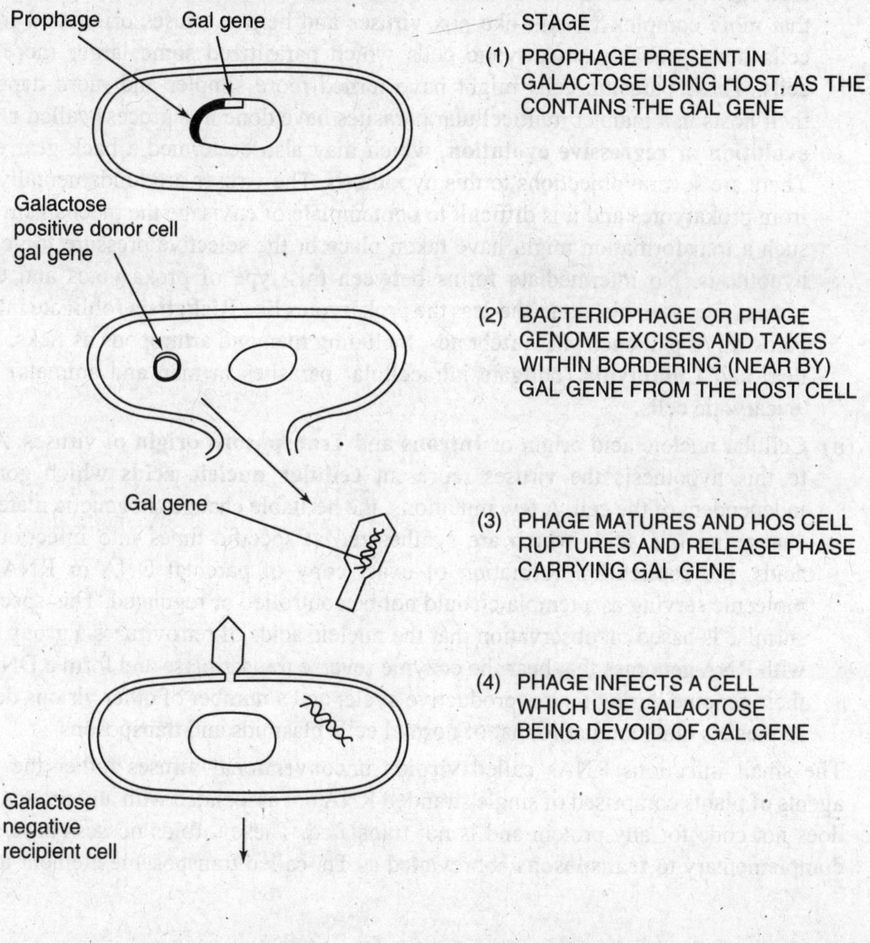

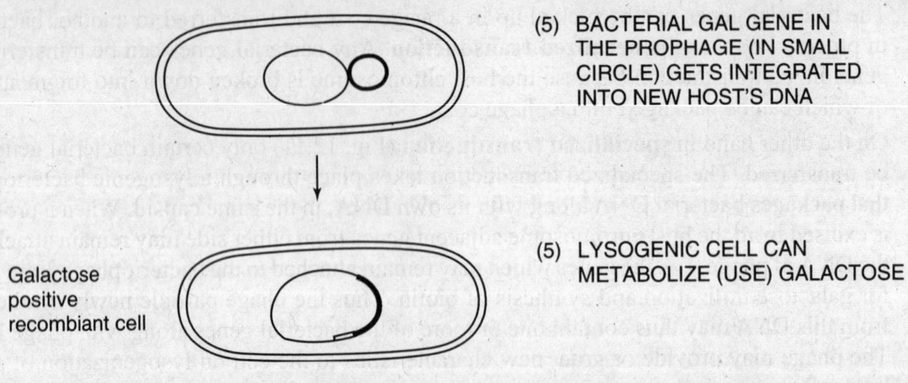

Fig. 12.11. Diagram showing **specialized transduction** in which **prophage** (phage DNA inserted into host-cell DNA) is excised (removed by) from its host chromosome along with a little of the adjacent or neighbouring DNA from the chromosome of the host bacterium.

Origin of Viruses

Q.7. What do you speculate of the origin of viruses? Discuss.

Ans. It is difficult to speculate or ruminate about the origin of viruses in absence of fossil record of the viruses yet there are the following major hypotheses, theorized by the virologists.

(i) **Retrograde** or **retrogress evolution** or **back-gear evolution** theory: It has ruminated that more complex viruses like pox viruses and herpes viruses originated from small cells most probably prokaryotic cells which parasitized some larger more complex cells. These parasitic cells might have turned more simpler and more dependent on their hosts in a manner multicellular parasites have done in a process called **retrograde evolution** or **regressive evolution**, which may also be termed a back-gear evolution. There are several objections to this hypothesis. The viruses are fundamentally different from prokaryotes and it is difficult to contemplate or envisage the mechanism by which such a transformation might have taken place or the selective pressure to gear up this hypothesis. No intermediate forms between this type of prokaryotes and enveloped viruses are available, none-the-less the prokaryotes like **Rickettsia** (obligate intracellular parasites or pathogens in vertebrates, including man and arthropods as ticks, mites and flea) and **Chlamydia** (obligate intracellular parasites in man and animals) parasitize eucaryotic cells.

(ii) Cellular nucleic acid origin or **Introns** and **Transposons origin** of viruses: According to this hypothesis the viruses represent **cellular nucleic acids** which got partially independent of the cell. A few mutations, the heritable changes in genetic material could change nucleic acids which are synthesized at specific times into infectious nucleic acids, the **replication** formation of exact copy of parental DNA or RNA, parental molecule serving as a template could not be controlled or regulated. This speculation or surmise is based on observation that the nucleic acids of retroviruses a group of viruses with RNA genomes that bear the enzyme reverse transcriptase and form a DNA copy of their genome during their reproductive cycle; and a number of other virions do not have sequences quite similar to that of normal cells plasmids and transposons.

The small infectious RNAs called **viroids** unconventional viruses rather the infectious agents of plants comprised of single stranded RNA not associated with any protein, the RNA does not code for any protein and is not translated. These viroids possess base sequences complementary to **transposons** abbreviated as Tn, called transposable element the regions around the boundary of mRNA, **introns**-the interrupted gene or one to many intervening sequences and portions of host DNA. Thus it has been conjectured that viruses have originated from introns and transposons. It is believed that cellular proteins spontaneously (naturally) assembled into icosahedra shapes around nucleic acids.

Cultivation of Animal Viruses

Q.8. How are animal viruses cultivated?

Ans. The animal viruses are cultured in the following way:

(i) The animal viruses can be cultured in laboratory animals. As the viruses are not able to penetrate skin the animals are incubated by infecting mucous membranes. A successful infection demonstrates disease symptoms which may include generalized body responses, tissue and cellular changes and particularly immunological responses.

(ii) The viruses are also cultured on embryonated chicken egg. To prepare egg for cultivation of virus the fertile egg is incubated in humid atmosphere for about 5 to 14 days at 37 to 39°C. Essentially an egg is **candled** for observing the contents of an intact egg by transmitted light. The embryonated egg in placed over an aperture in a box containing an electric lamp to see the internal details of the embryo and blood vessels which facilitates the inoculation of particular membrane or cavity (Fig. 12.12).

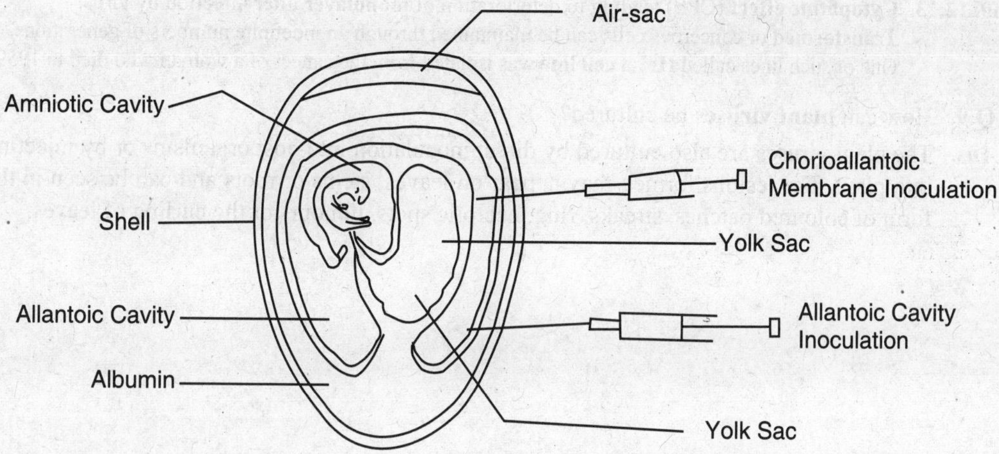

Fig. 12.12. Cultivation of virus with embryonated egg. The **chorioallantoic membrane** and the **allantoic cavity** the sites often used to grow animal viruses are also shown in the nine days old chicken embryo.

The cell surface is first cleaned and disinfected with iodine and penetrated by a small drill. The egg is inoculated by injecting virus into proper region where they grow well. After inoculation the drill hole is sealed with gelatin and the egg is incubated. The myxoma virus grows well on the **chorioallantoic membranes (CAM)** while mumps virus grows well in **allantoic cavity**. The infection with the virus produce lesions called pocks (pox). The embryonated eggs are regularly used by some pharmaceutical companies. The replicated viruses are isolated from embryo cultures and attenuated for use as vaccines to protect against diseases, like mumps, rabies and influenza.

Recently animal viruses have been grown in a thin single layer or monolayer prepared by cell and tissue culture. African green monkey kidney cells and human fibroblasts are widely used as viral hosts. Some antibiotics are also added to the growth media for cell/tissue culture to prevent bacterial and fungal contamination. A layer of animal cells is prepared in a petriplate and covered with virus inoculum and the viruses are allowed to settle and attach to cells. The cells are covered with a thin layer of agar to limit the spread the virion so that only adjacent cells are infected by newly produced virions. Consequently, the localized areas of cellular destruction or **lysis** take place and the clear zones so formed are called **plaques.**

Such plaques can be easily detected if stained with appropriate dyes like neutral red or trypan blue which differentiates living cells from dead cells.

It is not necessary that the viral growth always result into lysis of cells and form plaques. The animal viruses can also cause some microscopic or macroscopic degenerative changes or abnormalities in the host cells and tissues referred to as **cytopathic effects**. (Fig. 12.13).

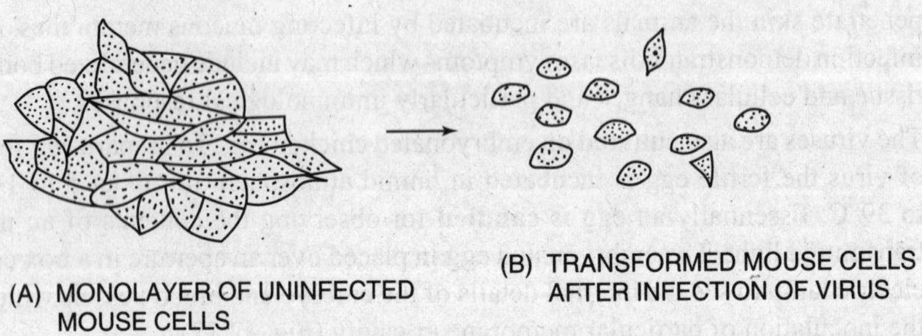

Fig. 12.13. Cytopathic effect (CPE) leading to deterioration of **monolayer** after infection by virus. **Transformed** or cancerous cells can be maintained through an indefinite numbers of generations. One of such lines called **HeLa cell line** was isolated from the cancer of a woman who died in 1959.

Q.9. How can **plant viruses** be cultured?

Ans. The plant viruses are also cultured by direct inoculation into host organisms or by injecting into plant. The lesions formed may appear on leaves, stems or roots and can be seen in the form of coloured patches, streaks, ring, necrotic spots, tumours or the curling of leaves.

13 CHAPTER

RIBOSOMAL RNA IN MICROBIAL TAXONOMY

Q.1. What is rRNA?

Ans. The rRNA is ribosomal RNA of various sizes and forms the part of ribosomes. It comprises 90% of total RNA of the cell while the remainder part of the ribosomes comprises ribosomal proteins.

Q.2. Give types of rRNA found in bacterial and archaeal 70 S ribosomes.

Ans. Bacterial and archael 70 S ribosomes comprise 50 S and 30 S subunits. The 50 S subunit has 23 S rRNA and 5 S rRNA. The 30 S subunit has 16 S rRNA.

Q.3. Give types of rRNA found in eucaryotic 80 S ribosornes.

Ans. The 80 S ribosomes comprise 60 S and 40 S subunits. The 60 S ribosome contains 28 S rRNA and 5.8 S rRNA while the subunit 40 S possesses 18 S rRNA.

Q.4. What is 16 S rRNA?

Ans. The ribosomal subunits have several ribosomal rRNA molecules of different molecular weights, e.g., 30 S subunits in bacterial and archael cells possess 16 S rRNA.

Q.5. Give the use of 16 S rRNA sequence in microbial taxonomy.

Ans. No doubt rRNA analyses are poor indicators of species level differences. Nevertheless RNA sequencing has been used extensively in microbial taxonomy. The sequences of 5 S and 16 S rRNAs are isolated from 50 S and 30 S subunits respectively. Their occurrence and the source may be represented as

$$5\text{ S rRNA} \xrightarrow{\text{Isolated form}} 50\text{ S Subunit}$$

$$16\text{ S rRNA} \xrightarrow{\text{Isolated form}} 30\text{ S Subunit}$$

Purified radioactive 16 S rRNA can be analyzed by treating it with the enzyme Ti ribonuclease that cleaves it into fragments as represented here.

$$\text{Purified radioactive 16 S rRNA} \xrightarrow{\text{Ti ribonuclease enzyme}} \text{Fragments (each of 6 nucleotides).}$$

The fragments so obtained are separated and all fragments composed of at least six nucleotides are sequenced. The sequence of corresponding 6S rRNA fragments from different bacteria are then compared using a computer and their association coefficient are calculated.

Q.6. Match the following:

(a)	Serve as excellent molecules for discerning evolutionary relationships among micro-organisms.	(A)	Are ancient molecules which are functionally constant.

(b)	The rRNAs are moderately well conserved across broad phylogenetic distances	(B)	Ribosomal RNAs
(c)	50 S and 30 S subunits	(C)	Procaryotes
(d)	Ribosomal rRNAs of 5 S, 16 S and 23 S occur in	(D)	70 S ribosornes
(e)	2900 nucleotides occur in	(E)	16 S rRNA
(f)	1500 nucleotides occur in	(F)	23 S rRNA
(g)	120 Nucleotides	(G)	5 S rRNA

Ans. [a B], [b A], [c D], [d C], [e F], [f E], [g G].

Q.7. Match the following:

(a)	Is more experimentally manageable than 23 S rRNA	(A)	Eucaryotes
(b)	18 S rRNA is the counterpart of 16 S rRNA in	(B)	16 S rRNA
(c)	16 S rRNA as a phylogenetic tool was suggested in 1970s at the University of Illinois	(C)	Phylogeny
(d)	Cultures of organisms are necessary to study	(D)	Carl Woese
(e)	Evolutionary distance (ED) is calculated by recording rRNA sequences on.	(E)	Computer

Ans. [a B], [b A], [c D], [d C], [e E].

Q.8. What is polymerase chain reaction?

Ans. The polymerase chain reaction (PCR) is a method employed to amplify a specific DNA sequence *in vitro* by repeated cycles of synthesis using specific primers and DNA polymerase.

Q.9. Define a ribosome.

Ans. It is a cytoplasmic particle comprising ribosomal RNA and protein which is a part of the protein synthesizing machinery of the cell.

Q.10. What is a **'signature sequence'?**

Ans. These are short oligonucleotides of unique sequence found in 16 S rRNA of a particular group of procaryotes. They are often found in defined regions of the 16 S rRNA molecules but are revealed when computer scans of two aligned sequences are carried out.

14

CHAPTER

CULTIVATION (CULTURES) AND GROWTH OF MICROORGANISMS

Q.1. Match the following.

(A)	Temperature, pH and osmotic pressure	(a)	A group of chemical requirements for growth of microorganisms.
(B)	Water, sources of carbon and nitrogen, minerals, oxygen and organic growth factors	(b)	Physical conditions for microbial growth.
(C)	Cold loving microorganisms	(c)	Mesophiles.
(D)	Moderate temperature loving microbes	(d)	Psychrophiles.
(E)	Heat loving microorganisms	(e)	Thermophiles.
(F)	Microbes responsible for spoilage of food in refrigerators	(f)	Psychrotrophs.
(G)	Optimum temperature for many pathogenic bacteria	(g)	Organic compost piles in which the temperature can rise from 50 to 60°C.
(H)	Thermophiles are important for	(h)	37°C.
(I)	Extreme thermophiles or hyperthermophiles	(i)	Optimum temperature for growth 80°C or higher.
(J)	Highest temperature which permits bacterial growth near hydrothermal vents in ocean.	(j)	110°C.

Ans. [A b], [B a], [C d], [D c], [E e], [F f], [G h], [H g], [I i], [J j].

Q.2. What is the optimum pH for most of the bacteria?

Ans. Near neutrality between 6.5 and 7.5.

Q.3. What are acidophiles?

Ans. Bacteria which can tolerate acidity.

Q.4. What are buffers?

Ans. Chemicals used to neutralise the acids and bases and maintain pH are called buffers. In other words the substance which tends to stablize the pH of a solution.

Q.5. Why do plasmolysis or shrinkage of the plasma of cells occur?

Ans. Because of osmotic loss of water (also called exosmosis).

Q.6. A low molecular weight compound such as sodium chloride has a greater antimicrobial effect. Give a comparative example.

Ans. NaCl or Sodium chloride having low molecular weight in comparison to sugar possesses higher antimicrobial effect.

Q.7. What are **extreme halophiles?** Give one example.

Ans. The microorganisms adapted to high concentrations of salts which is needed for their growth, e.g., the bacteria isolated from saline waters of Dead Sea usually require 30 per cent salt.

Q.8. What are **facultative halophiles?**

Ans. The microbes that do not require high salt concentration but can grow at salt concentrations of upto 2 per cent.

Q.9. What is superoxide dismutase (SOD)?

Ans. In the presence of oxygen obligate anaerobes are known to form some superoxide free radicals. Therefore, all organisms attempting to grow in atmospheric oxygen produce an enzyme superoxide dismutase (SOD) to neutralize toxic free radicals.

Q.10. What is **catalase**? Give its function.

Ans. Catalase is an enzyme which converts hydrogen peroxide (H_2O_2) produced in aerobic microorganisms into water and oxygen to get rid of their toxic effects of H_2O_2.

The presence of catalase can be detected easily when a drop of H_2O_2 is added to a colony and it effervesces (gives out bubbles).

Q.11. Why agar, which is a complex polysaccharide obtained from marine alga, and has been used to thicken foods like jellies, soups and ice-cream, is also used to solidify the media used in cultivation of microorganisms?

Ans. Agar melts at about 100°C and therefore, does not melt while incubating even thermophilic microorganisms needing high temperature to grow (approaching 100°C). Once agar melts it remains in liquid form even while the temperatures come down to 40°C on cooling. It can therefore, be easily handled while pouring into petriplates and culture tubes. In the laboratory agar is held in a water bath at 50°C, as at this temperature agar does not injure most of the microorganisms/bacteria when poured over the bacterial inoculum.

Q.12. What are media? Describe.

Ans. A brief description of media is as given below.

Laboratory Media for Cultivation of Microorganisms

There are innumerable microorganisms and therefore, it is most essential to prepare a medium to study their morphological, biochemical, diagnostic and other types of characteristics of basic and applied nature. Generally any liquid or solid preparation made specifically for the growth, storage or transportation of micro-organisms or another types of cells is referred to as a medium by students of microbiology. The growth media may be used for initiation of a culture or a subculture for enrichment, identification or diagnostic tests. There are special media for the growth and maintenance of mammalian cells, plant cells, protoplasts and other types of cells in tissue culture.

There are many chemoautotropic bacteria which can be grown in simple aqueous media containing mainly or only a mixture of inoganic salts. Nutritionally simple (undemanding) heterotrophs can be cultured on a large range of culture media as nutrient broth and peptone water. But nutritionally 'fastidious' heterotrophs can be grown only on an enriched medium. Strict anaerobes are often cultured on pre-reduced media that means media poised at or below a particular redox potential (oxidation-reduction potential represented by E_h). The poising agents used in the media are ascorbic acid, cysteine hydrochloride and thioglycolate. Poising is analogous to buffering in the context of pH, i.e., the E_h produced depends on the ratio oxidized: reduced poising agent. The stability of E_h is influenced by the absolute amount of poising agent in the medium.

The solid media are prepared from liquid media that have been solidified or gelled with an agent such as agar or gelatin. The other gelling agents are alginate, gelrite and pluronic polyol F 127.

The medium which without supplement can support the growth of nutritionally undemanding species of bacteria is often referred to as basal medium. The medium in which all the ingredients including trace substances are known quantitatively is called defined medium. The solid medium on which different types of organisms can be distinguished by their different forms of growth is termed differential medium. A medium which contains substances that encourage growth of the required organisms and/or inhibit the growth of other types of organisms is called enrichment medium, e.g., selenite broth and tetrathionate broth.

The medium used for the initial growth and the subsequent storage under conditions of minimal growth is known as maintenance medium. The maintenance medium for microorganisms is used to prepare a culture of the given organisms that can be stored either at ambient temperature or under refrigeration. Its subcultures are required at intervals ranging from 1 to 12 months. The constituents of the maintenance medium are often of minimum consistence with the need to maintain viability of the organisms that is being cultured, e.g., cooked meat medium and Dorset's egg.

The common media for cultivation of bacteria, often required in a laboratory are:

1. **Chemically defined media:** used to grow chemoautotrophs and photoautotrophs and for microbiological assays.
2. **Complex media:** used to grow most chemoheterotrophic organisms.
3. **Anaerobic growth media:** used to grow obligate anaerobes.
4. **Selective media:** used to suppress unwanted microbes and encourage desired microorganisms.
5. **Differential media:** These are used to distinguish colonies of desired microorganisms from others.
6. **Enrichment media:** These are similar to selective media but are designed to increase the number of desired microorganisms to a detectable level.

Q.13. What is a culture?

Ans. The microorganisms grown and multiplied in or on a culture medium are referred to as culture.

Q.14. What is a **deep**? How does it differ from a slant?

Ans. The agar solidified in a vertical tube is called a **deep** while the agar solidified in a tube in slanting position to increase the surface area for growth of the microorganisms is known as **slant.**

Q.15. Give a chemically defined medium for the growth of chemoautotrophic bacterium capable of using ammonium ions for energy.

Ans. It comprises ammonium sulphate [$(NH_4)_2$ SO)$_4$] 0.5 g, potassium phosphate monobasic (KH_2PO_4) 0.2 g, calcium chloride (Ca Cl_2) 0.04 g, magnesium sulphate ($MgSO_4$) 0.04 g, ferric citrate (FeC_6H_5 07) 0.0005 g and water 1 litre.

Q.16. Give the composition of a chemically defined medium for the growth of a typical **chemoheterotroph** (e.g., *Escherichia coli).*

Ans. It will comprise glucose-5.0 g, ammonium phosphate monobasic ($NH_4H_2PO_4$) – 1.0 g, sodium chloride (NaCl)-5.0 g, magnesium sulphate $MgSO_4$ $7H_2O$ –0.2 g, potassium phosphate dibasic (K_2HPO_4) –1 g, and water 1 litre.

Q.17. Give the composition of a chemically defined medium to culture **fastidious chemoheterotrophs** (e.g., *Neisseria gonorrhoeae).*

Ans. The constituents are carbon source, salts, amino acids, organic growth factor, reducing agent and water and their composition is given below. Glucose 9.1 g, starch 9.1 g, sodium acetate 1.8 g, sodium citrate 1.4 g and oxalo acetate 0.3 g as sources of carbon and energy. The salts are potassium phosphate dibasic (K_2HPO_4) 12.7g, sodium chloride (NaCl) 6.4g, potassium phosphate monobasic ($KH_2 PO_4$) 5.5g, sodium bicarbonate ($NaHCO_3$) 1.2 g, potassium sulphate (K_2SO_4) 1. 1 g, sodium sulphate (Na_2SO_4) 0.9 91 magnesium chloride ($MgCl_2$) 0.5g, ammonium chloride (NH_4 Cl) 0.4 g, potassium chloride (KCl) 0.4 g, calcium chloride ($CaCl_2$) 0.006g, ferric nitrate [Fe $(NO_3)_3$] 0.006g. The amino acids are cysteine 1.5 g, arginine 0.3 g, proline 0.3 g, glutamic acid 0.2g, methionine 0.2 g, asparagine 0.2 g, isoleucine and serine (each) 0.2 g, cystine 0.06 g. Organic growth factors are calcium pantothenate 0.02 g, thiamine 0.02 g, nicotinamide adenine dinucleotide 0.01g, uracil 0.006 g, biotin 0.005 g, hypoxanthine 0.003 g. The reducing agent is sodium thioglycolate 0.00003 g and water 1 litre.

Q.18. Give the composition of nutrient agar a complex medium for the growth of heterotrophic bacteria.

Ans. A complex medium is a medium the exact chemical composition of which varies from batch to batch. They are made up of nutrients like extracts from yeast, meat or plants or digests of proteins from these and other sources. The generally used nutrient agar comprises **peptone** (partially digested protein) 5.0 g, **beef extract** (concentrated extract of beef which is a dark brown paste-like substance) 3.0 g, sodium chloride 8.0 g, agar 15.0 g and water 1 litre.

Q.19. Agar itself is not a nutrient. Why is then **nutrient agar** called so?

Ans. Because it contains nutrients.

Q.20. What is a nutrient broth?

Ans. The liquid medium of the aforesaid type without agar is called **nutrient broth**, i.e., nutrients without agar.

Q.21. Why are reducing media used for cultivation of anaerobes?

Ans. Media of a special type known as reducing media are used for the cultivation of anaerobic bacteria because they may be required on exposure to oxygen. For this an anaerobic container is used to cultivate anaerobic bacteria on Petriplates. Water is mixed with the chemical packets containing sodium bicarbonate and sodium borohydride which results in generation of hydrogen and carbon dioxide. The palladium catalyst is placed in a screened reaction chamber, where the hydrogen and the atmospheric oxygen present in the jar combine to form water, freeing the environment in the jar from oxygen. Methylene blue is used as an anaerobic indicator which is blue when oxidized and turn colourless when the oxygen is depleted (removed) from the container.

Q.22. Why is *Salmonella typhi* (typhoid bacterium) cultured on bismuth sulphite agar?

Ans. Bismuth sulphite inhibits the growth of Gram positive bacteria and most Gram negative intestinal bacteria other than *Salmonella typhi*. It is a **selective medium.**

Q.23. Why is blood agar used as a differential medium to distinguish *Streptococcus pyogenes*?

Ans. The *Streptococcus pyogenes* causative agent of strep throat or sore throat show a clear ring around their colonies as the bacterium causes lysis of the surrounding blood cells around the colony.

Q.24. What is a pure culture? Why is pure culture needed? How are pure cultures obtained?

Ans. A population of one strain or species of bacteria or any other microorganisms is referred to as a pure culture.

The need for pure cultures was first realized by the German physician Robert Koch to study some human pathogens. Infectious materials such as pus, sputum and urine contain many types of bacteria. In the same way samples of soil, water and food also contain many kinds of microorganisms. Pure cultures are needed to identify and differentiate them on the basis of morphological and biochemical characteristics. Different organisms form unique colonies which are believed to have been formed from a single spore or individual vegetative cell on being spread on a medium.

The most common method to obtain pure culture of a bacterium or microorganism is the streak plate method. An inoculation loop is well sterilized by incinerating the loop (by heating over a flame). The loop is dipped in a mixed culture containing more than one type of microorganisms and is streaked on the surface of nutrient medium. While streaking, the bacteria are rubbed off the loop onto the medium in paths of fewer and fewer cells. In this way the last cells in the last part of the streak are sufficiently apart to grow and form isolated colonies.

Q.25. How are microorganisms preserved by deep-freezing and freeze-drying?

Ans. In deep-freezing pure cultures of microorganisms are placed in a suspending liquid and are frozen quickly at a temperature ranging from –50°C to –95°C. The cultures can be melted and used for several years.

In freeze-drying or **lyophilization** a suspension of microorganisms is frozen rapidly ranging from –54°C to –72°C and water is removed by a high vacuum. Under vacuum the containers (vials) are sealed by a high temperature torch. The powder like residue which remains can be stored for many years and can be revived by hydration with a suitable liquid nutrient medium at an appropriate time.

Q.26. What is **generation time**?

Ans. It is the doubling time or the time required for a cell to divide into two.

Q.27. What is a typical bacterial growth curve? What are its four phases?

Ans. A typical bacterial growth curve is formed when a few bacteria are inoculated into a liquid growth medium and the population is counted at intervals recording growth of cells over time taken. It comes out to be like, an inverted cup.

The phases that could be demarcated on this curve are:

(i) **Lag phase**—the period in which there is little or no cell division.

(ii) **Log phase** or **exponential growth phase**–this is the phase of maximum growth.

(iii) **Stationary phase**–this is the period of equilibrium.

(iv) **Death phase** or **logarithmic decline phase**–the number of cells formed is less than the number of cells that die.

Q.28. List the methods for direct measurement of microbial growth?

Ans. These are: (1) Plate counts, (2) serial dilution, (3) pour plate and spread plate, (4) filtration, (5) most probable number or MPN, and (6) direct microscopic count.

Q.29. Give the indirect methods for the estimation of microbial numbers.

Ans. These are: (1) Turbidity (2) metabolic activity and (3) dry weight.

15 CHAPTER

CONTROL OF MICROBIAL GROWTH (PHYSICAL, CHEMICAL AND ANTIBIOTIC)

Q.1. Why do we need to control the growth of microorganisms?

Ans. To control the growth of microorganisms of different types for various purposes is one of the most applied aspects of microbiology. We particularly control the growth of microorganisms for preservation of food and the environment, particularly for disinfection of water, to prevent the spoilage of articles of economic importance and to control the diseases caused by the microorganisms.

Q.2. Give devices and methods to control the growth of microorganisms.

Ans. These are divisible into three main categories:

(i) Physical control of microorganisms (with temperature, radiation, filtration, desiccation and oxygen to destroy strict anaerobes.

(ii) Chemical control of microorganisms (using disinfectants, preservatives and antiseptics).

(iii) Control by antibiotics and antimicrobics.

Q.3. Differentiate between sterilization and disinfection.

Ans. Elimination of microorganisms of all types from an environment by physical or chemical means is called sterilization, while destroying, removing or inactivating pathogenic microorganisms from an environment is referred to as disinfection.

Q.4. List the high temperature techniques of destroying microorganisms.

Ans. These techniques are incineration, dry heat and moist heat. **Incineration** is the complete combustion of a substance by flaming of wires and loops. High temperatures are used for decontamination of disposable plastic ware, paper ware, dressings and other medical wastes. Dry heat application for 2 hours at 180°C will destroy endospores. Moist heat is far more penetrating than dry heat, e.g., in high temperatures 121°C of an autoclave will destroyendospores at 15 pounds per square inch (psi) pressure.

Q.5. What is ultra high temperature sterilization of milk?

Ans. An exposure of milk to 141°C for 2 seconds.

Q.6. What is D value?

Ans. It is the decimal reduction time (DRT), which is an exposure time at a given temperature to kill/reduce the number of viable microorganisms by 90 per cent, e.g., the endospores of *Clostridium botulinum* have a D value of 0.21 minutes at 121°C. Heating food for 2.52 minutes at 121°C reduces the probability of the survival of C. *botulinum* endospores to 10^{-12}.

If a food can have one spore each, the probability of contamination after processing should be reduced to one in every trillion cans. Therefore, heating at 121°C for 2.52 minutes ensures the safety of canned food to any possible contamination by C. *botulinum*.

Q.7. Name the three diseases against which pasteurization has been effective.

Ans. Pasteurization has protected us against brucellosis or undulant fever caused by *Brucella,* a typical pneumonia fever caused by *Coxiella burnetii* and tuberculosis caused by *Mycobacterium bovis.*

Q.8. How does low temperature help protect the food from microbial spoilage?

Ans. By lowering growth and enzymatic activity.

Q.9. Name the bacterium which is resistant to exposures of 0.3 to 0.4 megarads that usually cause tenfold reduction in the number of viable bacterial spores.

Ans. *Micrococcus radiodurans* is resistant to ionizing radiation.

Q.10. How do ionizing radiations kill microbes?

Ans. **Ionizing radiations** disrupt hydrogen bonds, double bonds and ring structures in various molecules. The radiations in the presence of oxygen form hydroxyl free radicals (OH) which are toxic to the microbial cells.

Q.11. Why are the ultraviolet (UV) radiations useful in destroying microorganisms only on the surface?

Ans. UV wavelength 4 to 400 nm does not have high penetrating power below the surface.

Q.12. What sort of substances are sterilized by filtration?

Ans. Gases and liquids are sterilized by filtration.

Q.13. Name the microorganisms which cannot be removed from solutions by filtration.

Ans. Viruses and small bacteria such as mycoplasmas cannot be removed by filtration.

Q.14. How is air filtered in a laminar flow?

Ans. By using high efficiency particulate air filters or HEPA filters.

Q.15. Name a bacterial pathogen that can surivive desiccation (drying).

Ans. *Mycobacterium tuberculosis* can survive for weeks in dried sputum.

Q.16. What for is lyophilization or freeze-drying used and how?

Ans. **Lyophilization** is used to preserve microbial cultures by using high vacuum and low temperatures, such that water from frozen state goes directly into gaseous state without disrupting cellular structures.

Q.17. Give a microbiological application in which **hyperbaric oxygen** (pure oxygen under pressures used.

Ans. Pure oxygen under pressure (hyperbaric oxygen) is used for the treatment of the disease gangrene caused by an obligate (strict) anaerobe *Clostridium perfringens,* by forcing 100% oxygen into infected tissue at 3 atmosphere-pressure.

Q.18. Give examples of the following chemicals.

(i) **Algicide** (that kills algae).
(ii) **Bacteriocide** (the agent that destroys bacteria).
(iii) **Biocide** (that kills living organisms).
(iv) **Fungicide** (that kills fungi).
(v) **Germicide** (that specifically kills pathogens)
(vi) **Sporocide** (kills bacterial endospores).
(vii) **Virucide** (that inactivates viruses terminating their replication).
(viii) **Bacteriostatic** (that inhibits the growth and reproduction of bacteria).
(ix) **Fungistatic** (that inhibits the growth of fungi).

Ans. (i) Copper sulphate
(ii) Chlorohexidien, ethanol
(iii) Hypochlorite
(iv) Ethanol
(v) Formaldehyde, silver and mercury
(vi) Glutarldehyde
(vii) Cationic detergents such as cepacol and zephriam

(viii) Sorbate and benzoate

(ix) Zinc oxide and calcium propionate.

Q.19. Name the broad categories of chemical agents used to control microorganisms.

Ans. These are phenolics, halogens, alcohols, heavy metals, dyes, surface active agents, acids and alkalies.

Q.20. Name two test procedures to evaluate the effectiveness of disinfectants.

Ans. These are: (1) Phenol coefficient (the classic test procedure used until a few decades ago) and (2) To use-dilution method.

Q.21. What are antiseptics?

Ans. Antiseptics are the antimicrobial agents which can be applied to living tissues, e.g., alcohol. Alcohol denatures proteins, takes out membrane lipids and works as a dehydrating agent, all that make it an antiseptic.

Q.22. What is **oligodynamic action**?

Ans. The inhibitory effect of heavy metals against microorganisms is called oligodynamic action.

Q.23. What are **antibiotics**?

Ans. Antibiotics are antimicrobial substances produced by microorganisms.

Q.24. What are **antimicrobics**?

Ans. A common term used for antimicrobial agents produced by microbes or those made by chemists.

Q.25. Name some antimicrobics (antibiotics) which inhibit bacterial cell-wall synthesis.

Ans. Penicillin G , ampicillin, methicillin and cephalosporins.

Q.26. Name some antimicrobics (antibiotics) which inhibit baciterial protein synthesis.

Ans. Streptomycin, neomycin, chloramphenicol, tetracycline, bacitracin and erythromycin.

Q.27. Name an antimicrobic (antibiotic) which inhibits bacterial RNA synthesis.

Ans. Rifampicin.

Q.28. Name the antimicrobics (antibiotics) which damage plasma membrane of fungi.

Ans. Nystatin, amphotericin B and ketoconazole

Q.29. Name the antimicrobic (antibiotic) which inhibits mitosis.

Ans. Griseofulvin.

Q.30. Name the antimicrobics (antibiotics) which interfere with DNA synthesis in fungi.

Ans. Flucytosine and pentamidine.

Q.31. Name the antimicrobics which interfere with DNA replication of Protozoa.

Ans. Metronidazole.

Q.32. Name an antimicrobic which inhibits viral reverse transcriptase and is used against HIV infections.

Ans. Zidovudine.

Q.33. Name antimicrobic (antibiotic) that interferes with penetration and causes uncoating of viral particles.

Ans. Amantiadine is used against influenza type infections.

Q.34. List some antimicrobics (antibiotics) and also give their uses.

Ans. (i) **Penicillin G** Is used against Gram positive bacterial infections, streptococcal sore throat, gonorrhoea and syphilis.

(ii) **Ampicillin** Is used against Gram positive and Gram negative bacterial infections, middle ear infections, urinary tract infections cause by *Enterococcus faecalis* and infections caused by *Escherichia coli*.

(iii) **Methicillin** It is used against Gram positive and Gram negative bacterial infections, penicilliase producing *Staphylococcus aureus* infections.

(iv) **Cephalosporins** Are used against Gram positive and Gram negative bacterial infections, urinary tract and other infections due to *Escherichia coli*, middle ear infections and meningitis caused by *Haemophilus influenzae.*

(v) **Streptomycin** Used against Gram negative bacterial infections, bubonic plague and tularemia.

(vi) **Neomycin** Used against Gram negative bacterial infections. Used as topical ointment for general cuts and abrasions of the skin.

(vii) **Chloramphenicol** Is used against Gram positive and Gram negative bacterial infections, meningitis caused by *Haemophilus influenzae* or *Veisseria meningitidis* and typhoid fever.

(viii) **Tetracycline** Effective against Gram positive and Gram negative bacterial infections, pneumonia due to *Mycoplasma,* in nongonococcal urinary tract infections.

(ix) **Bacitracin** Is used against Gram positive bacterial infections, tropical ointment for general cuts and abrasions of the skin.

(x) **Erythromycin** Is used against Gram positive and Gram negative bacterial infections, whooping cough, diphtheria, diarrhoea, caused by *Campylobacter.* It is also effective against pneumonia caused by *Legionella* or mycoplasma.

(xi) **Rifampicin** Is used in Gram positive and Gram negative bacterial infections, tuberculosis and **Hansen disease** (leprosy).

(xii) **Nystatin** Is used against yeast infections and *Candida albicans* infections of skin and vagina.

(xiii) **Griseofulvin** Is used against ringworm of hair and nails.

(xiv) **Amphotericin B** Is used in systemic mycosis, histoplasmosis and cryptococcal meningitis.

(xv) **Flucytosine** Is used against candidiasis, cryptococcosis and aspergillosis.

(xvi) **Ketoconazole** Is used in candidiasis, fungal skin infections and some systemic mycosis.

(xvii) **Pentamidine** Is used against *Pneumocystis carinii* pneumonia and the African sleeping sickness (protozoal).

(xviii) **Metronidazole** Is used in the infections caused by *Entamoeba histolytica* and Trichomonas vaginalis.

(xix) **Zidovudine** Is used against HIV infections.

(xx) **Amantidine** Is used against influenza type A infections.

(xxi) **Acyclovir** Is used against herpesvirus in herpeps simplex, cytomegalovirus and varicella-zoster virus infections.

Q.35. What is **microbiostatic**?

Ans. The agent or the substance that inhibits the growth of microorganisms without destroying them is called microbiostatic.

Q.36. What is the Kirby-Bauer Test?

Ans. The **Kirby-Bauer test**, also known as the Baur-Kirby test is a standardized procedure to determine antibiotic susceptibility. The diameter of zone of inhibition which can be seen as a clear area around antimicrobic impregnated discs, gives the sensitivity of the microorganisms to the antibiotic. This test was developed by Dr Kirby and Dr Alfred Baur. The agar diffusion test procedure is designed for rapidly growing bacteria. This test is not directly applicable to filamentous fungi, anaerobes or slow growing bacteria. Nonetheless modifications of the media composition and incubation conditions can be made for testing the antibiotic susceptibility of such microbes.

Q.37. What is **minimum inhibitory concentration**?

Ans. It uses the dilution of antibiotics or antimicrobics to determine the lowest concentration of the antimicrobic (the MIC) which is effective for prevention of the growth of microorganisms.

Q.38. What is minimal bacterial concentration (MBC)?

Ans. The MBC, also known as minimal lethal concentration is the lowest concentration of antibiotic which destroys the defined portion of viable organisms in a bacterial suspension during a specified exposure time. A 99.9% kill rate of bacteria at an initial concentration of $10^5 + 10^6$ cells per ml during an 18 to 24 hours exposure is taken to define minimal bacterial concentration.

Q.39. Why do penicillins (obtained from fungus *Penicillium)* and cephalosporins (produced from fungus *Cephalosporium)* inhibit the formation of bacterial cell walls?

Ans. Both antibiotics contain **β-lectam rings**. The β -lectam portions of cephalosporin and penicillin bind the transpeptidase enzyme and hence prevent the binding of the enzyme to normal substrate D-alanyl-D alanine. The bacterial cell walls lacking the normal cross linking peptide chain are subject to attack by autolysins which are the antilytic enzymes produced by the bacterium and degrade the cells own cell wall structures.

Q.40. How do antimicrobics like erythromycin and tetracycline make therapeutically antibacterial agents?

Ans. They specifically target 70 S ribosomes and prevent continuation of protein synthesis by the bacterial cells.

Q.41. How does chloramphenicol inhibit protein synthesis?

Ans. It binds to 50 S ribosomal subunits and prevents the binding of t RNA molecules to the aminoacyl and peptidyl binding sites of the ribosome. Erythromycin, another antibiotic too binds to 50 Sribosomal subunits. Likewise rifampicin a semisynthetic derivative of rifamycin B also inhibits protein synthesis.

Q.42. How do quinolones work as DNA inhibitors?

Ans. **Quinolones** interfere with DNA gyrase and thereby prevent the establishment of replication fork and replication of DNA. As quinolones inhibit the DNA gyrase this leads to blocking of bacterial cells reproduction.

Q.43. Name the antibiotics which have prevented premature births by treatment of women with bacterial vaginosis.

Ans: Erythromycin and metronidazole.

Q.44. How is **acyclovir** inhibitory to herpes simplex viruses?

Ans. Acyclovir is an analog of guanine containing nucleoside and *in vivo* gets converted to acyclovir triphosphate. The enzyme uses it to form a false nucleotide that cannot be incorporated into DNA.

Q.45. Name some **heavy metals** which are used as germicides.

Ans. Silver, mercury, copper and zinc.

Q.46. How do heavy metals cause inhibitory action on microbes?

Ans. The heavy metals exert the antimicrobial action through oligodynamic action (oligo = a few). Heavy metal ions combine with thio (–SH) groups. As a result the proteins are denatured.

Q.47. What is the historical contribution of Ignatz Semmelweis to reduce infections in hospitals?

Ans. He introduced handwashing with **chlorinated lime** by hospital personnel to reduce infections.

Q.48. What are **quats**?

Ans. The **quats** (quaternary ammonium compounds) are cationic detergents attached to NH_4^+, disrupting plasma membrane. It causes leakage of cytoplasmic constituents of the cell. The quats are most effective against Gram-positive bacteria.

16
CHAPTER

NORMAL MICROFLORA, SYMBIOTIC RELATIONSHIPS AND PATHOGENICITY

Salient Features

- There are four possible symbiotic relationships.
- The **resident flora** comprise of **endosymbionts** found inside the body and **ectosymbionts** found outside the body on the surface of skin.
- **Transient flora** is temporary on skin and mucous membrane.
- In the **host parasite relationship**, the **parasitism** is a type of **symbiosis** between two species in which the smaller organism is physiologically dependent on the larger one called the host. The parasitic organism is harmful to host.
- The parasitic organism may colonize a host which is called **infection**.
- **Infectious disease** is produced as a result of interaction between parasitic organism/ microorganism and its host and consequently causes the host to shift from a **healthy state** to a **diseased state**.
- The organism/microorganism which causes the disease is called a **pathogen**.
- Internal wall of **gnotobiotic** or germ free rats is much thinner than normal ones and are poor in vitamin k.
- The virulence can be measured experimentally by plotting LD_{50} and ID_{50}.
- In the **pathogenesis of viral** diseases the basic step of viral infection is the expression of viral replicative cycle in a host.
- To cause a viral disease a virus enters a host, comes in contact with susceptible cells, reproduces, spreads to neighbouring cells, cause cellular injuries, provokes or creates a host immune response. Thereafter, it either disappears from the body of the host or turn into a persistent infection, may kill the host and find entry into the environment for a fresh beginning.
- In pathogenesis of bacterial diseases, transmission of the pathogen or their products is the first requirement to establish a disease.
- The pathogen binds to the host cell by special **adherence factors**.
- The pathogen may adopt both **active and passive mechanisms** to enter the host. On finding entry the pathogen may secrete certain products or enzymes which upgrade or elevate the dissemination (spread) of the pathogen through out the body of the host. The enzymes and chemical substances which are instrumental in spread of the pathogen are known as **virulence factors**.

- The pathogen establishes in the part of the body which provides most favourable conditions for growth and multiplication of the pathogen.
- One important follow up of conservation of chromosomal genes in bacteria is that the bacteria are clonal. There are only a few clonal types which exist in the environment in the pathogenic bacteria.
- In the process of co-evolution with the human hosts some bacterial pathogens have evolved **complex signal transduction pathways** to regulate the genes required for virulence.
- Many bacteria are pathogenic because they possess large segments of DNA known as **"pathogenic islands"** and carry genes responsible for virulence.
- Two main categories of toxins are **EXOTOXINS** and **ENDOTOXINS**.
- The exotoxins are heat labile, powerful or effective toxic proteins produced by the pathogen as a result of its normal metabolism. These exotoxins are specific in action and are grouped as **neurotoxins** (poisonous or destructive to nervous tissue), **cytotoxins** (have specific toxic actions on type of cell/tissue or organ for which they are specific *e.g.* nephrotoxins toxic to kidney, hepatotoxins are toxic to liver and cardiotoxins toxic to heart), and **enterotoxins** the toxins specifically affect the intestinal mucosa causing vomitting and diarrhoea.
- The most exotoxins being **dimeric** (consist of two parts or components) obey the **AB model** in which **A subunit** or fragment is enzymatic and the **B subunit** or fragment is binding portion.
- There are **several mechanisms of action** by which the A component of dimeric (having two parts) exotoxin enters the target cell.
- Bacterial exotoxins cause human diseases in three was: (i) Ingestion of preformed exotoxins, (ii) Colonization of mucosal surface followed by production of exotoxin, and (iii) Colonization of wound followed by production of local exotoxin.
- **ENDOTOXINS** are heat stable which are part of cell was polysaccharide of some Gram negative bacteria,
- Most endotoxins work by initially activating **"Hageman factor"** which them activates one to four humoral systems including blood clotting cascade, compliment activation, fibrinolytic system, and kininogen system. Endotoxins also provoke macrophages to release cytokines like IL – 1, IL – 6 and TNF – α.
- Microorganisms have developed several mechanisms to escape host defences.
- The viruses have developed mechanisms which either reduce or avoid the host's immune response.
- Bacteria have evolved the mechanism to escape the complement system, phagocytosis and specific immune responses.

Normal Microflora of Human Body

Q.1. Which organisms are found in the normal micro flora of the skin? Name the factors which control the **normal flora of the skin**.

Ans. The microorganisms (abbreviated-term microbes) commonly found on the skin are: non-pathogenic strains of *Corynebacterium* also called diphtheroids (excluding *C. diphtheriae*); *Propionibacterium* a genus of Gram positive asporogenous chemoorganotrophic; anaerobic bacteria; nonhaemolytic Staphylococci; enterococci (*streptococcus faecalis*); Coliforms (*Escherichia coli* and *Enterobacter* sp); acid fast bacilli *Mycobacterium smegmatis*; *Proteus* sp, *Pseudomonas* sp; *Candida albicans*-the yeast-like fungi; other fungi and different viruses.

The factors which often control the skin microflora are: (1) **Dryness of the skin** and (2) **Low pH (4–6)** of the secretions of the skin. The dryners of the skin, and acidic nature of the fluid discourage the growth of microorganism. Moreover the secretions of the skin possess antimicrobials such as **fatty acids** and **lysozyme** (enzyme found in tears saliva and many other body fluids).

Q.2. What is **acne**? How does diphthereroid *Corynebacterium acne* lead to formation of acne? Discuss.

Ans. The **acne** is one of the most common infections of the skin and is a disease of the **sebaceous glands** which are the cutaneous glands which secrete an oily substance called **'sebum'**. The ducts of these glands are short and straight and open into the **hair follicles**.

In the normal **sebaceous follicle**, from Latin *sebaceous* = a tallow candle, pertaining to fat or suet; the skin scales are shed and carried out in a continuous flow. However, in acne the dead cells and oil do not move smoothly out of the **pore** but get deposited along the inside of the duct consequently this deposition forms a plug or **comedo** a sort of glutton called **blackhead**. A worm-like cast formed of sebum occupies outlet of a sebaceous gland in the skin. The **comedones** have a black colour due to pigmentation. Thus the normal flow of sebum is stopped and the anaerobic diphtheroid *Corynebacterium acne* found in the tissue hydrolyze the sebum into fatty acids and begin to destroy the wall of the gland. The swollen gland then ruptures under the surface of the skin. The phagocytes are mobilized to attack the bacteria leading to formation of **pus** usually yellowish in colour comprising of bacteria and leucocytes. The necrotic tissue accumulates to form the characteristic **whitehead** or **pustule**. The formation of acne is known to be correlated with hormonal changes during **puberty**, the age at which reproductive organs become functionally active. The common sites of acne are the face, neck and upper part of chest and back.

A number of products containing salicylic acid and sulphur which pul off the top layers of the skin and loosen the hardened sebum and keep the pores open. Recently 13-cisretinoic acid a derivative of vitamin A has been introduced, which reduces sebum production approximately 90 percent.

Q.3. How do *Staphylococcus aureus* comprising normal microflora of the skin cause **boils**? How do **coagulase** and **leucocidins** protect this bacterium from phagocytes? Explain.

Ans. The boils (from Latin *bullire* = to bubble up) or furuncles (Fig. 16.1) represent an acute inflammatory condition surrounding a hair follicle caused by *Staphylococcus aureus* usually accompanied by **suppuration** *i.e.* the formation of pus. It has one opening for drainage in

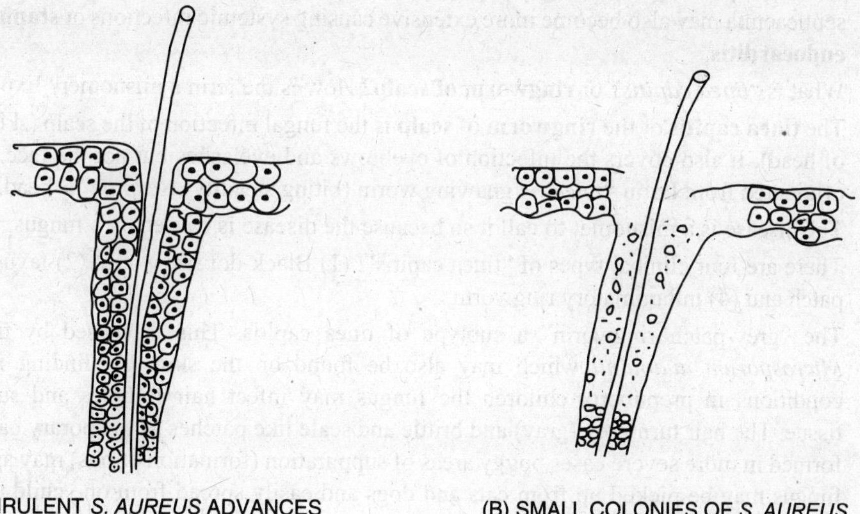

(A) VIRULENT *S. AUREUS* ADVANCES TO SUSCEPTIBLE HAIR FOLLICLE

(B) SMALL COLONIES OF *S. AUREUS* APPEAR UNCHECKED BY NORMAL DEFENCE MECHANISM

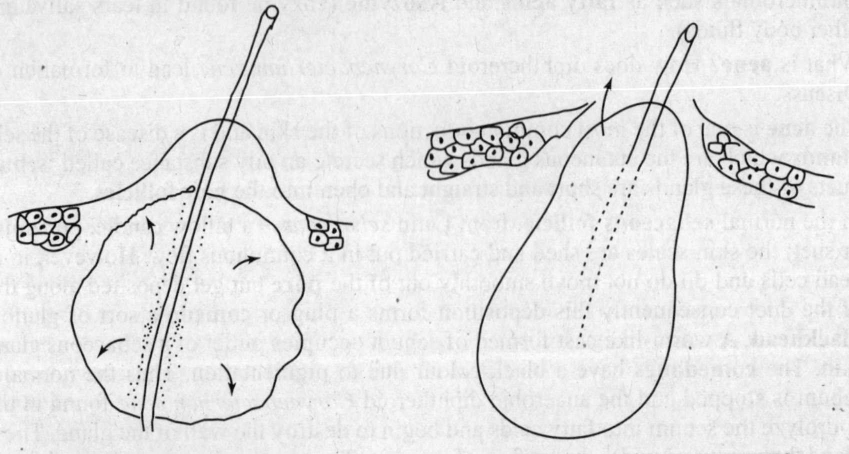

(C) RELEASE OF COAGULASE AND LEUCOCIDIN RESULT THE FORMATION OF PUS

(D) TAPERING UP BOIL RUPTURES RELEASING THE NECROTIC TISSUE LEAVING BEHIND MANY *S. AUREUS* CELL

Fig. 16.1. Normally found *Staphylococcus aureus* on the skin gets shifted to **parasitic relationship**. The **coagulase** (causes coagulation or clotting) and **leucocidin** (kills leucocytes) are staphylococcal toxins.

contrast to a **carbuncle** which in also caused by *Staphylococcus* involving several hair follicles and surrounding subcutaneous tissue forming an extensive **slough** – the septic tissue which becomes necrosed and separates from healthy tissue.

S. aureus secretes the enzymes **coagulase** and **leucocidins** which destroy phagocytes moving into the area. A boil should never be squeezed as the pathogens are deep within the lesion and may accidently be moved into the capillaries of the **dermis** the true skin or the cutis vera, the layer below the epidermis. In case the resistance of the host is poor it may develop **septicaemia** which is the presence of pathogens and their toxins in the blood. The septicaemia may also become more extensive causing systemic infections or **staphylococcal endocarditis**.

Q.4. What is "*tinea capitis*" or **ringworm of scalp**? How is the term a misnomer? Explain.

Ans. The **tinea capitis** or the **ringworm of scalp** is the fungal infection of the scalp (skin and hair of head). It also covers the infection of eyebrows and eyelashes but not the face. The term originated from Latin "*tinea*" = gnawing worm (biting steadily), "*capitis*" = head.

The disease is a misnomer to call it so because the disease is caused by a fungus.

There are four clinical types of "tinea capitis": (1) Black-dot ringworm, (2) favus, (3) grey patch and (4) inflammatory ringworm.

The "grey-patch ringworm" a subtype of tinea capitis. This is caused by the fungus *Microsporum audouinii* which may also be found on the skin. On finding favourable conditions in prepuberty children the fungus may infect hair follicles and surrounding tissue. The hair turn grey (gray) and brittle and scale like patches of temporary baldness are formed in more severe cases boggy areas of suppuration (formation of pus) may appear. The fungus may be picked up from cats and dogs and easily spread from one child to another. *Microsporum canis* also has been held responsible for causation of grey patch a clinical form of tinea capitis.

The drug **griseofulvin** and **Whitfield's ointment** (a preparation of benzoic acid and salicyclic acids) is used for therapeutic purpose.

Q.5. What is normal microflora of oral cavity and upper respiratory tract of human? Also give benefit of sneezing and harm of smoking.

Ans. The microflora of mouth before eruption (coming) of teeth comprise Streptococci (*S.viridans* and *S. salivarius*), staphylococci, Gram negative Diplococci (*Neisseria* sp.) Diptheroids (*Corynebacterium* sp) *Lactobacillus* (*L. casei*); and that after eruption of the teeth are represented by spirochaetes (*Leptospira* sp), *Bacteroides melaninogenicus, Fusobacterium, Vibrio, Actinomyces viscosus* and *Candida albicans*. The microorganisms generally found in the pharynx and trachea are *Streptococcus, Neisseria, Staphylococcus*, Diphtheroids, *Haemophilus, Mycoplasma* and *Bacteroides*. The microflora of the nose comprise *Corynebacterium, Staphylococcus* and *Streptococcus*.

The contaminating microorganisms moving towards mouth and nose are blown out by sneezing. The smoke of tobacco contains nicotine, which knocks out the cillia and permit trapped microorganisms to settle into the lungs. There are some other chemicals which stuns the cilia. The prolonged contact of such chemicals like nicotine with cilia may permanently debar them of motion and the probability of disease of lungs increases.

The oral cavity is usually sterile at the time of birth but within twelve hours a number of bacteria get established as normal microflora. *Streptococcus salivarius* is very common occurring on **gingiva,** the gum or the vascular tissue surrounding the neck of the teeth rather on the surface of the teeth. The disease which creates cavities in the teeth is called **dental caries**. The *Streptococcus sanguis, S. mutans, Actinomyces* and *Leptothrix* are found attached to enamel of the teeth.

Streptococcus mutans can ferment sucrose and produce an insoluble sticky polysaccharide called **dextran**. The sticky mat of the bacteria and dextran is called **dental plaque**. The sucrose is the most **cariogenic** or cavity producing carbohydrate. It is interesting that **fluoride** provides resistance to **acid decalcification** as the fluoride combines with calcium phosphate portion of teeth. Addition of fluoride to water supplies makes the tooth structure strong while the tooth is being formed, thus addition of fluoride is most effective in children. Adults do not incorporate compounds from drinking water since the growth of teeth has already stopped.

Q.6. Which microorganisms are common to **gastrointestinal tract**? What benefit bacteria gives to intestine? Discuss.

Ans. The common microflora of gastrointestinal tract consists of microorganisms as follows:

(i) In the **intestine of Breast – Fed Infants** the common microorganisms are *Lactobacillus bifidus* (*Bifidobacterium*) and Lactic Streptococci.

(ii) In **Normal Adult Microflora** (Bottle Fed Infants) the microorganisms (abbreviated as microbes) are Coliforms (*Escherichia coli, Serratia* and *Enterobacter*), Enterococci (*Streptococcus faecalis*), *Bacteriodes fragilis*, Lactobacilli (*Lactobacillus acidophilus*), *Clostridium perfringens, Pseudomonas, Proteus, Alcaligenes faecalis*, Yeasts, Fungi (*Penicillium*) and Enteroviruses. Oesophagus, the first portion of the tract contains only transient microflora. When the microbes enter the stomach many are destroyed by high acidity *i.e.* pH2 of the contents of stomach. There are approximately 10^3 to 10^5 bacteria per gram of contents in persons on regular diet. The number increases after a large meal since an open pyloric valve permits the mass of food to be neutralized more rapidly. A number of microorganisms comprising the resident flora exists that form the lining of the stomach, which are mainly aciduric streptococci and lactobacilli. They are not destroyed by the acids since are protected by the mucous lining inside the stomach.

The passage of food and microorganisms into the small intestine is facilitated by more alkaline conditions and consequently the number of microbes increases. The duodenum of an adult contains 10^6 bacteria per gram of the contents being close to stomach. The bacteria found in the duodenum are similar to that of stomach. The environment for the growth of microorganisms becomes more congenial in the **jejunum**, the middle part of the small intestine, and in the **ileum** the final portion of small intestine and the microflora increases to 10^7 per gram of the content. This count reaches 10^{10} in the large intestine or the **colon**. The microbial concentration goes upto 10^{11} in the discharged faeces (feces).

The bacteria in the intestine synthesize vitamin B_{12}, K, thiamin, pyridoxine (B_6) and riboflavin as a result of bacterial metabolic activities. Thus the normal microflora compete with and overcome pathogens and stimulate immune response when the pathogens accidentally find place in the tissue. Many microbes of normal flora produce organic acids (acetic, butyric and propionic and inhibits the growth of pathogens.

When the babies are shifted to cow's milk and solid foods to which they are not accustomed. The baby may experience cramping (sudden and painful tightening of muscles) emotional upset and changes. The intestinal flora shifts from typical breast fed *Bifidobacterium*-predominating to normal microflora of a bottle fed infant. Thus supplementing the diet with sweet acidophilus milk or **Lactinex**, the lyophilized concentrate of *Lactobacillus acidophilus* in the form of tablets has been found to reduce the trauma of **weaning** or make accustomed to changed food habit.

I. Mechnikov in 1908 attributed the longivity (over 100 years old) of Bulgarians to **yoghurt** (also spelt yogurt), the milk food prepared with fermentation of milk with a mixed culture of *Lactobacillus burganicus* and *Streptococcus thermophilus*).

Q.7. What is sudden infant death syndrome (SID)? Explain.

Ans. This is a disease of infants due to disturbed intestinal microflora which has also been called **"crib death"**. The sudden infant death syndrome (SID) has been regarded to be due to extremely small amounts of the toxin of *Clostridium botulinum*. A neurological problem also has been attributed because of *C. botulinum*. Under normal circumstances *Clostridium botulinum* does not pose any problem due to acidic nature of the contents of the intestine. The honey is not recommended for infants because it contains the spores of *Clostridium*.

Q.8. What is the normal **genitourinary tract microflora**. Also explain the terms **vaginitis** and **Döderlein's bacillus**.

Ans. The normal microflora of genital tract in the anterior **urethra**, the passage from the bladder through which urine is excreted, is similar to that of skin but in lesser number. The vagina after birth commonly contains **Döderlein's bacillus** which is an acidogenic lactobacillus which predominates for several weeks only. It later on changes to mixed microflora until puberty is reached. The normal microflora in the adult's vagina Döderlein's bacillus returns. The other microbes are *Clostridium, Peptostreptococcus, Streptococcus*, Coliforms, Spirochaetes (spirochetes), *Staphylococcus*, Diphtheroids and *Candida albicans* yeast-like fungus. After menopause Döderleins bacillus is lost and mixed flora returns.

The urinary tract infections in females are more in comparison to males because of shorter length of the duct and close proximity of the urethral meatus (opening) to the anus. The most common causative agents of urinary tract infections are *Escherichia coli, Proteus, Pseudomonas* and *Streptococcus faecalis* from the contaminating faecal matter. In healthy patients in voided urine collected there are about 10,000 bacteria per ml of the fluid. The counts in excess of 1,00,000 per ml are considered to be a state of urinary tract infection.

The infection of vagina is called **vaginitis** that may be caused by microorganisms as yeast-like fungus *Candida albicans*, and a flagellated protozoan *Trichomonas vaginalis*. **Döderlein's bacillus** (plural bacilli) are aciduric, Gram positive probably *Lactobacillus* observed by Döderlein in 1892 in human vaginal secretions.

Symbiotic Relationships

Q.9. What is **symbiosis**? Describe four symbiotic relationships occurring among organisms giving suitable examples.

Ans. A term used originally for any stable condition in which two different organisms called **symbionts** live in more or less close physical association, regardless of the nature of relationship. To sum up a close physical relationship is called **symbiosis**.

There are four basic types of symbiotic relationships.

(i) **Amensalism**, a relationship between different species in which one of the species is harmed. The typical classical example of an amensal relationships happen between the blue green mould *Penicillium* and the Gram positive bacteria *Staphylococcus*. The mould produces the antibiotic penicillin which bear an inhibitory effect on the bacteria. Thus bacteria are harmed as they cannot grow and reproduce but the presence or absence of bacteria have little or no effect on the mould. If the mould is benefitted it must be too little and insignificant.

(ii) **Commensalism** is the relationship in which one individual derives benefit but the other is not harmed e.g. relationship between *Streptococcus viridans* and human. *Streptococcus viridans* are found in the mouth of man and do not give harm. This environment suits them and they meet out their all natural needs easily. The relationship between the yeast-like dimorphic fungus *Candida albicans* (formerly the genus was known as Monilia) and human is another example of commensalism. The yeast-like fungus lives on the surface of the skin using glandular sections as nutrients. Humans receive no benefit from this relationship rather just provides support to yeast. It will be of interest to note that under normal circumstances the *Candida* is harmless but if conditions change it may produce the disease called **Candidiasis**. Besides on skin the *Candida* may also be found in the mouth, intestinal tract and vagina (cause **vaginitis**). The oral candidiasis or **thrush** forms a thick creamy coating on the tongue. The *Candida albicans* (Fig. 16.2) produces two type of spores, spherical thick walled asexual spores formed from hyphal cells are called **chlamydospores** while smaller asexual spores formed by budding are called **blastospores**.

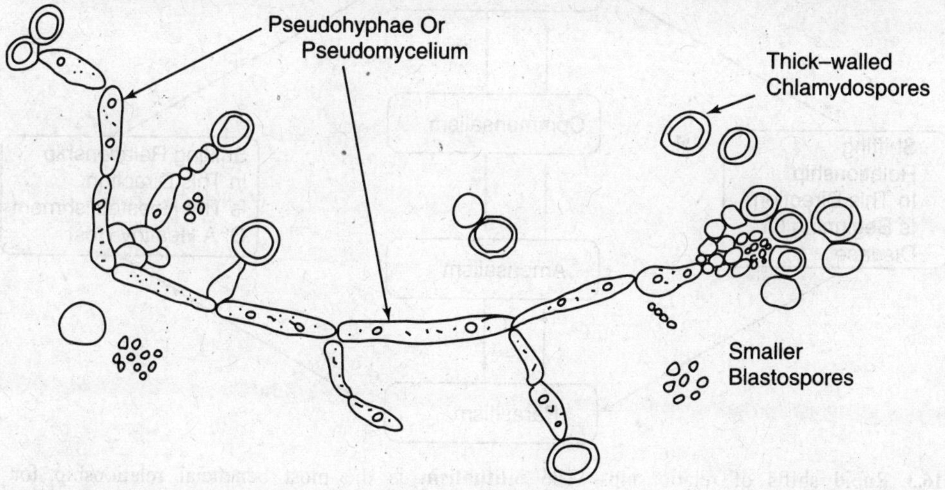

Fig. 16.2. Yeast-like dimorphic fungus *Candida albicans*. The scrapings scraped off the skin can be cultured on cornmeal agar. It is harmless under normal conditions, however, if conditions change this fungus may cause the disease called **candidiasis** (moniliasis). The **chlamydospores** are asexual spores formed from hyphal cells while **blastospores** are asexual spores formed by budding.

(iii) **Mutualism** is the condition of association in which both species of individuals is benefitted *e.g.* the rabbits eat grass and other plant material which is rich in cellulose despite of the fact that they cannot produce digestive enzymes that may break down cellulose into simple carbohydrates. However, the rabbits can get energy out of these molecules as they harbour special bacteria in their digestive tract; and these bacteria have cellulose digesting enzymes which hydrolyze cellulose into small carbohydrate molecules (sugars). Thus bacteria are benefitted because the intestine of rabbit provides them with moist, warm and nourishing habitat or niche. The rabbit in return, however, provide the food material to bacteria in the form of cellulose to which the rabbit was unable to digest itself. *Escherichia coli* and human are another example of mutualism.

(iv) **Parasitism** is the type of symbiotic relationship in which the lives in or on the host take benefit by obtaining nutrients and shelter and harm the host. The relationship among bacteria, rats, fleas and humans is a typical example of a complex type of parasitism. The fleas live in the fur of rats and feed on the blood of rats. The rat also may be host for other parasites like tapeworms, viruses and bacteria. One of the bacterium *Yersinia pestis* does little harm to rats but if carried to human may cause the disease **plague** known as "black death". Here the fleas spread the causative bacterium from one rat to another. The rat's flea may also hop or leap on humans and adopt or use them as hosts. Thus the plague can be spread to people. In this way in mid 14th centry the plague or "black-death" spread and killed millions of people. This disease in some of the countries killed 50 percent of population.

Role of Resident Flora

Q.10. What is **resident flora**? What sort of relationship resident flora have with human body? Discuss.

Ans. The symbiotic microorganisms of human body are divided into two main groups:

(i) **Resident flora** that generally occur inside the body called **endosymbionts** or on the surface of the body called **ectosymbionts**. The microorganisms comprising resident flora even if lost due to disease, chemical imbalance, or drug therapy they speedily return and reestablish the relationship. Generally they are not pathogenic but may turn up to be

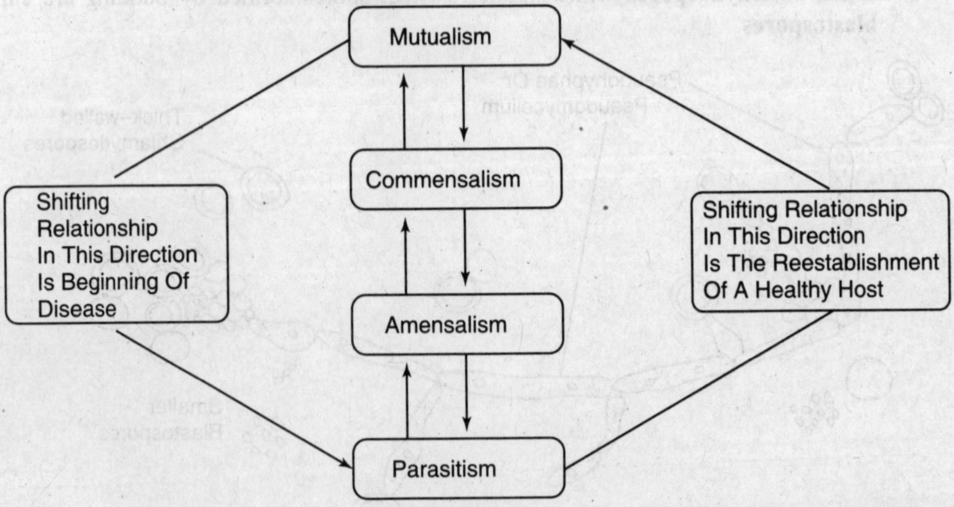

Fig. 16.3. Rapid shifts of relationships. The **mutualism** is the most beneficial relationship for both microorganisms and human and is farthermost or most distant to **parasitism** (*i.e.* antagonistic). A shift from mutualism or commensalism or amensalism to **parasitism** causes **disease** and a shift reverse to it returns **good health. Commonsalism** is a form of symbiosis in which one individual benefits while other is not harmed. In **amensalism** one species is harmed but the other is neither benefited nor harmed.

opportunistic and shift to a parasitic relationship if the situation appears. (Fig. 16.3). Majority of the residents enjoy an association with human called **commensalism** and rarely attack the tissue due to effective body defence mechanism.

(ii) **Transient flora** is generally found in association with skin and the mucous membrane. The microorganisms of transient flora are temporary and remain present for a few hours days or weeks. They do not establish permanent growing relationship with the body like residents. This is because of high competitive nature of the residents.

The competition between *Escherichia coli* and transients found in the intestine does not permit these outsiders (transients) to establish. No doubts, if residents are eliminated the transients may start working as opportunists and become established as parsites and cause disease.

The skin, eyes, ears, mouth, upper respiratory tract and genito urinary tract bear all these floras. However, the number and kind of microbes in each area is limited and kept under control in different manners. But because of exclusive nature of flora, each area may produce disease if these microorganisms get out of control.

Virulence

Q.11. What is **virulence**? Give the characteristics of the pathogen to determine virulence. How can virulence be measured in terms of $\mathbf{LD_{50}}$ and $\mathbf{ID_{50}}$? Explain.

Ans. The term was coined from Latin **virulentia** adapted from *virus* = poison. All symbiotic relationships which make a shift to parasitism depend on susceptibility of the host and virulence of the microorganism. The virulence is the degree to which the pathogen is able to cause damage or harm as a result of substances produced or associated with the pathogen in the course of **infection** which is the initial entry of the pathogen into a host. The characteristics which determine the virulence are:

(i) **Adherence**: A parasite must be able to adhere to an appropriate host cell, *e.g.* *Streptococcus sanguis* can adhere to gingiva but not to the enamel of the teeth. On the other hand *Streptococcus mutans* adheres to the enamel of the teeth when sucrose is available for the production of sticky dextran. The *Lactobacillus acidophilus* adheres to the cells lining the large intestine of human but not to cells in other animals.

(ii) **Focus of Infection**: The success of a parasite or a pathogen also depends on to establish a focal point of infection. The *Corynebacterium diphtheriae, Vibrio cholera, Bordetella pertusis*, and some other microbes remain attached on the surfaces and cause disease symptoms by releasing toxins or poisons which cause damage to the tissue deep within the host also referred to as **tissue damage**. Sometimes balance occurs between either the host defence mechanisms which keep the parasites in check (under control) or the genes responsible for the host damage are not being expressed. Thereby, the latency may continue for a considerable period before changing to active symptomatic infection. This happens with many strep throat infections and infections of *Mycobacterium tuberculosis* which have been latent in a host for several years. The degree to which the *M. tuberculosis* or any other pathogen causes damage or harm is known as *virulence*. When the microbes get established within the host the symptoms of virulence are expressed into specific ways *i.e.* toxin production or **invasiveness**.

(iii) **Pathogenic potential**: The pathogenic potential refers to the degree that the pathogen causes damage. A major aspect of pathogenic potential is toxigenicity. The toxigenicity is the pathogen's ability to produce toxins, the chemical substances which will damage the host and produce disease.

The virulence is generally measured experimentally by determining the lethal dose 50 (LD_{50}) or the infectious dose (ID_{50}). These experimental values allude to the dose or number of pathogen which will either kill or infect respectively, fifty percent of an experimental group of hosts within a definite time.

Intoxications (An Ill Adapted Term or Misnomer for Pathogen Caused Illness or Disease: Exotoxins and Endotoxins

Q.12. What are **intoxications**? Which are the main types of toxins which cause disease? Describe.

Ans. The literal meaning and usage of the term **"intoxication"** has been restricted or limited to **"alcoholic poisoning".** However, once in a blue moon *i.e.* very rarely the term has been used for disease caused by specific preformed toxin from Greek *toxikon* = poison, a metabolic product of microorganisms which changes normal metabolism of cells of the host causing harmful effect or damage to the host.

When the toxins find entry into the blood of the host it has been called **toxaemia** or **toxemia** from Greek *toxikon* = poison; *haima* = blood. It leads to generalized poisoning of the body by products of bacteria or damaged tissue which are **toxaemic** to the host.

EXOTOXINS: The exotoxins are soluble heat labile (from Latin *labilis* = apt to slip or get readily changed) proteins, sometimes enzymes which are released into the surrounding with the growth of the pathogen. The exotoxins can travel from the site of infection to the outer body tissue or the target cells to effect them. The major characteristics of exotoxins are:

(i) Certain bacteria having plasmids or **prophages**, latent form of temperate phage which remain integrated into the lysogenic bacterial chromosome, bearing genes for exotoxins, can synthesize exotoxins.

(ii) They are heat labile proteins which are inactivated at 60 – 80°C.

(iii) The botulinum is one of the most lethal toxin which is effective even in very small amounts of microgram per kilogram.

(iv) Exotoxins cause specific diseases adopting specific mechanism.

(v) They are highly immunogenic and provoke the formation of neutralizing antibodies known as **antitoxins**.

(vi) They can be usefully exploited by inactivating them with formaldehyde, iodine or other chemicals for the production of immunogenic **toxoids**.

(vii) They have been named after the disease produced by them as diphtheria toxin.

On the basis of structure and physiological habits or characteristics the **EXOTOXINS** have been differentiated into four main types, which are:

(i) AB Toxin: It has been named so because the portion of toxin B which binds to a host cell receptor is separate from portion A which bear enzymatic activity responsible for toxicity.

(ii) Second category of AB toxin comprises of those toxins which effect or target a specific host site as **neurotoxins** which effect nervous tissue, **enterotoxins** which effect the intestine and **cytotoxins** which effect the general tissues.

(iii) Haemolysins (Hemolysins) and other substances: These are disruptors of host cell membrane by attacking phospholipid layer or sterols in the host cell. This group of **EXOTOXINS** includes **leucocidins** which are microbial toxins which can damage or kill one or more types of leucocytes; **haemolysins** and **phospholipases**. The toxins of this group have been called haemolysins because the production of exotoxins is easily demonstrated by culturing bacteria on blood agar.

The **haemolysins** which completely destroy or disrupt red cells are known as **beta haemolysins**. **Streplolysin-S** (O_2 stable hence "S" in its name) is a kind of beta haemolysis produced by some pathogenic streptococci. The streptococci lead to causation of endocarditis an acute disease by releasing streptolysin S, and may cause severe heart valve ulceration and destruction, anemia and death within only a few days if not treated.

The second type of haemolysin will cause only partial destruction of red cells and has been called **alpha haemolysin**. These toxins produce a **green zone** around the colonies while cultured on blood agar.

The streptococci which release streptolysin O (gets inactivated by oxygen, therefore, O comes in its name, are alpha haemolytic and endocarditis due to such an infection is much less severe or subacute in comparison to one caused by streptococci which produce streptolysin *S*. The symptoms of subacute endocarditis include fever, weakness, gradual valve scarring with the development of a heart murmur and enlargement of spleen.

ENDOTOXINS: These are microbial toxins which are released only after lysis or rupture of the cell. The Gram negative bacteria possess lipopolysaccharide (LPS) in the outer membrane of their cell wall which are toxic to specific hosts under certain conditions. The LPS has been referred to as endotoxin since it is bound to the bacterium and is released after the cell undergoes lysis or gets ruptured. The toxic component of lipidpolysaccharide (LPS) is the lipid part known as **lipid A**. This is not a single macromolecular structure rather seems to be a complex array or protective costume of lipid residues. The major characteristics of endotoxins are:

(i) In contrast to exotoxins, the endotoxins are heat stable.

(ii) They are toxic only in high doses of miligram per kilogram.

(iii) They are poor immunogenically.

(iv) They are similar irrespective of the origin of their source.

(v) They can produce general systematic effects and are **pyrogenic** (produce pyrexia) from Greek *pyr* = fire, *i.e.* fever. The endotoxins may also cause shock, blood coagulation, work as fibrinolytic *i.e.* break down of fibrin, and may cause diarrhoea.

Mechanism of Action of Toxins

The mechanism of action of toxins differs from toxin to toxin. The membrane disrupting or damaging toxins kill phagocytic leucocytes known as **leukocidins** (from Greek *leukos* = white, *kytos* = cell, *caedere* = to kill). Most leucocidins are produced by pneumonococci, streptococci and staphylococci. The other toxins are **haemolysins** and **phospholipase enzymes**.

The mechanism of action of a dimeric exotoxin consisting of two parts or components can best be explained by **AB Exotoxin Model** of diphtheria toxin. (Fig. 16.4).

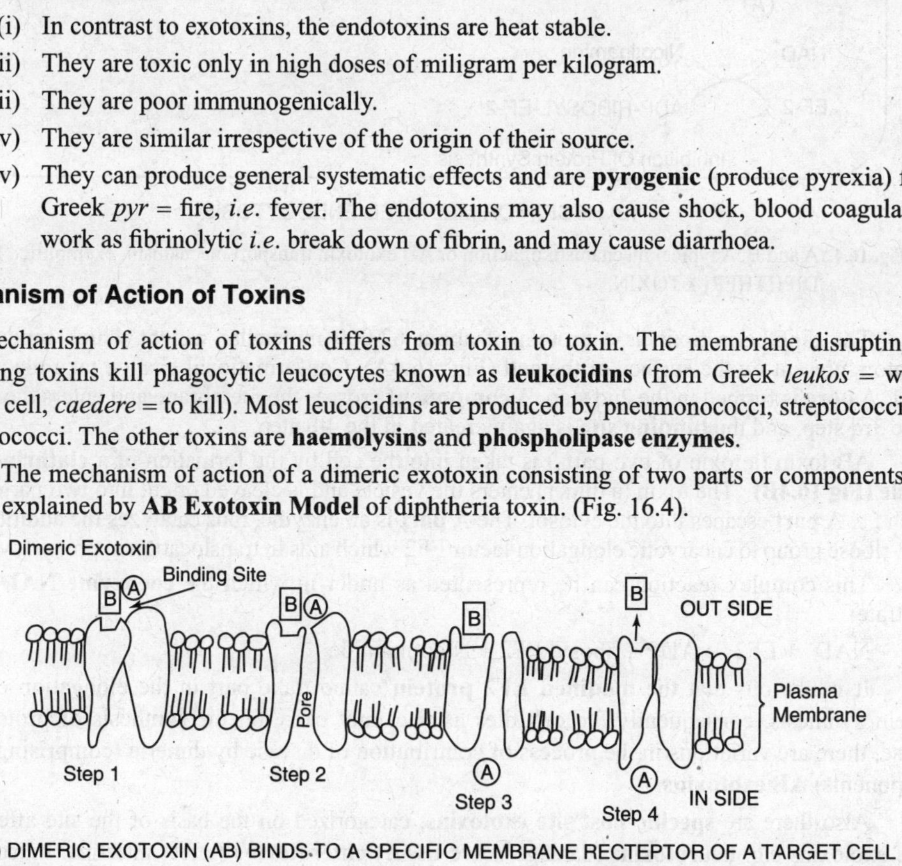

(A) DIMERIC EXOTOXIN (AB) BINDS TO A SPECIFIC MEMBRANE RECTEPTOR OF A TARGET CELL

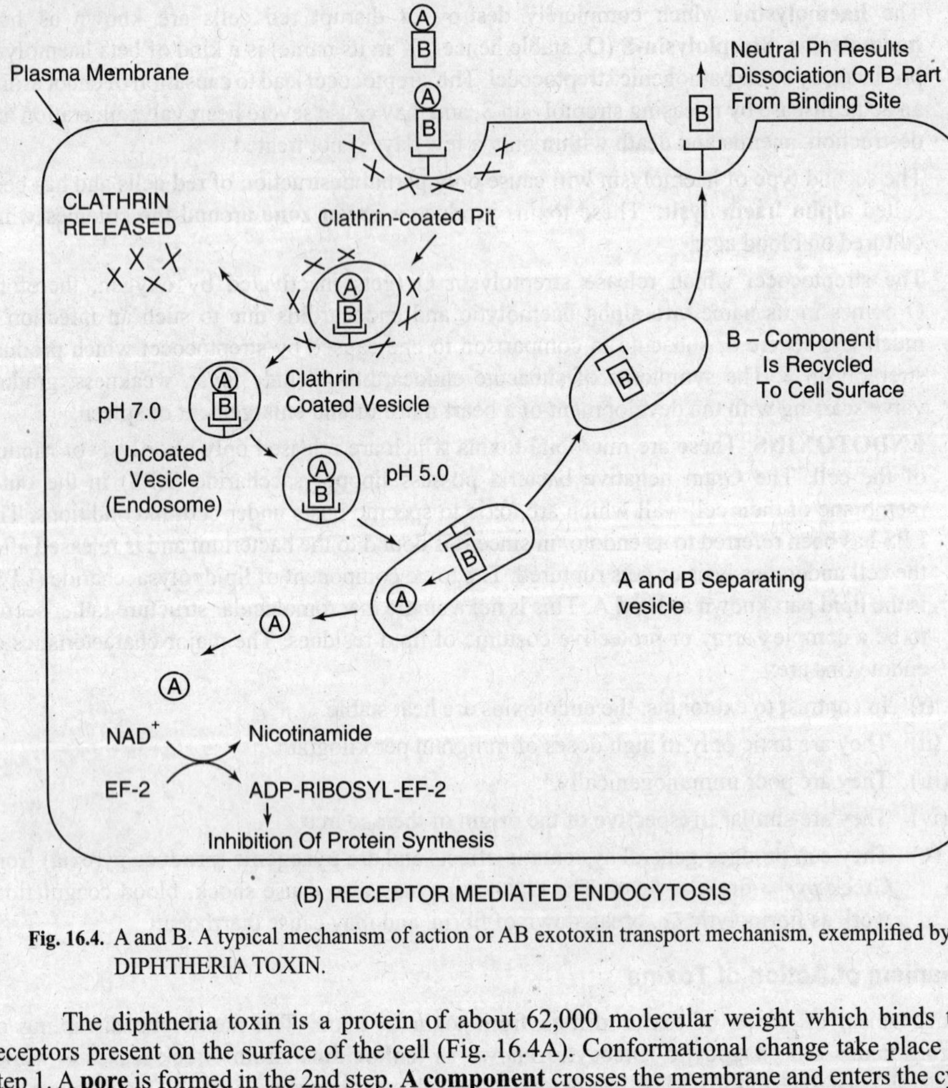

(B) RECEPTOR MEDIATED ENDOCYTOSIS

Fig. 16.4. A and B. A typical mechanism of action or AB exotoxin transport mechanism, exemplified by DIPHTHERIA TOXIN.

The diphtheria toxin is a protein of about 62,000 molecular weight which binds to the receptors present on the surface of the cell (Fig. 16.4A). Conformational change take place in the step 1. A **pore** is formed in the 2nd step. **A component** crosses the membrane and enters the cytosol in the 3rd step, and the **binding site** is again created in the 4th step.

AB toxin (a toxin of two parts) is taken into the cell by the formation of a **clathrin-coated vesicle (Fig 16.4B)** . The toxin (a dimer) enters the vesicle and is cleaved or cut into two parts one of which *i.e.* **A part** escapes into the cytosol. The **A part** is an enzyme, thus catalyzes the addition of an ADP-ribose group to encaryotic elongation factor EF2 which aids in translocation in protein synthesis.

This complex reaction can be represented as under in which the **coenzyme NAD$^+$** is the substrate:

$$NAD^+ + EF2 \rightarrow ADP - \text{ribosyl } EF2 + \text{Nicotinamide}$$

It is obvious that the **modified EF2 protein** cannot take part in the elongation cycle of protein synthesis, consequently the cell dies as it cannot carry on the synthesis of proteins. Of course, there are variations in the process of contribution of disease by dimeric (comprising of two components) **AB exotoxins**.

Also there are **specific host site exotoxins**, categorized on the basis of the site affected by these exotoxins *e.g.* **neurotoxins** (damage nerve tissue) **enterotoxins** (affect intestinal mucosa) and **cytotoxins** (general tissues). The neurotoxins are often ingested as preformed toxins in food as botulinum

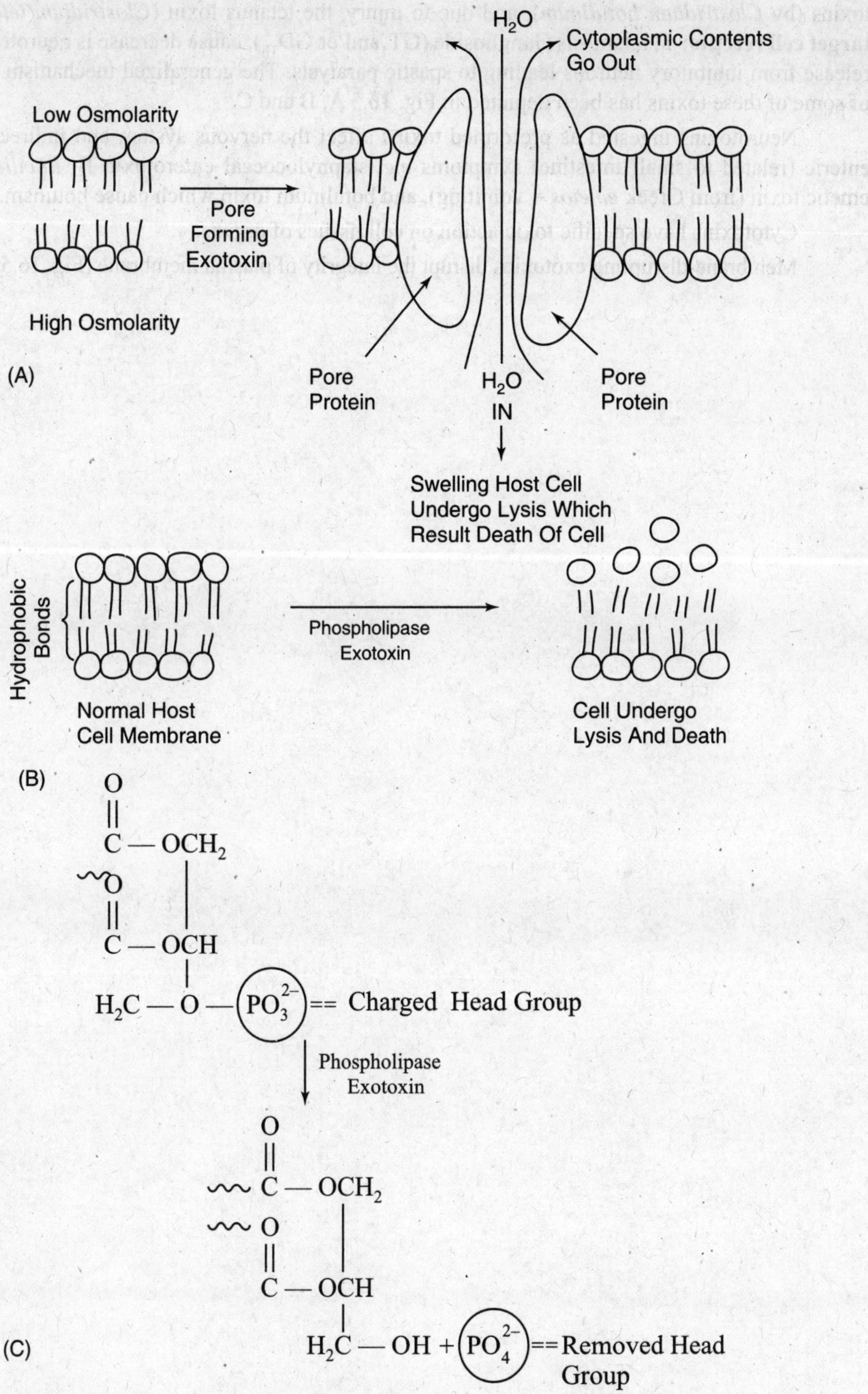

Fig. 16.5. A, B and C. Demonstration of **CELL LYSIS** by MEMBRANE DISRUPTING EXOTOXINS. A. Cytoplasmic contents go out, water comes in and cell dies. B. Phospholipase exotoxin damages integrity of membrane. C. Exotoxins removes **charged head groups**.

toxins (by *Clostridium botulinum*), and due to injury, the tetanus toxin (*Clostridium tetani*). The **target cell receptor** in tetanus is Ganglioside ($GT_1$and/or GD_{1b}), cause decrease is neurotransmitter release from inhibitory neurons leading to spastic paralysis. The generalized mechanism of action of some of these toxins has been depicted in Fig. 16.5 A, B and C.

Neurotoxins ingested as preformed toxins affect the nervous system and indirectly cause enteric (related to small intestine) symptoms *e.g.* staphylococcal enterotoxin B, *Bacillus cereus* emetic toxin (from Greek *emetos* = vomiting), and botulinum toxin which cause botulism.

Cytotoxins have specific toxic action on cell tissues of organs.

Membrane-disrupting exotoxins disrupt the integrity of plasma membrane (Fig. 16.5 A, B, C).

17

CHAPTER

MEDICAL MICROBIOLOGY

Q.1. Differentiate between *Staphylococcus* species and *Streptococcus* species.

Ans. There are two main distinguishing features between *Staphylococcus* and *Streptococcus* species:

- The *Staphylococcus* species produce catalase enzyme, whereas *Streptococcus* species are catalase negative.
- *Staphylococcus* species are usually found in grape like clusters, whereas *Streptococcus* species tend to form pairs and chains.

Q.2. How are the different clinically relevant species of *Staphylococcus* distinguished?

Ans. The three main clinically relevant Staphylococal species include: *S. aureus, S. epidermidis* and *S. saprophyticus. S. aureus* can be distinguished from the other species of *Staphylococcus* mainly by its ability to produce coagulase enzyme. The other two species which are coagulase negative are distinguished by susceptibility to antibiotic novobiocin: *S. epidermidis* is novobiocin susceptible and *S. saprophyticus* is novobiocin resistant.

Q.3. What are **toxins**?

Ans. A **toxin** is a poisonous substance, often a metabolic product of the organisms that damages the host.

Q.4. Describe the different types of toxins.

Ans. Toxins produced by microorganisms can be divided into two main categories: exotoxins and endotoxins.

Exotoxins: Exotoxins are soluble, heat labile proteins, which are either secreted by the bacterium or leak into the surrounding fluid as the pathogen grows. Exotoxins may act locally, or they may travel from the site of infection to other body tissue or target cells in which they exert their effect. Exotoxins are among the most lethal substances and they have specific mechanism of action. These toxins are usually characterized according to the tissues they adversely impact, e.g., neurotoxins, enterotoxins, or cytotoxins. They can be easily converted to toxoids and readily neutralized by antitoxins.

Endotoxins: Endotoxins are lipopolysaccharides (LPS), located in the outer membranes of the cell wall of many Gram negative bacteria. These toxins are released when the microorganism lyses. **Lipid A** is the toxic component of the LPS which is responsible for all the properties associated with the endotoxicity. These toxins are heat stable and are toxic only in large doses. Endotoxins cannot form toxoids and they are weakly immunogenic. They are capable of producing systematic effects.

Q.5. What do you understand by **dimorphic fungi**?

Ans. Those fungi which are capable of growing either as yeast like cells or as mycelia, depending on the environmental conditions are known as dimorphic fungi. When dimorphism is temperature dependent, the fungi are designated as thermally dimorphic. Many of the pathogenic fungi such as *Histoplasma capsulatum, Blastomyces dermatitidis* are dimorphic.

Q. 6. What is **satellitism**?

Ans. It is a phenomenon that helps in the recognition of *Haemophilus* species. These bacteria require both hemin (X factor) and NAD (V factor) for growth, and thus are unable to grow on 5% sheep blood agar media, which contains hemin but no NAD. Many bacterial species such as *Staphylococcus aureus*, are able to produce NAD. On sheep blood agar tiny colonies of *Haemophilus* species may be seen growing or engaging in "satellitism" around the V factor-producing organism; this phenomenon is known as **satellitism**.

Q.7. Mention the names of microorganism, for which the following culture media are selectively used.

(i) Mannitol salt agar
(ii) Thiosulfate bile salt sucrose agar
(iii) Modified Thayer Martin medium
(iv) Polymyxin lysozyme-EDTA-thallous acetate medium
(v) Tinsdale agar

Ans. Names of microorganisms suited to above media are :

(i) *Staphylococcus aureus*
(ii) *Vibrio cholerae*
(iii) *Neisseria gonorrhoeae or Neisseria meningitidis*
(iv) *Bacillus anthracis*
(v) *Corynebacterium diphtheriae*

Q. 8. Write down the full form of following terms :

(i) ELISA (ii) RIA (iii) STORCH
(iv) FIA (v) MIC (vi) VDRL

Ans.

(i) ELISA Enzyme Linked Immunosorbent Assay
(ii) RIA Radioimmunoassay
(iii) STORCH Syphilis, *Toxoplasma,* rubella, cytomegalovirus, and herpes
(iv) FIA Fluorescent Immunoassay
(v) MIC Minimal Inhibitory Concentration
(vi) VDRL Venereal Disease Research Laboratory Flocculation Test

Q.9. Mention the names of commonly used **anaerobic media.**

Ans. The primary plating media for inoculating anaerobic specimen are :

- Anaerobic blood agar
- *Bacteroides* bile esculin agar media
- Laked kanamycin-vancomycin blood agar media
- Anaerobic phenylethyl agar media

Q.10. Write a short note on universal precautions.

Ans. In 1987 CDC published guidelines, which have become known as universal precaution, to reduce the risk of Hepatitis B virus transmission in clinical laboratories and blood banks. These precautions are taken while handling samples like blood and body fluidis. Precautions do not apply to samples like faeces, saliva, sputum, sweat, urine or vomitus unless they are grossly bloody.

The essentials of universal precautions are as follows :

- Never eat, drink, smoke, or use cosmetics.
- Never insert or remove contact lenses.
- Do not bite nails or chew pens.
- Do not mouth-pipette.
- Wear gloves and aprons at all times, and if there is a risk of splashes, mask and goggles should also be worm.

- Every time when the gloves are removed, hand washing should be done.
- Avoid injuries with sharp objects such as needles and scalpels.
- Limit access to the laboratory to the concerned person only.
- All patients should be considred as infectious for blood borne pathogens.

Q.11. What is **Polymerase Chain Reaction (PCR)**?

Ans. PCR is a patented technology that generates multiple copies of a specific nucleotide sequence from a target organism. PCR is used to detect extremely low concentration of target organisms with high specificity. PCR is closely patterned after the natural principle of DNA replication. In its rudimentary form PCR is a three step process referred to as cycle, that is repeated a specified number of times. One PCR cycle consist of the following steps :

(a) Denaturation of target nucleic acid
(b) Annealing of primers
(c) Extension of primers

Q.12. What is normal flora?

Ans. A population of microorganisms routinely found in the normal healthy individual is known as **normal flora** of the body.

Q.13. Mention the advantages of normal flora.

Ans. **Advantages of normal flora are :**

Normal flora prevents colonization or invasion of the body by potential pathogens, either by simple competition with the pathogen for the nutrients or by producing inhibitory substances. Some suitable examples are as under :

- Vaginal Lactobacilli maintain an acidic environment by producing lactic acid, which suppresses growth of other microorganisms such as gonococci.
- Fatty acids produced by skin bacteria prevent growth of other pathogens.
- Some bacteria help in the disposal of waste matter.
- Some bacteria present in the gut are able to produce vitamin B and vitamin K and thus may contribute in the nutrition.
- The antigenic stimulation provided by the intestinal flora helps in increasing the immune status of the host.

Q.14. Name some media, which are commonly used, for the laboratory diagnosis of bacterial diseases.

Ans. Majority of the pathogens can grow on 5% sheep blood agar, which is the most widely used culture medium. Various selective media are used for cultivation of bacterial pathogens. These include :

Culture Media	**Microorganism**
MacConkey agar	for cultivation of coliform organisms and enteric pathogens
Chocolate agar	for cultivation of *Haemophilus* spp. and *Neisseria* spp.
Deoxycholate citrate agar	for cultivation of *Salmonella* spp. and *Shigella* spp.
Thiosulfate citrate bile salts agar	for cultivation of *Vibrio cholerae* and *V. parahaemolyticus*
Thayer Martin agar media	for the isolation of pathogenic *Neisseria* spp.
Mannitol salt agar	for selective isolation of *Staphylococcus* spp.
Lowenstein–Jensen media	for cultivation of *Mycobacterium* spp.
Robertson's cooked meat medium	for cultivation of anaerobes

Q.15. Name the primary plating media used for following specimens:

(i) Urine (ii) CSF (iii) Pus

Ans. Following media will be used for primary plating of the specimens

(i) Urine : Blood agar and MacConkey agar
(ii) CSF : Blood agar, chocolate agar and thioglycollate broth
(iii) Pus : Blood agar, chocolate agar and MacConkey agar

Q.16. Fill in the blanks.

(i) Erythrogenic toxin capable of causing scarlet fever is produced by a lysogenic strain of ....

(ii) Definitive identification of the *Corynebacterium diphtheriae* isolate as a true pathogen requires demonstration of .... by the isolate in question.

(iii) The genus *Bacteroides* can be divided into .... and .... species.

(iv) *Campylobacter* species that cause enteric illness are isolated from .... samples.

(v) .... pigment causes colonies of *Pseudomonas aeruginosa* to appear blue green on MacConkey agar.

Ans. (i) *Streptococcus pyogenes* (ii) Diphtheria toxin (iii) Bile tolerant and bile sensitive (iv) Stool (v) Pyocyanin

Q.17. Fill in the blanks.

(i) Full form of EPEC is ....

(ii) Whooping cough is primarily caused by ....

(iii) .... are the smallest free living organisms in the nature.

(iv) Infection with *Brucella* species (Brucellosis) is commonly referred to as ....fever.

(v) .... is mainly responsible for the moist, mucoid colonies characteristic of *Klebsiella pneumoniae.*

Ans. (i) Enteropathogenic *Escherichia coli* (ii) *Bordetella pertussis* (iii) Mycoplasmas (iv) Undulant (v) Capsule

Q.18. Define the following terms, which are used to analyze the performance of laboratory tests :

(i) Specificity (ii) Sensitivity

Ans. (i) **Sensitivity** is the percentage of individuals with the particular disease for which the test is used and in whom positive test results are found.

(ii) **Specificity** is the percentage of individuals who do not have the particular disease for which the test is used and in whom negative test results are found.

Q.19. Define the following terms :

(i) **Septicaemia** (ii) **Bacteremia**

Ans. (i) **Septicaemia :** It is a situation which indicates blood poisoning due to presence of actively multiplying bacteria or their toxins in the blood.

(ii) **Bacteremia :** This term indicates the presence of viable bacteria in the blood. Clinical signs of septicaemia are absent.

Q. 20. What is **botulism**?

Ans. Botulism is a paralytic disease caused by neurotoxin of *Clostridium botulinum*, which is an obligately anaerobic endospore forming, Gram positive rod.

Q. 21. Mention the names of pigments produced by *Pseudomonas aeruginosa*.

Ans. Two types of pigments are produced by *Pseudomonas aeruginosa*:

Pyoverdin - it is a fluorescent, water soluble greenish yellow pigment.

Pyocyanin - it is a non fluorescent phenazine blue pigment.

Q. 22. What is **plague** ?

Ans. The plague is a disease caused by *Yersinia pestis* (a small Gram negative rod), which infects rodents and is spread to humans either by the bite of rat flea (*Xeopsylla cheopsis*) or by inhalation of the droplets produced by the infected person.

Two major forms of infection are **bubonic plague** and **pneumonic plague**:

Bubonic plague is characterized by high fever and painful inflammatory swelling of lymph nodes in the groin and the armpit.

Pneumonic plague is caused by the inhalation of *Yersinia pestis* organism and is characterized by the malaise and pulmonary signs.

Q. 23. Explain the **Kirby-Bauer disk diffusion method** for determining drug susceptibility.

Ans. The **Kirby-Bauer disk diffusion test** is routinely used to determine the susceptibility of a given microorganisms to a variety of chemotherapeutic agents. The isolate to be tested is inoculated over the entire surface of the Mueller Hinton agar plates and filter paper disks containing the wide variety of antimicrobial agents are applied over the surface of these agar plates. After overnight incubation, the plates are examined for the presence of zone of inhibition around each antibiotic disk. The zone sizes are compared with established zone size ranges for individual antimicrobial agent and the results are recorded as 'S' (susceptible), 'I' (intermediate), or 'R' (resistant).

Q. 24. What are **coliforms**?

Ans. The coliform group of bacteria is defined as all Gram negative non spore forming rods that are able to ferment lactose along with the production of acid and gas. These organisms typically reside in the intestinal tract of warm-blooded animals and thus serve as a bacterial indicator of warm-blooded animals.

Q. 25. What are **prions**?

Ans. **Prions** are small proteinaceous infectious particles that contain only protein and no nucleic acid. They are able to cause a disease, which is known as **spongiform encephalopathies**. The main characteristics of prions are :

- Obligate intracellular parasites.
- Smaller size (< 100 nm).
- Lack of nucleic acid genome.
- Resistant to inactivation by UV light or nucleases, but are susceptible to chemicals that denature proteins.
- Slow replication.
- Cannot be cultured in the laboratory.

Q. 26. What are **Negri bodies** ?

Ans. Negri bodies are prominent cytoplasmic inclusion bodies, which are found in the nerve cells of human beings and animals infected with rabies virus. Negri first described them in 1903. They are spherical or oval, 2μm - 20μm in diameter. Their characteristic appearance and easy recognition made possible the rapid diagnosis of rabies infection.

Q. 27. What are the advantages and disadvantages of **oral polio vaccine (OPV)**?

Ans. The remarkable decline in poliomyelitis is due to the use of one or other of two polio vaccines currently available :

- Killed virus vaccine
- Attenuated live virus vaccine or oral polio vaccine (OPV)

The oral polio vaccine (OPV) has become the first choice in many countries. Its main **advantages** are :

- Easy administration
- Lower cost
- Immunity is induced at right place, the mucosal surface
- Confers "herd immunity" as well as individual immunity
- Its effectiveness approaches 100%.

The **disadvantages** of OPV are :

- Risk of reversion to increased virulence of the virus employed, particularly by type 2 and 3.

- Dissemination of virus to unvaccinated contacts.
- Can't be used in immuno-compromised persons.

Q.28. Write a note on mode of action of antibacterial agents.

Ans. Several potential antimicrobial targets exist within the bacterial cell, but those pathways or structures frequently targeted include :

- Inhibition of cell wall synthesis
- Damage to the cytoplasmic membranne
- Inhibition of nucleic acid
- Inhibition of protein synthesis
- Inhibition of specific enzyme systems.

Summary of mechanism of action of commonly used antibacterial drugs

Drug	Mechanism of Action
Beta-Lactams (e.g., penicillin, ampicillin carbenicillin, methicillin, cephalosporins)	Inhibition of the biosynthesis of the peptidoglycan cell wall structure by binding enzymes involved in peptidoglycan sysnthesis.
Aminoglycosides (e.g., streptomycin, gentamycin, kanamycin, tobramycin, neoymcin)	Inhibit protein systhesis by binding with 30S ribosomal subunit.
Chloramphenicol	Inhibit protein synthesis by binding with 50S ribosomal subunit.
Floroquinolones (e.g., ciprofloxacin, ofloxacin, norfloxacin	Inhibit bacterial DNA gyrase enzyme and thus interfere with DNA replication.
Polymyxin B	Disrupts plasma membrane.
Sulfonamides	Inhibit folic acid synthesis by binding the enzyme dihydropteroate synthatase.
Trimethoprim	Interfere with folic acid pathway by inhibiting the enzyme dihydrofolate reductase .

Q. 29. What are the major steps in the commercial **production of antibiotics?**

Ans. The major steps in the commercial production of antibiotics are :

- Preparation of inoculum
- Careful adjustment of the medium composition
- Sterilization of the medium.
- Inoculation of the medium in the fermenter.
- Forced aeration of the sterile air through out the media during incubation.
- Separation of the mould mycelium from the medium after fermentation.
- Extraction and purification of the antibiotic.

18 CHAPTER

MICROBIAL DISEASES (DISEASES CAUSED BY MICROORGANISMS)

Q.1. Name the microorganisms that form normal microbiota of the skin.

Ans. They are Gram-positive pleomorphic rods known as diphtheroids (e.g., *Propionibacterium acnes)* which produce propionic acid that helps to maintain low pH of the skin ranging from 3 to 5. The acidic effect produced by these bacteria safeguards skin against many harmful microorganisms.

The *Corynebacterium xerosis,* another diphtheroid is aerobic and inhabit the surface of the skin. The yeast *Pityrosporum* grows on oily skin secretions and causes scaling skin conditions known as dandruff.

Q.2. Give the main microbial diseases of the skin, causative agents, characteristics and treatment.

Ans. The common microbial diseases caused by bacteria, viruses, fungi and mites, and their cause, symptoms and treatment are listed in Table 18.1.

Q.3. Who was the last victim of **small pox**?

Ans. The last victim of small pox was an African in Somalia, who recovered from variola minor, in 1977.

Q.4. Which are the two main types of small pox?

Ans. Two main types of small pox are variola major and variola minor. Recovery from one type of disease provides immunity against both the types.

Q.5. Which are the two sites where the smallpox viruses are maintained?

Ans. One in USA and the second in Russia.

Q.6. What may be the after effect of use of **aspirin** to lower fevers in chicken pox and influenza?

Ans. Increases the chances of acquiring Reye's syndrome.

Q.7. What is **Reye's syndrome**?

Ans. Reye's syndrome is an occasional complication after chickenpox, influenza and other viral infections. In it a few days after the initial infection recedes the patient persistently vomits and shows the signs of brain disfunction. It may follow coma, fatty degeneration of liver and death. Death and brain damage in those who survive is due to swelling in the brain that prevents the circulation of blood.

Q.8. What is a **dermatophyte?**

Ans. A fungus that colonizes on skin including hair and nail is called dermatophyte. Dermatophytes grow on the **keratin**, the skin protein found in dermis, hair and nails and cause infections

Table 18.1 Common skin diseases caused by microorganisms

Disease	*Causal organisms*	*Symptoms*	*Therapy*
BACTERIAL			
Impetigo	*Staphylococcus aureus* occasionally *Streptococcus pyogenes*	Superficial skin infection forming isolated pustules	Penicillin (for *Streptococcus* only) and erythromycin
Erysipelas	*Streptococcus pyogenes*	Red patches accompanied with fever	Penicillin and erythromycin
Pseudomonas dermatitis	*Pseudomonas aeruginosa*	Superficial rash	Often self-limiting
Otitis externa	*Pseudomonas aeruginosa*	Superficial infection of external ear canal	Gentamicin and carbenicillin
Acne	*Propionibacterium acnes*	Inflammatory lesions	Benzoyl peroxide, tetracycline and isotretinoin
VIRAL			
Warts	Papillomavirus	Horny projection of the skin due to proliferation of cells	Liquid nitrogen cryotherapy, electrodesiccation, acids or lasers
Small pox	Small pox (variola) virus	Severe pustules, internal organs are also affected	None (except preventive vaccination)
Chickenpox	Varicella-Zoster virus	Vesicles often confined to face, throat and lower back	Acyclovir for immuno-compromised patients
Shingles (Zoster)	Varicella-Zoster virus	Vesicles like that of chickenpox but one sided on waist, face and scalp or upper chest	Acyclovir
Herpes simplex	Herpes simplex virus type 1	Cold sores-vesicles around mouth	Acyclovir
Measles	Measles virus	Rash on face spreading to trunk and other parts	None

(*Contd.*)

Disease	*Causal organisms*	*Symptoms*	*Therapy*
Rubella (German measles)	Rubella virus	Rash resembling measles but less and disappears in 3 days or less.	No therapy
FUNGAL			
Ringworm (tinea)	*Epidermophyton*, *Microsporum* and *Trichophyton*	Skin lesions of various types on various parts including loss of hair on head and beard	Griseofulvin (orally), miconazole and clotrimazole (topically)
Sporotrichosis	*Sporothrix schenckii*	Ulcer at the place of infection and spreads into nearby lymphatic vessels	Potassium iodide solution (orally) of appropriate dilution
Candidiasis	*Candida albicans*	Symptoms depending on the part of body affecting mucous membrane and moist parts of skin.	Miconazole, Clotrimazole and nystatin (topically)
MITES			
Scabies	*Sarcoptes scabiei*, a mite	Cause papules followed by hypersensitivity reaction to mites	Gamma benzenehexa chloride, and permethrin (topically)

Note: The treatment should be taken under supervision of a qualified physician.

called tinea or ringworms. **Tinea capitis** or ringworm of scalp results in bald circular patches on the head of school children.

Q.9. List the bacterial, viral and protozoal diseases of the eye. Also give their causal organisms, symptoms and therapy.

Ans. These are given in Table 18.2.

Q.10. List the microbial diseases of the nervous system.

Ans. These are listed in Table 18.3.

Q.11. List the microbial diseases of cardiovascular and lymphatic systems. Also give their causal microorganism mode of transmission and therapy.

Ans. These are listed in Table 18.4.

Q.12. Name the bacterial diseases of upper respiratory system.

Ans. These are given in Table 18.5.

Q.13. Name a viral disease of upper respiratory tract.

Ans. Common cold is the viral disease of upper respiratory tract which is caused by coranaviruses or rhinoviruses from respiratory secretions. There is no therapy for it.

Q.14. Give bacterial diseases of lower respiratory system.

Ans. Bacterial diseases of lower respiratory system are given in Table 18.6.

Q.15. Give fungal diseases of the lower respiratory system.

Ans. See Table 18.7.

Q.16. Name a protozoal disease of lower respiratory system.

Ans. *Pneumocystis* pneumonia is caused by *Pneumocystis carinii* which is transmitted through respiratory route. Trimethoprim-sulphamethoxazole and pentamidine isethionate are used for treatment.

Q.17. Give food borne bacterial diseases of digestive system.

Ans. See Table 18.8.

Q.18. Give viral diseases of digestive system.

Ans. See Table 18.9

Q.19. Name the fungal diseases of the digestive system.

Ans. See Table 18.10

Q.20. Give the protozoal diseases of the digestive system.

Ans. See Table 18.11.

Q.21. Give the common helminthic diseases of digestive system. What are their causal organisms and how are they transmitted? Give their generally followed therapy.

Ans. See Table 18.12.

Table 18.2 Diseases of eyes caused by microorganism

Disease	*Causal organisms*	*Symptoms*	*Therapy*
BACTERIAL			
Neonatal gonorrheal ophthalmia	*Neisseria gonorrhoeae*	Infection with pus, if not treated results in ulcers on cornea	Silver nitrate, tetracycline or erythromycin for prevention and penicillin for treatment
Inclusion conjunctivitis	*Chlamydia trachomatis*	Swelling of eyelid forming mucous and pus	Tetracycline
Trachoma	*Chlamydia trachomatis*	Conjunctivitis leading to scattering of eyelid may damage cornea mechanically	Tetracycline
VIRAL			
Herpetic keratitis	Herpes simplex Type 1 virus	Mild conjunctivitis may lead to corneal ulcers and severe damage	Trifluridine has been considered effective
PROTOZOAL			
Acanthamoeba keratitis	*Acanthamoeba* spp.	Commonly cause severe eye damage	Tropical application of propamidine isethionate, neosporin or eye removal surgery may be needed.

Table 18.3 Mirobial diseases of nervous system

Disease	*Causal organisms*	*Transmission media*	*Therapy*
BACTERIAL			
Meningococcal	*Neisseria meningitidis*	Contact with healthy carrier via respiratory tract	Third-generation cephalosporin
Hemophilus influenzae meningitis	*Hemophilus influenzae*	Respiratory route	Third-generation cephalosporin
Pneumococcal meningitis	*Streptococcus pneumonial*	Respiratory route	Third generation cephalosporin
Botulism	*Clostridium botulinum*	Ingestion of toxin	Antitoxins and respiratory support
Listeriosis	*Listeria monocytogenes*	Taking contaminated food	Penicillin
Leprosy or Hansen's disease	*Mycobacterium leprae*	Transfer of exudates	Sulphone drugs as dapsone, rifampin, clofazimine
Tetanus	*Clostridium tetani*	Deep anaerobic wound contamination	Immune globulin + respiratory support
FUNGAL			
Cryptococcosis	*Cryptococcus neoformans*	Inhalation of infected pigeon droppings	Amphotericin-B and flucytosine
VIRAL			
Poliomyelitis	Poliovirus	Ingestion of virus	No therapy except prevention by a vaccine
Rabies	Rabies virus	Rabid animal bite such as a mad dog bite	Immonoglobulin and vaccine
Arthropod-borne encephalitis	Arboviruses	Mosquitoes	No therapy

(*Contd.*)

Disease	*Causal organisms*	*Transmission media*	*Therapy*
PROTOZOAL			
African trypanosomiasis	*Trypanosoma brucei rhodesiense* and *Trypanosoma brucei gambiense*	Tsetse flies	Suramin, pentamidine isethiouate, melarsoprol and eflornithine
Naegleria microencephalitis	*Naegleria jowleri*	Nasal passage from fresh water	No treatment
PRION DISEASES			
Creutzfeldt-jacob disease	Prion	Ingestion, transplant or contamination of wounds	No treatment so far
Kuru	Prion	Ingestion or contamination of wound	No treatment

Table 18.4 Microbial diseases of cardiovascular and lymphatic systems

Disease	*Causal organisms*	*Transmission media*	*Therapy*
BACTERIAL			
Puerperal sepsis and infections causing abortions	Primarily *Streptococcus pyogenes*, *Clostridium* and *Bacteroides*	Unsanitory conditions at the time of delivery or abortion	Penicillin
Endocarditis (Subacute bacterial)	α-hemolytic strepto-cocci	Bacteremia	Penicillin for α-hemolytic cocci
Endocarditis (Acute bacterial)	*Staphylococcus aureus* *Streptococcus pyogenes*	Bacteremia from some focus of infection	Penicillin
Pericarditis	*Streptococcus pyogenes*	Bacterimia	Penicillin
Rheumatic fever	Group A β-hemolytic streptococci	Inhaling, causing sore throat	Penicillin
Tularemia	*Francisella tularensis*	Animal reservoirs, e.g., rabbit, skin cuts, bites, inhalation	Streptomycin and gentamycin
Brucellosis	*Brucella*	Animals, e.g., cows are reservoirs, taking milk, direct contact with skin abrasions.	Streptomycin and tetracyclin
Anthrax	*Bacillus anthracis*	Soil and animals inhalation or ingestion of spores	Penicillin
Gangrene	*Clostridium perfringens*	Contamination of undressed wounds	Debridement, amputation, hyperbarbic chamber, Penicillin
Rat bite fever	*Spirillum* or *Streptobacillus* spp.	Rat bite	Penicillin or tetracycline
Cat scratch disease	*Afipia felis* or *Rochalimaea henselae*	Cat	Rifampin

(*Contd.*)

Disease	*Causal organisms*	*Transmission media*	*Therapy*
Plague	*Yersinia pestis*	Rat flea (*Xenopsylla cheopis*)	Tetracyclin and streptomycin
Relapsing fever	*Borrelia* spp.	Soft ticks	Tetracycline
Lyme disease	*Borrelia burgdorferi*	A tick (*Ixodes* sp)	Penicillin, tetracycline
Epidemic typhus	*Rickettsia prowazekii*	Human louse	Tetracycline and chloramphenicol
Endemic murine typhus	*Rickettsia typhi*	Rat flea and ticks	Tetracycline and chloramphenicol
Rocky mountain spotted fever	*Rickettsia rickettsi*	Some ticks (*Dermacentor andersoni*)	Tetracycline and chloramphenicol
VIRAL			
Myocarditis	Coxsackievirus (enterovirus)	Air and food	No therapy
Infectious mononucleosis	Epstein Barr Virus	Oral secretions, kissing	No therapy
Yellow fever	Arbovirus (Yellow fever virus)	Mosquito *Aedes aegypti*	No therapy
Dengue	Arbovirus (**Dengue fever virus**)	Mosquito *Aedes aegypti*	No therapy
PROTOZOAL			
Toxoplasmosis	*Toxoplasma gondii*	Cat faeces and under cooked meats	Pyrimethamine in combination with sulphadiazine
Chagas disease	*Trypanosoma cruzi*	Bug bite & bug faeces	Nifurtimox
Malaria	*Plasmodium vivax*	Mosquito *Anopheles*	Chloroquine and mefloquine
HELMINTHIC			
Schistosomiasis	*Schistosoma*	Contaminated water	Praziquantel

Table 18.5 The bacterial diseases of upper respiratory system

Disease	*Causal organisms*	*Transmission media*	*Therapy*
Streptococcal pharyngitis or strep throat	*Streptococcus pyogenes*	Respiratory route	Penicillin
Scarlet fever	*Streptococcus pyogenes*	Respiratory secretions	Penicillin, erythromycin
Diphtheria	*Corynebacterium diphtheriae*	Respiratory secretions	Penicillin, tetracycline, erythromycin and antitoxins
Cutaneous diphtheria	*Corynebacterium diphtheriae*	Respiratory secretions	Antitoxin and penicillin, tetracyclin, erythromycin
Otitis media	*Staphylococcus aureus* *Streptococcus pneumoniae* β-hemolytic streptococci *Hemophilus influenzae*	Complications of cold, nose or throat infection	Penicillin and amoxicillin.

Table 18.6 Bacterial diseases of lower respiratory system

Disease	*Causal organisms*	*Transmission media*	*Therapy*
Pertussis or whooping cough	*Bordetella pertussis*	Respiratory secretions	Erythromycin
Tuberculosis	*Mycobacterium tuberculosis*	Respiratory exudates	Isoniazid and rifampin
Pneumococcal pneumonia	*Streptococcus pneumonia*	Respiratory from colonization of upper respiratory system	Penicillin
Hemophilus influenzae pneumonia	*Hemophilus influenzae*	Healthy carrier, colonization of upper respiratory system	Cephalosporin
Mycoplasmal pneumonia	*Mycoplasma pneumoniae*	Respiratory exudates	Tetracycline, erythromycin
Legionellosis	*Legionella pneumophila*	Aerosols, contaminated water	Erythromycin, rifampin
Psittacosis or ornithosis	*Chlamydia psittaci*	Aerosols of dried droppings and other exudates of birds, parrot	Tetracyclines
Clamydial pneumonia	*Chlamydia pneumoniae*	Respiratory secretions	Tetracyclines
Q-fever	*Coxiella burnetii*	Raw cow milk, aerosols	Tetracyclines

Table 18.7 Fungal diseases of lower respiratory system

Disease	*Microorganisms causing*	*Transmission media*	*Therapy*
Histoplasmosis	*Histoplasma capsulatum*	Birds, bats and air borne spores	Amphotericin B, itraconazole
Coccidioidomycosis	*Coccidioides immitis*	Soil organisms, airborne arthrospores	Amphotericin B, ketoconazole
Blastomycosis	*Blastomyces dermatidis*	Soil organisms and airborne spores	Amphotericin B

Table 18.8 Bacterial diseases of digestive system

Disease	*Causal bacterium*	*Therapy*
Staphylococcal food poisoning	*Staphylococcus aureus*	Replacement of lost water and electrolytes
Salmonellosis	*Salmonella*	Replacement of lost water and electrolytes
Typhoid fever	*Salmonella typhi*	Third generation cephalos porins/chloramphenicol
Bacillary dysentery (Shigellosis)	*Shigella* sp	Replacement of water and electrolytes fluoroquinolones
Cholera	*Vibrio cholerae*	Replacement of water and electrolytes
Vibrio parahaemo-liticus gastroenteritis	*Vibrio parahaemolyticus*	Replace lost water and electrolytes
Escherichia coli gastroenterilis	Enterotoxigenic Enteroinvasive Enteropathogenic, and Enterohaemorrhagic strains of *Escherichia coli*	Replacement of lost water and electrolytes
Campylobacter gastroenteritis	*Campylobacter jejuni*	Replacement of water lost due to dehydration
Helicobacter peptic disease syndrome	*Helicobacter pylori*	Bismuth containing drugs metronidazole
Yersinia gastroenteritis	*Yersinia enterocolitica*	Replacement of water lost due to dehydration and electrolytes
Clostridium perfringens gastroenteritis	*Clostridium perfringenes*	Replacement of water and electrolytes
Bacillus cereus gastroenteritis	*Bacillus cereus*	No therapy. Self control

Table 18.9 Viral diseases of digestive system

Disease	*Causal virus*	*Transmission*	*Therapy*
Mumps	Mumps virus	Saliva and respiratory exudates	None, but preventive vaccine available
Cytomegalovirus (CMV) inclusion disease	Cytomegalovirus (CMV)	Blood transfusion placental transfer	Ganciclovir
Hepatitis A	Hepatitis A virus (HAV)	Contaminated food and water	Vaccine anticipated
Hepatitis B	Hepatitis B virus (HBV)	Transfusion, sexual contact, blood contamination by needles	Preventive vaccine available
Hepatitis C	Hepatitis C virus (HCV)	Blood transfusion and contamination by needles	No therapy
Hepatitis D	Hepatitis D virus (HDV)	Blood transfusion contaminated needles sexual contact	Hepatitis B vaccine effective because coinfection required
Hepatitis E	Hepatitis E virus (HEV)	Contaminated food and water	No treatment
Viral gastro-enteritis	Norwalk agent or **rotavirus**	Contaminated food and water	Replacement of lost water and electrolytes

Table 18.10 Fungal diseases of digestive system

Disease	*Causative fungus*	*Transmission media*	*Therapy*
Ergot poisoning	Mycotoxin produced by *Claviceps purpurea*	Contaminated cereals/grains	Antispasmodic drugs
Aflatoxin poisoning	Mycotoxin produced by *Aspergillus flavus*	Contaminated food, e.g., Pea-nuts	No therapy

Table 18.11 Protozoan diseases of digestive system

Disease	*Causal Protozoan*	*Transmission*	*Therapy*
Giardiasis	*Giardia lamblia*	Consuming contaminated water	Metronidazole and quinacrine hydrochloride
Amoebiasis (Amoebic dysentry)	*Entamoeba histolytica*	Contaminated food and water	Metronidazole and iodoquinol both
Cryptosporidiosis	*Cryptosporidium*	Water borne	No therapy

Table 18.12 Helminthic diseases of digestive system

Disease	*Helminth*	*Transmission*	*Therapy*
Tapeworms	*Taenia saginata* (beef tapeworm) *T. solium* (pork tapeworm) *Diphyllobothrium latum* (fish tapeworm)	Consuming Contaminated meat and fish	Niclosamide
Hydatid disease	*Echinococcus granulosus*	Ingestion of eggs	Surgical removal, albendazole
Pinworms	*Enterobius vermicularis*	Ingesting eggs	Pyrantel pamoate and mebendazole
Hookworm	*Necator americanus*	Penetration of larvae into the skin from soil	Mebendazole
Ascariasis	*Ascaris lumbricoides*	Ingesting eggs	Mebendazole
Trichinosis	*Trichinella spiralis*	Ingesting meat containing larvae	

Q.22. How can *Neisseria gonorrhoeae* be microscopically recognized in the pus from a gonorrhoea patient?

Ans. A smear of the pus from a patient will show *Neisseria gonorrhoeae* as paired cocci contained within leukocytes which are Gram-negative.

Q.23. Give common bacterial diseases of the urinary system.

Ans. See Table 18.13.

Table 18.13 Bacterial diseases of the urinary system

Disease	*Bacterial species*	*Transmission media*	*Therapy*
Cystitis (Infection of bladder)	*Escherichia coli, Staphylococcus saprophiticus*	Opportunistic	Trimethoprim-sulphamethoxazole
Pyelonephritis (Infection of kidney)	*Escherichia coli* usually	Entry systemic or lower urinary tract infections	Cephalosporins
Leptospirosis (Infection of kidney)	*Leptospira interrogans*	Infected animals, their urine or contaminated water	Penicillin
Glomerulonephritis (Infection of Kidney)	Hypersensitive reaction to *Streptococcus pyogenes*	*Streptococcus pygenes* strains in any other part of body	No therapy Autoimmune

Q.24. Give bacterial diseases of the reproductive system in a tabulated form.

Ans. See Table 18.14 below.

Table 18.14 Bacterial diseases of reproductive system

Disease	*Bacterial species*	*Transmission media*	*Therapy*
Gonorrhoea	*Neisseria gonorrhoeae*	Sexual contact	Third generation cephalosporins
Nongonococcal	*Chlamydia*	Sexual contact	Tetracyclines
urethritis (NGU)	*Mycoplasma hominis, Ureaplasma urealyticum*	or opportunistic infection	Erythromycin
Pelvic inflammatory disease (PID)	*N. gonorrhoeae Chlamydia trachomatis*	Complication of STDs	Doxycline, cefoxitin sodium
Syphilis	*Treponema pallidium*	Often sexual contact	Benzathine penicillin
Gardnerella vaginosis	*Gardnerella vaginalis*	pH opportunistic vaginal pH 5 – 6.	Metronidazole
Lymphogranuloma venerium (LGV)	*Chlamydia trachomatis*	Most probably sexual contact	Doxychne
Chancroid or soft chancre	*Homophilus ducreyi*	Most often sexual contact	Erythromycin or ceftriaxone

Q.25. Give viral diseases of the reproductive system in tabular form.

Ans. See Table below.

Disease	*Virus causing disease*	*Transmission*	*Therapy*
Genital herpes	Herpes simplex virus 2 sometimes type 1	Usually sexual contact	Acyclovir
Genital warts	Papillomavirus	Contact	Interferon injection and removing by surgical and chemical methods.

Q.26. Name the most common fungal disease. Give its causative fungal species, transmission and therapy/treatment.

Ans. **Candidiasis** is caused by the fungus *Candida albicans* which is an opportunistic pathogen but may also be transmitted by sexual contact. Clotrimazole and miconazole are used.

Q.27. Name the disease caused by protozoan *Trichomonas vaginalis.*

Ans. It causes **trichomoniasis** (often **vaginitis**). The pathogen is usually transmitted by sexual contact and is generally treated with metronidazole.

19

CHAPTER

ALGAE, PROTOZOA AND MULTICELLULAR PARASITES

Salient Features

- The algae also occur on the hair of sedimentary South American Sloth (Common name for *Bradypus*, the eutherian (placentalia) mammals belonging to order Edentata) and the polar bear, and comprise a unique habitat.
- The algal genus *Sargassum* has been named so because it profusely grows in the subtropical Sargasso sea.
- All algae are eucaryotic.
- Some red, brown and multicellular green algae need rocks, snails or even other algal substrata for attachment.
- The body of algae are thalli (singular thallus) being not differentiated into root, stem and leaf.
- The body of some seaweeds are anchored or attached on rocks by means of structures called **hold fasts**, which have stem-like and usually hollow stipes and leaf-like structures called **blades**.
- The algal thallus is devoid of **xylem** the conductive tissue for taking water and minerals; and **phloem** for conduction of photosynthesised food.
- The macroscopic brown algae may grow many metres to kilometres in length, known as **kelp.**
- Recently algae are classified as **protists** or **plants** (autotrophic) depending upon their structure, pigments and other characteristics.
- The algin used as a thickner is obtained from cell wall of algae. The other uses of algin are formation of rubber tyres and hand lotions.
- The **agar** used for solidification of culture media is obtained from *Gelidium*, a red alga.
- **Carrageenan** a gelatinous material used for thickening evaporated milk, ice-cream and other pharmaceuticals, is obtained from red algae called **Irish moss**.
- *Gonyaulax* a genus of **marine dinoflagellates** (unicellular biflagellated) produce **neurotoxins** and cause paralytic shell fish poisoning. When excess of dinoflagellates are easten by mollusks like mussels and clams (bivalved molluscs e.g. *Unio* and *Lamellidens*). When these are eaten by humans the disease is passed on to them also. The large concentration of dinoflagellate *Gonyaulax* provides the ocean a deep red colour and because of this colour the tides therein have been called **red tides**. It is precautionary that mollusks should not be harvested for human consumption in a red tide.
- **Ciguatera** is a disease that occurs when dinoflagellate *Gambierdiscus toxicus* gets concenterate in large fish in the food chain.
- The great proportion of petroleum was formed from diatoms which lived over 300 million years ago following death of diatoms and other planktonic organisms and they were buried in sediments.

- The **Tridacna**, the giant clam have some unicellular algae as symbionts. The algae release glycerod into the blood stream of clam to meet out the clam's requirement for carbohydrates.
- The **malaria** (caused by *Plasmodium*), **leishmaniasis** (caused by *Leishmania*, Visceral leishmaniasis or the **Kala-azar** caused by *L. donovani*) and amoebiasis (caused by *Entamoeba histolytica*) are major protozoal diseases of the warm countries.
- Other protozoal diseases which are lesses but significant are *toxoplasmosis* (caused by *Toxoplasma gondii* an intracellular parasite of humans and dangerous to pregnant women because can cause congenital infection in utero); **giardiasis** (cuased by *Giardia lamblia*); **trichomoniasis** (caused by *Trichomonas* and **balantidiasis** (caused by *Balantidium coli*)
- The parasitic helminth's are parasitic flatworms of Phylum Platyheliminthes and parasitic roundworms of Phylum Nematoda.
- The tapeworm a Cestode comprise of **scolex** (head) and **proglottids**, a body segment of tapeworm containing both male and female organs.
- The nematodes that result infection of human through eggs are **pinworm** (*Enterobius vermicularis*) and a large nematode (*Ascaris lumbricoids*).
- The hookworm (*Nectar americanus*), *Trichinella spiralis* and anisakine or wriggly worms infect humans through their larvae.

Algae

Q.1. Are algae aetiological agents of disease? Explain.

Ans. The algae, often are not considered the aetiological (etiological) agents, the organisms which initiate or cause the disease. However, the dinoflagellate *Gonyaulax* (produce the toxin which cause paralytic shell fish poisoning. The **blooms**, the excessive growth of *Gonyaulax* also cause red tides in coastal marime **niches** or the habitats. Such algae and toxins produced in the duration of algal bloom get concentrated insite the bivalve shell fish like clams (bivalved molluses e.g. *Unio*, *Lamellidens* etc.) and oysters (edible, marine bivalve shell fish *Avicule margaritifera* in which pearl is formed). Therefore, the eating of shell fish possessing algal toxins by humans can lead to symptoms which resemble botulism. That is why to prevent the occurrence of blood poisoning the harvest and sale of shellfish is banned in areas of *Gonyaulax* bloom. The cell walls of dinolagellates (Fig. 19.1A) consist of many individual plates made of cellulose and silica.

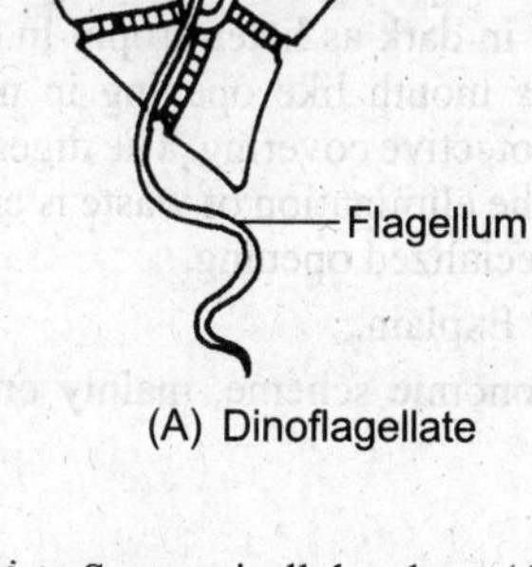

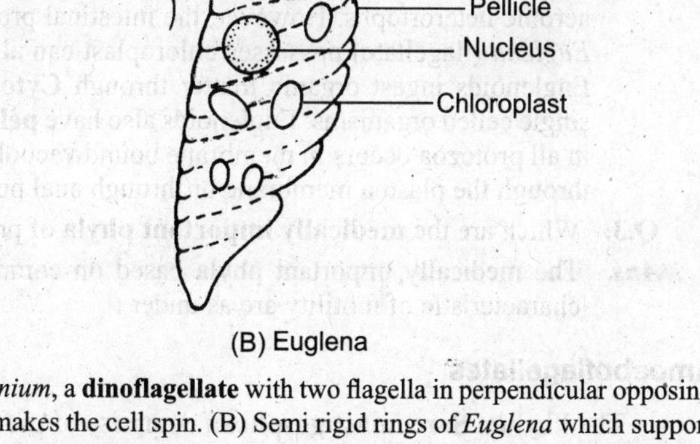

Fig. 19.1. Some unicellular algae. (A) *Peridinium*, a **dinoflagellate** with two flagella in perpendicular opposing grooves. Their simultaneous beat makes the cell spin. (B) Semi rigid rings of *Euglena* which support the **pellice**, the protective layer, enables the ***Euglena*** to change shape. Near opening of **Cytostome** (mount-like structure) there is a red **eyespot** at anterior end.

The bloom or excessive growth of **Cyanobacteria** or blue green algae may also cause poisoning of fish and live stock.

The euglenoids (Fig. 19.1 B) in the dark ingest organic matter through **cytostome**, the mouth-like opening.

Protozoa

Q.2. What are **protozoa**? Describe their **life cycle.**

Ans. Greek *protos* = first; *zoion* = living *i.e.* first formed animal life. Currently they are regarded as eucaryotic unicellular organisms of **Kingdom Protista**. The protozoa are **ubiquitous** seeming to be everywhere or at several places at the same time (occurring in soil, water, as normal microbiota of animals, as parasites and eaters of bacteria) and are of great significance in the economy of nature. Some parasitic forms are important medically e.g. **malarial parasite** (*Plasmodium vivax* the Apicomplexan or **Apicomplexa)**, cause of urinary and genital tract infections (*Trichomonas vaginalis* a **flagellate**) and cause of amoebic dysentery (*Entamoeba histolytica,* the only **pathogenic amoeba** of Subphyllum **Sarcodina.**

LIFE CYCLE OF PROTOZOA : The generalized account of life cycle of protozoa may be covered under subheadings asexual reproduction, sexual reproduction encystment and nutrition.

(i) **Asexual reproduction** in protozoa is by fission, budding and **schizogony**, in which there is nuclear fission. In nuclear multiple fission the nucleus divides into many nuclei prior to the division of the cell. The cytoplasm gets collected around each nucleus which emerges as a single cell and many daughter cells result.

(ii) **Sexual reproduction** has been observed in some protozoa as *Paramecium*; a **Ciliate** (Phyllum Ciliophora) undergo (i) **conjugation** a type of sexual reproduction in which contents of two morphological cells behave as **isogamous gametes**, or gametocytes and form a diploid zygote. (ii) But in protozoal conjugation there is variation in which two cells fuse and a haploid nucleus called **micronucleus** from each cell migrates to the other cell. This haploid micronucleus fuses with haploid macronucleus within the cell. Shortly thereafter the parent cells separate and each of the two is a fertilized cell. These cells produce daughter cells with recombined DNA when the cells divide.

(iii) Encystment is a process in which the protozoa form a protective capsule known as cyst which form of the organism enables it to survive at the time of deficiency of food moisture or oxygen. It also provided protection under adverse temperature and in presence of toxic chemicals, or when the parasite is outside the host's body. Since the parasitic protozoa lead life cycles involving more than one host the formation of cyst becomes highly significant to them.

(iv) **Nutrition**: They show a wide range of modes of deriving their nutrition. They may be aerobic heterortophs. However, the intestinal protozoa are capable of anaerobic growth. *Euglena* (flagellate) possesses chloroplast can also grow in dark as heterotroph. In dark Euglenoids ingest organic matter through **Cytostome,** a mouth like opening in many single celled organisms. Euglenoids also have **pellice** a protective covering. The digestion in all protozoa occurs in membrane bound vacuoles and the elimination of waste is either through the plasma membrane or through anal pore, a specialized opening.

Q.3. Which are the **medically important phyla** of protozoa? Explain.

Ans. The medically important phyla based on common taxonomic scheme, mainly on the characteristic of motility are as under :

Amoeboflagellates

The Phylum **Sarcomastigosphora** comprise of protozoa which involve either pseudopodia or flagella as means of motility. However, a few protozoa like *Naegleria* use both types of structures i.e. **pseudopodia** blunt the temporary projections and flagella whip-like structures.

AMOEBAS: The amoebas which are placed under Subphylum **Sarcodina** move with the help of **pseudopodia** the temporary protrusions of the cell associated with flowing movement of cytoplasm. It not only occurs in rhizopod and other protozoa rather takes place in white blood cells also. But all amoebas are not pathogenic. *Entamoeba histolytica* the cause of amoebic dysentery is the only pathogenic amoeba occurring as a pathogen in human intestline. The primary flood of this pathogenic amoeba is red blood cells (RBC) (Fig. 19.2).

It gets transmitted to healthy humans as a result of ingestion of **cysts** the protective capsules of *E. histolytica* which are excerted in the faeces of infected humans.

The amaebic forms of *Naegleria* and *Acanthamoeba* can infect blood, brain and spinal cord

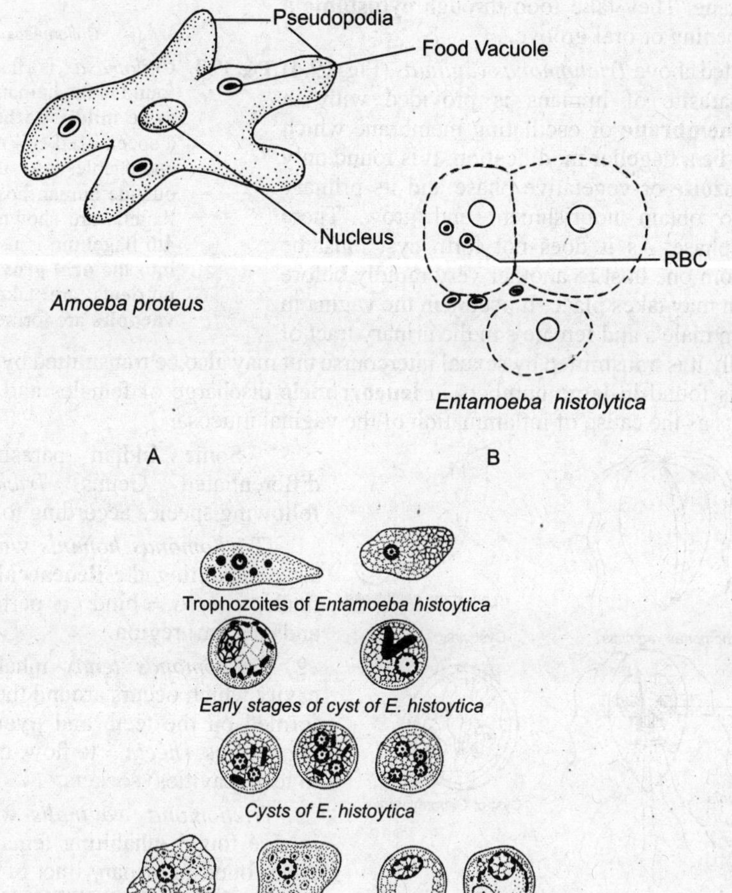

Fig. 19.2. Amoebas, (A) the organisms of Kingdom **Protista**, Subkingdom **Protozoa**, subphyllum **Sarcodina**. (B) *Entamoeba histolytica* with ingested red blood cells which serve as primary food of this protozoan (C) Trophozoites and Cysts of *Entamoeba.*

of human while swimming the waterborne cysts pass across **mucous membranes**, the moist epithelium.

FLAGELLATES : The flagellates (Subphylum Mastigophora) bear flagella which lash out like a whip and propel in their environment. Generally there are 2 or more flagella but some parasitic species bear as many as 8 flagella. The flagellates are spindle shaped. The flagella are projected from the front end of unicellular flagellates (Fig. 19.3) e.g. *Chilomastix* mildly pathogenic. They possess a flexible **pellicle**, a protective layer of outer tough membrane. They take food through **cytostome** a mouth like opening or oral groove.

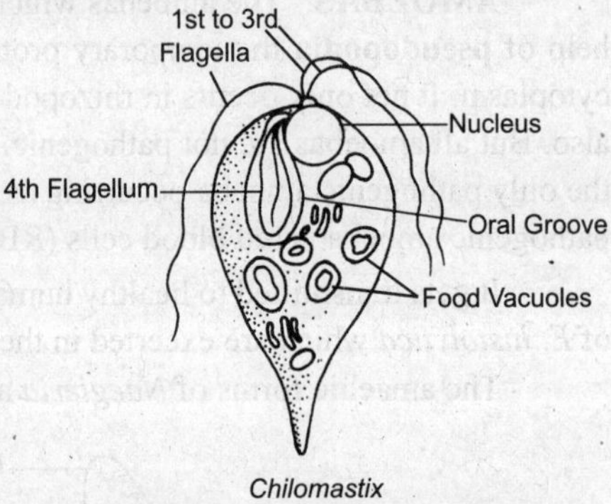

Fig. 19.3. *Chilomastix*, is a flagellate protozoan found in the human intestine known to be **mildly pathogenic**. The **cyst** a specialized cell makes it resistant and enables it to survive for months outside human host. The first three flagella are shown at the top. The 4th flagellum is used to move food into the **oral grove**, the site where nutrients are taken in and **food vacuoles** are formed.

As stated above *Trichomonas vaginalis* (Fig. 19.4) is another parasite of humans is provided with an **undulating membrane** or oscillating membrane which is believed to be a flagellar modification. It is found only in the **trophozoite** or vegetative phase and its primary activity is to obtain nourishment and grow. There is no cystic phase. As it does not form cyst must be transferred from one host to another very rapidly before its desiccation may takes place. It occurs in the vagina in females and in male's and female's in the urinary tract of both. Generally it is transmitted by sexual intercourse but may also be transmitted by toilet facilities. *T. vaginalis* is found in large numbers in **leucorrhoeic** discharge of females and is regarded by gynaecologists as the cause of inflammation of the vaginal mucosa.

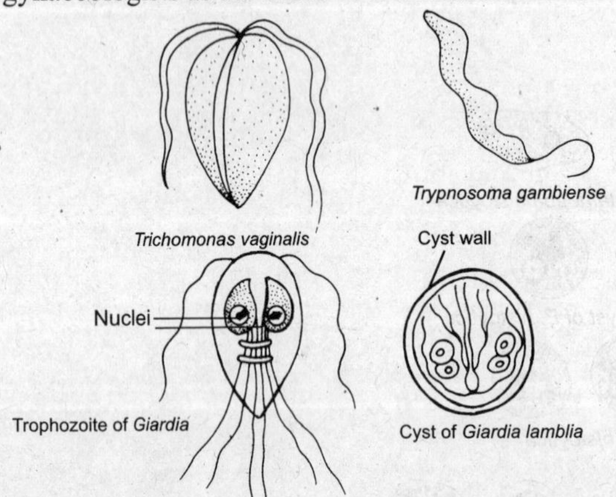

Fig. 19.4. Pathogenic members of Flagellates subphyllum Mastigophora, subkingdom Protozoa (singular Protozoan). *Trichomonas vaginalis* does not have cyst stage and causes urinary and genital tract infections.

Some Indian parasitologists have differentiated Genus *Trichomonas* into following species according to their habitats.

1. *Trichomonas hominis* which is 8 μm in size, inhabiting the **ileocaecal** (Latin *ilia* = flanks, *caecus* = bind *i.e.* pertaining to ileum and caecum) region.

2. *Trichomonas tenax* inhabiting the oral cavity which occurs around the **tartar** deposit formed on the teeth and **pyorrhoeic** (Greek *pyon* = pus, *rheein* = to flow, usually referring to teeth cavities) sockets.

3. *Trichomonas vaginalis* which is 13 μm in size found inhabiting female genital tract. Also found in urinary tract of both males and females. The **CYTOSTOME** (a cleft like depression at its side is less conspicuous.

Giardia lamblia (synonyms *Giardia intestinalis, Giardia duodenalis and Lamblia intestinalis)* was first observed by Leeuwenhoek in 1681 while examining his own stool. It occurs in **trophozoite** (vegetative form) and **cyst** forms, both (Fig. 19.4). It occurs in **small intestine** of humans and other mammals and is exerted in the) faeces as cyst phase. The cyst stage or phase enables it to survive in the environment before being ingested by some other host. It causes **GIARDIASIS** a common intestinal disease.

DIAGNOSIS : (i) It can be done by microscopic examination of freshly passed stool for the *Giardia* **trophozoites** and **cysts**.

(*ii*) *Giardia* trophozoites may also be recovered from bile A (aspirated from duodenum) and B (removed from bile duct) drawn by duodenal intubation.

HAEMOFLAGELLATES : These are parasitic flagellates which are found in the circulatory system of the host in blood. Besides blood they are also found in the tissue of human and other animals. They also pass through a cycle in the gut of insects which serve as intermediate hosts. the two genera which are of pathological importance are *Trypanosoma* and *Leishmania*.

They are transmitted by the bites of blood feeding bugs (insects) and occur in the circulatory system of bitten human and animals. The haemoflagellates possess long slender bodies and **undulating membrane** (oscillating membrane) to surve in the blood which is a viscous fluid. There are two species of the genus *Trypanosoma* (Fig. 19.4).

(1) *T. brucei gambiense.* This parasite was discovered by David Bruce in 1890, which causes African sleeping sickness which is transmitted by Tsetse fly. (2) *T. cruzi* cause **Chagas' disease** or **South American Trypanosomiasis** is transmitted by "kissing bug" named so because it bites on the face. *Trypanosoma* rapidly mutiply by fission. *T. brucei gambiense* cause **trypanosomiasis** in human. Bruce regarded it to be a cuase of **nagana** also in domestic animals. But Hoare (1970) after careful examination of illustrations of David Bruce was convinced that Bruce had an admixture of *T. congolense* which was responsible for the disease in the cattle in Zululand.

Kala – Azar

The **Kala** – **azar** is caused by *Leishmania donovani* a flagellated protozoan. This disease is also know as **visceral leishmaniasis**. The causative parasite has been named after Leishman and Donovan, both of these reported the organisms simultaneously. Leishman reported it from London in May 1903 and Donovan from Madras in July 1903. It is endemic in many places of India, China, Africa, Southern Europe, South America and Russia. In India Kala-ajar is common in Assam, Bengal along coasts of the rivers Ganges and Brahamputra. It is also known to be endemic in Bihar, Orissa, Madras and Eastern parts of Uttar Pradesh including Lucknow. It is an intracellular parasite of humans always found as intracellular in the **amastigote form** a form assumed by the cells of *Leishmania donovani* and many species of **TRYPANOSOMATIDAE**. Also have been called **Leishman-Donovan bodies**.

The parasite spends its life cycle in two stages:

(1) **Amastigate stage** : Earlier known as leishmanial form is found in human (in vertebrates).

(2) **Promastigate stage** : Earlier known as leptomonad form is found in the gut of sandfly, a small Dipteran insect *Phlebotomus*, having biting mouth parts.

Leishmnia donovani which causes visceral leishmaniasis or kala-azar involves reticuloendothelial system (RES). Thus essentially is a parasite of reticuloendothelial system. It causes intermittent fever and enlargement of spleen and liver. However, it is interesting that a **permanent immunity** is developed in the persons who gets recovered from the disease.

Ciliates

The species of Phylum Ciliophora are characterized by the presence of short hair like projections of processes called cilia. The cilia are arranged in precise rows and are shorter than flagella (Fig. 19.5). *Paramecium* is a typical **ciliate** which bear specialized structures for ingestion called mouth or **cytostome**, followed by **cytopharynx** and an opening for elimination of wastes called **anal pore**.

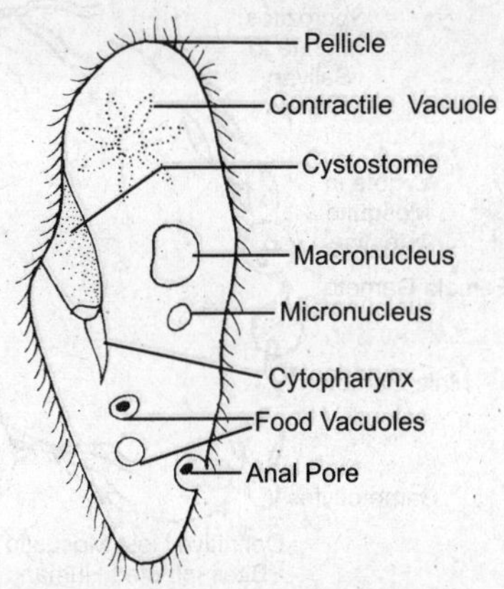

Fig. 19.5. *Paramecium* a unicellular Cilliate of Phyllum Ciliophora covered with cilia relatively short hair like projections. Macronucleus is associalted with synthesis of proteins while micronucleus participate in sexual reproduction. Contractile vacuole help in regulation of osmotic pressure.

There are two morphologically distinct nuclei, larger one called **macronucleus** meant for synthesis of proteins and smaller one called **micronucleus**. involved in sexual reproduction. The food enters through cytostome and is then passed to cytpopharynx, at the end of which food vacuoles are formed and after digestion of food, waste particles are released through the structure called anal pore.

The *Balantidium coli* is the only pathogenic species and is the known human parasite which causes severe although rare type of dysentery in human which has also been called **ciliate dysentery or balantidiasis**.

The parasite exists in two stages or phase: (1) **Trophozoite** stage and (2) **Encysted** stage. It perhaps is the largest protozoal parasite inhabiting the large intestine of human. The cysts on being ingested by the host enter the colon into which trophozoites are released. The trophozoites prey upon bacteria and eat them besides faceal debris as they multiply and the cysts formed are relased out with faeces. The cyst is slightly smaller than the trophozoite (trophic) form, measuring 50 to 60 μm in diameter.

Apicomplexans

The **Apicomplexa** is a **Phylum** of Protozoa. The organisms in this group are intracellular parasites. They are devoid of cillia and flagella, therefore, their adult forms are nonmotile. However, the **microgametes** i.e the male gametes are flagellated and motile in some species. The phylum has been named so because they possess a complex of special enzymes which assist them to penetrate the tissue of the host. They complete their life cycle in more than one host. A typical example of this phylum is the malarial parasite, the *Plasmodium vivax*.

Malarial Parasite: The *Plasmodium vivax* is the causative agent of malaria. The life cycle of malarial parasite is completed on two hosts :

1. **Definitive host** : which harbours sexually mature form. Here mosquito (*Anopheles*) is the definitive host. The mosquito on ingestion of gametocytes leads to sexual reproduction in its intestine.
2. **Intermediate host**: Human, is the intermediate host where asexual reproduction of the *P. vivax* takes place. It is by mutiple fission called **schizogony** in liver and red blood cells (RBC) of the human host (Fig. 19.6).

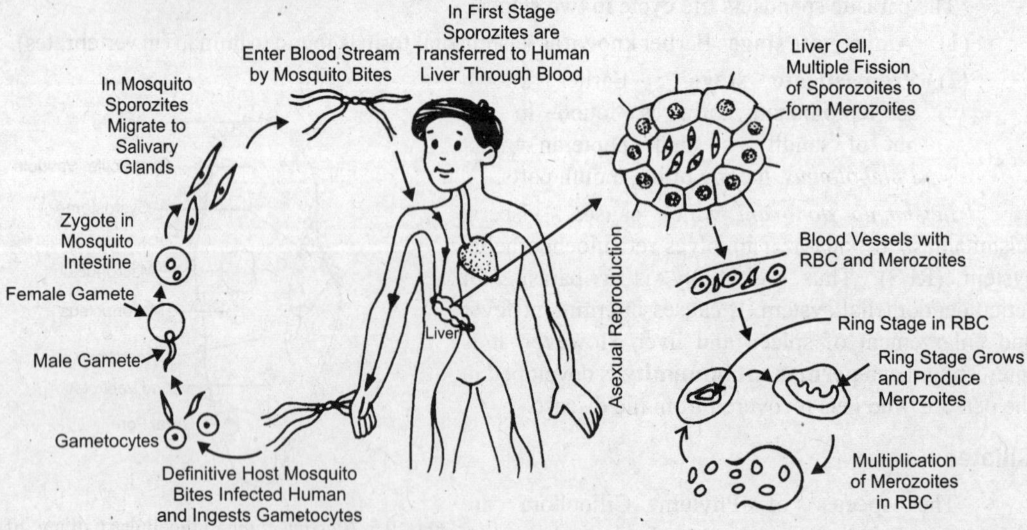

Fig. 19.6. *Plasmodium vivax.* Life cycle is completed on mosquito (Definitive host) and human (Intermediate host).

1st stage begins when mosquitoes carrying **sporozoites** (*i.e.* **trophozoites** vegetative form of small elongated cells which are produced in the mosquito and get localized in the salivary gland of the mosquito) bites the human being. Thus the sporozoites are transferred from salivary gland of infected mosquito to blood stream of human and from blood stream to liver of human. The sporozoites. In the 2nd stage undergo multiple fission called **schizogony** inside the liver cells and form thousands of progeny cells (say 12,000) known as **merozoites**. In the 3rd stage merozoites may enter the blood stream (alternatively to new parenchyma cell of the liver). In the 4th stage the young trophozoite (the technical term for the vegetative cell of the protozoa) takes the shape of a ring called **ring stage** in which nucleus and cytoplasm are distinct. In the 5th stage ring stage enlarges and divide a number of times repeatedly. In the 6th stage red blood cell (RBC) gets reptured and the large number of **merozoites** already formed in it are released. The release of merozoites and their wastes into the blood causes fever and chills to the host person. Most of the merozoites so released greatly infect the new red blood cells (RBC). Alternatively, some of the merozoites develop into male and female gametes called **gametocytes**. The gametocytes themselves do not cause any damage or harm to the human host.

In the 7th stage another mosquito comes to bite the infected human and sucks the male and female gametocytes which consequently enter the intestine of the mosquito and onward follow their **sexual cycle**.

In the 8th stage the male and female gametocyltes fuse to form a Zygote. The *zygote* gets encyted to form an **oocyst** in which cell division takes place and a large number of asexual **sporozoites** (the trophozoites of *Plasmodium vivax* formed in mosquitoes which are infective to humans) are produced. In the 9th stage the oocyst gets ruptured and sporozoites migrate to the salivary glands of the mosquito. Thus, from salivary glands these can be transmitted to a new host by the bite of the mosquito.

DIAGNOSIS OF MALARIA : The diagnosis of malaria is done by observing the presence of *Plasdmodium* in the thick smear prepared from the blood of a patient.

While developing the parasite the red blood cells (RBC) of the patient shows following changes:

(i) The infected red blood cell enlarges and becomes twice or double to its original size.

(ii) It becomes highly distorted in shape and becomes rhomboidal to irregular in outline.

(iii) It appears washed out and becomes pale to almost colourless.

(iv) The portion of the cytoplasm unoccupied by the malarial parasite gives a dotted or stippled look known as **Schüffner's dots** after the name of the discoverer. With Leishman's stain they appear as pinkish granules.

Toxoplasma

The *Toxoplasma gondii* the parasite discovered by Nicolle and Manceaux in 1908 in a small rodent, gondii (*Ctenodactylus gundi*) of Africa, is another apicomplexan intracellular parasite of humans. *T. gondii* poses a big threat to pregnant women as the cogenital infection in uterus are caused by it. A domestic cat is involved in the life cycle of this parasite. The trophozoites of the parasite known as **Trachyzoites** reproduce sexually and asexually in the infected cat. The oocytes each having 8 sporozoites are discharged out with faeces. When the oocysts are ingested by humans or other animals the sporozoites get raised as trophozoites which can infect tissues of new host.

DIAGNOSIS : Demonstration of *Toxoplasma gondii* is not always possible, however, can be carried out by microscopic examination of stained smears obtained from bone marrow punctures and splenic puncture or centrigused deposit of **C.S.F.** (Cerebro spinal fluid) and smears made from tissues removed by biopsy preferably lymph node or muscle) or at autopsy. Also the antibodies may be detected by ELISA or other tests.

Cryptosporium

The *cryptosporium* an intracellular parasite also is an apicomploxan which is the causative agent of diarrhoeal diseases in immunosuppressed people. It may also cause infection of gall bladder and respiratory tract.

DIAGONOSIS: Can be done by acid fast staining. The fluorescent antibody tests are also of diagnostic importance.

Microsporans

These recently have been placed under phylum Microscpora. These are also obligate intracellular parasites devoid of cillia and flagella. They are also obligate intracellular parasites devoid of cillia and flagella.They are known to cause chronic diarrhoea and inflammation of conjuctiva near cornea called **keratoconjuctivities** in the patients suffering from AIDS.

Nosema is a typical example of phylum Microspora which causes diarrhoea and keratoconjuctivities of animals.

Helminths

Q.4. What are **parasitic helminths**? How is the **laboratory diagnosis** of their eggs done ? Explain.

Ans. The helminths are multicellular eucaryotic animals. They generally possess digestive, circulatory, nervous, excretory and reproductive systems. There are also some **free living** (non parasitic) **helminths** in which one or the other of said system is either reduced or absent.

Most of them are placed under two phyla :

(i) Platyhelminthes (flat worms)

(ii) Nematodes (round worms)

The two classes of the parasitic flat worms are: (i) Trematodes, and (ii) Cestodes. The laboratory diagnosis of their eggs can be carried out adopting sedimentation techniques. These are as under.

(i) **Simple Sedimentation**: Sufficient amount of faecal material is thoroughly shaken with 10 to 20 times of its volume in tap water and allowed to settle in a cone shaped flask for an hour or two. The process is repeated several times till the **supernatant** fluid is clear. The sediment at the bottom is finally examined under microscope for the eggs. However, for examination of protozoal cysts this technique is not suitable.

(ii) **CONCENTRATION OF CYSTS AND HELMINTH OVA IN FAECES (FORMAL - ETHER CONCENTRAION)**: Concentration of all protozoal cysts and Helminth ova can be measured without distortion by Ridley and Hawgood (1956) method only in 5 minutes by dissolving in **formal saline** (formaldehyde in saline)

Take 10 g of faeces in approximately 30 ml of 4 per cent formaledyde in saline. The volume of formal saline depends on consistency of faeces. Pass the suspension through a wire sieve (of 250 μm diameter apertures). Pour 15 ml of the sieved suspension into a 25 ml centrifuge tube and add 3 ml of diethyl ether. Shake vigorously at least 50 times. Centrifuge at about 650g for 2 minutes. Loosen the fatty debris which collects at the liquid interface with a glass rod, keeping the glass rod in position invert the tube boldly to an angle of 45° with the horizontal, pouring away the supernate and the debris (supernatant fluid is decanted). Some of the debris remains attached to the inside of the tube. Clear this off with the glass rod by holding the rod under the jet to play on the rod and run back inside the tube.

Take samples of the deposit by means of a Pasteur pipette and examine them on a glass slide with a superimposed cover-slip. The factor of concentraion is 30 – 40 times. As the preparations are 'fixed' (by the presence of formalin) cysts will be killed and so eosin staining is valueless. However, iodine or sargeunt's stain are useful.

Q.5. Why are the helminths studied by microbiologists depsite of the fact that adult helminths are not microscopic ? Justify.

Ans. The adult helminths are not microscopic but the helminths show microscopic stages in their life cycle, and some of these stages cause diseases. That is why the microbiologists study helminths.

Platyhelminths

Q.6. What are Platyhelminths? Why is the digestive system of Playhelminths called incomplete? How do they obtain nutrition and parasitize host and complete their life cycle ? Explain.

Ans. The **Platyhelminths** comprise a group of small **flatworms** which are bilaterally symmetrical and dorsivetrally flattened from back to front. Their digestive system is regarded to be **incomplete digestive system** because their digestive system has one opening or mouth through which food enters and wastes leaves (go out). They include parasitic flat worms of two classes Trematodes (Trematoda) and Crestodes.

TREMATODES: They are called flukes with flat leaf-shaped bodies provided with oral and ventral suckers (Fig. 19.7). These small animals attach themselves on the host by means of these

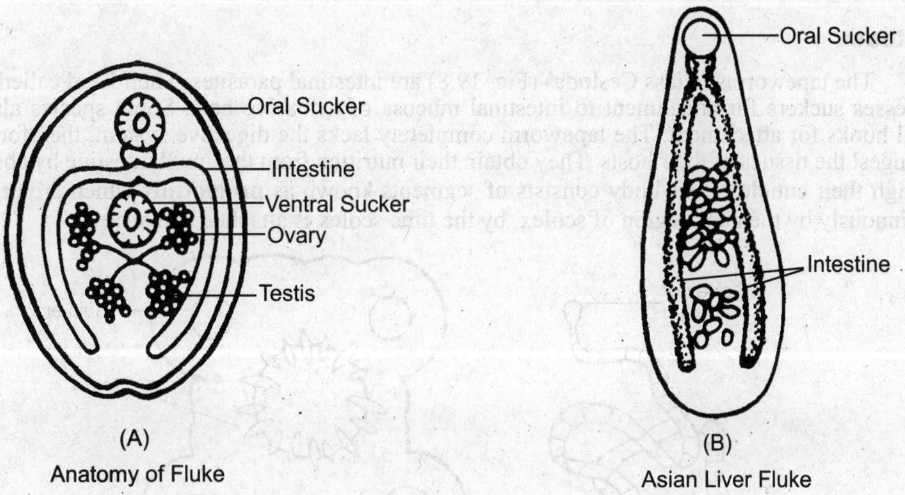

Fig. 19.7. **Flukes**, the parasitic flat warms (Trematodo of Phylum Platyhelminths). (A) Cutway view of an adult fluke with **oral** and **ventral suckers** which attach the fluke to the host. The mouth is situated in the centre of oral sucker. (B) An outline diagram of Asian liver fluke *Clonorchis sinensis*. The flukes are hermaphrodite and each animal has **ovary** and **testes** both.

suckers and suck fluids from the host. Besides these suckers the flukes can also derive their food by absorbing it their nonliving outer covering known as cuticle. Interestingly, they have been named according to the tissue of the definite host in which these adult creatures live *e.g.* lung fluke, liver fluke and blood fluke and parasitize the host. Extensive infestation of **Asian liver fluke** *Clonorchis sinensis* may block bile ducts. Its digestive system is incomplete and not well developed.

The life cycle of fluke has been best described in **lung fluke** *Paragonimus westermani*. This adult fluke 6 mm wide and 12 mm long lives in the branchioles of humans and other mammals. The life cycle of lung fluke may be divided into following 8 junctures or stages.

1. The hermaphrodite adults liberate eggs into the bronchi. The sputum which contains eggs is generally swallowed which consequently are exereted in faeces of definitive host. (2) The eggs in this stage find way to a water body. (3) The miracidium or miracidial larva develops in side the egg. (4) The larvae may enter the suitable snail which is the first intermediate host. (5) The **miracidium** develops into a **redia** one of the several types of larvae occurring in the life cycle of most trematodes including *Faciola* species). These redia develops into a **cercaria**. (6) The cercaria drills out (perforates) the snail and penetrates the cuticle of a crayfish. (7) The fluke which has parasitized crayfish gets encysted as a **metacercaria** in the muscles and other tissues of the crayfish. (8) The crayfish is consumed (eaten) by human. Thus metacercaria is freed in the humans small intestine. It drills out and wanders around till it penetrates the lungs, makes entry into the bronchioles and develops into an adult fluke inside the lung.

DIAGNOSIS: The sputum and faeces are examined microscopically in the laboratory. Eating undercooked crayfish may cause infection.

BLOOD FLUKE: *Schistoma* the blood flukes are not ingested rather they burrow through the skin of the human host and find entry in cculatory system. The adults have been reported from certain abdominal and pelvic views. The disease **schistosomiasis** from Greek *schistos* = cleft; *soma* = body; *iasis* = condition, is the infestation of human body by *Schistosoma* ('blood flukes') from drinking or bathing in infected water. The *Schistosoma haematobium* causes vesical schistosomiasis with ***haematuria*** (Greek *haima* = blood, *ouron* = urine ; means blood in urine symptom. *Schistosoma mansori* and *S. japonicum* produce intestinal **schistosomiasis** characterized at first by diarrhoea and latter by hepatosplenic disease.

Cestodes

The tapeworms (Class Cestoda) (Fig. 19.8) are intestinal parasites. Their head called scolex possesses suckers for attachment to intestinal mucosa of definitive host. Some species also bear small hooks for attachment. The tapeworm completely lacks the digestive system, therefore, does not ingest the tissues of their hosts. They obtain their nutrition from the small intestine by absorbing through their cuticle. Their body consists of segments known as **proglottids** which are produced continuously by the neck region of scolex, by the time scolex is attached and alive.

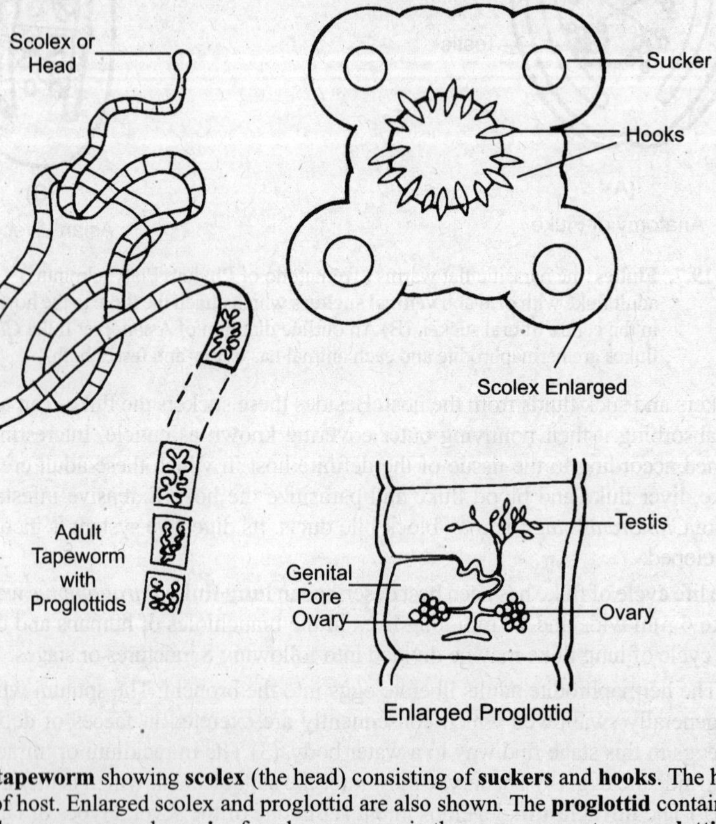

Fig. 19.8. An adult **tapeworm** showing **scolex** (the head) consisting of **suckers** and **hooks**. The hooks attach to tissue of host. Enlarged scolex and proglottid are also shown. The **proglottid** contains both **testes** male sex organs and **ovaries** female sexorgans in the same segment or **proglottid**.

The adults of **beef tapeworm** *Taenia saginata* live in humans and can gain a fascinating **length of six metres.** The head of this tapeworm called **scolex** is two mm in length and accompanied by a thousand or more proglottids. The presence of proglottids and thousands of eggs in each proglottid in faeces is indicative of parastization by tapeworm. Having twisted away from the faecal material they may spread and are ingested by grazing animals. On ingestion by the cattle the eggs are hatched into **larvae**, the immature stage of life cycle, may perforate or bore through the intestinal wall. The larvae finally reach muscle meat of the cattle and encyst as **cysticerci**. These when ingested by man are digested except the head of scolex, which anchors in the small intestine begins to generate or produce **proglottids**, the body segments of tapeworm having both male and female organs.

DIAGNOSIS OF TAPEWORM : The presence of mature proglottids and eggs in faeces is a diagnostic feature of infestation of tapeworm in man while the cysticerci can be seen macroscopically in meat and such meat is called **"measly beef"**.

Control Measures : (1) The meat meant for human consumption should be inspected for measly appearance. (2) the untreated human sewage should not be treated as a fertilizer in grazing pastures.

The **pork tapeworm** *Taenia solium* parasitizer human and is the only definitive host of it. The eggs are produced by adult worms in the intestine of human and are naturally released in the faecal matter. On being eaten by pig the larval helminth encysts in the muscles of pig. On eating undercooked Pork the humans get infected.

It is interesting enough that there is a wide range of size of tapeworms.

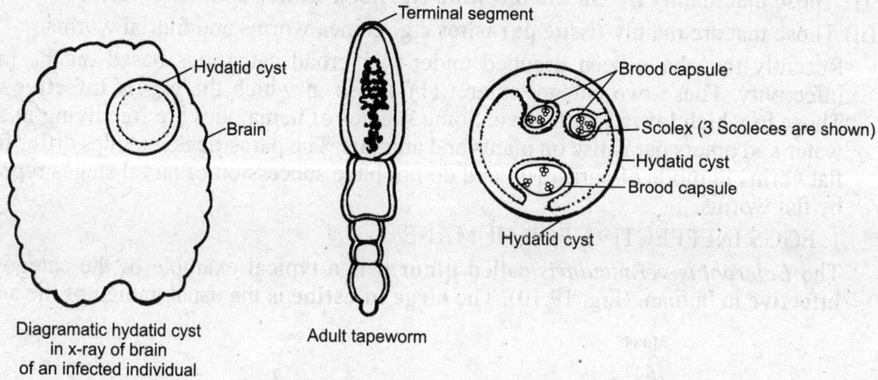

Fig. 19.9. A **minute (tiny) tapeworm** *Echninococcus multilocularis* which is adult of a closely related species *Echinococcus granulosis* found in dogs, wolves and foxes. A diagramatic view of **Hydatid cyst** formed by *Echinococcus granulosus* is also shown. Such cysts have been observed in X-ray of brain of infected human. The parasite can complete the life cycle provided the cysts are ingested by definitive host which eats the intermediate host. The human serves as an intermediate host but there is no question of promotion of parasite unless the human is eaten by the definitive.

Echinococus granulsus(Fig. 19.9) is minute 2 to 8 mm only. It comprises of **scolex** (head), short **neck** and **strobila** consisting of 3 segments (occasionally 4). First segments is immature, the second one is mature and the last one (4th one when present) is gravid. Terminal segements is biggest. The humans are intermediate hosts for it while dogs and **coyote**, the prairie - wolf *Canis latrans* small wolf-like animal found in western north America, are the definitive hosts. The eggs are exerted with the faeces which are ingested by deer, sheep or humans. The humans may get infected by infecting their hands with faeces or saliva licking from the mouth of dog. The eggs hatch in the humans small intestine and the larvae may travel to the liver and lungs. The larva mature and develop into a hydatid cyst i.e. a cyst containing **"brood capsules"**.

Each 'Brood Capsule' contains many scolces (singular scolex). Thus a single cyst containing many brood capsules may produce thousands of scolces. The cysts may be in a deer in natural (wild) conditions and the deer may be eaten by a wolf. The soleces attach themselves in the intestine of wolf and produce proglottids.

Diagnosis of hydatid cyst: The diagnosis of hydatid cyst is often done only on **autopsy**, from Greek *autopsia* = seeing with one's own eyes *i.e.* examination of a dead body for diagnostic purposes. One will feel relieved to know that now a days the presence of hydatid cysts can be done by X-rays. The hydatid cyst formed in the brain of infected individuals has recently been seen in X-ray.

Nematodes

Q.7. What are nematodes? Which type of digestive system do they have? What structure guides sperms to female's genital pore? Which are their two broad categories? Explain.

Ans. The nematodes are placed under Phylum **Nematoda** and are worm like creatures which have two sexes and an intestinal canal. These have been called **roundworms** and are cylindical and tapered at each end. They possess a **COMPLETE DIGESTIVE SYSTEM**, having distinct mouth, an intestine and an anus. Majority of species are dioecious. **Males** are smaller than **females**. The males have one or two hardened **spicules**, the hind most part i.e. posterior in position.

The structures of the male roundworm which guides sperms to female's genital pore are the spicules located at the posterior end.

They have been categorized in different ways. According to one of the way they have been divided into two categories on the basis of their **habitat** or **ecological niche**:

(i) Those that mainly **live in the intestine** e.g. hook worms and whip worms.

(ii) Those that are mainly **tissue parasites** e.g. guinea worms and filarial worms.

Recently they have been grouped under two broad categories based on the mode of infectivity. These two categories are: (1) Those in which the **egg is infective** and (2) Those in which **larva is infective**. Some species of nematodes are free living in soil and water and others parasitise on plants and animals. The parasitic nematodes differ from the flatworms in mode of parasitism and do not have succession of larval stages represented by flat worms.

1. EGGS INEFFECTIVE FOR HUMANS

The *Enterobius vermicularis* called **pinorm** is a typical example of the category **eggs infective** in human. (Fig. 19.10). The **large intestine** is the usual habitat of the adult

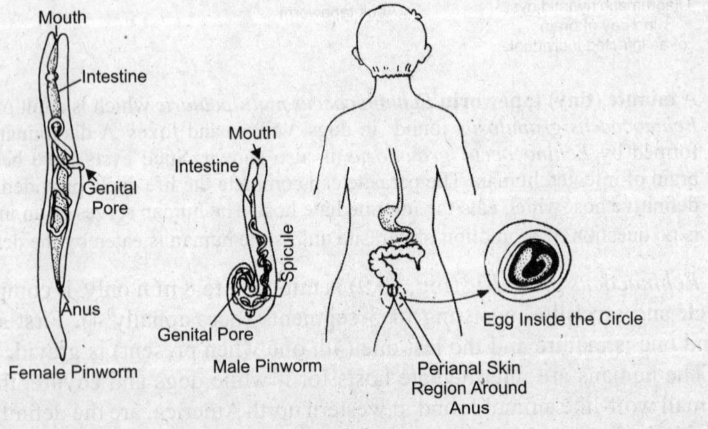

Fig. 19.10. *Enterobius vermicularis*, the pinworm which spends its complete life in large intestine of human. It exhibits **sexual dimorphism** (dioecious). Female pinworm is larger than male. The female deposits its eggs on perianal skin at night. Eggs are 55 μm long. The **spicule** in male guides sperms to female's genital pore.

pinworms. The female, 8 - 13 mm long migrates to the extreme termination of the rectum, the *anus*, to deposit her eggs on **perianal skin** (Greek *peri* = around ; Latin *anus* = anus) i.e. skin surrounding the auns. The eggs may be ingested by the host or any body else through contaminated cloths. Males are smaller 2 - 5 mm long and bear genital pore as well as spicules at posterior end.

DIAGNOSIS OF PINWORM: The infection caused by pinworm may be diagnosed using **Graham sticky - tape method** in which a piece of transparent tape is kept on the perianal skin so that the sticky side of the tape picks up eggs deposited by the female pinworm. The presence of eggs can be detected by microscopic examination of the tape.

An another nematode which infects human through eggs is *Ascaris lumbricoides* which is a large nematode 30 cm (one foot) in length. *Ascarius lumbricoides* exhibits **sexual dimorphism**. The male is smaller and has a curled tail. It lives in the intestine of human and animals as pigs and horses. It survives on semidigested food of the host. The eggs excreted with faeces of the host remain alive in the soild for a long duration by the time ingested by some other host. The eggs are incubated in the small intestine of the host, mature in the lungs and migrate to the intestine.

DIAGNOSIS: Diagnosis can be made on examining adult worms excreted along with faecal matter. The disease in humans can be prevented by living hygienically.

Larvae Infective for Humans:

The nematodes which infect humans by means of their larvae are hookworm *Nector americanus*, another worm *Trichinella spiralis* and aniksakine worms. The adult hookworm *Necator americanus* live in the small intestine of humans. *Ancyclostoma duodenale* is also called hookworm. *Nectar americanus* bear shooks (Fig. 19.11) around the mouth. These hooks are used for attaching to and removing food from the tissue of the host.

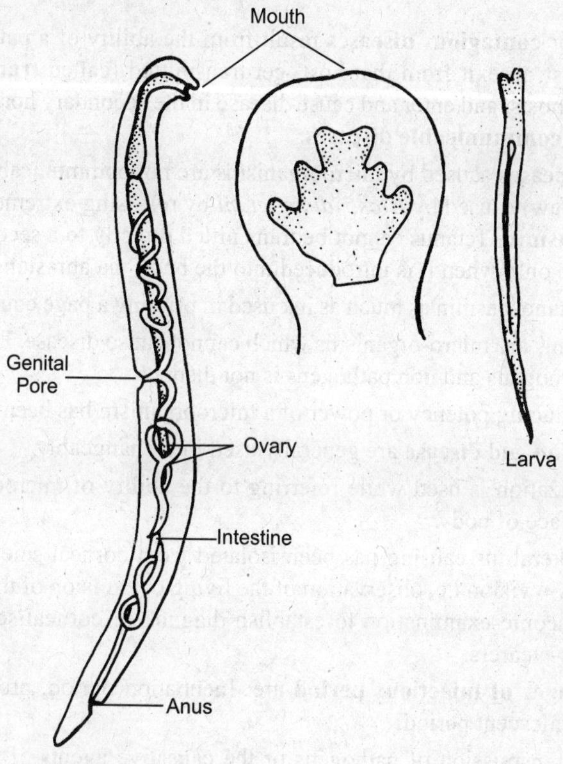

Fig. 19.11. *Necator americanus* (hookworm). The adult female showing complete digestive system with **mouth, intestine** and **anus**. Enlarged mouth with hooks around it meant for attachment. Free living larvae in the soil infect humans by penetrating the skin of definitive host (human).

The eggs are excreted in the faeces of the host. The larvae hatch in the soil and feed on bacteria. The larva find entry inside the host by penetrating the skin of the host. It enters the blood or lymph vessel which carried it to lungs. It is coughed up in sputum. The sputum is swallowed and the larva finally reaches small intestine.

DIAGNOSIS: The diagnosis can be done by observing eggs in the faeces. The infections can be prevented by wearing shoes.

20
CHAPTER

INFECTIOUS DISEASES AND DIAGNOSTIC IMMUNOLOGICAL TECHNIQUE

Salient Features

- The **infectious** or **contagious diseases** result from the ability of a pathogenic organism to multiply in a host, to exit from that host, get transmitted (called **transmission**) to one or more secondary hosts, and enter and cause disease in the secondary host. Such diseases have also been called **communicable diseases**.
- All infectious diseases caused by microorganisms are not communicable as **tetanus** (Greek *tetanus* = Lock-jaw) caused by *Clostridium tetani* by releasing extremely potent neurotoxin called **tetanospasmin**. Tetanus cannot be transmitted directly to a secondary host. *C. tetani* produces disease only when it is introduced into the body via abrasions or wounds.
- An amount of tetanospasmin as much as ink used in printing a page could kill 30 individuals.
- The non pathogens are micro-organisms which cannot cause disease. However, the dividing line between pathogens and non pathogens is not distinct.
- The disease producing potency or power of a microorganism has been called **virulence**.
- The term **infection** and **disease** are generally used interchangeably.
- The term colonization is used while referring to the ability of microorganism to establish itself on the surface of body.
- *Acanthamoeba* keratitus causing has been isolated from corneal smears or **biopsy** (Greek *bios* = life; *opsis* = vision i.e. observation of the living or excision of the tissue from a living body for microscopic examination to establish diagnosis), corneal scrappings or tissue in soft contact lens-wearers.
- The **clinical stages of infectious period** are: Incubation period, prodromal period, acute period and convalescent period.
- The means of transmission of pathogens or the causative agents of the disease is by: (1) **Contact** which may be **direct contact transmission** i.e. **person to person transmission** ; **indirect contact** i.e. **fometes** or non living objects. (2) **Vehicle transmission** i.e. through water (*Salmonella* infection), through food (tapeworm infection), through air as by droplet nuclei (as measles virus and *Mycobacterium*). Some fungal spores are also transmitted through air causing **histoplasmosis** (by *Histoplasma capsulatum*), **Coccidiomycosis** (by *Coccidiodes immitis*) and **blastomycosis** (by *Blastomyces dermatidis*). (3) **Vectors** in which insects including houseflies (spread pathogens of typhoid fever and bacillary dysentery from faeces of infected people to food) and other arthropods are counted.

- **Nosocomial infections** are hospital-acquired: caused by *Staphylococcus aureus*, *Escherichia coli*, and *Pseudomonas aeruginosa*.
- **Diagnostic immunology** can be used in diagnosis of diseases involving precipitation reactions, agglutination reactions, neutralization reactions, compliment fixation rections, radio immunoassay, immunofluorescence and fluorescent - antibody techniques; and enzyme - linked immunosorbent assay (ELISA). .
- The **Ouchterlony test** is based on precipitation reactions performed in agar gel medium, called **geldiffusion** (= immunodiffusion) procedure in which antibodies (usually in an antiserum) and antigens diffuse through a gel medium forming precipitate in the gel.
- Diagnosis of diseases can be done by a rising titer or **seroconversions** i.e. from no antibodies to the presence of antibodies.
- The **neutralization reactions** are used for elimination of harmful effects of bacterial exotoxin or virus.
- The presence of antibodies against a virus can be detected by the ability of the antibodies to avoid or prevent **cytopathic effects** (CPE) of viruses in cell culture e.g. prevention of the formation of **SYNCYTIUM** (giant cell).
- Ignatz Semmelweis (1816-1865) a Hungarian Physician working at the Obsterics Ward in the Vienna General Hospital, instructed all staff to wash their hands in **chlorinated lime**.
- Joseph Lister (1827–1912) treated surgical wounds with solutions of **phenol** (carbolic acid) and used spray of carbolic acid in surgery.
- In India more than three thousand years ago **Acharya Sushrut** used **ant-heads** for stiching surgical cuts.

Communicable Diseases or Infectious Diseases

Q.1. What is **communicable** or **contagious** or **infectious disease**? Define.

Ans. Any disease which can be transmitted from one individual to another either directly by physical contact between infected and uninfected individual e.g. many vinereal diseases, or indirectly by infection through a **fomite** (a non living object which can spread infection) like surgical instruments, bedding, toys and eating utensils, or **vectors** like (carriers of disease causing organisms) like arthropods – insects, ticks, mites, fleas or vertebrates – dogs, cats, bats and rats.

Q.2. Which are two main types of infections? Explain.

Ans. The infections are divided into two main types: Acute and chronic.

(1) The **acute disease** lasts for a short time and is severe. The symptoms usually last as long as the pathogen and its products are in the body of the patient.

(2) **Chronic disease** lasts for a long period of time. The chronic disease persists because the pathogen is adapted to defend itself against host defence mechanisms or because the host defences cannot respond normally. In either situation the microbes are not killed rather persist for a long time.

Q.3. Which are the **clinical stages** of infection of an acute **infectious disease**?

Ans. These are: Incubation period, prodromal period, acute period and convalescent period.

1. **INCUBATION**: this period starts after microbe makes a contact with the host and tries to establish a parastic relationship with the host which may last from two days to several month because the infecting microbe or the pathogen must rise to certain

population level before the symptoms become obvious. The virulent bacteria like *Streptococcus pneumoniae* have incubation period of one to three days. However, fungal diseases have more prolonged incubation period e.g. *Coccidiodes immitis* which causes fungal **coccidiomycosis** may take four weeks. The **lentiviruses** (*lenti* - = slow) are viruses causing disease slowly. Human Immunodeficiency Virus (HIV) which belongs to the group of viruses called lentiviruses causes disease very slowly. It may take from a months to a year to develop the full blown AIDS. Rabies caused by Rhabdovirus exhibit incubation period from 2 to 6 weeks to one year.

2. **PRODERMAL PERIOD:** The term originaed from Greek *prodromos* = running before, e.g. preceding as the transistory rash before the true rash of an infectious disease. This is the period in which patient feels the beginning of disease and it may take the form of **malaise** – a feeling of illness and discomfort. The patient may feel nausea head-ache or upset stomach. The microbes begin to increase in number and cause enough harm leading to production of symptoms of disease.
3. **ACUTE PERIOD:** This period is the culmination or peak point of the disease and specimen collections may be made. The patient shows all the typical symptoms of the disease as fever, chills, rash, nausea, vomiting, diarrhoea and others more characteristics of a *particular* infection. In many cases the acute period marks the time when the disease may be transferred directly or indirectly to other susceptible persons in the population. this is referred to as **communicable period**. However, in some diseases, the communicable period begins in the incubation period and extends through the acute phase as in diphtheria (from Greek diptheria = leather, membrane) caused by *Corynebacterium diphteriae* characterised by a grey, adherent, false membrane growing on a mucous surface usually that of upper respiratory tract.
4. **CONVALESCENCE:** This period occurs if the patient has not experienced enough damage to cause death. This period is also called the recovery period.

Infectious Disease Cycle, Reservoir and Carriers

Q.4. What do you mean by **infectious disease cycle**?

Ans. The infectious disease cycle is based on characteristics of the pathogen and is the chain of infection which support reproduction and dissemination of pathogen as given below:

Pathogen ⟶ Source of Pathogen ⟶ Transmission to Host ⟶ Suspectibility of the host ⟶ Exit of Pathogen from the Host

The first link or the connection in the infectious disease is the pathogen. The second link is the reservoir or source of pathogen. The 3rd link is the transmission to the host, the 4th one is the susceptibility of the host and the 5th or the last one is the exit from the host (Fig. 20.1).

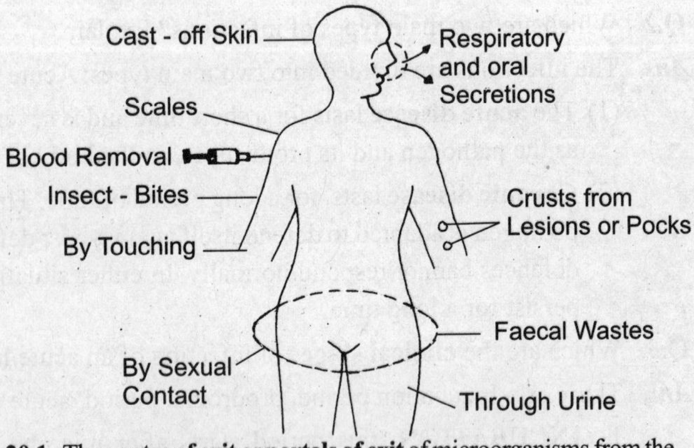

Fig. 20.1. The **routes of exit** or **portals of exit** of microorganisms from the body. The microorganisms are often aerosolized by vocalization (speaking and singing) sneezing, coughing, released by expectoration, kissing, drool and drivel, and runny nose.

Reservoir or the Source of Disease Causing Pathogen

Q.5. What do you mean by **reservoir** or source of a disease causing pathogen?

Ans. A source of the pathogen is the second link in the infectious disease cycle. A reservoir or source is the location from the pathogen is immediately transmitted to the host either directly through the environment or indirectly through an intermediate agent. The source may be either animate as humans and animals or inanimate as water, soil and food. The time in which the source is infectious or is disseminating the pathogen is referred to as **period of infectivity**. Manytimes human hosts are most important animate sources of the pathogen and have been called carriers. A carrier is an infected individual who is a potential source of infection. The carriers play an important role in the **epidemiology** (from Greek *epi* = on; *demos* = people ; *logos* = discourse), the scientific study of the occurrence and distribution of diseases. The main types of carriers are:

(i) **Active Carrier**: Individual having an **overt** or **symptomatic clinical case** of the disease. The term symptomatic has been derived from symptom [Greek *symptoma* = anything that happens] i.e. a subjective phenomenon or manifestation or apparent clarity of disease.

(ii) **Covalescent Carrier**: An individual who has recovered from the infectious disease but continues to port or shelter (harbour) large number of pathogen.

(iii) **Healthy Carrier**: The individual who shelters or harbours pathogen but is not sick.

(iv) **Incubatory Carrier**: An individual who is incubating the pathogen in large number but is not ill inspite of that.

The convalescent healthy and incubatory carriers shelter or harbour the pathogen for a short period only say hours, days or weeks and thus have been called casual, acute or **transient carriers**. In case they harbour or shelter the pathogen for long periods say months, years or life they are referred to be **chronic carriers**.

The infectious diseases found in animals are called **zoonoses** can be transmitted to humans. Thus these animals also serve as reservoirs. The humans may come in contact of these animals in several ways.

The disease tularemia caused by *Francisella tularensis* a Gram negative aerobic bacterium may be caught by human on coming in contact with flesh of animals. *Pasteurella multocida* causes **haemorrhagic septicaemia** of cattle, fowl cholera and pneumonia in **farm animal**. It also causes **Pasteurellosis** in fowl, cattle, sheep, swine, goats, mice, rats, rabbits and may be passed to human by biting of these animals.

The tuberculosis caused by the *Mycobacterium bovis* "complex" (*Mycobacterium tuberculosis* is said to be human pathogen) and brucellosis are passed to humans through cow's milk. The anthrax and trichinosis are passed to humans by eating insufficiently cooked infected meat. A number of diseases are transmitted by biting of arthropods working as vectors.

Transmission of Pathogens

Q.6. What is **transmission of pathogens**? Describe the important routes of transmission of pathogens.

Ans. The transmission is the act of mailing or sending of pathogen from one host to another or from the source to the host and is, the 3rd link in the 'infectious disease cycle'. The important routes of transmission are:

(i) **Airborne transmission:** The pathogen may be in the **droplet nuclei** which are small particles 1 to 4 μm in diameter which are created as a result of evaporation of larger particles of the size of 10 μm or more in diameter known as droplets. The droplet nuclei remain viable in the air for hours or days and are taken away to long distances. The humans and animals which serve as source may give out pathogens from the respiratory tract into the air by coughing, sneezing or vocalization. A large number of moisture droplets come out as an aerosol in a single sneeze and travel long distances in the air.

(ii) **Contact transmission**: The contact of the source of pathogen or reservoir and host may be direct or indirect. Among the direct contacts person to person contact by touching, kissing or sexual contact in case of the sexually transmitted diseases (STDs). **Direct contact** may also be through oral secretions or body lesions as herpes and boils, through nursing mothers (staphylococcal infections) and through the placenta for AIDS and syphilis.

Indirect contact may be through intermediate in animals or objects, thermometers, eating utensils, drinking cups, bedding and toys.

(iii) **Vehicle transmission**: The nonliving or inanimate objects that cause transmission of pathogen are known as vehicles. These spread the pathogen but do not support or back the reproduction of pathogen, e.g. surgical instruments, bedding and eating utensils. These utensils have been called **fomites** (singular fome or fomite) in epidemiology..

(iv) **Vector - borne transmission:** The living agents involved in transmission of pathogens are known as **vectors**. Generally the vectors which transmit pathogens are arthropods – insects, ticks, mites, fleas or vertebrates – dogs, cats, bats and rats. The vector-borne transmission may be either external or internal. In **external transmission** the pathogen is carried over the body surface of the vector and the vector keeps the process passive without adding to the growth of the pathogen, e.g. the Gram negative bacterium *Shigella* intestinal pathogen of man is carried by flies on their feet from a faecal source to the food plate.

The pathogen carried in the **internal transmission** remains inside the vector. The vector either serves as a portal or harbourage or go into the biological transmission phase, e.g. *Yersinia pestis* the causal organism of plague carried by the rat flea to human without permitting the pathogen to undergo any morphological or physiological change.

It should be kept in mind that in **biological transmission** the pathogen undergoes morphological and physiological change inside the vector as the *Plasmodium* a genus of protozoa in which reproduction occurs in mosquito and leads to formation of large number of fusiform or needle shaped uninucleate cells called **sporozoites** which are infective to man and cause malaria.

Q.7. What makes the host susceptible to pathogen in an infectious disease cycle?

Ans. The 'host' is the 4h link in an infectious disease cycle. The susceptibility of the host to pathogen depends on capacity of the pathogen to cause disease called **pathogenicity**; and non specific and specific defence mechanisms of the host.

Q.8. What is the Fifth and last link in an infectious disease? How do pathogen leave the host? Explain.

Ans. The release or exit of the pathogen in an infectious disease cycle is the 5th or last link. The escape or exit of pathogen from the host is as much important as the coming of pathogen in contact with the host and making entry inside the host.

The disease cycle will break and the pathogenic species will not everlast if the pathogen does not escape or exit successfully out of the host. The escape, however, may be active or passive, rather a combination of both. The microorganisms usually adopt the passive escape.

When the pathogen actively moves to a **portal** – the position or gateway to exit and leaves the host it is called **active escape** ; e.g. parasitic helminths causing helminthiasis (from

Greek *helmins* = worm, *iasis* = condition) migrate through the body of their hosts consequently reaching the surface to exit.

When a pathogen or its progeny leaves the host in faeces, urine, droplets, saliva or **desquamated cells** (from Latin *desquamare* = to scale of *i.e.* shedding or flaking off), it has been called **passive escape**.

Routes of Exit or Portals of Exit

The routes of exit of pathogens have also been called portals of exit (Fig. 20.1). The respiratory and gastrointestinal tracts are common portals of exit. The pathogens of diseases like tuberculosis, whooping cough, pneumonia, scarlet fever, meningococcal meningitis, measles, mumps, small pox and influenza are discharged through respiratory route. The pathogens responsible for causation of salmonellosis, cholera, typhoid fever, shigellosis, amoebic dysentery and poliomyelitis are released through gastrointestinal tract by discharge of faecal matter. Some pathogens are also released via saliva (e.g. rabies virus) The sexually transmitted diseases occur in secretion from the penis and vagina which thus serves as portals of exit.

The urine also may have pathogens associated with typhoid fever and brucellosis. The drainage from wounds and infected fomite also spread the diseases.

The yellow fever, rocky mountain spotted fever, tularemia (epidemic disease of rodents e.g. deerfly fever or rabbit fever caused by *Francisella* (=*Pasteurella*) *tularensis* and malaria are transmitted by insects. Some diseases as AIDS and hepatitis B are transmitted by contaiminated syringe-needles.

Nosocomial Infection or Hospital – Acquired Infection

Q.9. What are nosocomial or hospital-acquired infections? Which are the pathogens often involved in nosocomial infections? Also give the infection conrol policies which should be followed by an Hospital Epidermiologist (who practices epidemiology).

Ans. According to American Public Health Association (APHA) nosocomial or hospital-acquired infection is "An infection originating in a medical facility, e.g. occurring in a hospitalized patient in whom it was not present or incubating at the time of admission or is the residual of an infection acquired during a previos admission. Includes infections acquired in the hospital but appearing after discharge; it also includes infections among staff.

The term **nosocomial** has been coined from *nosos* = disease, and komeion = to take çare of. It is a necessity that period of hospitalization be reduced as much as possible. Approximately 3 to 5% of all patients admitted to hospitals annually develop nosocomial infections which either lengthen their confinement or result in additional treatment.

The four pathogens most commonly found causing infections are: *Escherichia coli, Staphylococcus aureus*

Streptococcus **(Group D) and** ***Pseudomonas.***

The infection control policies that are generally recommended are:

(i) Isolation of patients who are compromised and those who have infectious diseases.
(ii) Adoption of aseptic techniques.
(iii) Employee health programme.
(iv) In-service training to help reduce infections.
(v) Catheterization, intravenous procedures and other closed system care.
(vi) Review for appropriate antibiotic therapy for antimicrobial drug resistance of infectious agents.

(vii) To provide clean environment.
(viii) Research in infection control.
(ix) Evaluation of disinfectants, rapid test systems and other products.
(x) To encourage appropriate legislation related to infection control at the state level.
(xi) Maintenance of log of incidents related to infections and communicable diseases.

Q.10. What do you mean by a **compromised patient** or **host**?

Ans. Latin *compromissum* means mutual agreement. Thus an individual with decreased resistance is called a **compromised host**. Many infections encountered in medical practices today result from low virulence microorganisms which are normally found on the tissue of the host (normal flora) but are able to cause infection due to compromised conditions of a person's defence mechanism.

Diagnosis and Control of Communicable or Infectious Diseases

Q.11. What is **diagnosis**? Describe the common diagnostic techniques adopted to detect communicable or infectious diseases.

Ans. Diagnosis is the identification of diseases. The term has been coined from Greek *dia* = through; *gnosis* = recognizing. The diagnosis is art of distinguishing one disease from another and arriving at a correct decision between diseases presenting a similar clinical picture. There are many microbiological techniques used in diagnosis of infectious diseases.

1. CYATODIAGNOSIS: Greek *kytos* = cell ; *dia* - through ; and *gnosis* = recognizing. This is the diagnosis by microscopic study of cells.
2. DIAGNOSTIC IMMUNOLOGY: There are a number of diagnostic techniques as under:

(*i*) **Precipitation reactions**: These are as a result of reaction between soluble antigens with antibodies as IgG and IgM leading to formation of **lattices** due to interlocking of molecular aggregates.

The precipitation reactions takes place in two stages. In the first stage small antigen – antibody complexes are formed rapidly, within seconds. 2nd stage is slow that may take a few minutes to hours in which antigen antibody complexes form lattices which precipitate from solution. Of course, these precipitation reactions happen when the ratio of antigen and antibody is optimal. No visible precipitates are formed when either of the component is in excess. The optimal ratio can be produced when separate solutions of antigen and antibody are placed adjacent to each other and permitted to diffuse together. The precipitation curve is shown in Fig. 20.2. The maximum amount of precipitate is formed in the zone of equivalence where the ratio is optimal.

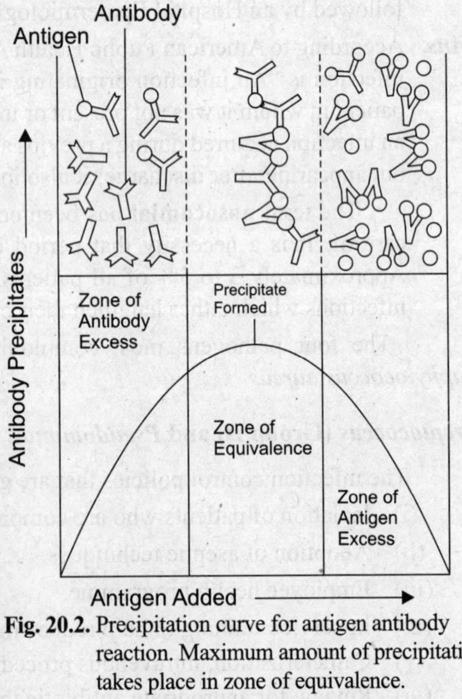

Fig. 20.2. Precipitation curve for antigen antibody reaction. Maximum amount of precipitation takes place in zone of equivalence.

A precipitation test is performed in a capillary tube in which a cloudy line of precipitation appears in the ring - shape which has been referred to as **precepitin ring test** (Fig. 20.3). This ring appears in the area in which the optimal ratio has been reached i.e. the **zone of equivalence**.

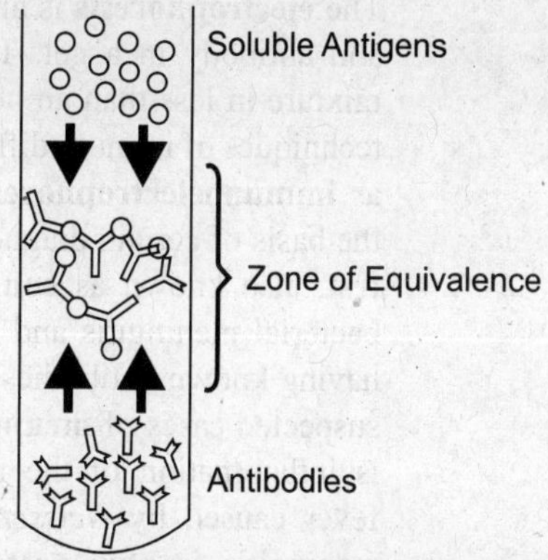

Fig. 20.3. Diagramatic **precipitin ring test.** The antigens and antibodies get diffused in equal proportions in zone of equivalence, leading to formation of a visible ring of precipitate.

There are immuno diffusion tests for precipitation reactions carried out in agar gel medium plated in a Petriplate or on a slide. One of types of such a test is **Ouchterlony test** or **Ouchterlony immunodiffusion precipitation test**. The wells are cut into purified agar gel in Petriplates. The serum containing antibodies i.e. antiserum is added to one well and soluble test antigens are added to each surrounding well (Fig. 20.4). The line of visible precipitate developes between the wells at the point where optimal antigen antibody ratio is reached. This test is employed to know whether the antigens are identicle, partially identicle or totally different.

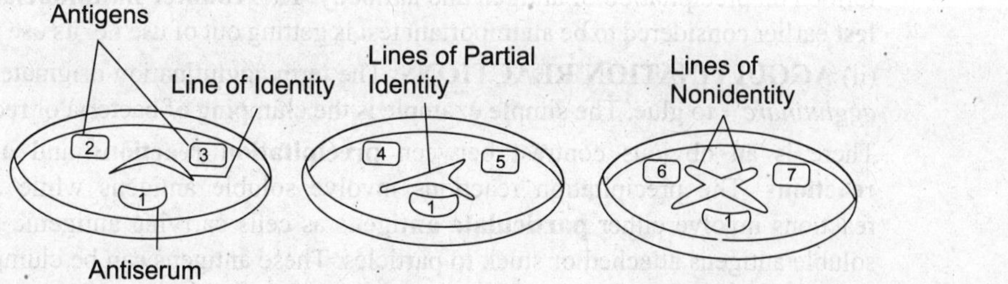

Fig. 20.4. Three Petriplates showing Ouchterlony immunodiffusion test used to calculate the degree of relatedness among antigens.

In the 1st Petriplate neither antigens in wells 2 and 3 nor antibodies in well numbered 1 are able to diffuse past each other as they react and precipitate to form the line (inverted boat shaped). Thus, antigens 2 and 3 are identical.

In the 2nd Petriplate, the wells numbered 4 and 5 are not identical but they share many antigenic determinant sites and are thus related. The spur shape of the line indicates that some of the antigens did not react with antibodies and diffused through the precipitation zone.

In 3rd Petriplate, the antiserum has antibodies against antigens in wells numbered 6 and 7. The antigens are unrelated as they diffused across each other's zones of precipitation.

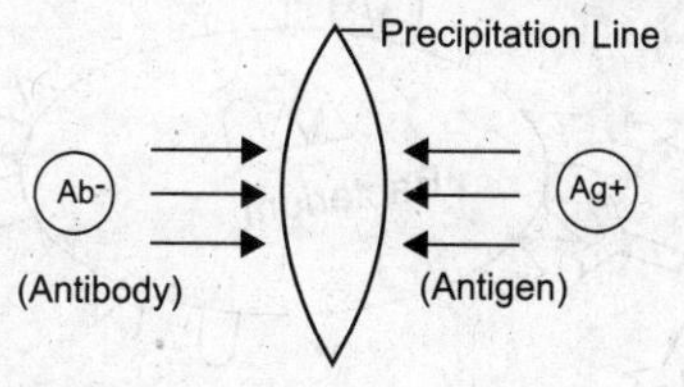

Fig. 20.5. Countercurrent immunoelectrophoresis, in which antigens (Ag+) and antobodies (Ab–) are placed in opposite wells and pH of surrounding medium are adjusted in such a way that antigen and antibdody have opposite charges. When current is passed antigen and antibodies move in opposite directions. On meeting the antigen and antibody form **precipitation line.**

The **electrophoresis** is an other type of test used to speed up the movement of antigen and antibody in a gel. This procedure can also be exploited to separate the protein mixture in less than an hour. This method has further been modified by combining the techniques of immunodiffusion and electrophoresis which process is collectively known as **immunoelectrophoresis** and is used to separate proteins in human serum and is the basis of certain diagnostic tests, as **countercurrent immunoelectrophoresis (CIE) test**, also known as **counter-immunoelectrophoresis** which is used in diagnosis of bacterial meningitis and some other diseases, (Fig. 20.5). This test is used for serum having known antibodies to identify an unknown antigen in certain body fluid, e.g. for suspected cases of **meningitis** (Greek *menigx* = membrane ; *itis* = inflammation which is inflammation of the meninges, epidermic form of which is called cerebrospinal fever caused by *Neisseria meningitidis* (meningococcus) The terms meningococcal meningitis is now preferred. There are also two more term – **leptomeningitis** (Greek *leptees* = thin ; *meningx* = membrane ; *itis* = inflammation) i.e. inflammation of the inner covering of membranes of brain and spinal cord, and **pachymeningitis** (Greek *pachy* = thick, *menigx* = membrane ; *itis* = inflammation) of dura matter. On passing electric current in the said test antigen and antibodies move towards the pole with the electric charge opposite to the one they carry. They pass through each other line. A line is formed on precipitation of antigen and antibody. The **counter immunoelectrophoresis test** earlier considered to be an important test is getting out of use i.e. its use is on decline.

(ii) **AGGULTINATION REACTIONS**: The term agglutination originated from Latin *agglutinare* = to glue. The simple example is the clumping of bacteria or red blood cells.

There is an obvious contrast between **precipitation reactions** and **agglutination reactions**. The precipitation reactions involve soluble antigens while agglutination reactions involve either **particulate antigens** as cells carrying antigenic molecules or soluble antigens attached or stuck to particles. These antigens can be clumped or linked by antibodies in the form of visible aggregates which can be seen with naked eye as well as under microscope and the reaction is known as agglutination (Fig. 20.6). The IgM is known as most efficient immunoglobulin for agglutination but IgG also takes part in agglutination reactions.

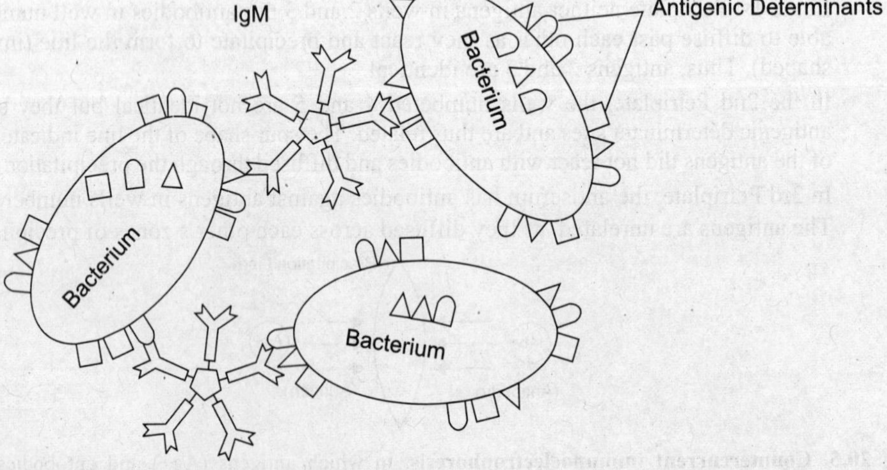

Fig. 20.6. Demonstration of agglutination reaction. The particulate antigens (here bacterial cells) agglutinate with IgM known as most efficient immunoglobulin for agglutination.

(a) DIRECT AGGLUTINATED TEST: Direct agglutination tests are used for testing brucellosis [Latin *Brucella* named after Dr. David Bruce ; Greek *osis* = condition] caused by one of species of *Brucella*. This test is also used to separate isolates of *Salmonella* into serovars. The amount of particulate antigen in each well of microtiter plate is same but the amount of serum containing antibodies is diluted in a way that each successive well has half the antibodies of the preceeding or the well means the well which comes before in order (Fig. 20.7).

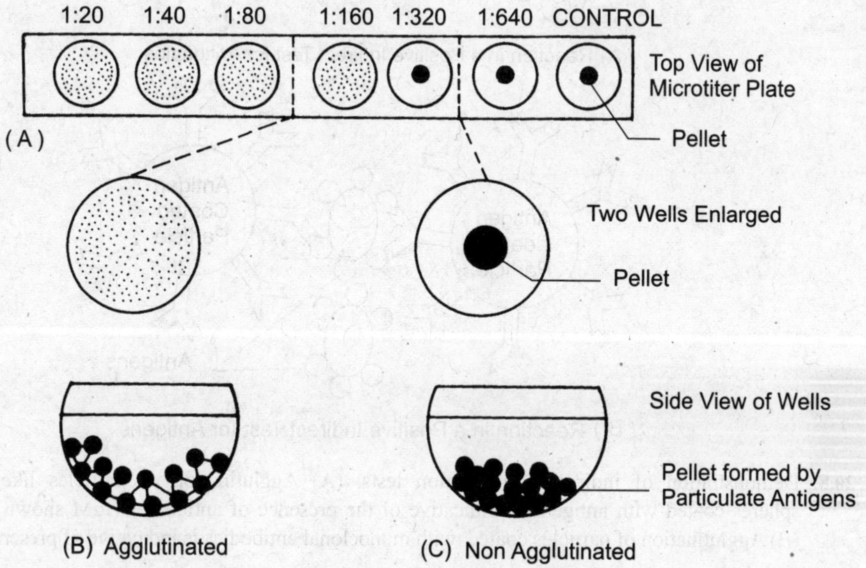

Fig. 20.7. Direct agglutination test for measurement of antibody titer which is the amount of antibody in the serum. (A) Each well of microtitler plate from left to right has the concentration just half the concentration of serum kept in the preceding well i.e. well coming before in order. (B) **Positive reaction** due to agglutination. (C) **Negative reaction** characterized by formation of **pellet**. The well with 1: 160 concentration is the most diluted concentration causing positive reaction or agglutination. Thus antibody titer is 160.

The dilution at which the agglutination takes place is the measure of **titler** – the concentration of serum antibody

In a **positive reaction** causing agglutination enough antibodies are present in the serum to link the antigen – antibody complexes on the bottom of the well. In a **negative reaction** it does not cause agglutination because of shortage of antibodies which are needed to link the antigens and consequently the particulate antigens as red blood cells roll down the sloping sides of the well and form a **pellet** at the bottom of the well. There may be rise of titer as the disease progresses. This change is known as seroconversion which is change in person's response to an antigen in a serological test and is of diagnostic importance. Such situations often come across while testing for infections of HIV.

(b) INDIRECT (PASSIVE) AGGLUTINATION TESTS: These are also called **latex agglutination tests** which are used for rapid detection of serum antibodies against many bacterial and viral diseases because of use of latex spheres. The antigens are absorbed or coated onto particles as red blood cells, bentonite clay, or most often latex spheres (Fig. 20.8). These tests are often employed for streptococci which cause sore throat. The diagnosis can be completed very rapidly say just in ten minutes short time.

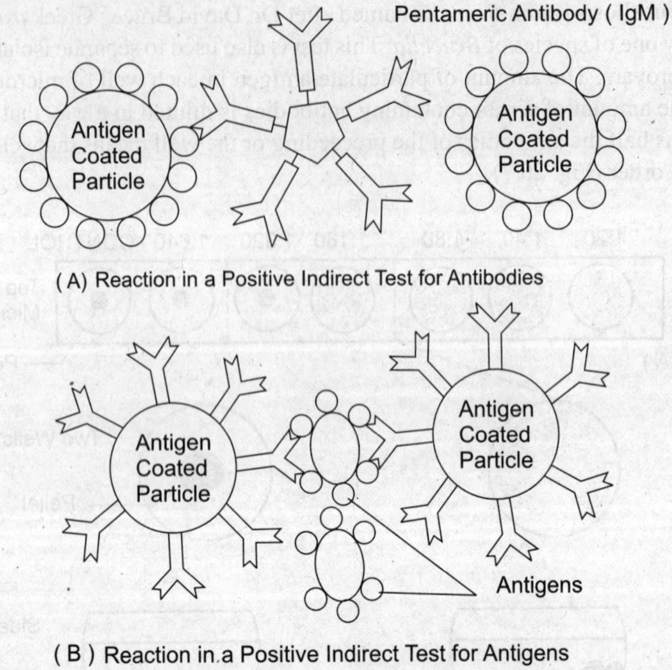

Fig. 20.8. Demonstration of indirect agglutination tests. (A) Agglutination of particles like minute latex spheres coated with antigens is indicative of the presence of antibodies (IgM shown in the figure). (B) Agglutination of particles coated with monoclonal antibodies is indicative of presence of antigens.

The antibodies as immunoglobulin M (IgM) react with the soluble antigens coated onto the latex particles and result agglutination. The same type of particles can be used in reverse (by using particles coated with antibodies) to detect the antigens against which they have specfcity.

Haemagglutination

The **haemagglutination** is the reaction in which clumping of ***erythrocytes*** (Greek *erythros* = red, *kytos* = cell) the red blood corpuscles or cells takes place. This reaction takes place when red blood cell surface antigens and their complimentary antibodies come in contact and get linked. The viruses which are causative agents of mumps, measles and influenza bear the ability to agglutinate red blood cells (Fig. 20.9) No doubt that this type of haemagglutination can be prevented by antibodies which can neutralize agglutinating virus.

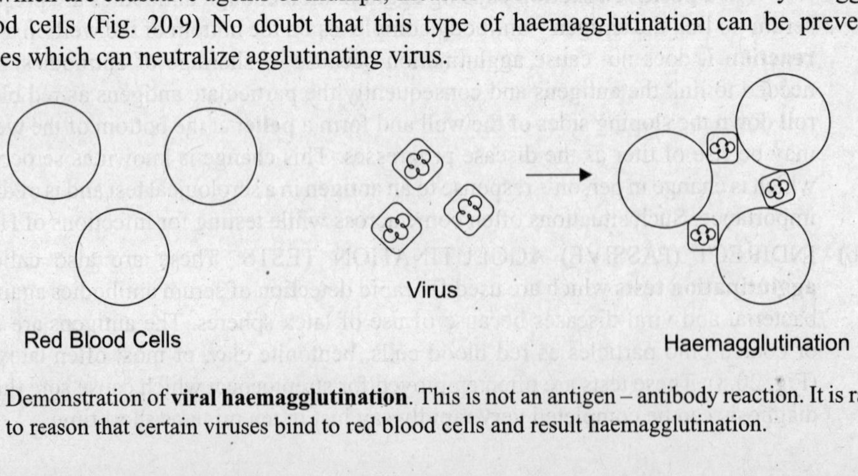

Fig. 20.9. Demonstration of **viral haemagglutination**. This is not an antigen – antibody reaction. It is rather due to reason that certain viruses bind to red blood cells and result haemagglutination.

Neutralization Reactions

Neutralization tests involve living microorganisms exposed to antibodies. These are antigen antibody reactions. The immune serum neutralize the toxic substances e.g. produced by *Corynebacterium diphtheriae*. The neutralization tests are also used in diagnosis of viral infections. One of the most often used neutralization test is viral haemagglutination inhibition test (Fig. 20.10) which is used in diagnosis of influenza, measles, mumps and number of other infections caused by viruses. The harmful effect of a bacterial exotoxin or virus is blocked by a specific antibody in neutralization reactions.

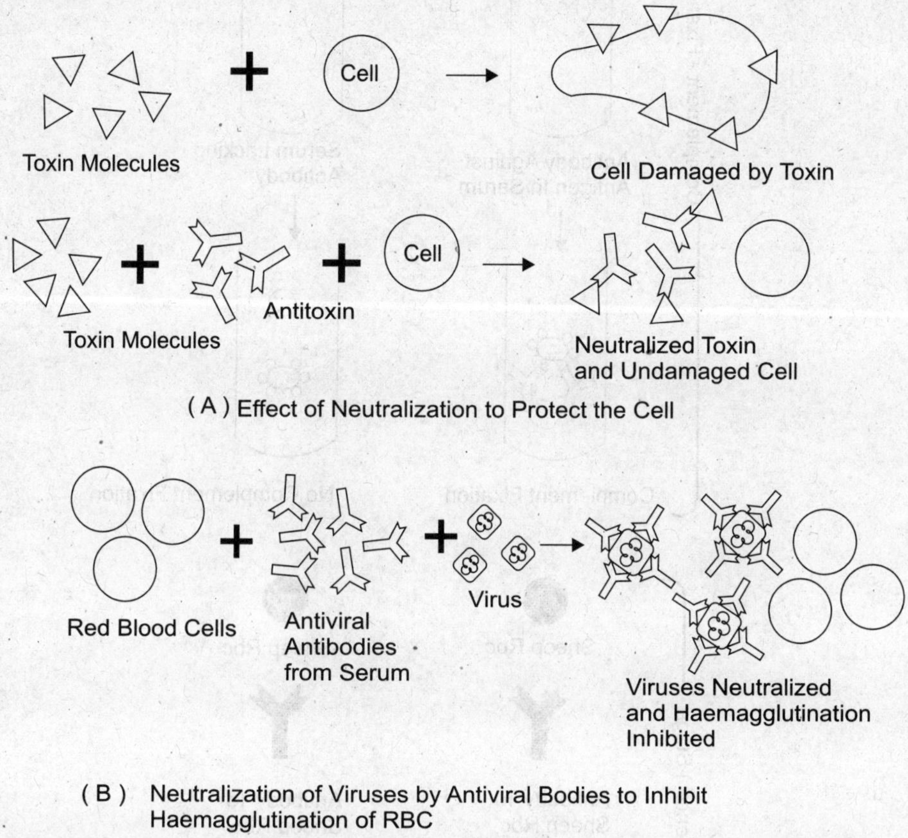

Fig. 20.10. Demonstration of reactions in **neutralization tests**. (A) Toxins damaged the susceptible cell and were protected by neutralization of toxin by **antitoxin** or antibodies to toxin. (B) Viral haemagglutination inhibition test.

Compliment – Fixation Reactions

The principle behind this reaction lies that antigen and antibody are are allowed to interact in presence of compliment used up. A group of serum proteins are collectively called **complement**. The complement binds to antigen antibody complex and is used up *i.e.* gets fixed (Fig. 20.11).

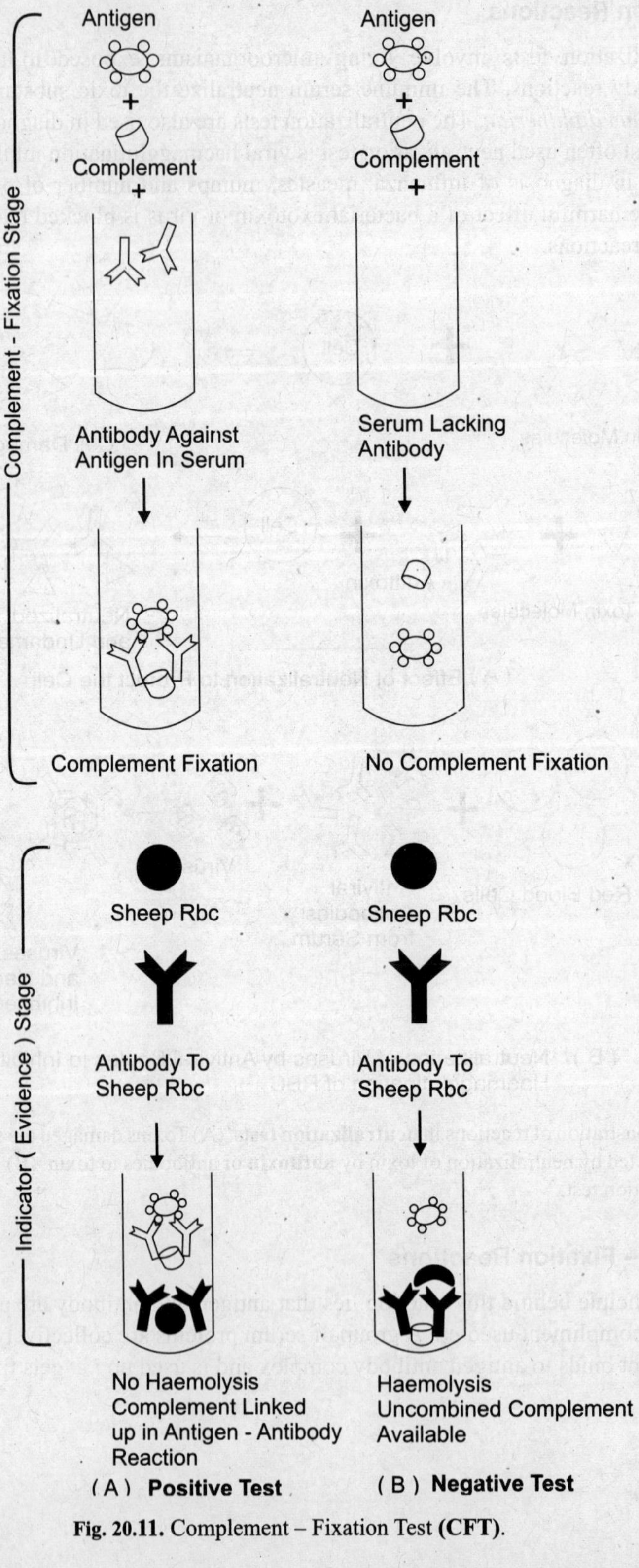

Fig. 20.11. Complement – Fixation Test **(CFT)**.

The complement fixation also called **Wassermann reaction** (also spelt as Wasserman reaction) is a **standard test for syphilis** in which the antibody (**reagin – the 'Wassermann antibody'**) undergo a complement fixing reaction with the **cardiolipin – lecithin – cholesterol antigen**. Complement fixation has also been used in diagnosis of certain viral, fungal and rickettsial diseases. To conclude this test is used to indicate the presence of antibodies to a known antigen. The complement will combine or get fixed with an antibody which is reacting with the antigen. It may be noted that if all the complement is fixed in the compliment fixation stage then none will remain to cause haemolysis (hemolysis) of red blood cells (RBS) is the indicator stage. In the tube shown on leftside (Fig. 20.10A) all ready (available) complement is tied up (fixed up) by the antigen – antibody rection and **no haemolysis** is seen. (Greek *haima* = blood ; *lysis* = a loosening) *i.e.* disintegration of RBC and liberation of contained haemoglobin do not take place. Thus the **test is positive** for the presence of antibodies. In the tube on the right side (Fig. 20.10B) the complement remains, the RBC are lysed in the indicator or evidence stage and haemoglobin contained in RBC is liberated, thereby the **test is negative**.

Radioimmunoassay

This method is used in detection of hormones, drugs and immunoglobulins. This is also used for diagnosis of several types of viral hepatitis e.g. to determine the amount of hepatitis B virus in a clinical sample, (Fig. 20.12). The sample to be assayed is combined with radioactive labelled virus and the antibodies to the virus. The radioactive virus and the virus in the sample will compete for antibody. The antigen – antibody complex can be isolated and analyzed for their radioactivity. Suppose, when antigen antibody complexes show high level of radioactive antigen, then the serum sample does not have much antigen i.e. the virus to combine with the antibody.

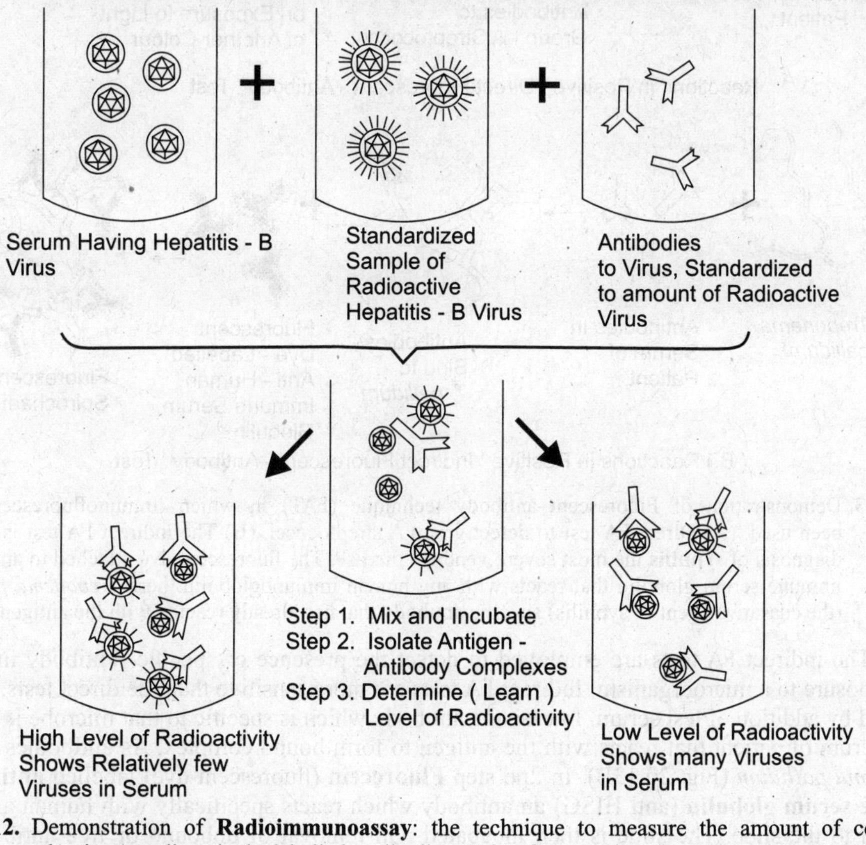

Fig. 20.12. Demonstration of **Radioimmunoassay**: the technique to measure the amount of compound adoptinmg radioactively labelled antibodies. It is often employed to detect hepatitis B virus in clinical samples.

Immunofluorescence and Fluorescent – Antibody Technique

The **fluorescent antibody techniques** (FAT) can be used to identify microorganisms in clinical specimens and can detect presence of specific antibody in serum. The fluorescent dyes as **fluorescein isothiocyanate** (FITC) are used in this technique along with antibodies to enable them fluoresce on exposure to ultraviolet light. These methods are rapid, sensitive and very specific to work. The fluorescent – antibody test for **rabies** [Latin madness] can be performed within few hours, with nearly cent per cent accuracy. The fluorescent – antibody tests also called **fluorescent – antibody technique** is the process in which immune fluorescence (Fig. 20.12) is employed for clinical lab diagnosis for counting or detecting specific microorganisms in water samples. These are of two types: (1) Direct and (2) Indirect.

The Direct FA tests (Fig. 20.13A) are done to identify microorganisms as Group A streptococci from patient's throat. The specimen having the antigen to be identified is fixed on a slide. Fluorescein – labelled antibodies are added and the slide is incubated. In the 2nd step slide is washed to remove any free antibody not bound or linked to antigen. On examining the slide under fluorescence microscope it emits **yellow-green fluorescence**.

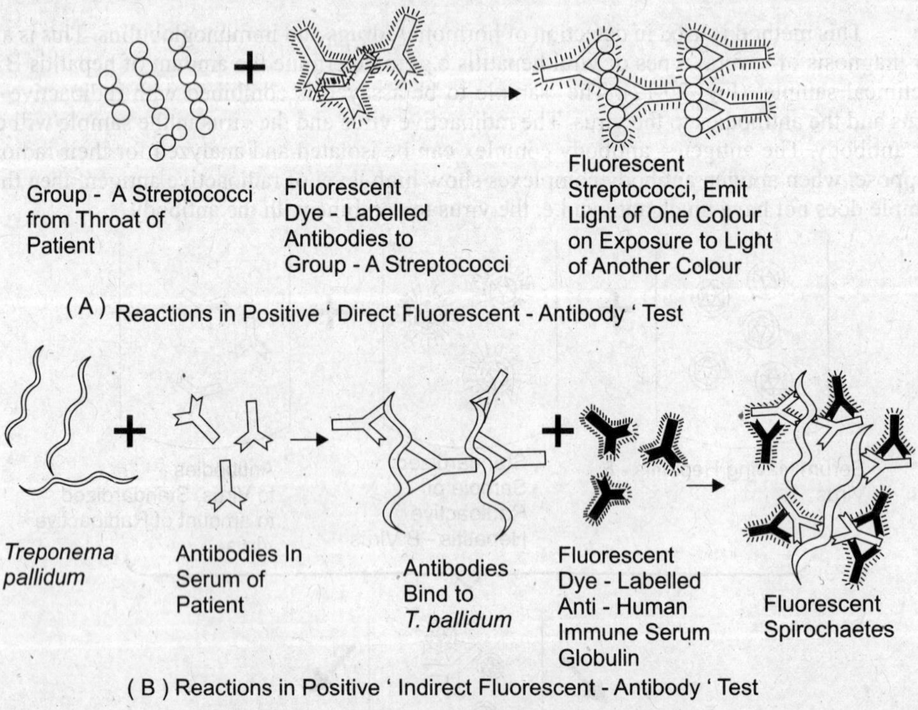

Fig. 20.13. Demonstration of Fluorescent–antibody technique (FAT) in which immunofluorescence has been used. (A) Direct FA test to detect group A streptococci. (B) The indirect FA test is done in diagnosis of **syphilis** the most severe venereal disease. The fluorescent dye attached to antihuman immune serum globulin that reacts with any human immunoglobulin, here *Treponema pallidum* (the cuasative agent of syphilis) specific antibody that has already reacted with the antigen is seen.

The indirect FA tests are employed to detect the presence of specific antibody in serum after exposure to a microorganism. Indirect FA tests are more sensitive than the direct tests. This is followed by addition of test serum. In case the antibody which is specific to that microbe is present in the serum of patient that reacts with the antigen to form bound complex. As antibodies bind to *Treponema pallidum* (Fig. 20.13B). In 2nd step **Fluorecein** (fluorescent dye) labelled **antihuman immune serum globulin** (anti HISG) an antibody which reacts specifically with human antibody is added to the slide. The slide is then incubated. On removal of unbound or free antibody the slide is seen under fluorescence microscope. Suppose, if known antigen fixed to the slide gives fluorescence, it is evident that antibody specific to test antigen is present.

Fluorescence – Activated Cell Sorter (FACS)

The **fluorescence – activated cell sorter** (FACS) is an adaptation or modification of **flow cytometer** which **sorts** (differentiates) cells besides separating and counting them. The flow cytometer, however, is an instrument which is capable to separate and count cells; and the detection is done by the presence of fluorescent tag on the surface of cell. The **fluorescence detector** identifies fluorescent cells by wavelength of fluorescene light emitted by the cell.

The **T-cells** or T-lymphocytes are characterized by antigenically specific receptors on their surface as $\mathbf{CD_4}$ and $\mathbf{CD_8}$ (CD stands for cluster of differentiation). As stated above FACS is modification of flow cytometer in which the suspension of cells **evacuates** (leaves) the **nozzle** in the form of **droplets** having not more than one cell each. The formation of such droplets is invited or encouraged by rapid vibration or **palpitation** of the nozzle (Fig. 20.14).

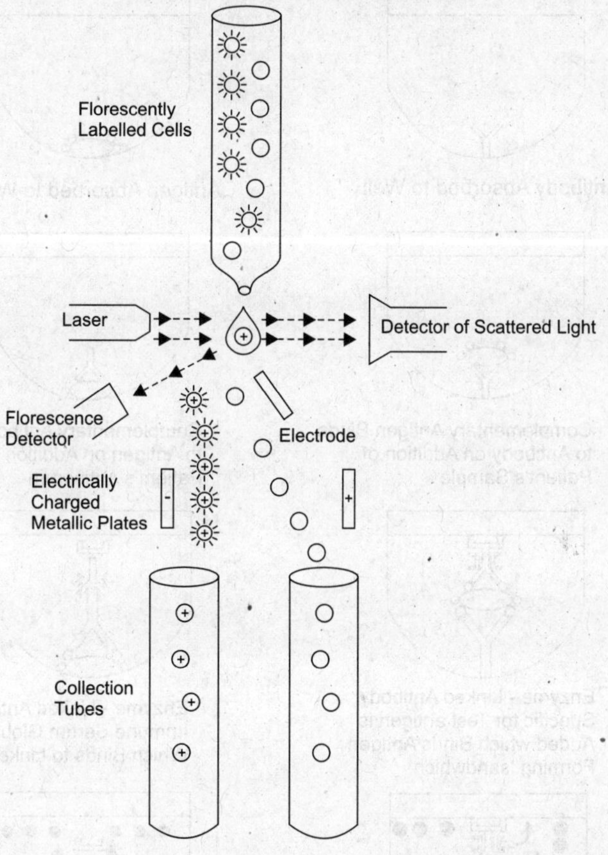

Fig. 20.14. Diagram of **fluorescence – activated cell sorter** (FACS).

As soon as the laser beam touch or hit the droplet containing a cell the beam is received by a **detector** (fluorescence detector) which measures certain characteristics like size. When the cells are laden with fluorescent antibody markers the detector of scattered light can measure the fluorescence to identify them as CD_4 or CD_8 T-cells. When a flow cytometer is provided with such a detector it becomes a fluorescence – activated cell sorter or FACS. When a cell of preselected size or fluorescence is detected by **laser beam**, the cell is provided either a positive or negative charge. The charged droplets pass between **electrically charged plates** and are collected into the receiving or **collection tubes** meant for the collection of either of plus charged cells or minus charged cells respectively. Thus millions of cells can be separated by this technique for further use for experimental purposes.

Enzyme-Likned Immunosorbent Assay or Elisa

It does not require radioactive materials like RIA test rather is a highly sensitive immunoassay for **specific antibodies** or **antigens**. Thus the enzyme - linked immunosorbent assay (ELISA) comprise a group of serological tests which use enzyme reactions as **indicators**, and is easy to perform for a technician and provide either **positive** or **negative result**. ELISA is mainly of two types: (1) Direct ELISA and (2) Indirect ELISA. Both the tests are generally performed in a microtiter plate with numerous **shallow** i.e. depthless or less deep wells may be 96 in number or less even. There may be variation in the method as the reagents can be bound onto tiny latex particles rather than on surface of microtiter plates.

The **direct ELISA** is employed for detection of antigens while indirect ELISA for detection of antibodies (Fig. 20.15).

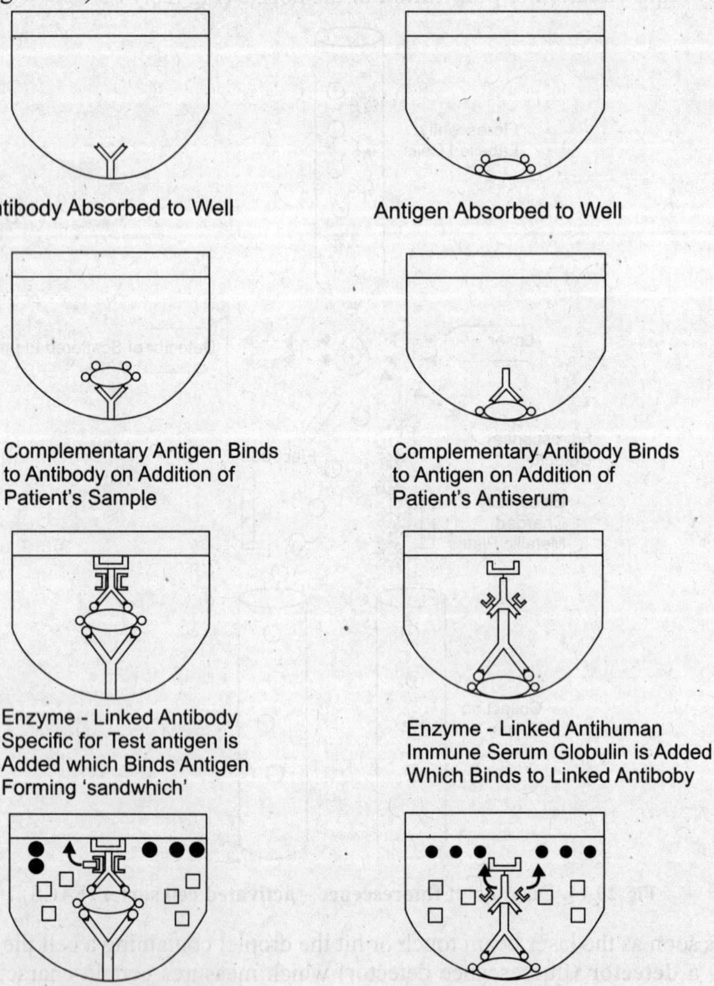

Fig. 20.15. Demonstration of **enzyme - linked immunosorbent assay** (ELISA) or **enzyme immuno assay** (EIA).

DIRECT ELISA: It may be described in 4 steps: In the 1st step, the antibody against the antigen to be studied for detection is adsorbed to the surface of the wells of mircotiter plate represented in the 1st well of the left column (Fig. 20.15). In the 2nd step the sample of a patient having undetermined antigen is then added to each well. In case the antigen interacts specifically with the antibodies absorbed to the well, then the antigen will be retained there when the well is washed free of unbound antigen. In the 3rd step second antibody sepcific for the antigen is added. If the both antibodies one absorbed to the wall of the well and the antibody known to be specific for the antigen have reacted with the antigen, a **"sandwich"**is formed with the antigen lying between two antibody molecules.

The conspicuity or prominenece of this reaction is just due to linkage or binding of second added antibody to an enzyme e.g. horseradish peroxidase or alkaline phosphatase. The free or unbound enzyme-linked body is washed from the well. This is followed by the 4th step in which enzyme's substrate is added to it. The colour change is obvious and can be seen easily or can be read by means of a special spectrophotometer. This colour change (Fig. 20.15) is shown in the lowermost well of the left column. The test is referred to be positive if the antigen has reacted with antibodies adorbed to the wall of the well in the first step. The test is said to be negative if the free or ubound antigen gets removed out of the well on washing.

INDIRECT ELISA: The first step is the addition of known antigen (instead of antibody as done in case of direct ELISA (Fig. 20.15) which is shown in the first well of right side column in the diagram. In the 2nd step patient's antiserum is added. The complementary antibody binds or gets reacted to antigen. In the 3rd step enzyme-linked **antihuman immuneserumglobulin (Anti HISG) is added which binds to the bound antibody.** All unreacted antiserum is washed from the well.

The anti-human immune serum globulin (anti - HISG) which is the antibody bearing capability to react with any human immunoglobulin (Ig) is then permitted to react or bind the antigen – antibody complex. The anti - HISG that already has been joined or bound to an enzyme reacts with the antibodies which are already bound to antigens in the well of microtiter plate.

In 4th or the final step unbound (free) anti - HISG is washed away and the appropriate substrate for the enzyme is added. The wells in which bound antigen has **reverberated or reacted** with the antibody in the serum sample exhibit coloured enzymatic reaction which can be seen easily. This technique bear resemblance to indirect fluorescent – antibody test.

What is your **mentation** (reasoning) to differentiate between IFAT and Indirect ELISA? Indirect assay or indirect ELISA employ linkage with an enzyme while in indirect fluorescent – antibody test a fluorescent dye is used (i.e. fluorecein labelled ani-human serum globulin or anti-HISG is used).

The ELISA find application to investigate or test the antibodies against HIV. Moreover, many ELISA – like tests are used in clinical laboratories automated for having printed results on computer. Some ELISA based testst have also been availed to public as **home pregnancy test** by using monoclonal antibodies. The home pregnancy test detects a hormone known as human chorionic gonadotropin (hCG) which is excreted only in the urine of a pregnant woman.

Q.12. Which are the two types of monoclonal antibodies in a **home pregnancy testing**? Also give its working system.

Ans. There are two types of monoclonal antibodies in the test area of the dipstick: (1) **Free monoclonal antibodies**, and (2) **Capture monoclonal antibodies** which are bound to the substrate in the test windows. The free antibodies are specific to hCG and are colour labelled. The capture antibodies are specific for hCG-free antibody complex. The dipstick on coming in contact with the urine having hCG, makes the free antibodies bind the hCG and are then bound capture antibodies forming a "sandwich". Consequently colour labelled antibodies undergo a visible colour change in the test windows (Fig. 20.16). Thus the hormone known as human chorionic gonadotropin (hCG) which is exerted in the urine of pregnant woman only gets detected. Human chorionic gonadotropin (hCG) is a hormone produced by viable placental tissue during pregnancy and is excreted in urine approximately 20 days after last menstrual period. The level of hCG rise rapidly reaching peak levels after 60-80days and then hCG levels falls suddenly and eventually 'plateau out'. The hCG molecule consists of two combined dissimilar subunits namely aplha and beta.

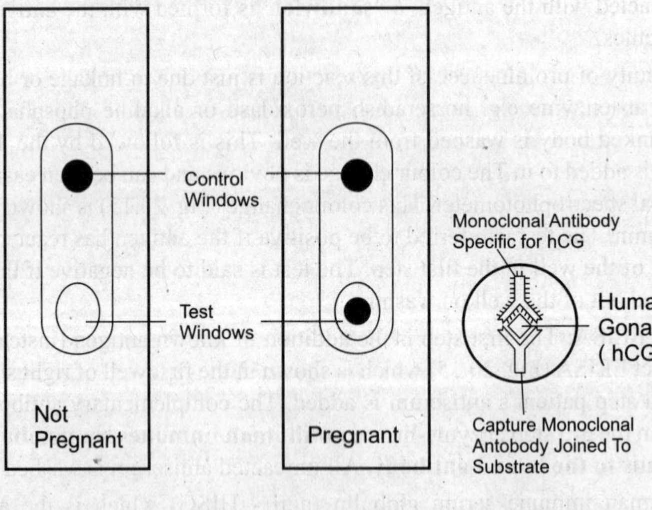

Fig. 20.16. Diagram of **home pregnancy** test using monoclonal antibodies. This test is done to characterize the presence of a hormone known as **human chorionic gonadotropin** abbreviated as **hCG** which is excreted in the urine of only a pregnant woman. The colour-labelled antibodies bring a visible colour change in the test windows while other bodies make the colour change in the control window. In this way the presence of hCG is detected in the urine of a pregnant woman.

Q.13. How can the **infectious** or **communicable diseases be controlled** ? Explain.

Ans. The measures to control infectious or communicable diseases varies from hospital to hospital. These are more or less same as the control policies for nosocomial infections described earlier. The important means to control disease are:

(i) Isolation of rooms and wards.

(ii) The tubs and other nonliving articles used should be disinfected.

(iii) The bandages and **intubations** *i.e.* insertion of tubes into organs such as trachea should be of single-use disposable or sterelized prior to use.

(iv) The antibiotics can be prescribed by the physician to improve the patient's resistance to infection.

(v) The immuno-suppressive drugs be given to the patient, the least.

(vi) The hospital should have an infection control committee.

21 CHAPTER

BACTERIAL PATHOGENS AND DISEASES

Salient Features

- The pyogenic (Greek *pyon* = pus; *genesis* = production) bacteria are pus forming cocci which include staphylococci, streptococci and members of genus *Neisseria*.
- Gram positive spore forming bacterial genera are *Bacillus* and *Clostridium*.
- The **Zoonosis** (Greek *zoon* = animal; *oxis* = condition), the diseases of animals which on occasions are transferred to human. The Zoonotic infections have been controlled by antibiotics.
- The clostridia which cause diseases such as gangrene, tetanus and botulism are also under control.
- The enteric (Greek *enteron* = intestine)
- Gram negative bacilli are subdivided into coliforms) *Proteus*, *Salmonella*, *Vibrio* and *Pseudomonas*.
- The Coliforms are bacteia which resemble *Escherichia* are divided into two groups: (1) *Escherichia*, *Enterobacter*, *Klebsiella* and *Serratia*; and (2) *Arigona*, *Edwardsiella* and *Citrobacter*.
- The *prophylaxis* (Greek *prophytaxis* = a guarding) is the prevention, from occurring diseases e.g. the use of antimicrobials to prevent occurrence of infection. The antibiotic rifampicin (rifampin) has been regarded a drug of choice in prophylaxis against infection by *Haemophilus influenzae*.
- The **empirical therapy** is the treatment given on the basis of experience and not as a result of *susceptibility testing* as measuring the diameter of zone of bacterial growth inhibition by placing the paper disc on the inoculated surface.
- The fluids which filters out of blood capillaries into spaces between tissues is known as *interstitial fluid* which is situated in the interstices on spaces of a part distributed through the connective structures.
- Interstitial fluid enters lymph capillaries called lymph (Latin *lympha* = water; *i.e.* fluid contained in the lymphatic vessels).
- **Lymphatics** (lymph containing) return lymph to blood.
- **Lymph nodes** contain fixed macrophages, β-cells and T-cells.
- The growth of microorganism in blood is known as **septicaemia**.
- The signs of septicaemia include **lymphangitis** (Latin *lympha* = water; Greek *aggeion* = vessel, *its* = inflammation *i.e.*, inflammation of lymph vessels).

- *Sterptococcus pyogenes* is the most frequent cause of **puerperal sepsis** i.e. inflection of genitlal tract happening within 21 days of abortion or child birth. However, puerperal sepsis is now not common because of modern hygienic techniques and antibiotics.
- The **strep throat** on **streptococcal pharangitis** is caused by Streptococus pyogenes a Gram positive bacterium of the **group A β haemolytic streptococci**.
- The 0.01 mg of the endotoxin produced by lysogenized cells of *Corynobacterium diptheriae* causative agent of **diptheria** is enough to kill a 91 kg (200 pound) weight person.
- The **pertussis** on whooping cough a disease of lower respiratory system is caused by *Bordetella pertussis*.
- The bacilli of *Mycobacterium tuberculosis* the causal organism of **tuberculosis** remains protected from defences of the body in **macrophages** inside their **phagosomes** and **phagolysosomes**.
- **Phagolysosome** is the vacuole formed in phagocytes in the body of the host by fusion of enzyme containing **lysosome** with **phagocytic (digestive) vacuole**.

Treatment

Q.1. How should a treatment for bacterial disease be designed? Explain.

Ans. A treatment be designed in a way that the chance of cure is maximal and the chance of developing a resistant microbial mutant is minimal. Therefore, it becomes imperative to know which chemotherapeutic agents are effective against the disease causing microorganism. The most common procedure of determining antibiotic sensitivity is to place an antibiotic impregnated disc on Mueller-Hinton agar surface which has been, 'inoculated with a film of test bacterium'.

A zone of inhibition of bacterial growth? Can be seen around the disc containing the effective antibiotics. This type of testing demands upto 24 hours. However, with critically ill patient it may be desirable to initiate treatment before conducting such tests.

This sort of **empirical therapy** is generally used under clinical situations where the time may be critical to the patient thus it becomes a need to have logical choice to give antibiotics based on previous experience.

PARENTERAL ADMINISTRATION OF MEDICINE : Many medicines cannot be given orally by mouth as they taste bad, are destroyed by stomach acid or are not absorbed from the intestine. These compounds are given parenterally by injection or by means of catheter. The antibiotics **aminoglycosides** comprise a class of braod spectrum antibiotics. The class is generally taken to and used include amikacin, apramycin, neomycin, kanamycin, netilmicin, gentamicin and tobramycin which are **bacteriostatic** that prevents the growth of bacteria but does not directly kill these microorganisms.

Because of the fact that aminoglycosides cannot be given orally. Thus the patient requires to be hospitalized for parenteral administration in order to prevent excessive accumulation of antibiotics, in patients receiving aminoglycosides. It becomes necessary to **monitor** (watch) the concentration of compound present-in body fluids. Such determinations are known as **antibiotic assays,** that are generally performed on the blood of patients for the presence of level of these or other toxic antibiotics. To conclude it is possible to monitor the amount of antibiotic given to the patient and maintain levels below **toxic concentration** *i.e.* the degree of being poisonous.

Prophylaxis

Q.2. What is **prophylaxis** ? Explain.

Ans. Greek **prophylaxis** = a guarding. The prophylaxis is the measures taken or treatment given usually disinfection, immunological or chemotherapeutic designed to protect an individual from a future attack by the pathogen. A term equivalent to prophylatic used in plant pathology is the 'protectant' for any chemical agent which prevents the occurrence of particular disease among plants treated with the agent.

Shibasaburo Kitasate and Email Vol Behring, assistants of Robert Koch published a joint paper on their results for **antitoxins** (antibodies) for **tetanus** and then von Behring on **diptheria**, in their efforts in developing an understanding of immune response. Another student in Koch's laboratory Paul Ehrilich also worked on developing an understanding of immune response. Enrlich in 1891 published a paper in which he differentiated active immunity from passive immunity. According to Ehrlich, the **active immunity** occurs when one's own body develops antibodies as a result of prior interaction with pathogens or their products and **passive immunity** occurs when one receives antibodies from another person or animal. The development of active immunity provides long term protection against disease while the passive immunity as the one transferred from mother to foetus (fetus) provides only temporary **prophylaxis**, the protection against infectious microorganisms.

Human Disease Causing Bacteria

Q.3. What is the criterion for grouping of the **bacterial disease**? Explain

Ans. Several criteria have been adopted for grouping or categorizing bacterial diseases. The bacterial disease most often have been categorized on the basis of body part infected by the bacterial pathogens as:

(i) Bacterial diseases of skin and eyes.
(ii) Bacterial disease of nervous system.
(iii) Bacterial diseases of cardiovascular and lymphatic system.
(iv) Bacterial disease of respiratory system. (Diseases of upper respiratory system and lower respiratory system).
(v) Bacterial diseases of digestive system.
(vi) Bacterial disease of urinary and reproductive system.

The bacterial diseases have also been categorized or grouped on the basis of source of derivation (capture) of the pathogen leading to causation and transmission of disease. This pattern has been adopted in this text. Thus the sources from which the diseases are generally caused, are : (1) Air-borne diseases, (2) Anthropod-bore disease, (3) Direct contact diseases and (4) Food and water-borne diseases.

Air-Borne Diseases or Air-Generated Diseases

Q.4. Describe the Common air-borne bacterial diseases giving aetiologic (also spelt etilogic) agent, symptoms, prophylaxis (prevention), therapy and laboratory diagnosis.

Ans. The most common air-borne diseases (diseases caught from air) caused by bacterial pathogens are: (A) Bacterial diseases of upper respiratory system, (B) Bacterial disease of lower respiratory system, and (c) Skin diseases.

The important air-borne diseases are described below:

Bacterial Diseases of Upper Respiratory System

The important bacterial diseases of upper respiratory system are named and described below:

Streptococcal Pharyngitis or Strep Throat or Sore Throat

The aetiologic agent of sore throat is Streptococcus pyogenes. It is a Gram positive bacterium of group A β – haemolytic streptococci. The streptococci are generally called by abbreviated name strep. The pathogenic characteristics of different serotypes are because they produce: (i) Extracellular enzymes which break down the molecules of host. (ii) They produce streptokinases or plasminogen activator (activates profibrinolysis). Plasminogen is a precursor of plasmin. The activation of host blood factor by streptokinases dissolved blood clots. It is interesting that fibrinolytic effect of the streptokinase has been used with thrombolytic therapy to speed up removal of intravascular fibrin. (iii) The cytolysins - streptolysin O and streptolysin S causes death of host leucocytes, and capsules enable the bacterium to escape phagocytosis. This bacterium of course, is also known to cause some skin and soft tissue infections as impetigo, erysipelas and acute bacterial endocarditis. It should be noted that pharangitis is also caused by viruses.

SYMPTOMS: Local infection and fever are common. Generally tonsilitis occurs and the lymphnodes in the neck become enlarged and tender.

The common complications may be Otitis media (Greek ous = ear, itis = inflammation, That means inflammation of middle ear cavity.

PROPHYLAXIS: There is no vaccine but prevention and control measures are to avoid contract-with objects contaminated with nasal secretions and not to take unpasterized milk.

THERAPY: The penicillin is the drug of choice. The failure of treatment with penicillin is due to production of β-lactamases by other members of microbiota of throat. Treatment of streptococcal pharyngitis is required to prevent complications like rheumatic fever (an autoimmune disease associated with haemolytic streptococci in the body. It is called rheumatic fever because two common symptoms are fever and pain in joints like that of rheumatism), and glomerulonephritis (from Latin glomus = a ball, Greek nephros = kidney, itis = inflammation i.e. bilateral non-suppyrative inflammation of glomeruli of the kidneys).

LABORATORY DIAGNOSIS: (i) Earlier the laboratory diagnosis was done by throat swab. (ii) Currently it is done using indirect agglutination diagnostic tests using microscopic latex particles coated with anti-bodies against group A streptococci. Such tests can be performed just in 10 minutes. However, it does not give surity because Streptococcus pyogenes can be detected in the throats of some asymptomatic (symptomless) carriers of this bacterium. (iii) A rise in I_g^M anti-body titer is the best diagnosic feature or indicator of sore throat of genuine streptococcal origin but a negative result be checked by culturing a throat swab.

The other streptococcal and other bacterial diseases of upper respiratory tract are as under:

Scarlet Fever or Scarlatina

Aetiologic Agent: It is caused by *Streptococcus pyogenes* strain carrying lysogenic bacteriophage i.e. genetic information of bacteriophage (bacterial virus) is incorporated into the chromosome of the bacterium which consequently codes for production of an *erythrogenic* (rash inducing) *toxin*, which causes shedding of skin.

Symptoms: The scarlatina occurs mainly in children. It is characterized by high fever and appearance of pink-red skin rash due to hypersensitivity reaction of skin because of toxins circulating in the circulatory system. The tongue has a spotted straberry-like appearance as it loses its upper membrane and becomes very red and enlarged. As the disease advances in its course, the affected skin gets peeled off giving look of sun burned skin which resembles scalded (as an injury caused by moist heat) skin syndrome due to *Staphylococcus aureus.*

Prophylaxis: The scarlet fever is a communicable disease. The patients should be kept isolated to avoid inhalation of infective droplets from an infected person particularly in school children.

Therapy: Penicillin is effective in treating original streptococcal infection.

Diagnosis: By observing clinical symptoms of typical red skin rash and red tongue. Indirect agglutination diagnostic tests using microscopic latex particles coated with antibodies and rise in I_g^M titer may be of additional help.

Diptheria

Aetiologic Agent: The diptheria is caused by *Corynebacterium diptheriae* a Gram positive non-endospore forming rod. The bacterium is pleomorphic frequently clubbed, from Greek *corynes* = clubbed.

Symptoms: The term diptheria originated from Greek *diphthera* = leather or membrane. The disease is characterized by grey, adherent false membrane growing on mucous surface usually that of **upper respiratory tract**. The greyish membrance contains fibrin, dead tissue and bacterial cells which completely block the passage of air to the lungs. Though the bacteria themselves do not invade the tissue but the cells which have been lysogenized by bacteriophage (carrying tax gene) can produce a powerful endotoxin. The toxin circulating in the bloodstream interferes with protein synthesis. It is interesting that 0.01 mg of this highly virulent toxin is enough to kill a 91 kg (200 pound) weight bearing person. The organs as heart or kidney are affected by the toxin and the disease can turn fatal. The disease begins with a sore throat and fever followed by general malaise and swelling of the neck.

Prophylaxis: Immunization with **DPT Vaccine** which is a combined vaccine used to provide active immunity containing diphtheria and tetanus toxoids and killed *Bordetella pertursis* cells, which provides protection. Thus **DPT** stands for diphtheria pertussis – tetanus.

Therapy: The antibiotics penicillin and erythromycin controls the growth of bacteria but the antibiotics do not neutralize the diptheria toxin. Therefore, the antibiotics along with antitoxins are used to treat diptheria.

Diagnosis: The laboratory diagnosis by identification of bacteria is difficult. It is done using several types of selective and differential media. The identification becomes complicated because both toxin producing and non toxin producing strains occur in the same patient.

Cutaneous Diphtheria

Aetiologic Agent: *Corynebacterium diphtheriae* causes infection through a wound or skin lesion and there is a minimal circulation of the toxin.

Prophylaxis: Immunization.

Therapy: Antibiotics.

Diagnosis: Slow healing ulcerations covered by a grey membrane.

OTitis Media (Inflammation of Middle ear Cavity) The term otitis originated from Greek *ous* = ear; *itis* = inflammation.

Aetiologic Agents. *Streptococus pneumoniae*

Hemophilus influenzae, Moraxella (=Branhamella*)*

Catarrhalis, (10 to 15%), *Streptococcus pyogenes* 8 to 10%. and *Staphylococcus aureus* (1% to 2%).

Therapy: Broad spectrum antibiotics such as penicillins like amoxicillin are the first choice for children under the age of 8 years, and penicillin in case of older patients.

Symptoms: The inflecting microorganisms cause the formation of pus which build up pressure against the eardrum and make it inflammed and painful. More common in early childhood because the auditory tube connecting the middle ear to throat is small.

Prophylaxis: Avoid bath in swimming pools of contaminated water.

Diagnosis: Culture pus from ear.

Bacterial Diseases of the Lower Respiratory System

The bacterial diseases of lower respiratory system are pertussis, tuberculosis and many types of pneumonia caused by bacteria. Some uncommon diseases like psittacosis (Ornithosis) and Q fever also come under this category.

Pertussis (Whooping Cough)

Aetiologic Agent: *Bordetella pertussis* is the causative agent of pertussis. The causative bacterium is non-motile Gram negative coccobacillus. The virulent strains form the capsule. The bacterium is an obligate aerobe and produces an **endotoxin** as part of its cell wall. The bacterium also produces **exotoxin** in the cytoplasm which is released along with **endotoxin**. When cell of bacterium undergoes autolysis (self destruction of plasma membrane in diseases) and death. The bacteria do not attack tissues but it attaches the cillia in the trachea and interfere with their action leading to accumulation of mucous. Consequently, **tracheal cytotoxin** (a toxic fragment of bacterial cell wall) causes loss of ciliated cells. The pertussis or whooping cough is disease of children with attacks of coughing which reach peak of violence ending in an inspiratory whoop.

1. The initial stage is known as **catarrh** (nasal catarrh or cold), has also been called **catarrhal stage** resembling a common cold. 2. Prolonged sieges of coughing characterise the **second paroxysmal** (coming on in attacks or paroxysms). The name pertussis is based on Latin *per* = thoroughly and *tussis* = cough. The mucous gets stored when ciliary action gets compromised and the infected individual desperately attempts to cough up the mucous that got stored. The violence of the coughing in small children may cause broken ribs. The coughing episodes may be repeated many times a day and last for one to six weeks. 3. The third stage is convalescence stage that may last for months together.

Symptoms: A childhood disease with catarrh (Greek *katarrheein* = to flow down) a sysmptom of common cold i.e. inflammation of mucous membrane with constant flow of mucous. The **catarrhal stage** culminating into paroxysmal stage with whooping sound in its characteristics.

Prophylaxis: The vaccination with heat inactivated whole bacteria with **DPT vaccine** where P in DPT stands for pertussis. The vaccination is generally recommended at about two months of age when the titer of mother's antibodies in the baby start falling (declining) and child's own antibodies have not taken their place. The **DPT vaccine** is a combined vaccine meant to provide active immunity and comprises of diptheria and tetanus toxoids and killed *Bordetella pertussis* cells. "Acellular pertussis" (aP) vaccines a new generation of vaccines came by 1990 and two "aP" vaccines have been licenced in USA.

Therapy: The antibiotics although do not provide rapid improvement, non-the-less render the patient noninfectious.

Diagnosis: Besides clinical symptoms the diagnosis of pertussis can be done by: (1) Morphology and staining of *Bordetella* which show ovoid Gram negative cocco-bacilli mostly uniform in size but longer bacillary forms appear particularly on **repeated cultures**. Capsules may be seen in freshly isolated cultures. In film form cultures loos clumps of bacilli with clear spaces in between, give the so called **"thumb-print"** appearance. (2) Primary cultivation requires an **enriched medium** containing catalase and a substance e.g. albumin or charcoal which absorbs toxic products possibly unsaturated fatty acids on **Bordet-Gengou** medium which is potato-glycerol - agar medium with a high concentration of blood (33 per cent). The medium is made more selective by addition of penicillin (0.25 units per ml) and diamidine M and B 938. Typical whitish highly refractile colonies appear after 2-3 days incubation. The colonies have been compared to small mercury drops or bisected pearls which are 0.5 to 1.0 mm in diameter, smooth raised and cohesive and may be picked off entire for film preparation or slide agglutination. It is interesting that **subsequent cultures** can be obtained on non-blood containing nutrient agar to which charcoal or starch has been added.

(3) The biochemical and biological activities are that *B. pertussis* do not show significant fermentative activities. It produces heat labile pertussis toxin, endotoxin (lipopolysaccharide) histamine sensitivity factor and lymphocytosis factor. All of these or some may be associated with the pathogenesis of the bacterium. (4) Antigenic characters are due to haemagglutinin and protective antigen.

Q.5. Which are three major **VIRULENCE FACTORS** of *B. pertussis*?

Ans. The three major virulence factor of *Bordetella pertussis* are: (1) Pertussis toxin (2) Filamentous haemagglutinin and (3) Lipopolysaccharide LPS) endotoxin in the cell wall of the bacterium.

Pertussis toxin and filamentous haemagglutinin are proteins. Pertussis toxin is responsible for most of the systemic pathological changes characteristic of whooping cough. It is made of 5 different polypeptides joined to form two functional subunits. One of the subunits binds the toxin to the host tissue. The second subunit inactivates eucaryotic cell proteins involved in regulating, cellular metabolism. Filamentous haemagglutinin is a large fibrous protein on the surface of *B. pertussis* which mediates the attachment of *B. pertussis* to cillia of epithelial cells lining the respiratory tract. On getting attached to cillia *B. pertussis* immobilizes them and this proceeds to destroy cilliated epithelial cells. Since there are no cilia to sweep mucous away the only alternative to clear the respiratory tract is intense coughing.

Tuberculosis

Q.6. Describe the **aetiology** (eliology), therapy (treatment), prophylaxis and laboratory diagnosis of tuberculosis.

Ans. **Aetiologic Agent:** Greek *aitia* = cause, *logos* = discourse. The aetiology is the science dealing with causation of the disease. The term aetiology is also spelt as etiology. The term tuberculosis has been coined from Latin *tuberculum* = small swelling, and Greek *osis* = condition. The tuberculosis is an infectious disease of **lower respiratory tract** caused by *Mycobacterium tuberculosis* an acid fast, slender road which is an obligate aerobe (Latin *obligatus* = bound, means no alternative i.e. strictly require oxygen to maintain life). These are slow growing rods with a generation time (doubling time) of 20 hours or more. They sometimes form filaments and show the tendency to grow in clumps. Their growth appears mould-like (Fig. 21.1) fungus-like characteristic of the bacterium has been used in nomenclature of the genus *Mycobacterium* from Greek *Mykes* = fungus, bacterium = small rod. Under microscope it can also be seen in shape of "serpentine cord" (Fig. 21.2).

Since *Mycobacterium tuberculosis* cause the disease it is also called a **pathogen** and the origin and development of the disease is known as **pathogenesis** (Greek *pathos* = disease; *genes* = producing).

Staining of the Bacterium: The *Mycobacterium tuberculosis* is stained with Ziehl-Neelson or Kinyoun technique being resistant to normal staining technique. In this method the cells are stained with brilliant red stain '**carbolfuchsin**'. The specimen is flooded with carbolfuchsin and heated gently for 5 minutes. It is safe to heat by steaming.

Fig. 21.1. Mould-like or fungus-like growth of *Mycobacterium tuberculosis* is generally common in liquid growth media. The organism has been given the generic name because of this fungus-like growth.

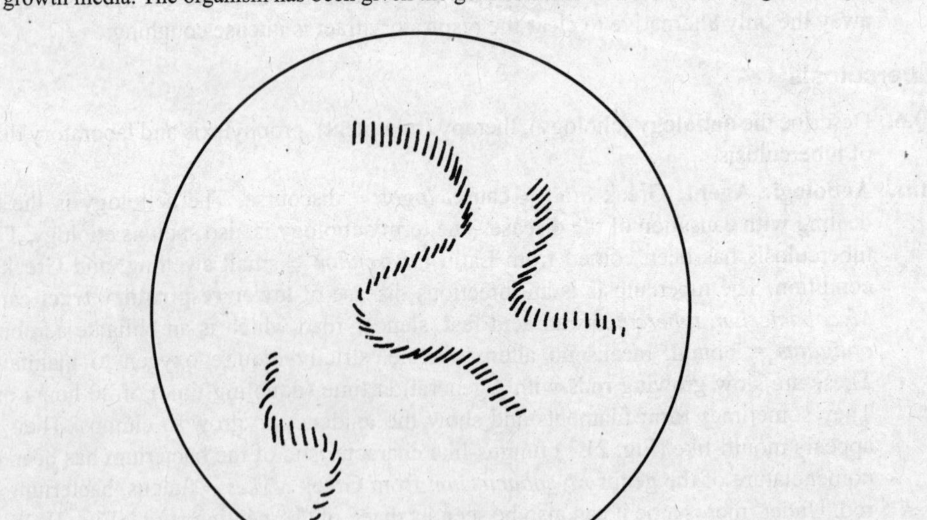

Fig. 21.2. Side by side 'serpentine cords' of *Mycobacterium tuberculoris* as observed in the field of microscope.

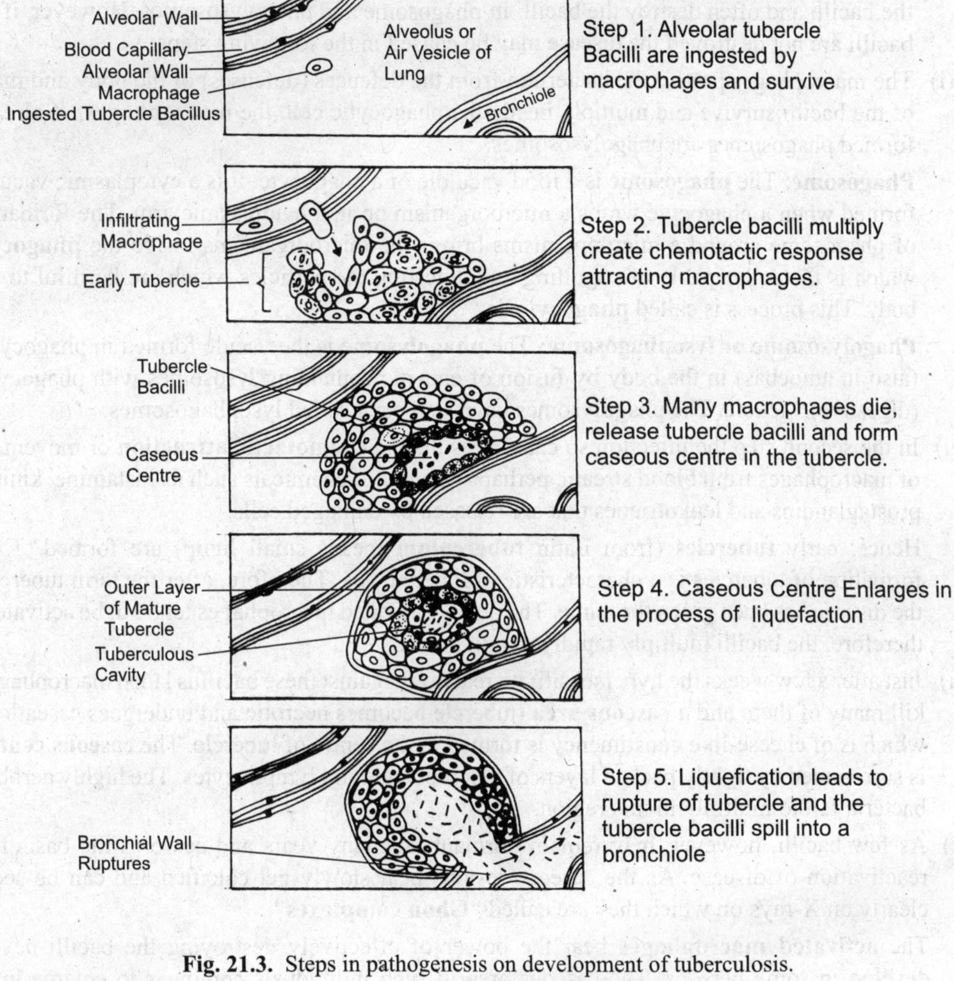

Fig. 21.3. Steps in pathogenesis on development of tuberculosis.

Heating is done to enhance penetration and retention of the dye. The Mycobacteria have a cell wall of unique composition. It has a high content of complex wax-like lipids called **mycolic acids**, which provide mycobacteria with a very protective permeability barrier that shields them from harsh environmental conditions, and can survive for weeks in dried sputum and are very resistant to chemical antimicrobials.

The cells stained with carbolfuchsin dye cannot be decolourized with a mixture of acid and alcohol. Therefore, these bacteria have been called **acid fast**.

The tuberculosis is a typical example of the importance of the ecological balance between host and parasite infectious disease. The factors that may affect host resistant levels are: the presence of other illness and physiological and environmental factors like malnutrition, overcrowding and stress.

Causation and Transmission of Disease or Pathogenesis:

The tuberculosis is most commonly acquired by inhaling the bacilli. Very fine particles containing 1 to 3 bacilli enter the lungs where they are phagocytized by a macrophage in the **alveoli** (singular alveolus) the air vesicles of the lungs (Fig. 21.3).

The **macrophages** [from Greek *makros* = large, *phagein* = to eat], the *phagocytic* cell which are the enlarged monocytes, in healthy individual become activated because of presence of the bacilli and often destroy the bacilli in phagosome and phagolysosomes. However, if the bacilli are not destroyed the disease may be caused in the following steps:

(i) The macrophages protect the bacterium from the defences (defenses) of the body and many of the bacilli survive and multiply inside the phagocytic cell, the **macrophages**, in already formed phagosomes are phagolysosomes.

Phagosome: The **phagosome** is a food vaculole of a phagocyte. It is a cytoplasmic vacuole formed when a phagocyte wraps a microorganism up in a cytoplasmic arm. The formation of phagosome around a microorganisms brings the microorganisms inside the **phagocyte** which is the cell capable of engulfing and digesting the particles which are harmful to the body. This process is called **phagocytosis.**

Phagolysosome or lysophagosome: The **phagolysome** is the vacule formed in phagocytes (also in amoebas) in the body by fusion of enzyme containing lysosomes with phagocytic (digestive) vacuole. The phagolysomes have also been called lysophagosomes.

(ii) In the second step the infection so caused results the **chemotactic attraction** or movement of macrophages from blood stream, perhaps because of chemicals such as histamine, kinins, prostaglandins and leukotrienes that are released by damaged cells.

Hence, early **tubercles** (from Latin **tuberculum** means small lump) are formed. Thus formation of tubercles is a characteristic of tuberculosis. Therefore, after the term tubercle, the disease has been given the name. The newly activated macrophages tend not be activated, therefore, the bacilli multiply rapidly within them.

(iii) Just after a few weeks the hypersensitivity reactions against these bacillus laden macrophages kill many of them and a **caseous area** (tubercle becomes necrotic and undergoes caseation) which is of cheese-like constituency is formed in the centre of tubercle. The **caseous centre** is surrounded by tightly packed layers of macrophages and lymphocytes. The highly aerobic bacteria fail to multiply in this region.

(iv) As few bacilli, however, may remain dormant far many years and serve as the basis for reactivation of disease. As the caseous lesions heal slowly get calcified and can be seen clearly on X-rays on which they are called **"Ghon complexes"**.

The **activated macrophages** bear the power of effectively destroying the bacilli never develop in some persons. The caseous area in such individuals continues to enlarge in a process which is referred to as *liquifaction* leading to formation of an air-filled tuberculosis cavity. The conditions within the cavity begin to favour the prodiferation of the tubercle bacilli. These bacilli set on multiplying outside the macrophages for the first time.

(v) As the bacilli row in an extraordinary large number, the lesions rupture and the bacilli are released into the respiratory system and then to blood and lymphatic systems. This disseminated infection is known as **'miliary tuberculosis'** (from Latin *miliarius* = millet) i.e. resembling a millet seed. The formation of numerous millet seed sized tuberculous nodules are widely disseminated throughout organs and tissues of the body.

The condition overwhelms the body's remaining defences and is characterized by a loss of weight, coughing generally with blood, and general loss of vigour. A popular term consumption has also been used for **pulmonary tuberculosis**, which consume the body.

Therapy (Treatment) and Vaccine:

(i) **Sanatoriums:** Before 1940s fresh air therapy for tuberculosis was popular. The patients were often taken to hospital roofs or sanatoriums also spelt sanitarium and sanatarium (care cottages). The first Sanatorium was made by Edward L. Trudeau in a small village of Saranc Lake in the Northern region of Adirondack Moutains in Upper New York State with 300 care cottages.

(ii) **Streptomycin:** The streptomycin was the first effective antibiotic used in the treatment of tuberculosis. (1) Today, however, isoniazid (INH) is the most commonly used synthetic chemical. (b) A combination of INH and rifampin is generally used. (2) There is another treatment in the above combination of INH and refampin, plus pyrazinamide which is effective against bacteria inside the macrophages. (3) Ethambutol pyrazinamide and streptomycin are given in cases of intolerance or resistance to the primary drugs.

This is to be kept in mind that chemotherapy for tuberculoris must be continued for one to two years. The prolonged treatment of tuberculois is required because tubercle bacillus grows very slowly and many antibiotics have been considered effective against growing cells. Moreover, the bacilli (rods) of tuberculosis remain hidden for long period in macrophages or other locations making it difficult for antibiotics to reach. (4) DOT : One approach to high risk patients to their medication routinely is called DOT (Direct observed therapy) in which a patient comes to a clinic and takes his or her medication in presence of a health care worker.

PRophylaxis: (1) A six to twelve months course of INH is given to people exposed to tuberculosis. (2) The BCG (Bacillus Calmette and Guerin vaccine) was developed by Albert Calmette and Camille Gurein who were two French scientists at the Pasteur Institute, Paris, France. The BCG vaccine is live culture of attenuated *Mycobacterium bovis* that has been made avirulent through a long cultivation on artificial media.

Laboratory Diagnosis: (1) **Tuberculin skin test** is the skin test used in diagnosis of tuberculosis. Tuberculosis patients respond with cell mediated immunity against the bacterium. This type of immune response instead of humoral immunity develops because the pathogen is located mostly within macrophages. Such an immunity involving sensitized T cells (T-lymphocytes) is the basis of the **tuberculin skin test** in which **purified protein derivation** (PPD) in prepared from filtration of streamed culture grown on a synthetic medium. The filtrate is treated with substances like trichloroacetic acid to precipitate **tuberculoprotein** which is then washed and prepared in standard dilutions. The tuberculin may refer specially to PPD. The **mammalian tuberculin** used in the comparative single intradermal tuberculin test in PPD. The **mammalian tuberculin** used in the comparative single intradermal tuberculin test is PAD prepared from human or bovine tubercle bacilli while avian tuberculin is PPD from the avian tubercle bacillus.

In this test the PPD is injected cutaneously. In case the person has been infected with tuberculosis in the past sensitized T-cells react with these proteins and a **delayed hypersensitivity reaction** (a cell mediated hypersensitivity) appears in 48 hours. The reaction can be seen as an induration i.e. hardening and reddening of the area around the site of injection.

(iii) **Mantoux test:** In this test dilutions of 0.1 ml of antigen are injected and reacting area of skin is measured. This test is considered to be the most accurate.

(iv) **Moro test:** In Moro test tuberculin is smeared over a small area of skin.

(v) **Vollmer patch test:** A modification of tuberculin test in which a patch of tuberculin-impregnated gauze is fastened to the skin with adhesive tape.

In the young individuals, the positive tuberculin test provides the probable indication of an active case of tuberculosis. However, in older individuals it may indicated only hypersensitivity resulting from a previous infection or vaccination, but not a current active case of tuberculosis yet it is suggestive of further examination as the x-ray of chest to detect the lesions in the lung and isolate the bacterium for further examination. As the growth of the bacterium is slow, the confirmation of tuberculosis by isolating bacterium is a tedious task. It may take 3 to 6 weeks for the formation of colonies and additional 3 to 6 weeks to reach the reliable identification series.

The bacteria grow more rapidly in broth cultures, therefore, some clinical laboratories grow bacteria in a medium containing radio-labelled palmitic acid as substrate. To conclude, the release of carbon dioxide is indicative of the presence of mycobacteria. This technique involves 7 to 14 days and additional time is required for further positive identification of the species.

To amplify the bacterial DNA by PCR and follow the assay using DNA probe can be a good process.

22

CHAPTER

FUNGAL DISEASES OR MYCOSES

Salient Features

- The fungal kingdom has been raised on the basis of their nutritional habit of deriving nutrients by means of absorption.
- The **yeasts** (unicellular) **moulds** (filamentous fungi) and **fleshy fungi** (pulpy fungi mushrooms, puffballs and coral fungi are important groups of applied nature.
- Generally most of the fungi are decomposers but some of them are parasitic on plants and animals.
- The fungal **thallus** (body) comprise either of nonseptate tubular structures which are **coenocytic** i.e. with many nuclei or filaments of cells that form a tuft of these **hyphae** (singular hypha) is called **mycelium**.
- The yeasts are unicellular fungi and often make a powdery coating on fruits and leaves.
- The *Schizosaccharomyces,* the fission yeasts divide symmetrically while *Saccharomyces*, the budding yeasts divide asymmetrically, and the buds do·not separate from mother cells which result the formation of **pseudohyphae** (false hyphae) or **pseudomycelium** false mycelium.
- There are some pathogenic **dimorphic fungi** i.e. represent two growth forms and yeast-like at 37ºC and mould-like at 25ºC.
- The fungi are classified on the basis of their asexual and sexual spores.
- The fungi can grow in acidic, low moisture and aerobic environments.
- The **dimorphism** in *Mucor rouxii* depends on concentration of CO_2 . *M. rouxii* shows a yeast-like growth on the surface of agar in the Petriplate while it is mould-like (filamentous) in the agar of Petriplate (Fig. 22.1) Recently a filamentous form of baker's yeast *saccharomyces cerevisiae* has been indused by starvation.

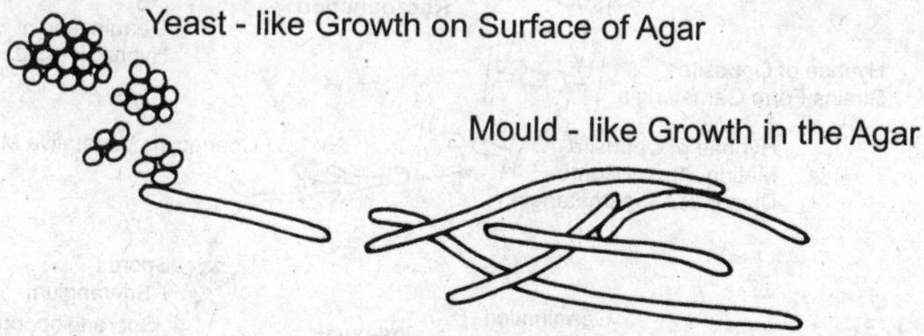

Fig. 22.1. The **dimorphism** or two growth forms of fungus *Mucor rouxii* that depends on concetration of CO_2. The fungus shows **yeast-like growth** on the surface of agar while the growth is a **mould-like form** in the agar.

- They mitabolize complex carbohydrates.
- Deuteromycota reproduce asexually by one or more types of spores, called **Chlamydospores,** the thick-walled spores in *Candida albicans*. The **arthrospores** are asexual spores formed as a result of fragmentation of seplate hyphae as in *Coccidiodes immitis*.The conidiospores

or conidia are asexual spores which are formed on sterigmata formed on vesicle formed by swelling of tip of conidiophores (the spore bearing stalk as in *Aspergillus flavus* (Fig. 22.2).

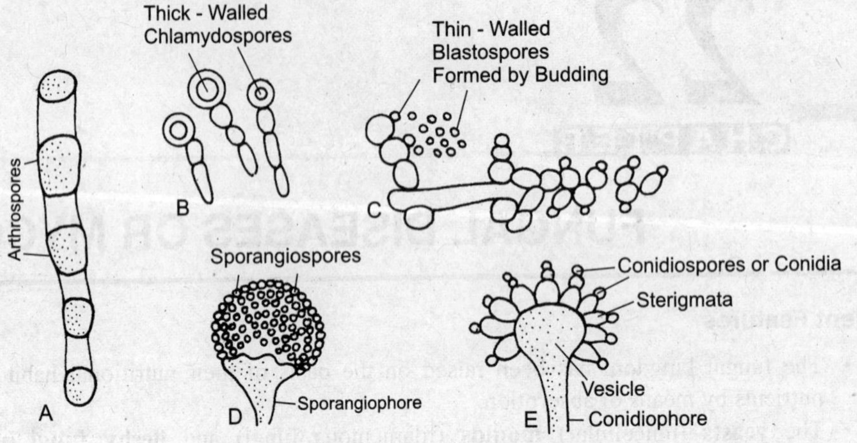

Fig. 22.2. Different types of asexual spores. (A) **Arthrospores** formed by fragmentation of hyphae in *Coccidiodes immitis*. (B) Thick-walled *chlamydospores*, asexual spores within the hyphae of *Candida albicans*. (C) Thin-walled **blastospores** formed from buds of parent cells in *C. albicans*. (D) Sporangiospores formed within the spore sac or **sporangium** of *Rhizopus*. (E) Chains of **condiospores** or conidia sterigmata (singular sterigma) formed on vesicle of conidiophore in *Asperigillus*.

The other asexually formed spores are blastospores formed by budding from parent cell as in *Candida albicans* (Fig. 22.2).

- The *sporangium* is the asexual spore formed within a wall.
- Some fungi are placed under Deuteromycota because sexual spores have not been observed in them.
- Zygomycota bear coenocytic (devoid of septa and are multinucleate) hyphae and produce sporangiophores with sporangia in asexual reproduction and zygospore, thick walled spore in sexual reproduction (Fig. 22.3).

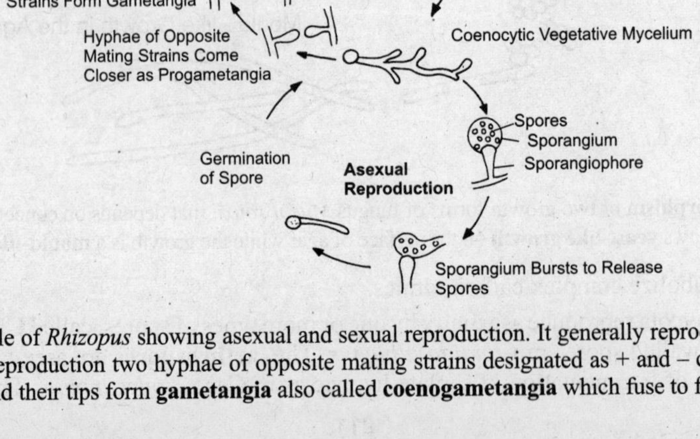

Fig. 22.3. Life cycle of *Rhizopus* showing asexual and sexual reproduction. It generally reproduces asexually. In sexual reproduction two hyphae of opposite mating strains designated as + and – come close to each other, and their tips form **gametangia** also called **coenogametangia** which fuse to form **zygospore**.

- The classification of *Pneumocystis* whether a fungus or protozoan is yet undecided and controversial .
- Most musrooms belonging to genus *Amanita* are poisonous. *Amantia phalloides* has been called the "death stool" or "death caps" in Europe. The *Amanita* sp, the pileus of which is flat when fully extended establish mycorrhizal association with Sal (*Shorea robusta*) in Uttarakhand, in India. Bird's nest fungus, *Scleroderma cepa* and fly-fungus (Fig. 22.4) belong to Basidiomyctes.

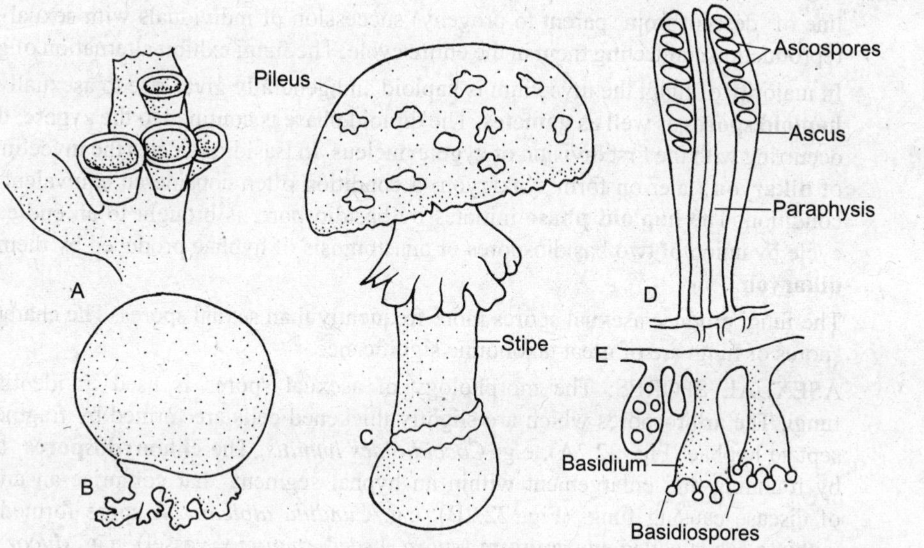

Fig. 22.4. Typical Basidiomycetes and spores of typical Ascomycetes and Basidiomycetes. (a) Bird's nest fungus *Crucibulum vulgare* growing on a branch of a tree. (b) *Scleroderma cepa* showing sporophore, the fruiting body with subsesile (small attachment) rooting base. It is a mycorrhizal fungus. (c) *Amanita muscaria* called **fly-fungus** for its use to kill flies. (D) Asci of *Morchella* with 8 *ascospores*, each. (E) **Basidia** of a Mushroom with 4 **basidiospres**, each.

- Ascomycota possess septate hyphae and form **ascospores** in *asci* (singular **ascus**) which are sac like structures, during sexual reproduction. More frequently they form conidia on conidiospores in sexual reproduction (Fig. 22.4D).
- Basidiomycota possess septate hyphae and produce **basidiospores** on club-shaped structure called **basidia** (Fig. 22.4E).
- The **systemic mycosis** (fungal infection in deep tissues) affect many tissues and organs.
- **Subcutaneous mycoses** are fungal infections below the skin.
- The **cutaneous mycoses** occur on keratin containing tissues like hairs, nails and skins.
- The superficial mycoses are localized on hair shaft and superficial skin cells.
- The opportunistic mycoses are caused by normal microbiota which usually are non pathogenic.
- Some fungi are plant pathogens.
- Not all fungi are harmful. *Saccharomyces* and *Trichoderma* are used in production of foods. There are also some **comestible or edible fungi**. Some fungi are also used as single cell protein (SPC). Edible fungi include mushrooms, puffballs and coral fungi.
- **Histoplasmosis** is a subclinical respiratory infection caused by *Histoplasma capsulatum*. The disease is acquired by inhalation of air borne conidia. It was first recognized in USA when X-ray showed lung lesions in people who gave tuberculin-test negative.

- **Coccidiomycosis** is caused by inhalation of airborne arthrospores of *Coccidiodes immitis*.
- **Blastomycosis** is caused by *Blastomyces dermatitidis* which infection begins in lungs and can spread causing extensive abscesses.

Life Cycle

Q.1. What is life cycle? Describe the life cycle of fungi.

Ans. The life cycle comprise the progressive changes undergone by an organism or lineal (direct line of descent from parent to progeny) succession of individuals with sexual or asexual reproduction connecting them in the entire cycle. The fungi exhibit alternation of generation.

In majority of fungi the mycelium is haploid and generally gives rise to asexually produced **haploid spores** as well as **gametes**. The diploid phase is confined to the **zygote**, the meiosis occurring with the first division of zygote nucleus. In Basidiomycetes the mycelium consists of **dikaryon** (later on form synkaryon) a condition often considered equivalent to diploid condition. The **haploid phase** initiates by basidiospore, is brought to an end early in life cycle by union of two basidiospores or anastomosis of hyphae produced by them to form a **dikaryon**.

The fungi produce asexual spores more frequently than sexual spores. The characteristic of spores of fungi are of great taxonomic significance.

ASEXUAL SPORES: The morphology of asexual spores is used in identification of fungi. The arthrospores which are slightly thickened cells are formed by fragmentation of septate hyphae (Fig. 22.2A), e.g. *Coccidioides immitis*. The **chlamydospores** are formed by rounding and enlargement within an hyphal segment, that comprise an another type of disease causing fungi (Fig. 22.2B), e.g. *Candida albicans*. A spore formed asexually within a sac is called **sporangium** (*sporo* = seed, *angion* = vessel), e.g. *Mucor* one of the causative agents of **mucormycosis**. The clinical symptoms of the disease are production of purulent nasal discharge that may be accompanied by gangrenous (death of part of tissue) mucosa. Mucormycosis is caused by fungi belonging to order Mucorales. The other genera which cause mucormycosis are **Absidia** and **Rhizopus**. These also form sporangium (plural sporangia). The stalk or the aerial branch which produces sporangium is called **sporangioshore**. An another important type of spore is conidiospore which may be unilocular or multilocular with horizontal septa or both horizontal and vertical septa. Chlamydospores are not enclosed in a sac (Fig. 22.2E). They are produced on branch or stalk called **conidiophore**. The conidiophores may further branch into **sterigmata** which bear conidiospores or conidia at their tips. In some species of Ascomycota **primary sterigmata** produce **secondary sterigmata** which in turn form conidia on them. e.g. *Aspergillus*. The blastospores are also a type of asexual spores which are thin walled (Fig. 22.2C) formed as a result of budding e.g. *Candida albicans* and *Cryptococcus neoformans*, non-fermentative imperfect yeasts.

SEXUAL SPORES: The sexual spores in fungi are formed in three stages. (1) A haploid nucleus of plus strain or donor cell penetrates the cytoplasm of minus strain or recepient cell. (2) The plus and minus nuclei fuse to form diploid zygote nucleus. (3) The diploid nucleus undergoes meiosis and result the formation of haploid nuclei or the sexual spores.

The formation of sexual spores is favoured by special circumstances e.g. **plant pathogenic ascomycetes** form ascospores when host plants are at the end of growing season. This shows that ascospores are initiated by changes in moisture, temperature or nutrient availability. The ascospores survive by the time plant hosts are growing again. These sexual spores are of great taxonomic significance and have been used to divide or classify fungi into different divisions, or subdivisions.

NUTRITIONAL ADAPTATIONS: The fungi are chemoheterotrophs and like bacteria they absorb nutrients rather than ingesting them like animals. The fungi are adapted better to grow under adverse conditions in comparison to bacteria as given below:

(i) The fungi can grow at pH 5.0 which is too acidic for the growth of most of the common bacteria.

(ii) The moulds generally are aerobic. Therefore, they grow on the surface rather than inside the medium or substrate.

(iii) The fungi can tolerate the osmotic pressure thus they can grow under higher concentration of sugar or salt.

(iv) The fungi can grow under low moisture conditions where bacteria fail to grow.

(v) The nitrogen requirement of fungi is less in comparison to bacteria.

(vi) The fungi can decompose the complex carbohydrates like lignin to which the bactéria cannot break down or decompose.

Medically Important Divisions (P hyla).

Q.2. Which are the medically important Divisions (Phyla) of fungi?.

Ans. The following are medically important divisions (phyla):

(i) DEUTEROMYCOTA: This group has been called so because the fungi of this division do not produce sexual spores and are also called "Fungi Imperfect". They reproduce asexually by forming chlamydospores, arthrospores and conidiospores. Budding is also seen in them and the hyphae of members are septate. When a perfect stage bearing sexual spores is observed that is classified either under Ascomycetes or Basidiomycetes.

(ii) A CASE OF MISTAKEN IDENTIFIY: The position of *Pneumocystis* is controversial. The *Pneumocystis* is a common opportunistic agent to cause pneumonia in immunosuppressed individuals and is the major cause of death in patients of AIDS. By 1970s *Pneumocystis carinii* was not regarded a human pathogen. It was discovered in 1908 and its taxonomical position has been uncertain. It was provisionally classified as a protozoan. However, on comparing the RNA sequence of *Pneumocystis* with that of *Protozoa*, *Euglena*, cellular slime moulds, plants, mammals and fungi it was revealed that they are members of fungal Kingdom.

(iii) ZYGOMYCOTA: They have also been called "Conjugation Fungi". They are saprophytic moulds bearing coenocytic hyphae e.g. *Rhizopus nigicans*, the common black bread mould. The *Rhizopus* produces dark coloured sporangiospores enclosed inside the sporangium in asexual reproduction. The sporangium breaks open to release the sporangiospores which germinate on coming in contact with a suitable medium. The **zygospores** after which Zygomycetes have been named are the sexual spores. The zygospores are formed as a result of fusion of nuclei of two cells which are morphologically similar (Fig. 22.3).

(iv) ASCOMYCOTA: The sexual spores in this group are formed in sacs called asci (singular **ascus**) (Fig. 22.4D) There are 8 ascospopres in a typical ascus. The moulds bearing ascospores and yeasts come under this division. The asexual spores are conidiospores or **conidia** which are produced in long chains on **conidiosphores** the spore bearing stalks or branches which in turn may produce **sterigmata** (secondary sterigmata also in some species). The sterigmata (singular sterima) are minute-stalks bearing spore or chain of spores.

(v) BASIDIOMYCOTA: They are characterized by production of **basidiospores** which are formed on club-shaped structures called basidia (singular basidium) Fig. 22.4E. There are four basidiospores on a single basidium.

Fungal Diseases

Q.3. What are the fungal diseases? Explain.

Ans. A disease caused by a fungus is called mycosis (plural mycoses). The fungal diseases are generally chronic and last for a long time. They have been categorised into following five types:

(i) **Systemic Mycosis**: The fungal infections which are deep within the tissue are called systemic mycoses. They can affect different tissues and organs of the body. They are caused by fungi which are soil dwelling (live in soil). They are caused by **inhalation** of spores which is the route of transmission. The infection originates in the lungs and spreads to other tissues of the body. Fortunately, they are not contagious from animal to human and humam to human. Two important systemic mycosis are : (a) Histoplasmosis and (b) Coccidiomycosis.

(ii) **Subcutaneous Mycosis**: These are the infections beneath the skin and are caused by saprophytic fungi living in soil and on vegetation. These are caused by direct implantation of fungal spores or fragments of mycelium into a wound or puncture in the skin.

(iii) **Cutaneous Mycoses**: These are the infections of epidermis, hair and nails. The fungi which cause cutaneous infections are called **dermatomycoses** or dermatophytoses or cutaneous mycoses. The mechanism of infection involves the secretion of enzyme **keratinase** by dermatophyte which breaks down kertain, the protein of which the skin, hair and nails are made of. The cutaneous mycoses can be transmitted from human to human, animal to human either by direct contact or through the tools in barber's shop.

(iv) **Superficial Mycoses**: The fungi which cause superficial mycoses are localized along hair shafts, surface of skin or the epidermal cells. These infections occur in tropical climates.

(v) **Opportunistic Pathogens**: The opportunistic pathogens are generally harmless in its normal habitat but can turn pathogenic in a host who is seriously debilitated (weak) or traumatized seeking treatment of broad spectrum antibiotics. Also in those individuals whose immune system is suppressed by drugs or by immune disorder. Therefore, the patients suffering from AIDS are quite susceptible to opportunistic pathogens.

The normal microbiota of a person may turn to be opportunistic pathogens under such immuno supporessed conditions. Thre common examples of the diseases caused by opportunistic pathogens are :

(a) **MUCORMYCOSIS**: The opportunic mycosis caused by *Rhizopus* and *Mucor* belonging to order Mucorales has been referred to as mucormycosis. The infection occurs mostly in patients with **ketoacidosis** caused due to diabetes melitus, leukemia or treatment with immunosuppressive drugs.

(b) **ASPERGILLOSIS**: The aspergillosis also is an opportunistic mycosis caused by *Aspergillus*. The aspergillosis is caused in individuals who have debilitating lung diseases or cancer, and have inhaled spores of *Aspergillus*.

(c) **CANDIDIASIS**: This opportunistic mycosis is caused by *candida albicans* as **vulvovaginalis candidiasis** during the period of pregnancy. The thrush is a mucocutaneous candidiasis which is an inflammation of mouth and throat occurring frequently in new borns and people suffering from AIDS.

Fungal Capsule

Cryptococcus neoformans is a fungus which causes a type of meningitis. The fungus produces a capsule which enables fungus to resist phagocytosis.

Diseases Caused by Fungal Toxins

Some diseases are caused by toxins produced by fungi. These are given below :

1. **ERGOTISM**: The disease ergotism was common in Europe during the Middle Ages. The ergotism is caused by *Claviceps purpurea*, a member of Ascomycetes which grows

on grains. The toxin is present in **sclerotia** which are highly resistant portions of fungal mycelium which may even get separated. The toxin of this fungus is called **ergot** which is an alkaloid and can cause hallicinations resembling to those produced by LSD. This is because that ergot is a natural source of LSD. Ergot may also cause **gangrene** by constructing capillaries and preventing proper blood circulation. *C. purpurea* though occassionally grows on grain but modern milling often removes the sclerotia.

2. **AFLATOXINS** : The aflatoxins are the fungal toxins which are carcinogenic (cancer causing). Aflatoxin is produced by growth of the mould *Aspergillus flavus.*
3. **MUSHROOM TOXINS**: A few poisonous mushrooms produce toxins called mycotoxins prominent mushroom **mycotoxins** are **phalloidin** and **amanitin** produced by *Amanita phalloides* which has been called "death stool". Its neurotoxins are so powerful that eating *Amanita* mushroom may cause the death of the consumer.

Cutaneous Mycoses or Dermatomycoses

Q.4. What are **cutaneous mycoses** or **dermatomycoses** ? Which are the three genera of fungi involved in the causation of the disease? Describe its clinical types? Also give the therapy.

Ans. The **cutaneous mycoses** or **dermatomycoses** or **tineas** or **ringworms** are the synonyms of fungal infections of the outer layer (stratum corneum) of the epidermis. The causative fungi are called **dermatophytes**. The dermatophytes grow on the protein keratin present in the skin. The name ringworm is a misnomer, the term got coined from ancient Greek belief that infections that tend to expend in a circular style were caused by a worm. The Romans wrongly linked ringworm with lice. Therefore, the term **tinea** was coined from Latin word for worms or insect larvae cutaneous mycoses has also been called **dermatophytoses**.

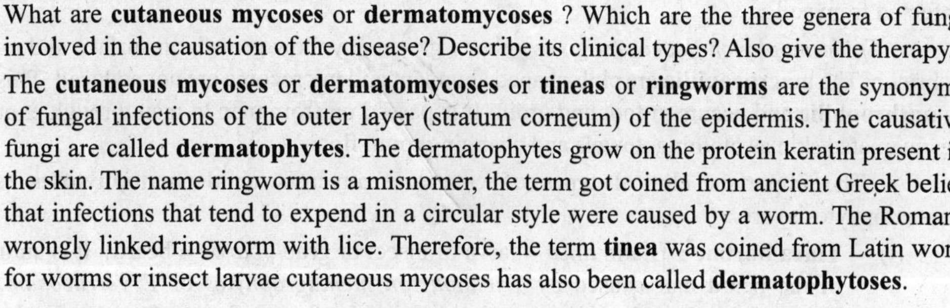

Fig. 22.5. Macroconia and microconidia of common dermatophytic fungi. A. Macroconidia and microconidia of *Trichophyton mentagrophytes*. B. Macroconidia and microconidia of *Microsporum gypseum*. C. Macroconidia or macroaleuriospores of *Epidermophyton floccosum* which does not produce microconidia.

Three genera of fungi held responsible for causation of cutaneous mycosis (Fig. 22.5 A, B and C) are:

1. *Trichophyton* can infect hair, skin or nails.
2. *Microsporum* often involves hair and skin only.
3. *Epidermophyton* infects skin and nails.

Trichophyton mentagrophytes has two distinct forms. The **zoophilic** subspecies var *granulare* produces a fast growing colony with a flat powdery or granular surface which is an indicator of abundant sporulation and an irregular edge. The anthrophilic subspecies var. *interdigitale* is often indistinguishable from *T. rubrum* on primary isolation since it produces a **white puffy aerial mycelium** that may become yellowish or tan with age. The underside of both colony types may be pink, red, yellow or rose-brown. The sporulation is much more obvious with granular form which also shows the presence of spiral hypae the **macroconidia** which are generally upto 6-celled and smooth thick walls. The macroconidia are rare in the **anthrophilic** form. Both forms are known to produce **microconidia** either borne along the sides of vegetative hyphae (Fig. 22.5A) or in grape-like clusters specially common in zoophilic isolates. *T. mentagrophytes* degrades the urea in **Christensens urea medium** supplemented with glucose 10g/litre sufficiently to change the indicator in the medium.

Microsporum gypseum produces granular light to mid-brown colony which grows rapidly with an irregular fringed border and cinnamon undersurface. There are numerous **macroconidia** (Fig. 22.5B) formed with upto 6 cells in each. The walls of conidia are thick and roughened. It is geophilic infection in humans is rare and often as a result of contact with animals like rodents infected from soil.

Epidermophyton floccosum gives velvety greenish yellow growth on surface and underside is brown. The **macroconidia** are club-shaped abundant and microconia are not poduced. This fungus is **anthrophilic** and occurs in foot and groin. It is passed on from man to man through exposure to sesquamated skin as clothing, towels and changing room floor.

The important clinical types of cutaneous mycosis or **dermatophytoses** (singular dermatophytosis) are :

(i) **Tinea barbae** : [Latin *tinea* = larva of insect or grub ; *barba* = beard]. It is an infection of the beard hair caused by *Trichophyton mentagrophytes* or *T. verrucosum*. It is a disease of males of rural areas who acquire the fungus from infected animals in their surrounding.

(ii) **Tinea capitis**: [Latin *capita* = head]. It is an infection of hair of head. It can be characterized by loss of hair, inflammation and scaling. This disease is found in children which is caused by *Trichophyton* or *Microsporum*. Person to person transmission of the fungus takes place in poor hygienic conditions in over crowded areas. The fungus also occurs in domestic animals from which the fungus may be transmitted to human coming in their contact.

Diagnosis: The infection can be diagnosed by Wood's light i.e. an ultra violet light because the fungus infected hair fluoresces on illumination with UV radiation.

(iii) **Tinea corporis** : [Latin *corpus* = body]. It is a dermatomycoses of any region of human skin. The disease may be characterized by appearance of circular, red, well demarcated scaly, vesiculopustular, lessions accompanied by itching. The causative agents of tinea corporis are *Trichophyton rubrum Trichophyton mentagrophytes* or *Microsporum canis*. The disease may be transmitted as a result of direct contact with the infected animals or indirect contact through fomites, the infected articles.

(iv) **Tinea cruris**: [Latin *crura* = leg]. It is a dermatophytic infection of groin region. *Epidermophyton floccosum*, *Trichophyton mentagrophytes* or *T. rubrum* are the causative agents of the disease. The pathogenesis and clinical manifestations of tinea cruris are same as that of tinea corporis. The factors which favour the disease are: moisture, occlusion (the closure of an opening especially of ducts and blood vessels) and skin trauma (injury). The disease is also supported by wet bathing suites, athletic supporters (infection also called jock itch or **jockey itch**), tight fighting-slacks, panty hose and obesity.

(v) **Tinea pedis**: (Latin *pes* = foot). It has also been called **athletes foot**. The tinea manuum (Latin *mannus* = hand) is fungal infection of hand. Thus, Tinea pedis is the dermatophytic infection of the feet and tinea mannum is that of hand. Clinical symptoms of disease vary from fine scale to a vesiculopustular eruption. Itching is common. The disease is favoured by warmth, humidity, trauma (injury) and occlusion (closure of the ducts and blood vessels). The infections are generally caused by *Trichphyton rubrum*, *T. mentagrophytes*, or *Epidermophyton floccosum*. The tinea pedis and tinea mannum are found in adults and both of these increase in frequency with the advancing age in humans.

(vi) **Tinea unguium**: [Latin *unguis* = nail]. It is the dermatophytic infection of the nail bed. This disease characterized by the changes in the nails which get discoloured and then get thickened. The causative agents are the *Trichophyton rubrum* and *T. mentagrophytes*.

THERAPY OR TREATMENT: The fungal diseases of cutaneous mycoses can be treated with topical ointments like miconazole (Monistat-Derm), tolnaftate (Tinactin) or clotrimazole (Lotrimin) for 2 to 4 weeks. Griseofulvin (Grifulvin V) and itraconazole (Sporanox) are the only oral antifungal agents approved by appropriate authorities for the therapy of dermatophytoses or cutaneous mycoses.

Subcutaneous Mycoses

Q.5. What are **subctuaneous mycoses**? Which are the clinical types of subcutaneous mycoses ? What are their causative fungi ? Also give the treatment for subcutaneous mycoses.

Ans. Latin *sub* = under ; *cuticula* = cuticle ; which collectively means beneath the skin. Thus the fungal infection beneath the skin are called subcutaneous mycoses. These are more serious than cutaneous mycoses. The dermatophytes which cause subcutaneous mycoses are usual saprophytic fungi of soil and decaying vegetation. But they are not able to enter the skin, therefore, must be introduced into the subcutaneous tissue by a puncture wound which has been contaminated with soil bearing the fungi. On finding way to subcutaneous tissue the disease develops slowly usually over a period of years. during this period fungi produce a nodule which consequently ulcerates and the fungus spread along lymphatic channels forming more subcutaneous nodules. Such nodules sometimes may come on the skin surface, occasionally. The clinical types of subcutaneous mycoses are as under.

(i) **Chromoblastomycosis**. The nodules in this type are pigmented to dark brown colour. The causative fungi are black moulds *Phialophora verrucosa* or *Fonsecaea pedrosoi*. These fungi are universal in distribution particularly in tropical and subtropical zones. Infections on legs and feet are very common.

(ii) **Maduromycosis** : It is caused by the fungus *Madurella mycetomatis* which occurs worlduride in general and in tropics in particular. The infection caused at maturity is called **eumycotic myceloma** or fungal tumour as the invading fungus damages subcutaneous tissue and cause serious malformation. The abrasions caused while walking bare-footed on contaminated soil may result a form of myectoma which has been called **Madurafoot**.

(iii) **Sporotrichosis** : It is caused by *Sporothrix schenckii* named after a mycologist. Schenck. This fungus is dimorphic (having two growth forms). Although the disease has been found through out the world yet is most common of subcutaneous fungal disease in the USA. The infection may be caused by a puncture wound of a thorn or splinter contaminated with the fungus. The fungus inhabits in the soil, on living plants like barberry shrubs, roses, in plant debris as *Sphagnum* moss and mulch of pine-bark. This is precarious for florists, gardeners, and workers in forestry. A small red papule appears and begins to ulcerate shortly after an incubation period of 1 to 12 weeks in the infected persons. This may be followed by the appearance of new lesions along lymph cannels and the disease may remain localized or spread throughout the body and thus **extracutaneous sporotrichosis** proliferates and becomes obvious.

Therapy: The administration of oral 5-fluoro-cytosine, iodides, amphotericin B and surgical removal of diseased parts are the common therapeutic measures. The subcutaneous fungal infections can also be treated with a dilute solution of potassium iodide but surprising enough the fungus *Sporothrix schenckii* is not affected *in vitro* by a 10 per cent solution of potassium iodide (KI) even.

Candidiasis

Q.6. What is **candidiasis** ? What suppresses the growth of causal fungus? Which factors favour the disease ? Also give treatment of the disease.

Ans. The **candidiasis** is the mycosis or the fungal disease caused by a dimorphic fungus *Candida albicans*. The disease has also been called **moniliasis** in the past because of an invalid synonym of *Candida albicans* (= *Monilia albicans*). The growth of the causal fungus *Candida albicans* is often suppressed by the bacterial microbiota of mucous membranes in the genito urinary tract and the mouth.

The fungus itself is not affected by antibacterial drugs. Therefore, sometimes the fungus overgrows mucosal tissues when the antibiotics suppress the normal bacterial microbiota. The changes in the normal mucosal pH may also favour the growth of *Candida albicans* because it can grow even in acidic conditions. Such overgrowths by the yeast-like *Candida albicans* are called candidiasis. This sort of white growth called **thrush** can be seen specially in the oral cavity of new borne infants because the normal microbiota has not been established in them. Therefore, it has also been called an opportunistic pathogen. The immunosuppressed individuals also are prone to infections of skin and mucous membrane. The diabetes, antibiotic therapy, oral contraceptives, pregnancy or any other factors that compromises the female host may lead to causation of **candidal vagnitis**. The lactobacilli which were observed by **Döderlein's bacilli**, being **aciduric** (tolerant of acidic conditions) can control *Candida* in vagina by the low pH created by them. If the number of Gram positive Döderlein's bacilli decreases because of any of aforesaid factors the Candida may grow excessively resulting a curd-like yellow white discharge from the vagina. During intercourse, the *Candida* may be transmitted to males and cause the fungal infection called **balanitis**. Therefore, it may also be called a sexually transmitted disease (STD). To conclude the balanitis is the infection of male **glans** (bulbous termination of penis). The disease begins as vesicle on the penis which turn into patches. A severe itching and burning is also accompanied with appearance of these patches.

THERAPY: Cutaneous lesions may be treated with topical agents like sodium caprylate, sodium propionate, gentian violet, nystatin, miconazole and trichomycin. Ketoconazole, amphotericin B, fluconazole, itraconazole, and flucutosins are used in 'Systematic candidiasis' treatment.

Histoplasmosis

Q.7. What is **histoplasmosis**? How does temperature affect its forms? Describe the causation of disease by the fungus. Also give treatment of the disease.

Ans. Greek *histos* = web ; *plasma* = anything formed ; *osis* = condition. It is the fungal disease caused by *Histoplasma capsulatum* var. *capsulatum*, a facultative parasitic fungus which grows **intracellularly**. Histoplasmosis superficially looks like tuberculosis. The dimorphic fungus is seen as small budding yeast in humans and on culture media but at 25°C it grows as mould and produce small microconidia, 1 – 5 μm in diameter, produced singly at the tips of short conidophores. The same fungus has also been observed to produce macroconidia 8 to 16 μm in diameter. That is how the *Histoplasma* behaves at different temperatures of 37°C and 25°C. The yeast-like cells of the *Histoplasma* are found in tissue and phagocytic cells of humans.

In soil *H. cappsulatum* var. *capsulatum* is found in mycelial form, particularly in the areas which have been contaminated with excrement of bats. The bats carry the fungus because they have lower body temperature than birds. The microconidia are easily spread by air.

The disease histoplasmosis is endemic to Minnesota, Mississippi, Kentuck, Tennessee, Ohio and Rio Grande River basis, and antibodies against *Histoplasm* are found in 75% of population living in these areas.

DIAGNOSIS: The Gram-stained dry smears as used generally in bacteriology are satisfacory for specimens of yeast-like fungi. But small yeast forms of *H. capsulatum* are seen best in alcohol-fixed Giemsa stained preparations. Identification in tissue specimen are required for proper diagnosis of histoplasmosis.

THERAPY: The amphotericin B or itraconazole are counted among most effective chemotherapy of the disease.

Otomycosis

Q.8. What is **otomycosis**?

Ans. Greek *ous* = ear; *mykes* = fungus; *osis* = condition. It is a fungal infection of the external auditory meatus, most often caused by *Aspergillus* and *Candida*.

23 CHAPTER

HUMAN IMMUNODEFICIENCY VIRUS AND AIDS

Q.1. How and when was **HIV** discovered?

Ans. By 1992, two cancer research laboratories were involved in research to find out the cause of a new disease, later called **Acquired Immune Deficiency Syndrome (AIDS)**. In 1993, research teams headed by Luc Montagnier and Robert C. Gallo in Pasteur Institute, Paris and National Cancer institute, USA, respectively, published reports hinting that they have found the cause of AIDS. The virus was referred to as HTLV-III (human T-cell lymphotropic virus) / LAV (lymphoadenopathy associated virus) by the scientific community. The name was finally changed to **Human Immunodeficiency Virus (HIV)** in 1986. A new strain of HIV was identified in 1985, referred to as HIV-2 relatively common in West Africa. It appeared relatively milder than HIV-1.

Q.2. How did the HIV virus originate?

Ans. Origin of virus is not yet certain. Evidence suggest that it existed in chimpanzees in West Africa and then infected humans possibly through a bite or scratch from a chimpanzee and from there it may have spread into a wider population.

Q.3. What is **AIDS?**

Ans. **AIDS** is primarily a disease of the immune system characterized by the severe depression of the immune system and the acquisition of life-threatening secondary infections. It is caused by the Human Immunodeficiency Virus (HIV).

Q.4. When were cases of HIV reported in India?

Ans. No country has been left unscathed by HIV including India. First cases of HIV infection were reported in sex workers in Tamilnadu, India in 1986. In 1987 a **National AIDS Control Programme** was launched to deal with the crisis by surveillance, blood screening, and health education. By the end of 1987, out of 52,907 who had been tested, around 135 people were found to be HIV positive and 14 had AIDS. The infection rates continued to rise and in 1992 the government set up **NACO (National AIDS Control Organization)**. Its functions were formulation of policies, prevention work and control programmes to combat the HIV and AIDS. HIV had now extensively spread in the country.

Q.5. What is **HIV**?

Ans. **HIV** stands for **Human Immunodeficiency Virus**. Like all viruses, HIV cannot grow or reproduce on its own. In order to make new copies of itself it must infect the cells of a living organism. It belongs to the class **Retroviruses** and family **lentiviridae**. On the basis of geographical distribution, biological and molecular characterization and extent of transmissibility, two types of HIV are recognized- **HIV1** and **HIV2.**

Q.6. What does HIV consist of?

Ans. HIV is 120 nm icosahedral, enveloped virus which stores its genetic information as **ribonucleic acid (RNA)**. RNA must be converted to DNA by a special enzyme **reverse transcriptase**. The outer envelope of virus consists of a **lipid bilayer** with 72 spikes or knobs of **gp 120** and **gp 41**. The protein core inside the envelope contains two copies of RNA and the virus enzymes **reverse transcriptase**, **integrase** and **protease** which are essential for viral replication and maturation.

Q.7. What steps are involved in the **life cycle** of HIV?

Ans. The **life cycle** of HIV consists of following steps.

Entry: Virus binds to the host cells by attaching its **Gp 120** to the host cell receptor. **CD4** molecule present on **T lymphocytes** is a predominant receptor. Entry of virus into host cell requires certain co-receptors e.g. **CCR-5**, **CXCR-4**, **CCR-2** and **CCR-3** etc. Once the **gp 41** of the virus fuses with the host cell membrane, the **capsid** is uncoated. The contents of the HIV particle are then released into the cell, leaving the envelope behind.

Reverse Transcription and Integration: A **ribonucleoprotien complex** capable of reverse transcription is formed and forms **cDNA** under the effect of viral enzyme, the reverse transcriptase. The nucleoprotein complex formed after transcription comprises of linear **double stranded DNA**, the **gag** matrix protein, the accessory **vpr** protein and the viral integrase. This is called **pre-integration complex** and is transported into the host cell nucleus. Viral integrase leads to a complex series of enzymatic steps and integration of virus occurs with the host cell genetic material forming the **provirus.**

Transcription and Translation: HIV **provirus** may lie dormant within a cell for a long time. But when the cell becomes activated, it treats HIV genes in much the same way as human genes. First it converts them into **messenger RNA** (using human enzymes). Then the messenger RNA is transported outside the nucleus, and is used as a blueprint for producing new HIV proteins and enzymes.

Assembly, Budding and Maturation: The strands of messenger RNA produced by the cell containing HIV genetic material gather together with newly made HIV proteins and enzymes to form new viral particles, which are then released from the cell. The enzyme **protease** plays a vital role at this stage of the HIV life cycle by chopping up long strands of protein into smaller pieces, which are used to construct mature viral cores. The newly matured HIV particles are ready to infect another cell and begin the replication process all over again. In this way the virus quickly spreads through the human body. HIV replicates rapidly with several billion new viruses made every day in a person infected with HIV. What makes HIV so difficult to stop, is its ability to mutate and evolve.

Q.8. How does HIV Devastate the Immune System?

Ans. Every day, HIV destroys billions of **CD4+ T cells** in a person infected with HIV, eventually overwhelming the immune system's capacity to regenerate or fight other infections. HIV can kill the cells in different ways:

Direct killing: When a large number of virus is produced and buds out from the surface of CD4+ T cell surface, the cell membrane is disrupted and this causes the cell to die. Also, the virus excessively uses the cell's machinery for its own purposes, disrupting normal activities needed for the survival of the cell.

Apoptosis: The disregulated infected cell may commit suicide by a process known as programmed cell death or apoptosis.

Destruction of un-infected Cells: Cells that are not infected with HIV may also die as a direct result of the effects of HIV infection.

- HIV may bind to the cell surface, making it appear as if the cell is infected. After antibodies attach to the virus on the cell, **killer T cells**, which serve to protect the immune system by killing infected cells, may mistakenly destroy the cell. This process is called **antibody-dependent cellular cytotoxicity**.
- **CD8 T cells**, also known as "killer T cells," may mistakenly destroy uninfected cells that have consumed HIV particles and display HIV fragments on their surfaces.
- Because some HIV envelope proteins bear some resemblance to certain molecules on CD4+ T cells, the body's immune responses may mistakenly damage these cells.
- Uninfected cells may undergo apoptosis. HIV envelope alone or when bound to antibodies sometimes sends an inappropriate signal to CD4+ T cells. This can cause the cells to undergo apoptosis, even if not infected by HIV.

Destruction of Immune Precursor Cells: HIV may also destroy young cells that have not yet fully developed and will that later mature into cells with special immune functions. HIV can also damage the bone marrow and the **thymus**, which are needed for developing precursor cells.

Q.9. How does HIV hide from the immune system?

Ans. The virus can hide within the **cytoplasm** or integrate into the infected cell's genetic material. Protected from the immune system, HIV can lie dormant in an infected cell for a long time, sometimes for years. These cells serve as a latent reservoir of the virus.

Q.10. What are the **different stages** of HIV infection?

Ans. HIV infection can be divided into following different stages:

Acute Primary Infection: Once HIV enters the body, the virus infects a large number of CD4+ T cells and replicates rapidly. During this acute phase of infection, the blood has a high number of HIV copies (**viral load**) that spread throughout the body, propagating in various organs, particularly the **lymphoid organs** such as the thymus, spleen, and lymph nodes. During this phase, the virus may integrate and hide in the cell's genetic material. In this phase of infection, up to 70 percent of HIV-infected people suffer flu-like symptoms.

The Immune System Strikes Back: Two to 4 weeks after exposure to the virus, the immune system fights back with **CD8+ T cells** and **antibodies**. At this point, HIV levels in the blood are significantly reduced. At the same time, CD4+ T cell counts rebound, and for some people the number rises to its original level.

Clinical Latency: During this phase, a person infected with HIV may remain free of HIV-related symptoms for several years despite the fact that HIV continues to replicate in the lymphoid organs where it initially seeded.

Progression to AIDS: The immune system eventually deteriorates to the point that the human body is unable to fight off other infections. The HIV viral load in the blood dramatically increases while the number of CD4+ T cells drops to dangerously low levels. An HIV-infected person is diagnosed with AIDS when he or she has one or more **opportunistic infections**, such as **pneumonia** or **tuberculosis**, and has fewer than 200 CD4+ T cells per cubic millimeter of blood.

Q.11. What are the **routes of transmission** for HIV?

Ans. The HIV can transmit by following routes.

Unprotective sexual inter-course: Whether heterologous or homologous, is a major route of transmission

Blood borne infection: Use of HIV-infected blood, transplanted organs or tissues, improperly sterilized needles and syringes that have been in contact with infected blood is an efficient way of transmission

Mother to child: HIV can transmit from an HIV-infected woman to her foetus or infant before, during, or after birth.

Q.12. What are the common **opportunistic infections** in AIDS?

Ans. Almost 80% of patients with AIDS die from infections other than human immunodeficiency virus (HIV). These infections usually occur late in the course of disease when CD4 T-cell count has fallen below 200 cells per cubic milliliter. Most of these infections are caused by organisms that do not normally afflict healthy individuals and are thus considered to be **opportunistic**. Common important pathogens are: ***Mycobacterium tuberculosis, Candida albicans***, ***Pneumocystis carinii*** (usually non-invasive pathogens), ***Cryptosporidium parvum*** (invades the cells but remains localized in the gut), ***Toxoplasma gondii*** (penetrates through the gut to cause systemic infection), **Cytomegalovirus** and **Herpes simplex virus**.

Q.13. What **Bio-safety** measures should be taken in laboratories while handling HIV-infected samples?

Ans. There is no vaccine for prevention of HIV/AIDS; treatment is expensive and not easily available. Therefore, prevention of infection is important in health care settings. **Universal precautions** should be taken while handling samples in laboratory. These include

Barrier protection like use of gloves, laboratory gowns, facial masks and protective glasses

Frequent and thorough hand washing

Safe techniques including use of **biological safety cabinet class 2**

Safe handling of sharp items and safe disposal of used sharps in puncture-proof containers

Safe handling of specimen

Safe techniques including mechanical pipetting device.

24 CHAPTER

IMMUNOLOGY

Q.1. What is immunology?

Ans. Study of a host's specific defence to a pathogen or a foreign substance is called immunology.

Q.2. What is an **antigen**?

Ans. A substance which on entering a body causes formation of antibodies and reacts only with a specific antibody is called antigen.

Q.3. What is an **antibody**?

Ans. A protein formed by the body in response to an antigen and is capable of combining specifically with that antigen often leading *in vivo* to its destruction or inactivation.

Q.4. What is **acquired immunity**?

Ans. It is the protection which an animal (vertebrate) develops over a period of time against certain types of microbes or foreign substances.

Q.5. Give types of acquired immunity.

Ans. There are two main types of acquired immunity: (1) **Naturally acquired** and (2) **Artificially acquired**. Naturally acquired immunity **further consists** of two types: (a) Natural Active immunity, in which antigens enter the body naturally, and the body produces antibodies and specialized lymphocytes. (b) Naturally acquired Passive immunity, as the antibodies pass from mother to foetus via placenta or to the infant through her milk.

Likewise, artificially acquired immunity also is of two types: (a) artificially acquired active immunity, e.g., the antigens are introduced in vaccines. The body produces antibodies and specialized lymphocytes. (b) artificially acquired passive immunity, e.g., preformed antibodies in immune serum are introduced into body by injection.

Q.6. What is **colostrum**?

Ans. The first secretion of breast milk of mammals after birth of infant. It helps to establish natural active immunity.

Q.7. What is **serum** (plural sera)?

Ans. The remaining fluid after the blood has clotted or coagulated.

Q.8. What is immune serum globulin or gamma globulin?

Ans. On subjecting the serum to an electric current during electrophoresis the proteins in it move at different rates. One of the components separated in this way contains most of the antibodies. This antibody rich component of serum is called immune serum or gamma globulin.

Q.9. What is **humoral immunity**?

Ans. It is an antibody mediated immune system found in vertebrates.

Q.10. Which cells of the blood are responsible for production of antibodies?

Ans. B-cells or B-lymphocytes.

Q.11. Production of which types of cells is involved in cell mediated immunity?

Ans. Specialized lymphocytes called T-cells or T-lymphocytes.

Q.12. What are **epitopes**?

Ans. The epitopes are specific regions on the surface of an antigen against which antibodies are formed.

Q.13. Name the common antigens.

Ans. These are proteins or large polysaccharides. The lipids and nucleic acids often are antigenic only when combined with proteins and polysaccharides. The antigenic components are often components of invading microbes, e.g., capsules, cell walls, flagella, fimbriae and toxins of bacteria, protein coats of viruses, and surface of other types of microbes.

Pollen grains, egg white, blood cell surface molecules, serum proteins from other individuals or species and surface molecules of transplanted tissues and organs are the examples of nonmicrobial antigens.

Q.14. What are the substances against which specialized B-cells defend in humoral or antibody mediated system?

Ans. Extracellular bacteria, bacterial toxins and viruses.

Q.15. Which are the substances against which T-cells defend in cell mediated immune systems?

Ans. Intracellular bacteria and viruses, fungi, protozoa, helminths, transplanted tissue and cancerous cells.

Q.16. What are **haptens**?

Ans. These are very small molecules which are not antigenic by themselves but work as an antigen on combining with larger carrier molecules, e.g., penicillin works as an hapten and combines with serum proteins causing allergy to some people.

Q.17. Which are the five classes of **immunoglobulins**?

Ans. The five classes of immunoglobulins (Igs) are IgG, IgM, IgA, IgD and IgE.

Q.18. What is common and similarity in **structure** of Ig G, Ig D and Ig E?

Ans. They are **monomers**.

Q.19. How are the structures of Ig M and Ig A closely related?

Ans. Both of them comprise two or more monomers joined together by monosulphide bonds. Ig M is a **pentamer** while Ig A is a **dimer**.

Q.20. Which is the most common of all immunoglobulins?

Ans. Ig G, which has 80% of all antibodies in serum.

Q.21. How does Ig G provide passive immunity to a foetus (fetus)?

Ans. They being monomer, readily cross the walls of blood vessels and also cross the placenta.

Q.22. Why does Ig M often remain inside blood vessels?

Ans. Their large size (Ig M from macro means large) prevents moving about as freely as Ig G, and does not enter the surrounding tissue of blood vessels.

Q.23. Where does Ig A (also called secretory Ig A) generally occur?

Ans. It occurs in secretions such as tears, saliva, mucous, intestine and milk. However, it also occurs in blood and lymph.

Q.24. How does occurrence of T-cells and B-cells differ?

Ans. The lymphocytes T cells and B cells originate from stem cells in adult bone marrow or in fetal liver. It is important to note that red blood cells, macrophages, neutrophils and other white blood cells also originate from these same stem cells. Some cells pass through the thymus and form into as mature T cells. The other cells remain in the bone marrow and become B cells. Both types of cells then migrate to lymphoid tissues, i.e., lymph nodes and spleen.

Q.25. How do B cells recognize antigens?

Ans. B-cells can recognize antigens by antigen receptors present on their surfaces.

Q.26. What is **clonal selection** in activation of antibody producing cells?

Ans. When an appropriate antigen binds to the antigen receptors of B-cells, the B-cells proliferate into a large clone of cells which is referred to as clonal selection.

Q.27. How many molecules of antibody are produced by each plasma cell which live for a few days?

Ans. 2000 molecules per second.

Q.28. The B and T cells that interact with self antigens are destroyed during fetal development. What is it called?

Ans. **Clonal deletion**.

Q.29. What is immunological memory?

Ans. The second response to an antigen is called **immunological memory** or anamnestic response.

Q.30. What are memory cells?

Ans. Some activated B lymphocytes do not result into antibody producing plasma cells. They rather persist as long-lived memory cells for many years.

Q.31. What is **hybridoma**?

Ans. Hybridoma is a clone of cells produced from the fusion cell. It is formed as a result of fusion of the antibody producing B-cell with a cancer cell.

Q.32. What is CD type of T-cells?

Ans. Cluster of differentiation.

Q.33. What are two types of CD receptors?

Ans. CD4 and CD8 cells.

Q.34. What are CD4 cells?

Ans. The CD4 cells are helper T cells (which also work as receptors of HIV in causation of AIDS).

Q.35. What are CD8 cells?

Ans. They are cytotoxic and suppressor T cells.

Q.36. What is **septic shock?**

Ans. Sudden drop in blood pressure due to infection by Gram-negative bacteria (even infection due to cell walls of population of Gram-negatives killed by antibiotics).

Q.37. What are helper T or (TH) cells?

Ans. Specialized T cells that generally interact with an antigen before B-cells start to interact with the antigen.

Q.38. What are cytotoxic T (Tc) cells?

Ans. These are specialized T cells which destroy infected cells possessing antigens. Therefore, they provide protection against viral infections and intracellular bacterial infections.

Q.39. What are delayed **hyper**sensitivity T (TD) cells?

Ans. These are specialized T-cells which produce lymphokines in type IV hypersensitivities. They are therefore, associated with certain allergic reactions. TD cells are the cells from which cell-mediated immunity was originally named.

Q.40. What are suppressor T (Ts) cells?

Ans. Ts cells are the cells which are believed to end an immune response after an antigen is no longer present.

Q.41. Match the following:

(A)	Natural killer cells are not immunologically specific but attack and destroy target cells by causing lysis.	(a)	Are soluble chemical messengers by which cells of immune system communicate with each other. These are produced by lymphocytes as well as macrophages.
(B)	**Cytokines** (formerly called lymphokines.	(b)	Virus infected and tumour cells.
(C)	Cytotoxic T (Tc) cell discharges a protein known as **perforin**.	(c)	Granulocyte—macrophage colony stimulating factor abbreviated as GM-CSF.
(D)	Stimulates the formation of red and white blood cells from stem cells.	(d)	Lyses the infected cell.
(E)	The population of CD4 cells is decimated.	(e)	Antibody producing spleen cells and myeloma cells (cancerous B cells) fuse to form.
(F)	Hybrid cells, the proliferation of which into clones is called **hybridoma**.	(f)	By infection of AIDS virus.

Ans. [A b], [B a], [C d], [D c], [E f], [F e],

Q.42. Fill in the blanks.

(i) ———— in 1975 discovered the process to fuse an "immortal" cancerous B cell a mutant that lacked the capability to produce antibody with a noncancerous antibody secreting plasma B cells from a mouse immunised with a particular antigen.

(ii) Humoral immunity involves bone marrow-derived ———— .

(iii) Cell mediated or cellular immunity primarily involves thymus derived ————.

(iv) In nonspecific defence mechanism the skin produces fatty acids which ———— proteins of cell membrane of the microorganism.

(v) In nonspecific immunity trypsin ———— proteins of cell membrane and wall of the microbe.

(vi) In nonspecific immunity a pH dependent polyamine found in sperm, seminal fluid and other tissues inhibits the growth of Gram positive bacteria is called ————.

(vii) Natural killer (NK) cells kill their targets by causing the formation of ———— in their cell membrane.

(viii) Perforins, serine proteases, nucleases TNF (tumour neurosis factor and FAS ligand (CD 95) are major ———— of activated NK cells.

(ix) Perforins of NK cells polymerize the cell membrane causing pores which allows ———— and toxic molecules inside the target cell.

(x) Serine proteases of NK cells degrade ———— in cell membrane of the target cell.

(xi) The ———— of NK cells degrade nucleic acid in target cell.

(xii) Tumour neurosis factor (TNF) of the NK cell ———— protein synthesis and lead to ———— of toxic free radicals in target cell.

(xiii) On maturing a thymocyte acquires certain ———— antigens in its membrane and loses others hence permitting identification of T-cell subsets by means of complex of these surface proteins.

(xiv) CD proteins and T-cells receptor for antigen are molecules on T-cell membrane that work in ———— acceptance.

(xv) Phagocytosis is the ———— immunity.

(xvi) Neutrophils (polymorphonuclear leucocytes or PMNs) monocytes and macrophages are major types of ———— cells involved in nonspecific immunity.

(xvii) The membrane bound bags called lysosomes fuse with the phagosome to form ____.

(xviii) The phagocytic vacuole or food vacuole of a phagocyte is called a ———— .

(xix) Aryl sulphatase that act on sulphate esters, cathepsin that act on protoglycans, cholesteryl esterase that act on lipoproteins, glycosidase that act on carbohydrates, lysozyme that act on ————, phosphatase that act on phosphate esters and phospholipase are important **lysosomal enzymes**.

(xx) ———— are the substances that bind to particle making it more susceptible to phagocytosis.

Ans. (i) Georges Köhler and Cesar Milstein

(ii) B lymphocytes or B cells

(iii) T lymphocytes or T-cells

(iv) denatures

(v) hydrolyzes

(vi) spermine

(vii) pores

(viii) cytotoxic products

(ix) cytosol

(x) proteins

(xi) nucleases

(xii) depresses, production

(xiii) CD (cluster differentiation)

(xiv) epitope (antigenic determinant)

(xv) nonspecific

(xvi) phagocytic

(xvii) Phagolysosome

(xviii) phagosome

(xix) peptidoglycan

(xx) Opsonins

Vaccines

Q.43. What is a vaccine?

Ans. A vaccine is an antigenic preparation administered in order to stimulate the recipient's protective immunity to one or more particular pathogens and/or toxins. Some vaccines e.g., Sabin vaccine (antipoliomyelitis) are given orally, others e.g. BCG are given parenterally.

To be effective a vaccine must stimulate the formation of antibodies in part of the body where they can efficiently counteract infection. There are six main types of vaccines.

(a) Inactivated vaccines e.g. those used against cholera and typhoid.

(b) Attenuated vaccines e.g. Those used against tuberculosis and yellow fever.

(c) **Toxoids** e.g. in the triple vaccine.

(d) Solutions or suspensions of antigenic extracts from specific pathogens e.g. polysaccharide capsular material from *Streptococcus pneumoniae*.

(e) Synthetic vaccines e.g. synthetic peptides antigenic determinants of the pathogen/toxin, e.g. foot and mouth disease, influenza virus and plasmodium.

(f) Genetically engineered vaccines, e.g. vaccinia virus.

Q.44. Match the following:

(A)	The individuals immunised with oral polio vaccine are example of	(a)	Inactivated (killed) or alive but attenuated microorganisms are used to form
(B)	Whole agent vaccines	(b)	Herd immunity
(C)	Crude fraction of *Vibrio cholerae* (Cholera vaccine)	(c)	Every 10 years for adults
(D)	Purified diphtheria toxoid which works against diphtheria	(d)	Works as a booster for every six months
(E)	Pertussis (whooping cough)	(e)	Plague
(F)	Crude fraction of *Yersinia pestis*	(f)	Killed whole or fragments of *Bordetella pertussis*
(G)	Pneumococcal pneumonia	(g)	Tetanus
(H)	Purified tetanus toxoid	(h)	Purified polysaccharide from *Streptococcus pneumoniae*
(I)	Killed or attenuated *Salmonella typhi*	(i)	Hepatitis B virus vaccine
(J)	Comprises of a portion of viral protein coat produced by a genetically engineered yeast	(j)	Is used to prepare vaccine against typhoid fever

Ans. [A b], [B a], [C d], [D c], [E f], [F e], [G h], [H g], [I j], [J i]

Q.45. Choose the correct (More than one option may be correct).

(i) **DPT** (Diphtheria, Pertussis, tetanus) is used for childhood immunisation.

(ii) **OPV** is the oral polio vaccine for children.

(iii) The **precipitin ring test** show a cloudy ring of precipitation which is also called **zone of equivalence**.

(iv) The **countercurrent immunoelectrcophoresis (CIE) test** also known as **counter-immunoelectrophoresis** can be used in the diagnosis of bacterial meningitis and some other diseases.

(v) The progress of disease in **HIV infection** is not followed by **seroconversion**.

(vi) The viruses causing mumps, measles and influenza can agglutinate and the process is called **viral haemogglutination**.

(vii) **Radioimmunoassay technique** can be used in detection of hormones, drugs and immunoglobulins.

(viii) The fluorescein isothiocyanate (FITC) a fluorescent dye with antibodies cause them emit fluorescence on exposure to ultraviolet light.

(ix) The **fluorescent-antibody technique** can detect rabies in a few hours with nearly cent per cent accuracy.

(x) Fluorescence-activated cell sorter is used to separate different classes of T-cells.

(xi) **Enzyme-linked immunosorbent assay (ELISA)** is also called **enzyme immunoassay (EIA).**

(xii) The **direct ELISA** detects antigens.

(xiii) The **indirect ELISA** detects antibodies.

(xiv) The ELISA tests are either clearly positive or clearly negative.

(xv) In positive direct ELISA to detect antigen a sandwich is formed with antigen between two antibody molecules.

(xvi) Indirect ELISA is used to test for **antibodies against HIV** and in **home pregnancy test**.

(xvii) The home pregnancy test cannot detect the hormone known as **human chorionic gonadotropin (hCG)**.

(xvii) The hormone called human chorionic gonadotropin is excreted only in the urine of a **pregnant woman**.

(xviii) In ELISA the antibodies are linked to an enzyme such as **horseradish peroxidase** or **alkaline phosphatase**.

(xix) The **ELISA** differs from radioimmunoassay **(RIA)** in the sense that **enzyme** is used in place of radioisotope in ELISA.

(xx) Radioimmunoassay (RIA) is an extremely sensitive method (to the picogram level). Therefore, can be used for quantitation of any substance which can be labelled with a radioactive isotope e.g. Iodine-125.

Ans. i, ii, iii, iv, vi, vii, viii, ix, x, xi, xii, xiii, xiv, xv, xvi, xviii, xix, xx Correct; v, xvii Wrong.

Immune System and Immunity

Q.46. Elaborate the following.

(i) History of immunology.

(ii) Structure, composition and function of cells and organs involved in immune system.

(iii) Host parasite relationships.

(iv) Microbial infection.

(v) Virulence and host resistance.

(vi) Immune response.

(vii) Naturally acquired immunity.

(viii) Artificially acquired immunity.

(ix) Immunohaematology.

(x) Blood groups.

(xi) Blood transfusion and Rh incompatibility.

Ans. (i) The Chinese physicians were the first to make children inhale powder of dry crusts or scabs of small pox to protect them from small pox.

In 1700s in Turkey Lady Montague, the wife of the British ambassador reported of the practice prevailing at that time that an old woman with a nutshell full of matter of small pox would come to place a needle full of the matter on the opened vein of a child. The protection provided was probably because of two reasons.

(1) The chosen portal of entry for the pathogen which normally infects body by entering through respiratory system.

(2) Probably the use of less virulent form of the virus.

This process called **variolation** can be a useful form of protection but virulence of the pathogen being uncertain sometimes may kill the recipient.

Edward Jenner was one of such persons who had received treatment at the age of eight. Jenner on interaction with a milk maid who was free of tension of disease small pox she being in contact with cow pox and motivated with his childhood memories of variolation Jenner conducted a series of experiments in 1798 and deliberately inoculated people with cow pox.

(ii) The three major types of cells are **erythrocytes** (red blood cells) and **leucocytes** (white blood cells) and platelets. The RBCs or erythrocytes which are rounded oval, biconcave biconvex in structure are 4.8 to 5.4 million per cubic millimeter (mm^3) perform the function of transportation of O_2 and CO_2. Here we are concerned with leucocytes which by themselves or derivatives their of function as **phagocytes**.

The leucocytes (WBC) can be further categorised into **granulocytes** characterised by presence of granules which get stained and can be seen under light microscope; and **agranulocytes** which lack granules. The granulocytes are neutrophils, basophiles and eosinophils. The **neutrophils** are 60 to 70% of leucocytes and their granules stain red and blue with a mixture of acidic and basic dyes respectively. The neutrophils are also called polymorphonuclear leukocytes (PMNs) or polymorphs. They are highly phagocytic and motile.

The granules of basophils (0.5 to 1% of leucocytes) stain blue with basic dye methylene blue and that of eosinophils (2 to 4% of leucocytes) stain red with acidic dye eosin. The neutrophils are also called polymorphonuclear leucocytes (PMNs) or polymorphs since the nuclei of polymorphs are contained in two to five lobes. The neutrophils are highly phagocytic and motile have got the ability to leave the blood and enter the infected tissue and destroy microorganisms and foreign particles. The **basophils** release substances like histamine and are important in inflammation and allergic responses and release heparin which is an aticoagulant of blood. The **eosinophils** are somewhat phagocytic and possess the ability to leave the blood. They produce toxic proteins against some parasites. Being small in size they cannot ingest helminths but attach themselves on the outer surface of the parasites and discharge peroxide ions and consequently destroy them.

Among the agranulocytes are 3% to 8% **monocytes** which lack granules are not actively phagocytic until they leave the circulating blood and mature into macrophages. The maturation and proliferation of macrophages along with lymphocytes is taken as one of the factors leading to swelling of lymph nodes during an infection.

The **lymphocytes** which are 20% to 25% are without granules are not phagocytic rather are involved in specific immunity. The lymhocytes occur in lymphoid tissues of the lymphatic system: in the tonsils, lymph nodes, spleen, thymus gland, thoracic duct, bone marrow, appendix, Peyer's patches of small intestine and lymph nodes in the respiratory, gastrointestinal and reproductive tracts. The lymphocytes are mainly of two types **B lyphocytes** or **B cells** carry on production of antibodies while **T lymphocytes** or **T cells** take part in cell mediated immunity.

(iii) There exist four types of relations among microorganisms which are **amensalism** in which one species is harmed while other remains unaffected; **commensalism** in which one organism is benefitted while the second is neither benefitted nor harmed; mutualism in which both interacting species are benefitted; and **parasitism** which is the association in which the parasite not only depends on host for nutrients rather is harmful to the host. Many disease causing bacteria are parasites.

Once the parasite or the pathogen finds entry inside the host after breaking nonspecific resistance or defence a number of changes are brought to each other by the parasite as well as host. The nonspecific defence by host also include **phagocytosis** involving the mechanism of chemotaxis, adherence, ingestion and digestion. The entry of the parasite also leads to inflammation of the damaged tissue, vasodilation and increased permeability of blood vessels, fever and complement system. The serum proteins in complement take part in lysis of foreign cells, inflammation and phagocytosis. In case of viral infections host animal produce antiviral proteins called **interferons**. The human interferons are of three principal types called **alpha interferon**, **beta interferon** and **gamma interferon**.

The infection by a pathogen parasite also leads to establishment of specific defence mechanism, through naturally acquired immunity (**active**—by forming antibodies against antigen, and **passive** antibodies passed from mother to foetus through placenta or mother to infant in her milk) or **artificially acquired active immunity** from vaccination.

(iv) The **infection** is invasion or colonization of the body (host) by pathogenic microorganisms. When the infection leads to a change from a state of health it is called **disease**.

(v) The capacity of a pathogen to cause disease is **virulence**. It is defined broadly in terms of severity of symptoms in the host. A highly virulent strain may cause severe symptoms in a susceptible individual while a less virulent strain would produce relatively less severe symptoms in the same individual. The severity of the disease which actually develops may be regarded as a result of joint expression of the degree of infectivity and virulence of the pathogen and the state of immunity either natural or acquired of the host in relation to the pathogen.

The ability of the host to ward off disease by offering defences is called **resistance**. The resistance is of two types: First non-specific resistance and the second specific resistance or immune responses. Under nonspecific resistance the first line of defence is presented by skin and mucous membranes and second line of defence is presented through phagocytes, inflammation, fever and production of antimicrobial substances. The specific resistance or immune response is the third line of defence offered by the host against the pathogen.

(vi) The **immune response** involves the production of specialised lymphocytes which are responsible for **cell mediated immune system**, and special proteins called **antibodies** responsible for antibody mediated or **humoral immunity**. The organisms or substances that provoke the immune response are called **antigens**. In **cellular immune response** the pathogen invades inside the host cell called target cell. This response is carried out by highly specialised cells that show presence of receptors which recognise target cells and cause their lysis. Thus the functions of immune system are to locate foreign cells, viruses and microorganisms; to neutralise and destroy them and to eliminate these after destruction.

(vii) **Naturally acquired immunity** is mainly of two types:

(1) Naturally acquired active immunity, and

(2) naturally acquired passive immunity.

In **naturally acquired active immunity** antigens enter the body in a natural way consequently the body produces antibodies and specialised lymphocytes, one acquired the immunity is life long for some diseases e.g. measles, chicken pox and yellow fever. The **naturally acquired passive immunity** is the natural transfer of antibodies, e.g. from mother to her infant, and from pregnant woman to foetus through the placenta.

(viii) **Artificially acquired immunity** is acquired artificially either by vaccination or injecting immune serum. In vaccination or providing oral drops (in case of polio drops) the antigens are introduced, consequently body produces antibodies and specialised lymphocytes. In **artificially acquired passive immunity** antibodies present in immune serum are injected. Immune serum can be collected from immune animal (e.g., immunized horse) or individuals.

(ix) **Immunohaematology** is the immunological haematology, that deals with blood, its composition, formation of blood, functions of blood and diseases of blood in relation to immunology. It includes cell mediated and humoral effector responses. The effectors of the humoral branch are secreted antibodies which can neutralise soluble antigens on microorganisms and also can activate the complement system leading to lysis of the invaders, **opsonization** (i.e., deposition of opsonins, substance e.g. an antibody or C3b that binds to an antigen and enhances its phagocytosis), and viral neutralisation. Antigen specific and nonspecific cells, both contribute to cell mediated immune response. The specific cells include cytokine—secreting $CD4^+$ T_H1 and T_H2 cells and $CD8^+$ T cytotoxic lymphocytes (CTLs) while the non-specific cells include macrophages, neutrophils, eosinophils and natural killer cells. Major types of cells have already been described above.

(x) **Blood groups** which is classification of is also referred to as **ABO blood group system**. The four principal types of blood A, B, AB and O were discovered by Karl Landsteiner in early 1900s. Since then at least 15 other blood group systems have been discovered, but ABO blood group and Rh system are the two best known of all the systems. The ABO blood type of an individual depends on the presence or absence of carbohydrate antigens present on the cell membrane of red blood cells also called RBCs. The cells of blood type O are devoid of A and B antigens.

The persons with type AB cells have plasma with no antibodies to either A or B antigens therefore, can receive transfusions with either A or B antigens and so can receive transfusions with A or B blood without any reaction. Type O individuals have antibodies against both A and B antigens. Being devoid of antigens their blood can be transfused without difficulty to others but they themselves can receive transfusions of type O blood only. In case of an incompatible transfusion, e.g. when type B blood is transferred to type A blood, the antigens present on the type B blood cells will react with anti-B antibodies present in the recipient's serum. Such antigen antibody reaction activates complement leading to donor's red blood cells as they enter the recipient. Among about 80% of the population, the individuals which are known as **secretors**, the soluble antigens of the ABO type occur in saliva and other bodily fluids. The saliva residues from a cigarette can be used by the investigators to detect the criminals.

(xi) **Blood transfusions and Rh incompatibility** takes a serious turn if blood from a Rh^+ donor is given to an Rh^- recipient. In such a case the donor's red blood cells stimulate the production of anti-Rh antibodies in the recipient. If after this the recipient receives Rh^+ red blood cells in a subsequent transfusion a serious hemolytic reaction develops rapidly. Besides blood transfusions, **the hemolytic disease of the newborn** is an another example in which an Rh^- person can become sensitized to Rh^+ blood. When an Rh^- woman and Rh^+ man produce a child then there are 50% chance that the child will be Rh^+. Since the child is Rh^+, the Rh^- mother can become sensitised to this antigen during birth. When the placental membranes tear and the foetus Rh^+ red blood cells enter the mothers circulation system, lead mothers body to produce anti-Rh antibodies of the IgG type. So, if the foetus in the later pregnancy is Rh^+ mothers anti-Rh antibodies will cross the placenta and destroy the foetal red blood cells. Now-a-days hemolytic disease of the newborn (HDNB) is prevented by immunisation of Rh^- mother with anti-Rh antibodies available in the market.

Antigens

Q.47. Fill in the blanks.

(i) If the response for an antigen in a functioning immune system leads to immunity then the antigen is called ——— and the antigen which produces a state of specific tolerance is called ——— .

(ii) The antibodies produced in response to the antigen react with or bind to only certain portions of the antigens which are known as ——— .

(iii) A low molecular weight molecule which is not immunogenic by itself but can serve as part of an antigenic determinant on an immunogenic molecule (the carrier) is called a ——— .

(iv) A common example of **hapten** is ——— .

(v) Penicillin causes allergic reactions to some people is a good example of ——— .

(vi) The antibodies are ——— which while introduced into an animal for which they are foreign, like any protein elicit the formation of antibodies more appropriate to call "anti-antibodies". Hence antibodies may work as antigens.

(vii) Original home pregnancy test kit found use of antihapten antibodies in a **hapten-inhibition assay**. This test revealed if a woman's urine had **human chorionic gonadotropin (HCG)** which is a sign of ——— .

(viii) Mitogens are capable of inducing ——— in high percentage of T or B cells.

(ix) The common examples of lectins with mitogenic activity are ———, ——— and ——— .

(x) The mitogen other than lectin is ——— component of Gram negative bacterial cell wall that function as a B-cell mitogen.

Ans. (i) immunogenic, tolerogen

(ii) antigenic determinants or epitopes

(iii) **hapten**

(iv) dinitrophenol (DNP)

(v) **hapten**

(vi) proteins

(vii) pregnancy

(viii) cell division

(ix) Concanavalin A (Con A), phytohaemagglutinin (PHA) and pokeweed mitogen (PWM)

(x) lipopolysaccharide (LPS)

Q.48. Give the **phylogeny of immunity**.

Ans. The immunity is a vertebrate trait which is found in all classes of vertebrates including the most primitive of living vertebrates, e.g. lampreys and hag fish but humoral immunity is greatest in birds and mammals. There are some interesting reports of memory responses in annelid worms (segmented animals with ring or ring-like structure, as earthworms, leeches etc.) and in echinoderms (animals with rough skin because of presence of spines, plates, oxides etc., strictly marine and nonparasitic, as star-fishes, sea-stars, feather stars, brittle-stars and sea-lillies. However, sometimes the second response in these animals is slower, i.e., not faster than the first.

Though a variety of effective defence mechanisms as secretions of antimicrobial substances, tissue reactions and phagocytic cells to engulf such invaders, are found in invertebrates but these are not specific and fail to qualify as immune.

Q.49. Give the diversity of antibodies.

Ans. The antibodies are proteins when introduced into an animal for which they are foreign they like any other protein induce the formation of antibodies that may be called antiantibodies. So the antibodies can also work as antigens. Their 3 main types are **isotypes, allotypes** and **idiotypes**.

(i) The human immunoglobulins while injected into rabbit, are recognised as foreign bodies by the rabbit. The rabbit antiserum precipitates with a number of distinct components in human serum. Each of the line of precipitate is called an **isotype** (iso = same). The dominant isotype is IgG.

(ii) The antibody **allotypes** (*all* = other) are genetically distinct types of antibody molecule which are found in some members of a species but not in others. They are inherited as simple Mendelian traits, e.g., rabbit IgG molecules exist in two allotypic forms, d11 and d12.

(iii) An **idiotype** (*idio* = one's own) is usually found only on that subset of antibodies that share an identical antigen binding site. Niels Jerne in 1973 proposed a conceptual theory called the **network theory** which predicted the consequences of immune-system recognition of self-antibody, for which Jerne was awarded Nobel Prize in 1984. According to the network theory as antibody is produced in response to an antigen it in turn induces the formation of antibodies to its unique variable region sequences. Jerne called each of the individual antigenic determinant of the variable region as an **idiotope**. Each antibody contains multiple idiotopes and sum of the individual idiotopes is known as an **idiotype** of the antibody.

Q.50. What are **adjuvants**? Describe.

Ans. The adjuvants are substances which on mixing with an antigen and injected with it increase the immunogenicity of that antigen. The adjuvants show one or more of the following effects.

(i) They prolong antigen persistence.

(ii) The adjuvants enhance co-stimulatory signal.

(iii) They induce granuloma (macrophage-rich mass of cells) formation.

(iv) Stimulate lymphocyte proliferation nonspecifically.

Some of the commonly used adjuvants are given below:

(a) Aluminium potassium sulphate (alum).

(b) Freund's incomplete adjuvant (comprises of aqueous solution of antigen, mineral oil and an emulsifying agent) which disperses the oil into small droplets surrounding the antigen.

(c) Freund's complete adjuvant which contains heat-killed *Mycobacterium* in water in oil imulsion.

(d) *Mycobacterium tuberculosis.*

(e) *Bordetella pertussis.*

(f) Bacterial lipopolysaccharide (LPS).

(g) Synthetic polynucleotide (Poly IC/poly AU).

Q.51. Fill in the blanks.

(i) A **vaccine** is the preparation of antigenic material to induce ———— against pathogenic organisms.

(ii) The attenuated strain of *Mycobacterium bovis* known as **Bacillus Calmette-Guerin (BCG)** was developed by growing *M. bovis* on medium containing increased concentration of ———— in the medium.

(iii) The poliovirus employed in the preparation of Sabin polio vaccine was attenuated by growing on monkey kidney ———— cells.

(iv) The major disadvantage of attenuated strain is the probability of their ———— to a virulent form.

(v) The vaccines for cholera, pertussis and plague are prepared by using inactivated whole ————.

(vi) Tuberculosis vaccine is prepared from ———— bacterial cells.

(vii) The whole inactivated ———— particles are used to prepare influenza vaccine.

(viii) The attenuated viral particles are used to prepare the vaccines of ————, ————, and ————

(ix) ———— viral particles are used to prepare polio (Salk).

(x) Varicella zoster (chickenpox) is prepared from ———— viral particles.

(xi) Yellow fever vaccine is prepared from attenuated ————.

(xii) A toxin which has been altered to eliminate its toxicity but still functioning as an ———— is known as a **toxoid**.

(xiii) Inactivated exotoxins are used to prepare ———— and ———— vaccines.

(xiv) **Recombinant surface antigen** (Hbs Ag) is used to prepare ———— vaccine.

(xv) The vaccines of *Haemophilus influenzae* type b, *Neisseria meningitidis* and *Streptococcus pneumoniae* are prepared from ————.

Ans. (i) immunity
(ii) bile
(iii) epithelial
(iv) reversion
(v) bacterial cells
(vi) attenuated
(vii) viral
(viii) measles mums and polio (Sabin)
(ix) Inactivated
(x) attenuated
(xi) viral particles
(xii) immunogen
(xiii) diphtheria and tetanus
(xiv) Hepatitis B
(xv) Capsular polysaccharide

Immunoglobulins

Q.52. Describe the **basic structure** of immunoglobulins.

Ans. Identification of antibodies in the serum protein fraction was demonstrated in the classic experiments of A. Tiselius and E.A. Kabat in 1939. However, the basic structure of immunoglobulin molecule was revealed by Rodney R. Porter and Gerald Edelman in 1950s and 1960s for which they received Nobel Prize in 1972. Porter and Edelman using ultracentrifugation differentiated γ -globulin (gamma globulin) fraction into high molecular weight fraction bearing sedimentation constant of 19 S and low molecular weight fraction with a sedimentation constant of 7 S. The 7 S fraction bearing a 1,50,000 MW gamma (γ) globulin was designated as immunoglobulin G or IgG by them. On subjecting brief digestion of IgG with the enzyme **papain** two similar fragments each with MW of 45,000 were obtained by Porter which were called **Fab fragments** since they retained the "antigen binding" activity and the one fragment with MW 50,000 was called Fc fragment for it crystallised during cold storage. Similar experiments were performed by Alfred Nisonoff using the enzyme **pepsin** and on brief digestion a single 1,00,000 MW fragment composed of two Fab-like fragments to which he designated $F(ab')_2$. Like Fab fragments the $F(ab')_2$ fragments precipitated antigens. The Fc fragment, however, was not recovered on digestion with pepsin, which got digested into multiple fragments. The experiments of Porter revealed that IgG has two 50,000 MW polypeptide chains designated as **heavy (H) chains** and two 25,000 MW chains designated as **light (L) chains**. By using antisera from goats it was concluded that Fab consists of portions of a heavy and a light chain and that Fc contains only heavy chain components. Hence, Porter and Edelman proposed the prototype structure for IgG according to which the IgG molecule consists of two identical H chains and two identical L chains which are linked by disulphide bridges (Fig. 24.1).

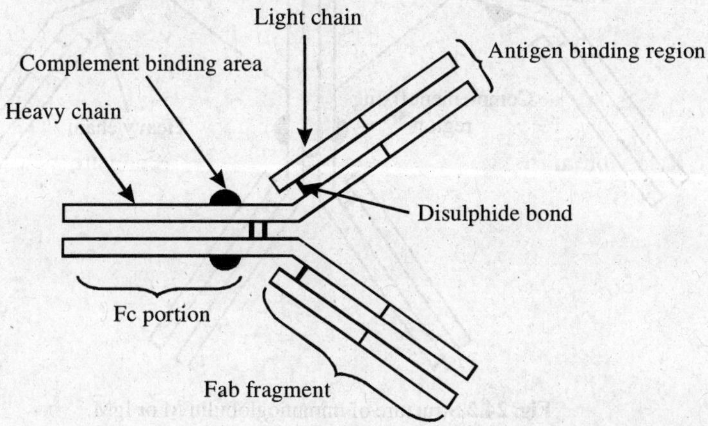

Fig. 24.1 The structure of immunoglobulin G. The prototype structure of IgG was proposed by Rodney Porter in 1962. The chain structure and disulphide bonds have been shown in the structure above. Two light chains are identical as are in the two heavy chains.

Q.53. Fill in the blanks.

(i) The four **subclasses** of IgG in human according to their decreasing average serum concentration are ———, ———, ——— , and ——— .

(ii) IgG1, IgG3 and IgG4 readily cross the placenta and protect the developing ——— .

(iii) IgG3 is the most effective complement ———, followed by IgG1.

(iv) IgG1 and IgG3 bind with high affinity to ——— on phagocytic cells.

(v) Fc binding permits IgG to function as an ——— .

(vi) IgM is the first **immunoglobulin class** produced as a primary response to ——— .

(vii) IgM is the ——— immunoglobulin synthesised by the neonate.

(viii) An IgM molecule can bind ——— small hapten molecules.

(ix) One hundred to one thousand times more molecules of ——— will be needed than that of IgM to achieve the same level of agglutination.

(x) The IgM due to its large sized molecules (Fig. 24.2) do not diffuse well, therefore, is found in very low concentration in the intra-cellular tissue ———— .

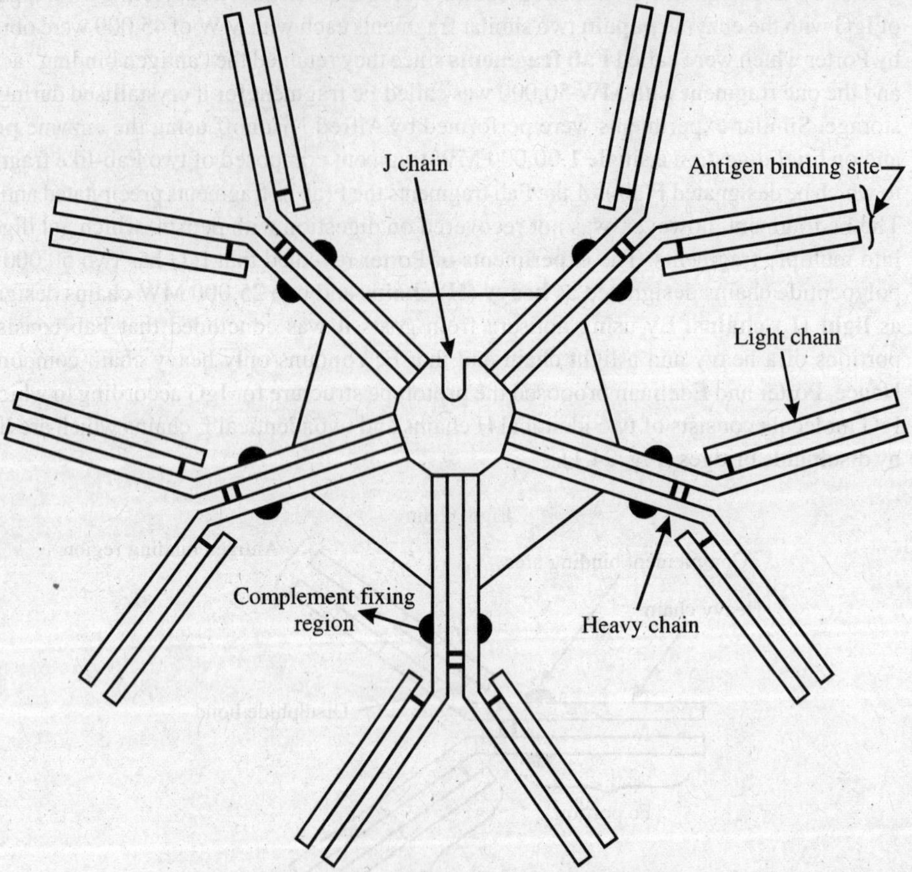

Fig. 24.2 Structure of immunoglobulin M or IgM.

(xi) The ———— is an important secretory immunoglobulin.

(xii) IgA is a predominant **immunoglobulin class** is external ———— like breast milk, saliva, tears and mucous of the bronchial, genetourinary and digestive tracts.

(xiii) IgA provides an important line of defense against bacteria such as *Salmonella, Vibrio cholerae, Neisseria gonorrhoeae,* and virus such as ————, influenza, and reovirus.

(xiv) The **P–K reaction** was the basis for earliest **biological assay** for ———— activity.

(xv) Most of the IgA in the serum form monomers having 2 H and 2 L chains. The L chains are like other classes of antibodies either ———— or lambda.

(xvi) A substantial **biological benefit** of IgA is that it is resistant to bodies own ———— as a secretory IgA in the intestine.

(xvii) One of the **biological activity** of IgE is that it mediate the immediate hypersensitivity — ———— responsible for symptoms of hay fever, asthma, hives and anaphylactic shock.

(xviii) The presence of a serum component that cause allergic reactions was first demonstrated by K. Prausnitz and H. Kustner in 1921 while they injected serum from an allergic person intradermally into a ———— individual.

(xix) The **biological activity** of IgE is accounted because of its property to bind ———— and to their tissue equivalent, mast cells.

(xx) Georges Köhler and Cesar Milstein devised a method for preparing **monoclonal antibody** which was derived from a single ————.

Ans. (i) IgG1 (9 mg/ml), IgG2 (3 mg/ml), IgG3 (1 mg/ml) and IgG4 (0.5 mg/ml)

(ii) foetus

(iii) activator

(iv) Fc receptors

(v) **opsonin**

(vi) antigen

(vii) first

(viii) ten

(ix) IgG

(x) fluids

(xi) IgM

(xii) secretions

(xiii) polio

(xiv) IgE

(xv) Kappa

(xvi) proteases

(xvii) reactions

(xviii) nonallergic

(xix) basophils

(xx) clone

Q.54. Give the **clinical uses** of **monoclonal antibodies**.

Ans. The clinical uses of monoclonal antibodies (Fig. 24.3) are as under:

(i) They are used for *in vitro* diagnostic reagents e.g., for detecting pregnancy; in diagnosis of pathogenic microorganisms, measuring the blood levels of various drugs; matching histocompatibility antigens; and detecting antigens shed by tumours.

(ii) Radiolabelled monoclonal antibodies too can be used for detecting tumour agents *in vivo*.

(iii) Radiolabelled monoclonal antibodies can also be used for early diagnosis of some primary or metastatic tumours, e.g. monoclonal antibody to breast cancer cells labelled with iodine-131 has been introduced into the blood to detect spread of tumour to regional lymph nodes. This is known as **monoclonal imaging technique** that can detect breast cancer metastases which remains undetected by other scanning techniques that are often used.

A monoclonal antibody that exhibit catalytic activity is called **abzyme**.

Antigen-Antibody Interactions

Q.55. Choose the correct.

(i) **Agglutination** cannot be carried *in vitro*.

(ii) The strength of multiple interactions between multiple antibody and antigen is called **avidity**.

(iii) Secretory **pentameric IgM** often show higher affinity than IgG.

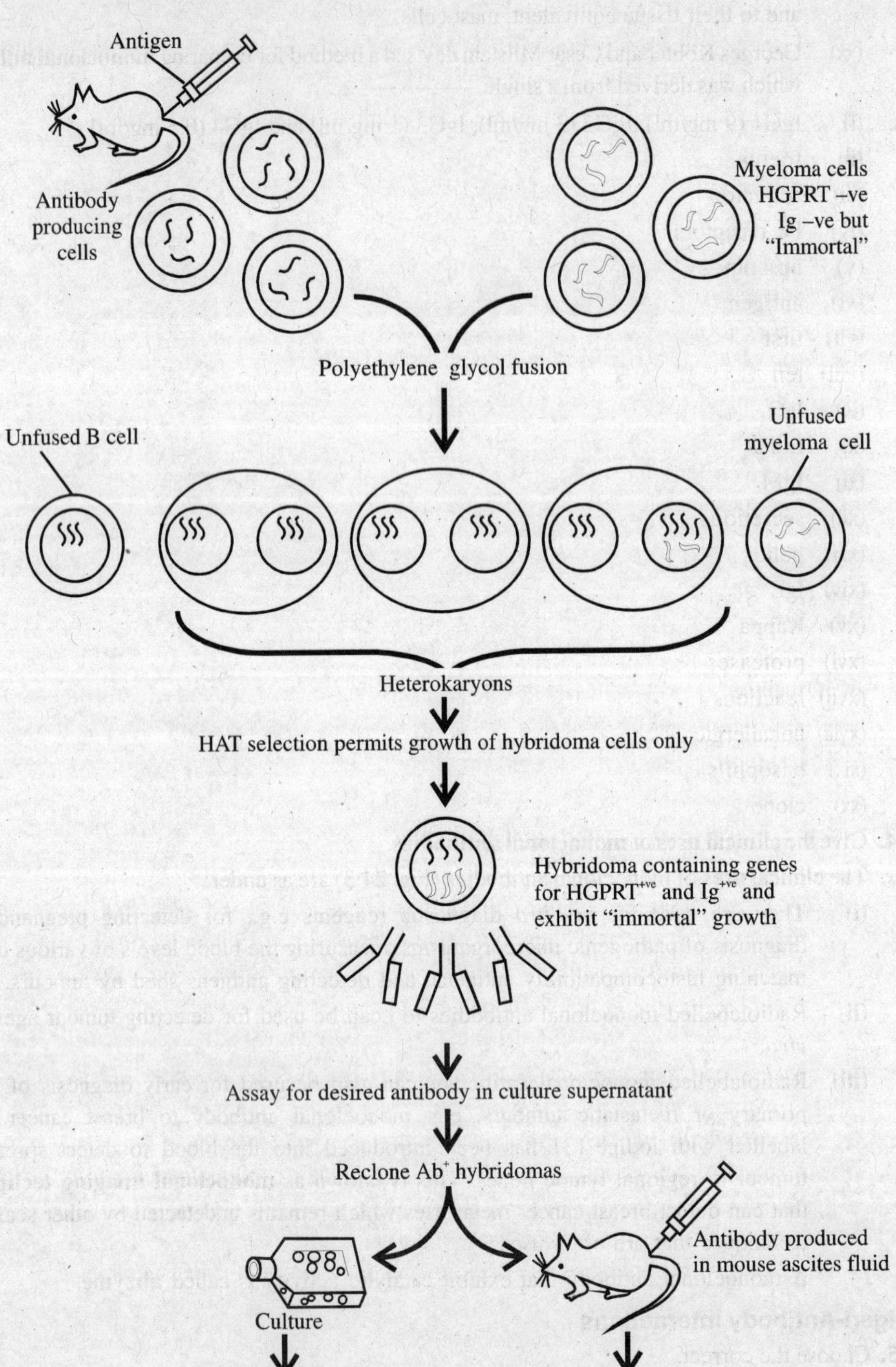

Fig. 24.3 Production of a monoclonal antibody.

(iv) The interaction between antibody and a particular antigen results visible clumping which is known as **agglutination**.

(v) The inhibition of agglutination reactions due to excess of antibody is not **prozone effect**.

(vi) At neutral pH red blood cells are surrounded by negative ions which makes the cells repel each other. This repulsive force is known as **zeta potential**.

(vii) **IgM** because of its **large size and pentameric nature** cannot overcome zeta potential.

(viii) **IgG** because of its **smaller size and bivalency** is less able to overcome zeta potential.

(ix) The agglutination titer of an antiserum cannot be used to diagnose a bacterial infection.

(x) The **Widal test** is the example of an agglutination test which is used for detecting serum antibodies to salmonella which cause human enteric fevers including **typhoid**. In it the serial dilutions of serum are tested for the presence of agglutinins to the O and H antigens of salmonellae.

Ans. Odd numbers, wrong; even numbers, correct.

Q.56. What are **precipitation reactions**? Explain.

Ans. The precipitation reactions occur as an interaction between an antibody and an antigen in aqueous solution forming a lattice which develop into a visible **precipitate**. The antibodies which aggregate soluble antigens are referred to **precipitins**. The precipitation reactions can also be used as a rapid test for the presence of antibody or antigen. The **interfacial (ring) precipitin test** is done by adding antiserum to a small tube and layering antigen on the top. The formation of a visible line of precipitation in the tube is indicative of positive reaction.

The precipitation reactions may also be in the gels. These **immunodiffusion reactions** can be used to determine relative concentration of antibodies or antigens. The immunodiffusion techniques are of following types:

(i) **Radial immunodiffusion** or **Mancini method** which is routinely used to serum levels of IgM, IgG and IgA. But this method cannot detect antigens concentration below 5–10 μg/ml.

(ii) **Double immunodiffusion** or **Ouchterlony method** in which both antigen and antibody diffuse radially from wells towards each other distinguish each other establishing a concentration gradient. The pattern of lines drawn or concluded from possible precipitin pattern indicates whether the two antigens have identical isotopes (**identity**), partially identically epitopes (**partial identity**) or no epitopes in common (**nonidentity**).

(iii) **Immunoelectrophoresis**, which is often used in clinical laboratories to detect the presence or absence of proteins in the serum.

(iv) **Rocket electrophoresis** in which the precipitate formed between antigen and antibody appear in the shape of a rocket. This technique gives quantitation of antigen levels as low as 0.2 μg/ml.

Immunofluorescence

Q.57. Fill in the blanks.

(i) The antibodies bound to cell or tissue sections can be seen by tagging the antibody molecule with a fluorescent dye and the technique is called **immunofluorescence**. The fluorescent dyes are ———— and ———— .

(ii) Immunofluorescence is used to identify a number of subpopulation of lymphocytes, e.g., CD4$^+$ and ————, T-cells.

Immunoelectron Microscopy

(iii) In immunoelectron microscopy an electron dense label is either integrated or an indirect staining technique is employed and the electron dense label is conjugated to immunoglobulin region. The technique has been employed in demonstrating that class I and class II MHC (——— ——— ———) molecules may be sequestered on different intracellular processing routes.

Enzyme-Linked Immunosorbent Assay

(iv) Enzyme-linked immunosorbent assay is often abbreviated as ——— or EIA.

(v) ELISA can be used to detect the presence of antibody or ———.

(vi) Three major types of ELISA are ——— ELISA, ——— ELISA and ——— ELISA.

(vii) The serum antibodies to HIV can be detected by indirect ELISA within ——— weeks of infection.

Ans. (i) fluorescein, rhodamine
(ii) $CD8^+$
(iii) major histocompatibility complexes
(iv) **ELISA**
(v) antigen
(vi) indirect, sandwich, competitive
(vii) six

Western Blotting

Q.58. What is **Western blotting**? Give its uses.

Ans. The technique **Western blotting** is used to identify a specific protein in a complex mixture of proteins. It has been named so because of its similarity to **Southern blotting** which is used to detect DNA fragments, and **Northern blotting** which is employed to detect mRNAs.

In Western blotting, the separated antibody bands can be seen using a labelled antigen. This technique has been used to identify the envelope and core protein of **HIV** and the antibodies to these components that develop in the serum of HIV infected patient.

Radioimmunoassay or RIA

Q.59. What is radioimmunoassay or RIA? Also give its applications in diagnosis of microbial diseases.

Ans. The **radioimmunoassay** a very sensitive technique to detect the antigen or antibody. This technique was developed by endocrinologists S.A. Berson and Rosalyn Yalow, in 1960 to determine levels of insulin-anti-insulin complexes in **diabetics**. This technique has also been used for quantitating hormones, serum proteins, drugs, and vitamins at a low concentration of 0.001 μg or less. After death of Berson significance of the technique was acknowledged when Rosalyn Yalow was awarded Nobel Prize in 1977.

The RIA involves competitive binding of radiolabelled antigen and unlabelled antigen to a high affinity body. The antigen is generally labelled with a gamma emitting isotope as ^{125}I.

The microtiter RIA has been widely used to detect the presence of **hepatitis B virus** in the blood of a donor which consequently has reduced incidence of spread of hepatitis B infections in the recipients of blood during blood transfusions.

Major Histocompatibility Complex and Tumour Immunology

Q.60. What is **major histocompatibility complex** MHC?

Ans. The mammalian species that have been investigated so far possess a tightly linked cluster of genes which has been called the **major histocompatibility complex (MHC),** the products of which are associated with intercellular recognition and to self and nonself discrimination. It is of two types.

(i) Histocompatible (determine that transplanted tissue be acceptible as self).

(ii) Histoincompatible (determine that transplanted tissue be rejected as foreign).

Q.61. What is HLA complex?

Ans. The MHC in human is known as **HLA complex**. However, the MHC in mice is referred to as **H-2 complex**.

Genomic Map of MHC Genes

Q.62. Fill in the blanks.

(i) The MHC in human is located on ——— ———.

(ii) In mice the MHC is located on chromosomes ———.

(iii) The MHC contains nearly 100 genes spanning some 2000 kb of mouse DNA and — ——— kb of human DNA.

(iv) Class I MHC genes encode ———, expressed mainly on the surface of nearly all nucleated cells where they present peptide antigens of altered self cells required for activation of Tc cells.

(v) Class II MHC genes encode glycoprotein D on antigen presenting cells as macrophages, dendritic cells and B cells to present processed antigenic peptides to ———.

(vi) Class III MHC genes encode ——— ——— as serum proteins, components of the complement system and tumour necrosis factors.

(vii) The class I MHC in human is about 2000 kb DNA and bear about ——— genes.

(viii) The class II MHC in both mice and human possess 2 genes LMP2 and ——— that encode proteasome subunits of two genes TAP1 and TAP2 that encode peptide transporter subunits.

(ix) In human as well as mice the class III region of the MHC have a heterogenous collection of more than ——— genes.

(x) The **cellular distribution of MHC molecules** revealed that low level of class I MHC molecules on liver cells may reduce the likelihood of graft recognition T cytotoxic — ——— of the recipient which is indicative of considerable ——— of such transplants.

Ans. (i) chromosome six

(ii) 17

(iii) 4000

(iv) glycoproteins

(v) T_H cells

(vi) secreted proteins

(vii) 20

(viii) LMP 7

(ix) 36

(x) lymphocytes, success

Regulation of MHC Expression or Gene Regulation

Q.63. Give the regulatory mechanism of expression of MHC genes.

Ans. Although it is the beginning of research in this area yet it has been revealed that both class I and class II MHC genes are flanked by 5′ promoter sequences that bind sequence specific transcription factors. The transcriptional regulation of MHC is mediated by both positive and negative elements. The MHC II transactivator referred to as **CIITA** and another transcription factor known as **RFX** have been noticed to bind the promoter region of class II MHC genes. A fault with these transcription factors cause a form of **bare lymphocyte syndrome**. The patients of this category do not have class II MHC molecules or their cells, therefore, suffer a great **immunodeficiency** since class II MHC molecules play a central role in maturation and activation of T-cells.

Besides the gene regulation the expression of MHC molecules is also regulated by various **cytokines** and **interferons** and tumour necrosis factors each of which lead to increase of class I MHC molecules on the cells.

The MHC expression is also increased or often decreased by some **viruses** as humancytomegalovirus (CMV), hepatitis B (HBV) and adenovirus 12 (Ad 12).

MHC and Immune Responsiveness or IR Genes

Q.64. Comment on genes concerned with immune responsiveness or **IR genes**.

Ans. The variability of **immune responsiveness** is various haplotypes of MHC has been accepted on the basis of below given models.

(i) **Determinant-selection model**, according to which various class II molecules differ in their capacity to bind processed antigens. It means that MHC polymorphism within a species will lead to generation of different patterns of responsiveness and nonresponsiveness to different antigens.

(ii) **Holes-in-the repertoire model**, in which the T-cells bearing receptors recognise foreign antigens that closely resemble the self antigens, which may be eliminated during thymic processing.

Q.65. What is the HLA **tissue typing**? Explain.

Ans. Besides **ABO blood group compatibility typing, HLA typing** is also a device for tissue typing. That is done on the basis of antibody mediated microcytotoxicity which can show the presence or absence of different MHC alleles. However, if a fully HLA compatible donor is not available the transplantation can be done after doing **one way mixed lymphocyte reaction** which is abbreviated as **MLR**. The MLR is done to assess quantitatively the degree of class II MHC compatibility between potential donors and a recipient. The drawback of MLR is that it takes about 6 days to run the assay for MLR purpose.

Q.66. Give immunologic basis of **graft rejection**.

Ans. The degree of immune response varies with the type of graft, which are given below.

(i) Autograft: The tissue is transferred from one site of the body to the another in the same individual, e.g., transferring healthy skin to the burned areas in the patients with burns.

(ii) Isograft: Transfer of tissue between genetically identical individuals, e.g., preparation of isograft between genetically identical twins (monozygotic) and in inbred strains of mice from one mouse to another syngeneic mouse.

(iii) Allograft: Transfer of tissue between genetically different members of the same species, e.g. in mice transfer of tissue or an organ from one inbred strain to another. In human transfer of organs from one individual to another but not between identical twin.

(iv) Xenograft: It is the transfer of tissue between different species, e.g., the graft of heart of a baboon into a human.

The mechanism involved in graft rejection is by a cell mediated immune response to **alloantigens** (primarily **MHC molecules**) expressed on cells of the graft. The process of graft rejection can be divided into two stages, which are: **sensitisation stage** in which $CD4^+$ and $CD8^+$ T-cells recognise alloantigens expressed on cells of foreign grafts and leading to their proliferation in response, and effector stage which involve cell mediated reactions involving delayed type hypersensitivity and CLT mediated cytotoxicity and less common mechanisms are antibody plus compliment lysis and destruction by antibody dependent cell mediated cytotoxicity (ADCC). The culmination of graft rejection involves cell mediated reaction is an influx of T cells and marophages into the graft.

Q.67. What is **autoimmunity**? Define. Also give its major types.

Ans. An expression of immunity in which the body directs immunological mechanism against one or more of its own components is called autoimmunity. The major types or categories of which are listed below.

(i) Disease mediated by direct cellular damage.

(ii) Disease mediated by stimulating or blocking autoantibodies.

(iii) Systemic autoimmune diseases.

Q.68. Which are the autoimmune diseases mediated by direct cellular damage?

Ans. The **autoimmune diseases** which are caused when lymphocytes or antibodies bind to cell membrane resulting into cellular lysis or an inflammatory response, which are given below.

(i) **Hashimotos thyroiditis**, is seen in middle aged woman leading to formation of goiter and visible enlargement of thyroid gland, because of production of autoantibodies and sensitised T_{DTH} cells.

(ii) **Autoimmune anemia**, is caused when autoantibodies bind to **intrinsic factor** (membrane bound intestinal protein) blocking the intrinsic factor mediated absorption of vitamin B_{12} from small intestine. It can be treated by injecting vitamin B_{12}.

(iii) **Goodpasture's syndrome**, is caused because of binding of basement membrane antigens to basement membranes of kidney glomeruli and alveoli of lungs leading to damage of kidney and haemorrhage in lungs.

(iv) **Insulin-dependent diabetes mellitus**, which is caused because of an autoimmune attack against insulin producing beta cells.

Q.69. Name the diseases mediated by stimulating or blocking antibodies.

Ans. The diseases of this type are given below.

(i) **Graves disease**, which is caused by binding of autoantibodies to the receptor for thyroid stimulating hormone (TSH) leading to unregulated activation of thyroid resulting the overproduction of hormones.

(ii) **Myasthenia gravis**, is caused by production of autoantibodies to acetylcholine receptors on the motor end plates of muscles which inhibits activation of muscles.

Q.70. Enlist the **systemic autoimmune diseases**.

Ans. These are as under :

(i) **Systemic lupus erythematosus (SLE)**, which is characterised by fever, weakness, skin rashes and disfunction of kidney in women between 20 and 40 years of age. It is because of attack of autoantibodies to tissue antigens like DNA, histones, RBCs, platelets, leukocytes and clotting factors.

(ii) **Multiple sclerosis (MS)**, which is accompanied by formation of inflammatory lesions which destroy myelin.

(iii) **Rheumatoid arthritis**, in which chronic inflammation of joints is caused in women of 40 to 60 years old. The individuals with rheumatoid arthritis produce a group of antibodies referred to as **rheumatoid factors**.

Cancer and the Immune System

Q.71. Match the following:

(A)	The term cancer mainly refers to	(a)	Tumour arising from endodermal or ectodermal tissues.
(B)	Carcinoma	(b)	**Malignant tumour**
(C)	Derived from mesodermal connective tissue e.g. bone, fat and cartilage.	(c)	Sarcomas are.
(D)	The leukemias and lymphomas	(d)	Retinoblastoma
(E)	A rare childhood cancer in which tumours develop from neural precursor cells in the immature retina	(e)	Are from hemato poetic cells of the bone marrow.
(F)	**Alpha-fetoprotein**	(f)	An increased level of carcinoembryonic antigen (CEA) is indicative of.
(G)	Resumed tumour growth	(g)	One of the two well studied oncofetal tumour antigens.
(H)	Nude mice lacks thymus.	(h)	Lacks T cells.
(I)	Monoclonal antibodies	(i)	Tumour growth.
(J)	MHC expression is accompanied by progressive	(j)	Have been used to make tumour specific immunotoxin.
(K)	Show elevated expression of the oncogene encoded Neuprotein.	(k)	Human breast cancer cells.
(L)	**Tumour antigens**	(l)	AFP (alpha-fetoprotein) and CEA (carcinoembryonic antigen).
(M)	Two well studied **oncofetal antigens** are	(m)	Hypoxanthine, aminopterin and thymidine containing medium.
(N)	**HAT medium**	(n)	This defect leads to their inability to survive in a medium having hypoxanthine, aminopterin and thymidine
(O)	The myeloma tumour cells used for hybridoma production are deficient in enzyme hypoxanthine phosphoribosyltransferase.	(o)	Have been noted on **tumours** induced with chemical or physical carcinogens and some **virally induced tumours**.
(P)	The antibody producing **spleen cells**, possess the enzyme hypoxanthine phosphoribosyltransferase.	(p)	Enables **fused hybridoma cells** survive in selective medium and can be recognised by their ability to grow indefinitely in the selective medium.

Ans. [A b], [B a], [C c], [D e], [E d], [F g], [G f], [H h], [I j], [J i], [K k], [L o], [M l], [N m], [O n], [P p].

25
CHAPTER

MICROBIAL GENETICS

Q.1. What is an **auxotroph**? Define.

Ans. Auxotroph is an organism which has developed a nutritional requirement through mutation.

Q.2. Define **conjugation.**

Ans. Conjugation is the transfer of genes from one procaryotic cell to another by a mechanisms involving cell to cell contact and plasmid.

Q.3. Define the term **diploid.**

Ans. An eucaryotic cell or organism containing two sets of chromosomes.

Q.4. What is electroporation?

Ans. The use of an electric pulse to induce cells to accept free DNA or plasmid.

Q.5. Define gamete.

Ans. The haploid germ cells resulting from meiosis in eucaryotes.

Q.6. What is genetic map?

Ans. Genetic map is the arrangement of genes on a chromosome.

Q.7. Define genotype.

Ans. The precise genetic make up of an organism.

Q.8. What is haploid? Define.

Ans. It is a cell or organism that has only one set of chromosomes.

Q.9. What are **mutagens**?

Ans. The causative agents of mutations.

Q.10. Define mutation.

Ans. An inheritable change in the base sequence of the genome of an organism.

Q.11. Define phenotype.

Ans. The observable characteristics of an organism comprise phenotype.

Q.12. Define plasmid.

Ans. An extrachromosomal genetic element which has no extracellular form.

Q.13. Define point mutation.

Ans. A mutation which involves one or only a very few base pairs.

Q.14. What is **recombination**?

Ans. Recombination is the process by which parts of or all of the DNA molecules from two separate sources are exchanged or brought together into a single unit.

Q.15. Define selection.

Ans. It is the placing of organisms under conditions where the growth of those with a particular genotype will be favoured.

Q.16. What is **transduction**?

Ans. Transduction is transfer of host genes from one cell to another by a virus.

Q.17. What is **transformation**?

Ans. Transformation is transfer of bacterial genes involving free DNA.

Q.18. What is a **transposable element**? Define.

Ans. It is a genetic element which has the ability to move (transpose) from one site on a chromosome to another.

Q.19. Define the term **transposon**.

Ans. It is a type of transposable element which transfers genes in addition to those involved in transposition.

Q.20. How will you phenotypically differentiate a mutant of *Aspergillus nidulans* by change of colour of pigment?

Ans. The colour of unmutated wild type is due to green pigment, therefore, any change of colour from green to other colour will indicate mutation.

Q.21. What is the importance of microbial genetics? List the main points.

Ans. The importance of microbial genetics is due to the following reasons:

(i) The function of the gene is a base for the function of the cell. It is therefore indicative of the functioning of microorganisms.

(ii) Microorganisms are simple to study through their genetic phenomena.

(iii) Microorganisms can be exploited for isolation and duplication of specific genes from other organisms, which is referred to as molecular cloning.

In molecular cloning genes are manipulated and are incorporated into microorganisms where they can be induced to increase in number.

(iv) The use of genetically improved microorganisms for large scale industrial production comprise an important part of biotechnology.

(v) By understanding the genetics of disease causing microorganisms we can control the diseases caused by them.

(vi) By studying processes such as conjugation we can understand how the genes are transferred from one organism to another.

Q.22. Name the technique by which nutritional mutants can be detected.

Ans. Replica plating.

Q.23. What is a nutritional mutant, with a requirement for growth factor often called?

Ans. **Auxotroph**.

Q.24. Name the one word technical term for the wild type parent from which an auxotroph is obtained.

Ans. **Prototroph**.

Q.25. What is a histidine auxotroph?

Ans. *Escherichia coli* or any other microorganisms with a His phenotype is called histidine auxotroph.

Q.26. What indicates the development of antibiotic resistant colonies within the inhibition zone of an antibiotic assay disc?

Ans. Development of antibiotic resistant mutants.

Q.27. Tabulate the common kinds of mutants and their characteristics for detection.

Ans.

Type	*Change*	*Characteristics for detection*
1. Nonmotile	Loss of flagella or flagella are nonfunctional	The colonies formed will be compact instead of being flat and spread out
2. Noncapsulated	Loss or modification of surface capsule	Small rough colonies instead of larger smooth ones.
3. Rough colony	Loss or change in lipopolysacc-haride outer layer	Granular, irregular colonies instead of smooth glistening colonies.
4. Auxotroph	Loss of enzyme in biosynthetic pathway	The mutant does not grow on medium lacking nutrient.
5. Sugar fermentation	Loss of enzyme in degradative pathway.	Lack of colour change on agar containing sugar and on pH indicators.
6. Drug resistant	Alteration of permeability to drug or drug target or detoxification drugs	Show growth on medium having growth inhibitory concentration of the drug.
7. Virus-resistant	Loss of virus receptors	Growth in the presence of large amounts of viruses.
8. Temperature sensitive	Change in some essential proteins	The mutant being unable to grow at temperatures that usually support growth (e.g., 40°C). but are still able to grow at lower temperatures (e.g., at 30°C).
9. Pigmentless	Loss of enzyme in biosynthetic pathway leading to loss of one or more pigments	Shows in different colour from original or lacks colouring.
10. Cold sensitive	Change in some essential proteins and therefore inactivated at low temperatures	Does not grow at low temperatures (e.g., at 20°C) that usually supports growth.

Q.28. Differentiate between screening and selection.

Ans. Screening is to examine a large population of organisms for selectable mutants from among nonselectable mutants while selection is the isolation of a single mutant from a population containing millions or billions of parental organisms.

Q.29. What is the purpose of the penicillin selection method?

Ans. It is to isolate mutants that need amino acids or other growth factors.

Q.30. How does the penicillin selection method work?

Ans. Penicillin destroys only the growing cells. If penicillin is added to a population with wild parents and mutants growing in a medium which lacks growth factor, the parent cells would be destroyed while the non-growing mutants would survive. Thus after preliminary incubation in the absence of growth factor the penicillin is washed away leaving the mutant population free from penicillin which can be transferred to plates containing growth factor. The colonies that finally grow are growth factor mutants.

Q.31. Define the following terms: (a) Silent mutation (b) Missense mutation (c) Nonsense mutation (d) Temperature sensitive mutation (e) Conditionally lethal.

Ans. (a) **Silent mutation.** When a change in base pair substitution causes no apparent effect it is called silent mutation, e.g., a change in RNA from UAC to UAU bears no apparent effect as UAU is also a tyrosine codon.

(b) **Missense mutation.** A change in base pair substitution that leads to change in formation of a protein, e.g., a single base change from UAC to AAC results in a change in the protein from tyrosine to asparagine.

(c) **Nonsense mutation.** A base pair substitution leading to an incomplete protein. The change has been named so as the change is from a codon for amino acids (sense codon) to a stop codon (nonsense codon).

(d) **Temperature sensitive mutation.** A missense mutation can lead to an enzyme which is temperature sensitive, hence called temperature sensitive mutation, e.g., the temperature sensitive bacterial mutants function usually at 30°C but not at 40°C at which temperature the wild type works well.

(e) **Conditionally lethal.** The temperature sensitive mutations have also been called conditionally lethal mutations as the mutant cannot grow under one condition but can grow under another.

Q.32. What are frameshift mutations?

Ans. Any deletion or insertion of base pairs results in a reading frame shift and this completely upsets the translation of the gene.

Q.33. What is back mutation or reversion? Define.

Ans. The point at which mutations are reversible and in which the wild type phenotype that was lost is restored is back mutation. These processes result in the formation of two types of revertants: (1) Same site revertants, and (2) Second site revertants.

Q.34. Name the types of mutations involving many base pairs.

Ans. (i) Deletion (elimination of a region of DNA),

(ii) Insertion (addition of new bases to DNA),

(iii) Translocation (in which large section of chromosomal DNA gets moved to a new location and in eucaryotes even to a different chromosome),

(iv) Inversions (in which the orientation of a particular segment of DNA gets reversed with respect to surrounding DNA).

Q.35. List the rates of mutations of some important types.

Ans. (i) Spontaneous mutations occur at a frequency of 10^{-6} per generation.

(ii) Transposition events occur more frequently, i.e., at about 10^{-4}

(iii) Nonsense mutations are less frequent, i.e., at 10^{-6} to 10^{-8}.

Q.36. What is the rate of mutation in ribonucleic acid (RNA) genomes? Why is it higher than in DNA genomes?

Ans. The mutation rate in RNA genome is about 1000 fold higher than in DNA genome. RNA mutation is a comparatively short process. The reasons being that the RNA replicates do not seem to have proofreading activities like those of DNA polymerases. Also there are no comparative RNA repair mechanisms. Instead there are many repair systems for DNA that correct many a changes before the changes get fixed as mutations.

Q.37. Can point mutations spontaneously revert?

Ans. Yes.

Q.38. Do missense mutations occur in genes encoding transfer RNAs (t RNAs)?

Ans. Yes.

Q.39. What are mutagens?

Ans. Mutagens are chemical, physical or biological agents that enhance the mutation rate by making alterations in DNA. However, the alteration is a mutation only if it is inherited.

Q.40. Name the common chemical mutagens and describe how they cause mutations.

Ans. There are following types of chemical mutagens:

(i) Base analogs: Many of them are base analogs resembling DNA purine. Pyrimidine bases in structures may consequently lead to faulty pairing, e.g., (1) 5-bromouracil is an analog of thymine. Consequently 5-bromouracil can pair with guanine causing AT to GC substitution. (2) 2-aminopurine serves as an analog of adenine, hence 2-aminopurine can pair with cytosine causing AT to GC substitution.

(ii) Alkylating agents: Alkylating agents such as nitrosoguanidine are powerful mutagens. They may produce direct change even in non replicating DNA. Like base analogs they tend to induce base pair substitutions.

(iii) Acridines: The acridines are intercalating agents leading to microinsertions or microdeletions. The acridines (mutagens) get inserted between DNA base pairs.

Q.41. Which are the physical mutagens?

Ans. Several types of radiations count among physical mutagens. They are mainly of two types:

(a) Nonionizing radiations, e.g., UV radiations in 260 nm region.

(b) Ionizing radiations: They are short wavelength rays like X-rays, cosmic rays and gamma rays. Among the free radicals the hydroxyl radical (OH) reacts with and inactivates the macromolecules in the cell, of which DNA is the most important.

Q.42. Name the complex cellular mechanism associated with DNA repair.

Ans. SOS regulatory system.

Q.43. What is biological mutagenesis?

Ans. It is the process of transposon mutagenesis. Transposons are widely used by microbial geneticists as mutagenic agents.

Q.44. What is site directed mutagenesis?

Ans. It is the production of specific mutations in specific genes using recombinant DNA technology with the use of synthetic DNA.

Q.45. What is **Ames test**?

Ans. This test developed under the direction of Bruce Ames at University of California at Berkeley employs a sensitive bacterial assay system for detecting chemical mutagens in the environment. It was primarily conducted for carcinogenic screening.

Q.46. Name the organisms often used in Ames test.

Ans. Histidine auxotroph of *Salmonella typhimurium* and tryptophan auxotrophs of *Escherichia coli.*

Q.47. How does the mechanism of recombination occur in procaryotes?

Ans. It involves DNA transfer during the process of transformation, transduction and conjugation.

Q.48. Name the protein found in all procaryotes which assists in pairing required for homologous recombination.

Ans. Rec A protein.

Q.49. What is complementation test?

Ans. Complementation test is used to detect mutations in the same genetic (functional) unit.

Q.50. What is the extreme length of DNA in *Bacillus subtilis?*

Ans. 1700 μm.

Q.51. Who was the first to show evidence of bacterial transformation?

Ans. British scientist Frederick Griffith in 1928 was the first to prove it with *Streptococcus pneumoniae* (pneumococcus).

Q.52. What were the two strains of *S. pneumoniae* (pneumococcus) used by F. Griffith and what did he conclude?

Ans. Griffith used two forms, i.e., smooth (virulent) and rough (nonvirulent) forms of the bacteria. He found that a mixture of smooth virulent (pathogenic) killed by heat and live, rough nonvirulent (nonpathogenic) while injected to mouse proved fatal. It indicated that genetic material (DNA) formed by heat killed smooth type pneumococci brought transformation into the genetic make up of rough nonvirulent form.

Q.53. What is competence? On what factors does competence depend?

Ans. A cell which can take up a molecule of DNA to be transformed is said to show competence. Competence depends on special proteins and may include membrane associated DNA binding proteins, a cell wall autolysin and various nucleases. In *Bacillus subtilis* and *Streptococcus pneumoniae* induction of competence depends on the medium and growth stage of the culture.

Q.54. Give variations in uptake of DNA.

Ans. Gram-negative bacterium *Haemtophilus* takes up double-stranded DNA into the cell while in the Gram-positive bacteria *Streptococcus* and *Bacillus* only a single DNA strand is taken up while the complementary DNA is simultaneously degraded. However, generally double stranded DNA binds more effectively to the cells. Competent cells may bind 1000 times more DNA than noncompetent cells.

Q.55. Give the mechanisms of transformation (Integration of transforming DNA).

Ans. During the process of transformation competent bacteria first bind DNA reversibly, soon however, making it irreversible. The DNA undergoing transformation is bound at the surface of bacterial cells by a DNA binding protein which remains attached to DNA perhaps to prevent it from the attack of enzyme nuclease till the time it reaches the chromosome of the bacterial cell where Rec A protein takes over. It would be worthwhile to add that once the transforming DNA is bound at the surface by the DNA binding protein either the entire double stranded fragment is taken up or one strand is degraded by a nuclease and the other strand is taken up.

Q.56. What is transfection? What is its use?

Ans. It is the transformation of bacteria with DNA extracted from a bacterial virus instead of taking DNA from another bacterium. Transfection is a very useful tool in studying the mechanisms of transformation and recombination as the small size of phase genomes allows for the isolation of a nearly homogenous population of DNA molecules.

Q.57. What do you mean by the term '**transfection**' in relation to eucaryotes?

Ans. Eucaryotic microorganisms and animal and plant cells can take up DNA in a process which resembles bacterial transformation. Because the word transformation in mammalian cells is used to describe the conversion of cells to the malignant state. Therefore, introduction of DNA into mammalian cells has been called transfection.

Q.58. Name the bacterial genera which possess high efficiency natural transformation.

Ans. High efficiency natural transformation occurs in *Azotobacter, Bacillus, Streptococcus, Haemophilus, Neisseria* and *Thennus,* which are easily transformed

Q.59. How can artificial competence be induced in Gram-negative bacterium *Escherichia coli?*

Ans. Competence can be introduced in *E. coli* by treating it with high concentration of calcium ions and then stored in the cold. On doing so it becomes transformable at low efficiency. *Escherichia* coli treated in this way takes up double stranded DNA and therefore, transformation by plasmid DNA is more efficient because no recombination is required. How does calcium treatment work is not well understood but this method is also used with some other Gram-negative bacteria.

Q.60. How is DNA transferred by electroporation?

Ans. Small pores are formed on the membranes of bacterial cells exposed to pulsed electric fields. The DNA molecules present outside the cells during electric pulse, can enter the cells through these pores. By this method plasmid can be directly transferred from one cell to another if both the types of cells are present during electroporation.

Q.61. How can yeast, algae, plant cells, mitochondira and chloroplasts be transfected by particle gun?

Ans. A small steel cylinder containing a gun-powder charge is used to fire nucleic acid coated particles at the target cells. The nucleic acid enters the cell by piercing the wall without damaging the cell and combines with the host DNA.

Q.62. Which are the two main types of transduction?

Ans. Transduction is of two types: (1) Generalized transduction (2) Specialized transduction.

Q.63. What is an **episome**?

Ans. Plasmids having the ability to integrate into host chromosomes are known as episomes.

Q.64. What is the main mechanism of cell to cell transfer of plasmids?

Ans. Conjugation.

Q.65. What are conjugative plasmids? Name the gene that controls it.

Ans. Plasmids which govern their own transfer by cell to cell contact are called conjugative plasmids. The set of genes called *tra* region that controls the transmission of the plasmids.

Q.66. Some plasmids with gene function other than replication not known, have been detected by gel electrophoresis of cell extracts.

Ans. Cryptic plasmids.

Q.67. What are the main types of plasmids? Name the organisms in which they are found.

Ans. The following are the main types of plasmids.

(i) Conjugative plasmids, e.g., F plasmid in *Escherichia coli,* pfd m, K in *Pseudomonas,* P in *Vibrio cholerae,* and SCP in *Streptomyces.*

(ii) R plasmids, which show resistance to a wide variety of antibiotics, resistance to mercury, cadmium, nickel, cobalt, zinc and arsenic, e.g., in Enteric bacteria, *Staphylococcus* and *Pseudomonas.*

(iii) *Plasmids for bacteriocin and antibiotic production:* The bacteriocins are ribosomally synthesized peptides but the *Escherichia coli* produces colicins coded col plasmids, and *Bacillus subtilis* produces subtilisin. They are also produced by *Clostridum*, enteric bacteria and *Streptomyces*.

(iv) Plasmids to control physiological functions include:

(a) Utilization of lactose, sucrose, urea and fixation of nitrogen, e.g., enteric bacteria and *Rhizobium*.

(b) Degradation of octane, camphor, naphthalene and salicylate, e.g., *Pseudomonas*.

(c) Production of pigments, e.g., *Erwinia* and *Staphylococcus*.

(d) Nodulation and symbiotic nitrogen fixation *Rhizobium*.

(v) Virulence plasmids

(a) Enterotoxin, K antigen, endotoxin in *Escherichia coli*.

(b) Tumorigenic plasmid *Agrobacterium tumefaciens*.

(c) Adherence to teeth (dextran), e.g., in *Streptococcus mutans*.

(d) Coagulase, hemolysin, fibrinolysin and enterotoxin, e.g., in *Staphylococcus aureus*.

Q.68. What is generalized transduction?

Ans. It is the genetic transfer of host genes (host DNA derived from any portion of host genome) becoming part of DNA of the mature virus particles in place of the virus genome. It was first discovered and reported in *Salmonella typhimurium* with phage P 22. It has also been studied in *Escherichia coli* with phage P1.

Q.69. What is specialized (restricted) transduction?

Ans. The specialized (restricted) transduction occurs only in some temperate viruses where DNA from a specific region of the host chromosome is integrated into the virus genome usually replacing some of the genes. It was first discovered in *Escherichia coli* bringing transduction of the galactose genes by the temperate phage lambda. The region in which lambda integrates is immediately adjacent to the cluster of host genes which control the enzymes involved in galactose utilization, and the DNA of lambda is inserted into the host DNA at that site.

Q.70. Why is altered phage particle lambda dgal or λ dgal so called?

Ans. Because it is defective for galactose. It is a defect of the phage genes lost and does not make mature phage. But a helper phage can provide those functions missing in the defective particle.

Q.71. What is abortive transduction?

Ans. The fate of transduced DNA in the recipient cell (now called a transductant) depends on various factors. If DNA is a complete replicon (e.g., a plasmid) it may be stably inherited by the transductant. But if the DNA is a fragment of a chromosome or plasmid it may meet one of the three possible fates.

(i) It may be completely degraded by the recipient cell's restriction endonuclease system.

(ii) It may undergo recombination with a homologous region of the parents chromosome (or plasmid) such that at least some of the genes it carries can be stably inherited also called complete transduction.

(iii) It may persist in the cell in a stable but not replicating form which is referred to as abortive transduction. The **transduced DNA** in an abortive transductant may exist as a circular DNA protein complex.

Q.72. **What are the plasmids with the ability to integrate into the host chromosome, called?**

Ans. Episomes.

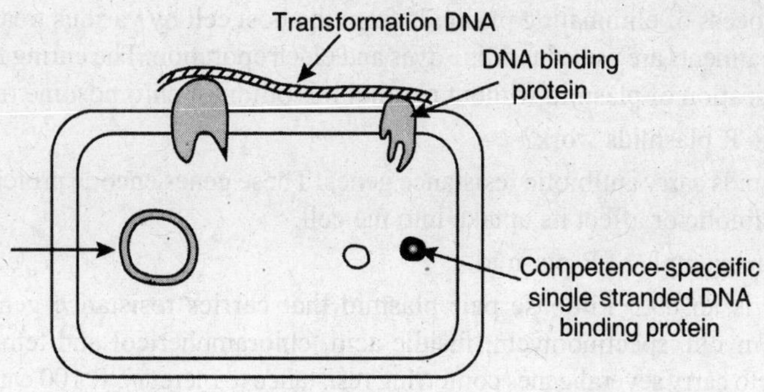

Step 1. **Binding of free DNA by a membrane bound DNA binding protein.**

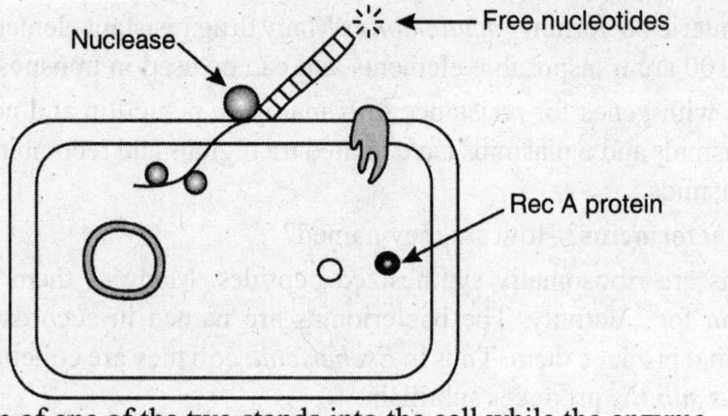

Step 2. **Passage of one of the two stands into the cell while the enzyme nuclease degrades the other strand.**

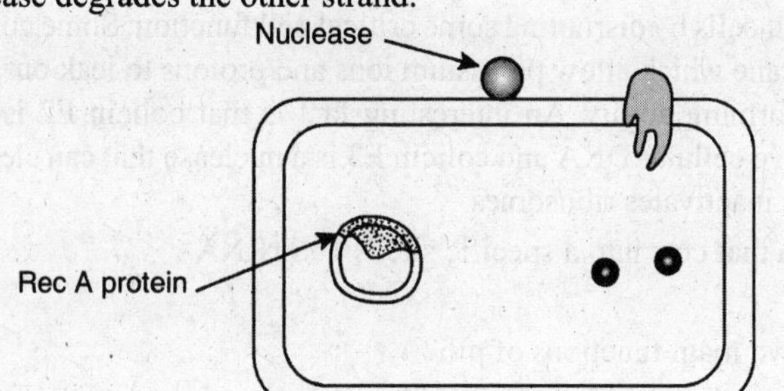

Step 3. **The single strand in the cell is bound by specific proteins and the recombination with homologous regions of the bacterial chromosome mediated by Rec A protein.**

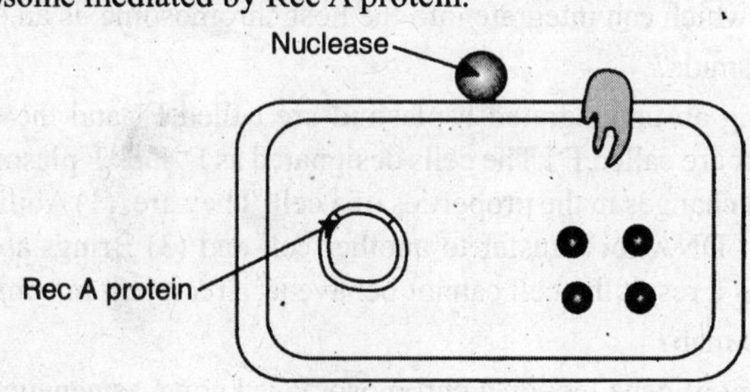

Step 4. **A transformed cell.**

Fig. 25.1. **Steps in mechanism of DNA transfer by transformation in Gram +ve bacterium.**

Q.73. What is the curing of plasmid?

Ans. The process of eliminating plasmids from the host cell by various treatments is called curing. The treatments are use of acridine dyes and electroporation. The curing takes place by inhibition of replication of plasmid without parallel inhibition of chromosome replication.

Q.74. How do R plasmids work?

Ans. R plasmids carry antibiotic resistance genes. These genes encode proteins that either inactivate the antibiotic or affect its uptake into the cell.

Q.75. Give an example of R plasmid.

Ans. R 100 is an 89.3 kilobase pair plasmid that carries resistance genes for sulphonamides, streptomycin, spectinomycin, fusidic acid, chloramphericol and tetracycline. R 100 is also known to carry several genes conferring resistance to mercury. R 100 can be transferred between enteric bacteria of the genera *Escherichia, Klebsiella, Proteus, Salmonella* and *Shigella* but not to nonenteric bacterium *Pseudomonas* Many drug resistant elements in R plasmids such as those in R 100 are transposable elements and can be used in transposon mutagenesis.

R plasmids with genes for resistance to Kanamycin penicillin and neomycin too are known. Many R plasmids and F plasmids have related tra regions and recombination can occur between F and R plasmids.

Q.76. What are **bacteriocins**? How are they named?

Ans. Bacteriocins are ribosomally synthesized peptides. Many of them require postranslational modification for ' activity. The bacteriocins are named in accordance with the species of organisms that produce them. Thus in *Escherichia* coli they are colicins coded by Col plasmids and *Bacillus subtilis* produces subtilisin.

Q.77. How are colicins harmful?

Ans. The colicins kill cells by disrupting some critical cell function. Some colicins form channels in the cell membrane which allow potassium ions and protons to leak out resulting in loss of the cell's energy forming ability. An interesting fact is that colicin E2 is a DNA endonuclease which can cleave cellular DNA and colicin E3 is a nuclease that can cleave at a specific site in 16S rRNA and inactivates ribosomes.

Q.78. Name a colicin that cuts into a specific site is 16S rRNA.

Ans. Colicin E3.

Q.79. What are the two main functions of pili?

Ans. They are: (1) to make specific contact with the receptor on the recipient bacterial cell, and (2) to retract after that to bring the two cells together such that a conjugation bridge is formed.

Q.80. Is F plasmid an episome?

Ans. The F plasmid which can integrate into the host chromosome, is an episome.

Q.81. What are F plasmids?

Ans. The cells having an unintegrated F plasmid are called F^+ and the cells which can work as recipients of Hfr are called F^-. The cells designated as F^- lack F plasmid. The F plasmids result in three distinct changes in the properties of a cell. They are: (1) Ability to produce F pilus, (2) Mobilization of DNA for transfer to another cell and (3) Brings about alteration of surface receptors and as a result the cell cannot behave as a recipient in conjugation.

Q.82. What is genetic map?

Ans. The arrangement of gene loci on a chromosome is known as genetic map.

Q.83. What is transposition? Give its importance.

Ans. Transposition is the process by which a gene moves from one place to another in the genome. It is important for genetic evolution and in genetic analysis.

Q.84. What are transposable elements?

Ans. They are the genetic elements on which the movement or transposition of genes depends. These have also been called "jumping genes".

Q.85. What are different classes of transposable elements?

Ans. The transposable elements in bacteria are of three types. They are: (1) Insertion sequences (IS) (2) Transposons Tn), and (3) Some special viruses (e.g. Mu).

In eucaryotes transposable elements are in yeast (e.g., sigma and Ty), in fruit fly (e.g., copia, P), in maize (e.g., Ac) and in humans (e.g., retrovirus sarcoma, human immunodeficiency virus or HIV).

Q.86. Give the mechanisms of transposition.

Ans. One of the essential features of transposition is the inverted repeats found at the ends of transposable elements. The second essential feature is an enzyme called transposase which recognizes the aforesaid repeats. The enzyme transposase on recognizing DNA, cuts it and then ligates the DNA.

While a transposable element becomes inserted into another DNA (referred to as target DNA) a short sequence in the target DNA at the site of interaction is duplicated. The target DNA sequence was not present in the transposon but the transposable element causes duplication of this DNA by the insertion process. The duplication of target sequence apparently arises, as single stranded breaks are generated by the transponase. There are two categories of mechanisms of transposition.

(i) Conservative, in which the transposable element is excised from one location in the chromosome and gets inserted at a second location, e.g., in Tn 5.

(ii) Replicative, in which the transposon is duplicated and a new copy is inserted at another location, e.g., in the transposon bacteriophage Mu.

Q.87. Name three often studied composite transposons.

Ans. These are Tn 5, Tn 9 and Tn 10 which carry genes for resistance to kanamycin, chloramphenicol and tetracycline respectively.

Q.88. What are the simplest transposons called?

Ans. The simplest transposons are called insertion sequences (IS).

Q.89. What is general recombination (or homologous recombination)?

Ans. In general recombination genetic exchange between homologous DNA sequences from two different sources takes place. As homologous DNA sequences have the same sequence, the base pairing takes place over an extended length of the two DNA molecules.

Q.90. What characteristics do insertion sequences (IS) and transposons have in common?

Ans. They show some common characteristics which are:

(i) Both have inverted repeats at their ends.

(ii) Both encode for 'transposes' which is an enzyme required in transposition.

Q.91. What are the use of transposable elements?

Ans. Transposable elements can be used as biological mutagens.

Q.92. How are locations of various genes mapped on a bacterial chromosome?

Ans. The genes in a bacterial genetic map are located using the mechanisms of transformation, transduction and conjugation, e.g., a circular reference map of *Escherichia coli* strain K-12 has been prepared by giving map distances in minutes of transfer with 100 minutes for the whole chromosome. In it "zero time" is set arbitrarily for threonine operon which is the first genetic transfer. This can be detected using the original Hfr strain.

Using a combination of conjugation and transduction over 1400 genes have been located on the circular chromosome of *E. coli.* It was found that some set of related genes, e.g., *trp* genes (genes involved in biosynthesis of amino acid tryptophan are tightly clustered while arg *genes* (involved in arginine biosynthesis) are scattered.

Q.93. What is **regulon**?

Ans. A set of operons controlled by the same regulatory proteins is a regulon.

Q.94. What is **operon**?

Ans. A cluster of genes (often in procaryotic cells) whose expression is controlled by a single operator is operon.

Q.95. Give the range and size of procaryotic chromosomes.

Ans. *Mycoplasma genitalium* a parasitic bacterium has 580 kilobase pairs in its chromosome and is the smallest chromosome while *Myxococcus xanthus* has an unusually large chromosome of 9500 kilobase pairs and that *Escherichia coli* is formed of 4700 kilobase pairs.

Q.96. What is the genetic proof that mitochondria and chloroplasts of cells of eucaryotes have evolved in the past from endosymbiotic bacteria?

Ans. The genomes of these organelles are usually circular like that of procaryotes.

Q.97. What is signal hypothesis?

Ans. Signal hypothesis means that the translocation of a protein from its site of synthesis into or through a membrane requires that the protein be synthesized in a precursor form (preprotein) having a specific N-terminal sequence of amino acid residues (= signal peptide, signal sequence and leader peptide) that is essential for initiation of translocation but is excised during translocation.

Q.98. What is a signal recognition particle (SRP)?

Ans. It is a particle in eucaryotes that is involved in translocation of proteins across the endoplasmic reticulum. An SRP contains 6 polypetides together with 7 S RNA that is essential for the function.

Q.99. What is **apoptosis**?

Ans. It is genetically programmed cell death.

26
CHAPTER

MICROBIAL BIOTECHNOLOGY

Biotechnology is the application of living organisms and their products in industrial processes on large scale. Microbial biotechnology is that aspect of biotechnology which involves the use of microorganisms or their products. Microbial biotechnology is sometimes also referred to as industrial microbiology which is an old field that has been given new dimensions because of the discoveries made in the field of **genetic engineering** *in vitro* manipulation of DNA molecules to generate new combinations of genes or sequences, to place the gene under the control of different regulatory systems, to introduce a specific mutation in a molecule, etc. Industrial microbiology came into existence initially by establishment of alcoholic fermentation processes to produce wine and beer. Thereafter, came the microbial production of antibiotics, and food additives such as amino acids, enzymes, butanol and citric acid.

Genetic engineering has enabled us to use microorganisms for production of new substances which the microorganisms could not have produced normally, such as the production of hormone **insulin** a pancreatic hormone which stimulates the transportation of glucose into cells. The production of insulin by bacterium has been possible due to genetic engineering techniques of inserting human insulin gene into bacterium. Microbial biotechnology can be divided under two subheadings: (1) Traditional microbial technology which is the large scale manufacture of products which are normally produced by microorganisms. (2) Microbial technology with genetically engineered microorganisms in which new genes have been inserted.

Industrial Microorganisms

Industrial microorganisms are those microorganisms which have been selected carefully to make one or more specific products. Industrial microorganisms are selected for their metabolic activities which are capable of specific products and give high yield of particular metabolites. To achieve the desired objectives of high metabolic specialization the industrial strains are genetically modified by mutation or recombination using techniques of genetic engineering. The minor metabolic pathways are either brought down or eliminated. Though the industrial strains may grow well under highly specialized artificial conditions of the fermentor they may show poor growth characteristics in natural conditions of competitive environments.

Source of Industrial Strains The initial and ultimate source of industrial strains has no doubt been nature but through the experience of years of large scale microbial processes perfection has been achieved for greater yield. The strains so developed have been deposited in culture collections. To get a new industrial process patented, the applicant is required to deposit a strain capable of carrying on the process, to a recognized culture collection. Although most industrial companies would be reluctant to deposit their best cultures with any recognized culture collection, yet these collections serve as a ready source of cultures. A very big list of culture collections is given in the "World Directory of Collection of Microorganisms" (1982), updated by V.F. Mc Gowan and V.B.D. Skerman.

Some of the general culture collections with their abbreviation, full name and locations are given below.

AMRC (FAO-WHO International Reference Centre for Animal Mycoplasmas) Institute of Medical Microbiology, University of Aarhas, Denmark.

ATCC (American Type Culture Collection) 12301.
Parklawan Drive, Rockvilla, Maryland 20852, USA.

CBS (Central Bureau Voor Schimmelcultures) Oousterstraat 1, Baarn, The Netherlands.

CCEB (Culture Collection of Entomophagous Bacteria) Institute of Entomology, Czechslovak Academy of Sciences, Femingovo N2, Prague 6, Czechoslovakia.

CDDA (Canadian Department of Agriculture) Ottawa, Canada.

CIP (Collection of Institute Pasteur) Rue due Dr Roux, Paris 15, France.

CMI (Commonwealth Mycological Institute presently known as International Mycological Institute) Kew, UK.

DSM (Deutsche Samn—dung von Microorganismen) Grisebachstrasse 8, Gottingen, Federal Republic of Germany.

FAT (Faculty of Agriculture, Tokyo University) Tokyo, Japan.

IAM (Institute of Applied Microbiology) University of Tokyo, Bunkyo-ku, Tokyo, Japan.

IFO (Institute of Fermentation) 4-54 Jusonishinocho, Osaka Japan.

IMI (International Mycological Institute) Bakeham Lane, Egham, Surrey TW 209 TY, UK.

IMY (Institute of Microbiology & Virology) Academy of Sciences of the Ukranian S.S.R., Kiev.

NCIB (National Collection of Industrial Bacteria) Aberdeen, Scotland.

NCTC (National Collection of Type Cultures), London, U.K.

NRRL (Northern Regional Research Laboratory) Peoria, IL USA.

UQM (Culture Collection, Department of Microbiology, University of Queensland), Herston, Brisbane 4006, Australia.

UWO (University of Western Ontario Culture Collection, Department of Plant Sciences), Ontario N6A 587, Canada.

Maintenance and preservation of cultures is also very essential for getting cultures from the culture collections.

Strain Improvement Microorganisms from original sources are highly modified in the laboratory. The modification is done to achieve a target of higher yield. One of the interesting examples of the progressive improvement is the antibiotic penicillin produced by the fungus *Penicillium chrysogenum.* The production of penicillin on industrial scale was for the first time 1 to 10 μg/ml. However, the programmes of strain development accompanied by changes in medium and the growth conditions after hard work of many years, increased the yield of antibiotic penicillin to about 50,000 μg/ml. Thus, almost **50,000** times increase was possible by mutation and selection without involving genetic engineering techniques. The addition of new genetic engineering techniques has brightened the chances for yet greater yield.

Requisites for an Industrial Microorganism

Not all but only a few selected microorganisms are suitable for application in industrial production. An industrial microorganism must grow readily in largescale manufacturing equipment. No doubt an industrial microorganism's first requisite is to manufacture the product of interest but it must also grow rapidly and on relatively cheap culture media. An industrial microorganism should have the possibility of undergoing manipulations genetically for its strain improvement. An industrial microorganism should neither be pathogenic nor it should produce any toxic products which may be harmful to humans, animals or plants.

Q.1. Whether the following are True or Flase.

(i) **DNA typing** is the analysis of DNA which is being accepted for the identification of individual criminals in murder and rape cases.

(ii) Agarose is not the gel which can be used to separate DNA molecule.

(iii) The production of cells with changed properties by inserting genes from other sources is known as **Genetic engineering**.

(iv) Ethidium bromide which is a fluorescent dye bind to nucleic acid molecule, can be used to detect DNA in gels.

(v) The common vectors employed for gene cloning are plasmids and viruses.

(vi) PCR is not the technique which can be used to produce billions of copies of a single piece of DNA

Ans. ii, vi, False; i, iii, iv, v, True.

Q.2. Fill in the blanks.

(i) ———, a bacterium multiplied through nitrogen and phosphorus fertilizers was manipulated in cleansing Alaskan beaches following Exxon Valdez oil spill.

(ii) A normally shy and reserved microbiologist of Indian origin, **Anand Mohan Chakraborty,** was the first to have a ——— of oil eating bacterial strain from the US Supreme Court in 1980.

(iii) Genetically engineered *Escherichia coli* produces gamma interferon, a human protein which promotes the ——— response.

(iv) **Herbert Boyer,** known to be one of the founders of Genetech was the first scientist to apply for a patent on ———.

(v) Addition of microorganisms to a hazardous waste site so that they may react with hazardous wastes and make the hazardous wastes harmless, is known as ———.

Ans. (i) *Pseudomonas*

(ii) **patent**

(iii) immune

(iv) genetic material

(v) **bioaugmentation**

Q.3. Give a brief account of new proteinaceous therapeutic products generated by development and advancement of biotechnology.

Ans. The recombinant DNA technology generally called **genetic engineering** has revolutionized our ability to isolate, manipulate and express genes (hence proteins) virtually at will. The first and prime achievement of this has been expression and isolation of many **therapeutic proteins** including antiboides. In most cases these therapeutic proteins furnish replacement therapy for defective, inactive or absent proteins in patients with well characterized biochemical defects. The typical examples are the use of **insulin** to treat diabetes; and **adenine deaminase** to treat 'boy in bubble syndrome. **Growth hormones** can be used to treat dwarfism.

On the other has **therapeutic proteins** may also be used to stimulate parts of or all the immune system to generate favourable response, i.g. the use of **colony-stimulating factors (CSF)** to stimulate platelet and neutrophil (blood cells) production in neutropenic patients and **interferons** and **interleukins** for stimulating anticancer response.

A large number of therapeutic proteins are now expressed in bacterial, yeast, insect and mammalian cell systems which can be manipulated by adopting techniques of genetic enginnering. However, large bioscale production in some of these may be difficult. Therefore, proteins are being expressed in **transgenic** or **genetically engineered animals** and **transgenic plants** which serve as **expression vats** or container to make proteins a process of therapeutic application called **pharming of proteins**.

Some proteins prepared through the said approach include **lactoferrin** and **protein C**, made in cow and used in infant formula blood clotting. The **cystic fibrosis transmembrane receptor (CFTR) protein** has been expressed in goats and used as a treatment to **cystic fibrosis**, a recessive genetic disease with a carrier frequency of about 4%, in caucasians. The same way alpha -1 antitrypsin has been expressed in sheep for the treatment of **emphysema** characterized by gaseous distension of tissues as alveolar distension in pulmonary emphysema Hence, here comes a solution to **diseases of our farm animals** and human being itself.

There are some proteins expressed in transgenic plants as **serum albumin** in potatoes has been used as a blood expander; **enkephalin** expressed in rape plants, as a painkiller; and **melanin** in tobacco plants as a sunscreen.

The process of farming using transgenic animals and plants may also assist in production of complex proteins. That could not be expressed in bacteria or yeast on industrial scale, i.g. **human collagen,** which bear application in **skin healing**, **tissue transplantation** and **dermatolgy**.

The biotechnologists working in the field of genetic engineering have made a significant contribution by providing information about mechanisms of cancer and working of both nervous and immune systems.

Q.4. What are **Bence - Jones proteins**?

Ans. These are free monoclonal antibody light chains found in the urine of patients with multiple myeloma.

Q.5. Define **stem cells**.

Ans. These are immature or precursor cells from which mature cells arise, typically the cells of the bone marrow which are the progenitors of the T cells and B-cells of the immune systems.

Q.6. Give a chronological account of new therapeutic products of biotechnology in the form of enzymes and proteins.

Ans. Some of the novel therapeutic products of biotechnology are tabulated below in **Table 26.1** in chronological order of development.

Table 26.1 A chronological account of production of new therapeutic products like therapeutic enzymes and other proteins from biotechnology (After Maulik and Patel, 1997).

Year	Therapeutic Product	Used for Disease manufacturer	Name of the company or
1982	Humulin, the synthetic insulin	Type I diabetes	Genetech, Inc.
1985	Protropin, a human growth hormone	Growth hormone defeciency	Genetech, Inc.
1986	Intron A (interpheron alpha	Hairy cell leukemia	Biogen, Inc and Scheringplough,
	Roferon A (interpheron alpha)	Hairy cell leukemia	Genetech, Inc. and Roche, Inc.

	Orthoclone OKT3 (monoclonal antibody)	Kidney transplant rejection	Orthobiotech, Inc. (Johnson & Johnson)
	Recombivax AV (synthetic vaccine)	Hepatitis B	Chiron Corp and Merck and Co
1987	Humatrope (Human growth hormone)	Growth hormone deficiency	Biogen, Inc and Eli Lilly, Inc.
1988	Intron A	Genital warts	Biogen, Inc and Schering-plough Inc.
1989	Epogen (Erythropoietin)	Renal anemia	Amgen, Inc
	Engerix B	Hepatitis B	Smith R Kline Beecham
	Alferon N	Genital warts	Interferon Sciences, Inc.
1990	Cyto Gram	Cytomegalo virus infection	Genetech, Inc
	Activase	Pulmonary embolism	
	Alpha Nine	Hemophilia B	
	ProCrit (erythropoietin alfa)	AIDS/AZT anemia	Orthobiotech, Inc
	Actimmune (interferon-gamma	Chronic gramulomatous disease	Genetech, Inc
	Adagen (adenosine deaminase)	Severe combined immunodeficiency (SCID)	Enzon, Inc
1991	Intron A	Hepatitis C	Biogen, Inc., and Schering Plough, Inc.
	Neopogen (colony stimulating factor)	Neotropenia from chemotherapy	Amgen, Inc.
	Leukine (GM - CSF i.e., Gramulocyte macrophage-colony-stiumulating factor)	Neutrophil recouery from bone marrow transplants	Immunex, Inc
	Ceredase (alglucerase)	Gaucher's disease	Genzyme Corp.
	Provisc	Ophthalmic surgery	Chiron Corp.
1992	Proleukin (interleukin-2)	Renal cell cancer	Chiron Corp.
	Alpha Nine SC	Factor IX viral infection	
	Intron A	Hepatitis B	Biogen Inc- and Schering-Plough, Inc.
	Recombinate (Factor VIII)	Hemophilia A	Genetics Institute, Inc and baxter Corp.
	Onco Scint OV 103 monoclonal antibody complex	Ovarian cancer imaging	Cytogen Corp.
	OncoScint CR 103 monoclonal antibody complex	Colorectal cancer imaging	Cytogen Corp.
1993	KoGEN (antihemophiliac factor)	Hemophilia A	Miles, Inc
	Procrit	Anemia from Chemotherapy	Ortho Biotech
	Orthoclone OKT3	Heart and liver transplant rejection	Ortho Biotech (Johnson & Johnson

	Betaseron (inter feron-beta)	Multiple sclerosis	Berlx Labs, Inc. and Chiron Corp
	Neutropin	Short stature (associated with chronic renal insufficiencies	
	Pulmozyme (D Nase)	Cystic fibrosis	Genentech. Inc
1994	Oncospar (Pegaspargase)	Acute hymphocytic leukemia	Enzon, Inc., and Rhone-poulenc Roser
	Neutropin	Growth hormone deficiency	Genentech, Inc.
	Cerezyme (imiglucerase)	Gaucher's disease	Genzyme, Inc.
	Neupogen	Bone-marrow transplants	Amgen Corp
	Albunex	Diagnosis of heart disease	(Mallinck rodt)

Q.7. What is **Southern blotting** ? Define.

Ans. It is a procedure for the identification of DNA by transmitting a fragment isolated on an agarose gel to a nitrocellulose filter where it can be hybridized with a complimentary 'probe' sequence. The technique was perfected by professor E.M. Southern in 1975.

Q.8. Define **Northern blotting.**

Ans. It is a technique to identify RNA molecules by hybridization which is analogous to Southern blotting.

Q.9. What is **Western blotting**?

Ans. It is a technique by which proteins are separated and immobilized on a paper sheet also called paper towels and then analyzed, generally by means of a labelled antibody.

Q.10. Have **'eastern' transfers** been reported ?

Ans. No, so far no one has come with 'eastern' transfers. So far this obscure term remains frolic and frisky.

Q.11. Which Nobelist is known as father of **test tube babies** ? Also give the success story of the Nobel Laureate.

Ans. Robert Edwards of Britain (= U.K.) who won 2010 Nobel Prize in physiology and medicine for the development of *in vitro* fertilization (IVF) is known as father of test tube babies.

Immediately after finishing Manchester Central High School on Whitworth Street in Central Manchester, heart of the historical industrial city (now industries have been closed), he served in British Army and then completed his undergraduate studies in Biological sciences at the University of Wales, Bangor. He obtained degree in Biology with major specialization in Zoology and minor specialization in Botany, and subsequently he studied at Institute of Animal Genetics and Embryology, at the faculty of Science at University of Edinburgh. He got Ph.D. in 1955 and joined University of Cambridge in 1963.

Edwards started to study human fertilization and he continued his work in collaboration with Patrick Steptoe, a gynaecologist surgeon from Oldham. Edwards developed human culture media to allow the fertilization and early embryo culture while Steptoe utilized laproscopy to receive ovocytes from patients with tubal infertility.

It was made an **ethical issue** connected with beliefs and principles about what is right and wrong. Their attempts faced significant hostility and opposition including a refusal of the British Government to fund their research and number of law suits. He developed technique in which egg cells are removed from a woman, fertilized outside her body and then implanted into the womb, together with British Gynaecologist surgeon Patrick Steptoe who died in 1988.

The birth of Louise Brown the world's first **'test tube baby'** at 11 = 47 p.m. on 25th July 1978 at Oldham General Hospital made medical history in *in vitro* fertilization, meant a new way to help infertile couples who formerly had no possibility of having a baby.

Refinement in the field of this technology has increased pregnancy rates and it is estimated that upto 2010 nearly 4 million children have been born by IVF with about 1,70,000 coming from donated oocyte and embryos. This was a major break through that also lead to prepare ground work for further innovations like **intracytoplasmic sperm injection, embryo biopsy** and **stem cell** (the cells which retain the capacity to self renew as well as to produce pregnancy which can generate cells of multiple lineage) research.

Edwards and Steptoe founded BOURN HALL CLINIC as place to advance their work and train new specialists.

Alas, Steptoe died in 1988. But their researches lead to **ectopic pregnancy,** fertilization and development of an egg which occurs outside the uterus.

Q.12. What is a **bacterial artificial chromosome** (BAC)?

Ans. A cloning vector which uses the origin of replication of *F* plasmid. It can take inserts of 100 – 300 kb and is believed to be less prone to cloning artefacts, such as chimeric inserts or rearrangements, upon propagation within the host.

Q.13. What is **yeast artificial chromosome** (YAC) ?

Ans. A cloning vector using a yeast host cell, can accept very large inserts of DNA (~ 1 Mb).

Q.14. What is **zoo blotting** ?

Ans. It is Southern blotting where human DNA is used to hybridise DNA from a range of other mammals.

Q.15. What is X-chromosome inactivation ?

Ans. It is clonal inactivation of one of the X-chromosomes in females during development.

27 CHAPTER

ANIMAL CLONING

Salient Features

- The cloning serves farmers and ranchers (as an American **ranch** or a large farm where people can go on holiday and do the activities of type that cow boys do) to accelerate the reproduction of their most productive live stock of choice in order to better produce safe and healthy food.
- The cloning helps in minimizing the use of antibiotics, growth hormones and other chemicals by reproducing healthy animals.
- The cloning is beneficial to consumers because the meat and milk from cloned animals and their offsprings will be more healthful, consistent and safe.
- The nucleus of a donor somatic cell can be transferred to 'nucleus-discarded-egg' of female animal.

Q.1. Define animal cloning.

Ans. It is a process by which an entire organism is reproduced from a single cell taken from the parent organism and in a genetically identical manner. This means the cloned animal is an exact duplication or replica in every way of its parent and bears same exact DNA.

Q.2. How can animal cloning be carried in lab? Describe the technique.

Ans. The attempts to clone animals have been made by the scientists since long. Many of the early attempts were futile without any result.

The first successful results in animal cloning were seen when **tadpoles**, (The form of a frog or toad from the time it leaves the egg to the time when it takes its adult form) were cloned from frog embryonic cells. The tadpoles so created did not survive to grow although in mature frogs yet it was a major breakthrough.

After that, using nuclear transfer scientists managed to clone mammals but cloned animals did not survive.

This was followed by first successful instance of animal cloning of the sheep named **Dolly** which not only lived but went to reproduce herself and naturally. Dolly was created by Lan Wilmut and his team at the Roslyn Institute of Edinburgh, Scotland, in 1997. This attempt was unlike or different from that of previous instances. She was not created out of developing embryonic cell but from a developed **mammary gland cell** taken from a full – grown sheep.

Since then the scientists have successfully produced a variety of other animals.

ANIMALS THAT HAVE BEEN CLONED: These are rats, cats, horses, bullocks, pigs, deer and buffalo.

Animals Cloned in India

Lohri born on 13th January 1987 was the first successful non-surgical **embryo transfer** cattle **calf**, which work was accomplalished by the National Institute of Immunology and the Indian Agricultural Research Institute (Fig. 27.1). But it was not a clone.

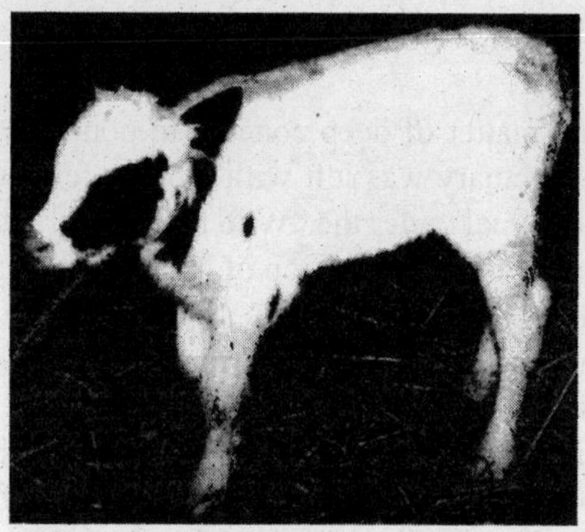

Fig. 27.1. Lohri, the first successful non-surgical embryo transfer cattle calf of India. Adapted from Biotech News 6 (1) : 17, 2011.

However, The first animal clone of India and the first buffalo clone of the world was produced at N.D.R.I. (National Dairy Research Institute) Karnal. Since then clone of the sheep 'PASMINA' a variety of sheep known for its wool – quality, has been cloned in India, at Shere Kashmir University of Agricultural Sciences and Technology.

Over a dozen animal clones have been prepared in India. Important among these are given below in a chronological order:

SAMRUPA: The world's first cloned buffalo calf was to be India's answer to Dolly, the sheep. But unlike Dolly, the first mammal cloned 13 years ago, who lived for seven years, Samrupa succumbed to a lung infection five days after it was born.

This first cloned calf was born on February 6, 2009 and died on February 11, 2009, just after 6 days. The scientists of NDRI had created a land mark technique by creating first buffalo calf.

GARIMA: The world's second cloned buffalo calf through the advanced hand guided technique was born at NDRI, Karnal on June 6, 2009. It survived for more than two years but she died of heart failure on August 18, 2011.

GAMINI: On Aug. 22, 2010 a cloned calf of female buffalo **'Garima-2'** nicknamed Gamini was born from embryonic stem cell ; she is mother to a calf named **'Mahima'**.

SHRESTHA: On Aug. 26, 2010 a cloned calf of male buffalo 'Shrestha' was born from somatic cell of an ear from an elite bull. It is producing good-quality fertile semen.

NOORI: On 9th March 2012 world's first Pashmina goat was produced by a process of cloning, using the nuclear transfer method. Noori was produced at the Centre of Animal Biotechnology at Shere–Kashmir University of Agricultural Sciences and Technology (SKAUST) in Srinagar, India, funded by World Bank, the project was jointly worked by SKAUST and National Dairy Research Institute, Karnal.

MAHIMA: On 15th January 2013 at 1.51 PM by normal parturition a female calf was born from cloned buffalo Garima. The NDRI has named the newborn female calf "Mahima". In the world it is the first calf born from cloned buffalo, produced through hand guided cloning technique. Mahima weighed 32 kg at the time of birth.

SWARAN: On March 18, 2013 a clone of male buffalo 'Swaran' was born from somatic cell. It is alive and healthy.

PURNIMA: On September 6, 2013 a clone of female buffalo 'Purnima' was born from high milk yielding animal. It survived for 21 days before dying.

LALIMA: It was born through the NDRI's 'hand-guided cloning technique', on May 02, 2014. It is a clone of **adult Murrah Buffalo**. Murrah is an elite breed of buffalo.

RAJAT: On July 23rd 2014 a male cloned calf 'Rajat' was produced through the "Hand guided technique".

DEEPASHA: It is a matter of deep concern to ponder over the environmental conditions that Chhatisgarh wild life Sanctuary was left with a clone female buffalo, popularly called **Asha**. The buffalo is a schedule - I animal under the "Wild life Conservation Act — 1972" and is in the red list of the "International Union for Conservation of Nature". It is an endangered animal and there is only one buffalo in the country. 'Deepasha' is the second female of its species. Earlier Chhatisgarh Wild buffalo delivered male calves after natural mating. It was a major concern of Chhatisgarh authorities to protect this buffalo due to old age and other risks. 'The Wild Life Trust of India' the technical partner of Chhatisgarh state approached NDRI for assistance. The team of scientists of NDRI visited the State of Chhatisgarh and collected some cells from the clone wild buffalo **Asha** and cultured it in NDRI which resulted the birth of the clone Deepasha on December 12, 2014. The calf took birth by normal parutrition and its weight at the time birth was 32 kg. The donor mother Asha lives at 'Udanti Wild Life Sanctuary' and as stated is the lone buffalo of its species.

It has been said that even human being can be cloned which has given rise to new ethical debate. The people argue, is it right to produce human clones? What would that do to the fabric of our society?

Process of Animal Cloning

Early attempts in artificially induced animal cloning were done using developing embryonic cells and transferred that into an unfertilized egg from which nucleus has already been removed. The process of fertilization was stimulated by giving an **electric shock** or by some **chemical treatment** method. The cells which developed from the artificially induced union were than implanted into host mothers. The cloned animal which resulted had a genetic makeup that was **identical genetic makeup of the cell**, which was transferred to nucleus removed egg cell.

Since Dolly, of course, it is now possible to create clones from **non-embryonic cells**.

The animal cloning now can be done both for **reproductive** and **therapeutic purposes**.

However, for the second case cloning is done to produces **stem cells** or other such cells which can be used for therapeutic purposes like healing or recreating damaged organs. The intention is not to duplicate the whole organism.

Technique

In animal cloning in the process of **somatic cell nuclear transfer** i.e. transfer of the nucleus of cell of the animal other than the germ cell, the scientists collect a cell from the animal that is to be cloned which is called the genetic donor. It is obvious that the somatic cell contains the DNA of **genetic donor animal**. In the second step the scientists collect an egg from a female animal called the **egg donor** and discards the nucleus of the egg cell which is the part of the cell which contains egg donor's genes. (Fig. 27.2).

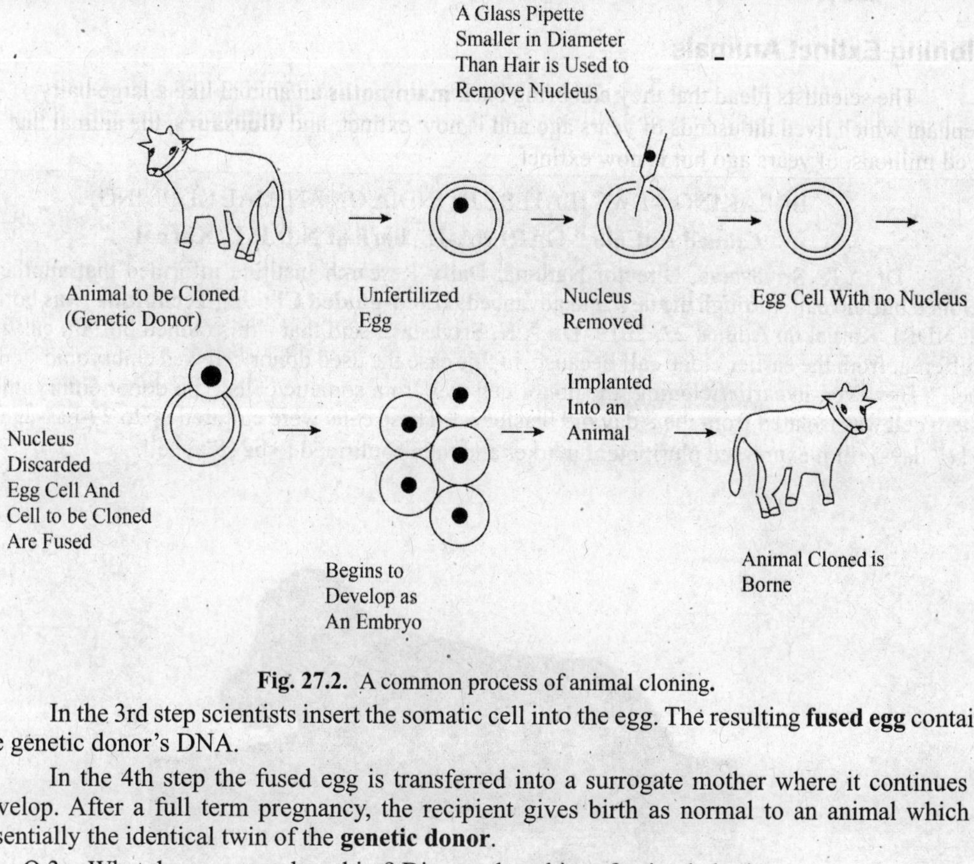

Fig. 27.2. A common process of animal cloning.

In the 3rd step scientists insert the somatic cell into the egg. The resulting **fused egg** contains the genetic donor's DNA.

In the 4th step the fused egg is transferred into a surrogate mother where it continues to develop. After a full term pregnancy, the recipient gives birth as normal to an animal which is essentially the identical twin of the **genetic donor**.

Q.3. What do you mean by ethics? Discuss the ethics of animal cloning.

Ans. The ethics is the branch of philosophy that deals will moral principles or a system of moral principles.

Despite of the fact that most of scientists count the process of animal cloning as a major breakthrough and see many beneficial possibilities in it but many people are feeling uneasy with this idea considering it to be against nature and ethically damming particularly in the instance of preparing clone of human being.

It is because most of general public are not aware of general details involved in cloning. Therefore, there are a lot of misconceptions about the entire matter.

Recently there has been a spurt of new laws banning or regulating around the world. In some countries animal cloning is allowed but human cloning is not permitted.

Some advocacy groups are seeking to ban cloning for therapeutic use even if this could potentially save from many debilitating illnesses.

Points Against Animal Cloning

The points that go against animal cloning are as under: The cloning process fails in the course of pregnancy in many cases and some sort of birth defects occur e.g. in a recent case a calf was borne with two faces. Sometimes defects manifest themselves later and kill the clone.

Points in Favour of Animal Cloning

1. The cloning providers a hope for intention to save certain endangered species of animals and protect them from becoming totally extinct.

Cloning Extinct Animals

The scientists plead that they can bring back **mammoths** an animal like a large hairy elephant which lived thousands of years ago and is now extinct; and **dinosaurs**, the animal that lived millions, of years ago but is now extinct.

BREAKING NEWS (LATEST IN INDIA ON ANIMAL CLONING)

Cloned Buffalo " GARIMA-II" born at N.D.R.I., Karnal

Dr. A.K. Srivastava, Director National Dairy Research Institute informed that another cloned buffalo calf, through the new and advanced **'Hand-guided Cloning Technique'** was born at NDRI, Karnal on August 22, 2010. Dr. A.K. Srivastava said that ''this cloned buffalo calf is different from the earlier clone calf because, in this case the used donor cell was embryonic stem cell". However, in earlier cloning, the donor cell was from somatic cells. The donor embryonic stem cell was isolated from the 8 day old blastocyst. These cells were cultured up to 29-passages (117 days) till it expressed pluripotent marker and then confirmed to be stem cell.

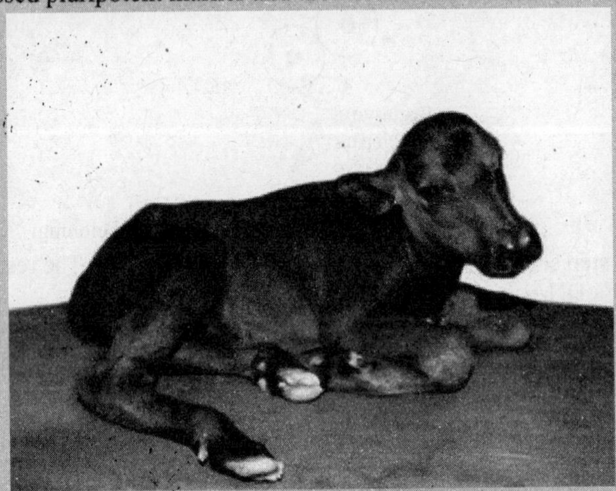

Dr. A.K. Srivastava, Director emphasized that this technology could go a long way in helping for faster multiplication of superior milk buffaloes in India. He said that although the world's largest population of buffaloes is in India and they are contributing about 55% of total milk production in country, but the percentage of elite animals is very low and there is an urgent need to enhance the population of these elite buffaloes. He further emphasized that there is an acute shortage of good bulls and the technology of cloning will decrease this gap between supply and demand of breeding the bulls in the shortest possible time.

The team of the jubilant scientists involved in the production of this cloned calf using embryonic stem-cell as donor cell are Dr. M.S. Chauhan, Dr. S. K. Singla, Dr. R.S. Manik,

Dr. P. Palta, Dr. Shiv Parsad, and Dr. Aman George of N.D.R.I., Karnal. The scientists are of the opinion that the embryonic stem cells have better cloning ability as compared to somatic cells, as such the epigenetic reprogramming of these cells is much more efficient than the somatic cells, which are already differentiated and lineage committed.

Earlier NDRI has produced the world cloned buffalo calf on 6th February, 2009. The hand-guided cloning technique developed at NDRI, is an advanced modification of the "Conventional Cloning Technique". In this technique, immature oocytes were isolated from ovaries and were matured *in vitro*. These were then denuded and treated with an enzyme to digest the outer layer of oocytes called '**zona pellucida**'. The oocytes were then treated with chemicals to push their genetic material to one side of the oocyte. This protruded side was then cut off with the help of "hand held fine blade" for removing the original genetic material of the oocyte. The enucleated oocyte was then electrofused with single cell taken from colony of embryonic stem cells. The resulting embryos were cultured and grown in the laboratory for seven days to develop them to the stage of blastocyst. The blastocysts were transferred to recipient buffaloes.

This cloned calf weighing 32 kg. was born through caesarian operation carried out by a team of doctors comprising Dr. R.S. Bisla from CCSHAU, Hisar and Dr. K.P.S. Tomar, Dr. Subhash Chand and Dr. Parveen Kumar and Dr. M.K. Srivastava from NDRI, Karnal. The animal is as alert as any other animal and is apparently normal and healthy.

Dr. S Ayyapan, Director General, Indian Council of Agricultural Research and Secretary DARE, Govt. of India, New Delhi congratulated the team and said that the new technology of "Hand-guided Cloning" of buffaloes may lead a new era in faster multiplication of elite germplasm to face the challenges of increasing demands of milk in view of the ever growing human population. Dr. K.M.L Pathak, Deputy Director General and Dr. C.S. Prasad, ADG, (AS) ICAR also congratulated the team.

Dr. A.K. Srivastava, Director, N.D.R.I., informed that another cloned male Buffalo calf named '**Shresth**', was born at N.D.R.I., Karnal during early Hours of August 26, 2010, through the new and advanced 'Hand-guided Cloning Technique'. This cloned buffalo calf is different from the earlier clone calves as, in this case, (I) the foster mother was provided opportunity for Normal delivery. However, earlier two calves were born through caesarian Operation, (II) the cloned calf was from ear somatic cell of 2 week old Buffalo calf; however, the earlier two calves were produced by using cells from foetus and embryonic stem cell, respectively. Furthermore, (III) the Embryo which led to successful pregnancy and normal delivery had remained Frozen at -196C for one week in liquid nitrogen and brought back to active Life upon thawing at room temperature.

The jubilant scientists are of the opinion that the cryopreservation of embryos will need to be made as part of technique, so that the embryos could be transported and used at several places.

This cloned calf weighing 41 kg was born through normal delivery with slight assistance carried out by a team of doctors.

Earlier NDRI has produced the **world's first cloned buffalo calf on 6th February, 2009** followed by on June 6, 2009 and August 22, 2010. The Hand-guided cloning technique developed at NDRI, is an advanced modification of the "Conventional Cloning Technique".

Dr. S. Ayyappan, Secretary DARE and Director General, Indian Council of Agricultural Research wished good luck and congratulated the team and said That the new technology of "Hand-guided Cloning" of buffaloes may lead a new Era in faster multiplication of elite germplasm to face the challenges of increasing demands of milk in view of the ever growing human population. Dr. K. M. L Pathak, Deputy Director General (Animal Sciences) and Dr. C. S. Prasad, ADG(AS) ICAR also congratulated the team and praised the efforts by NDRI, Karnal.

28

CHAPTER

INDUSTRIAL MICROBIOLOGY

Q.1. What is industrial microbiology?

Ans. The use of microorganisms for manufacture of industrial products is known as industrial microbiology.

Q.2. What is a bioreactor?

Ans. Vessels used for industrial fermentation are called **bioreactors**. These are designed for proper aeration with pH control and temperature control.

Q.3. What is the purpose of a series of impeller paddles and stationary wall **baffles** in a continuously stirred type bioreactor?

Ans. To keep microbial suspension agitated.

Q.4. Why is there a diffuser at the bottom of the bioreactor?

Ans. For molecular break up of incoming air for maximum aeration.

Q.5. Give one example each of a primary metabolite and a secondary metabolite.

Ans. Ethanol and penicillin.

Q.6. Give two characteristics of primary metabolites.

Ans. (i) Primary metabolites are produced almost at the same time as the cells.

(ii) The production curve of primary metabolites follows the curve for population of cells in parallel with only minimal lag phase.

Q.7. Give two characteristics of **secondary metabolites.**

Ans. (i) A secondary metabolite is not formed until the microbe has completed its growth phase also known as trophophase, and has attained the stationary phase of growth. The following period in which most of the metabolite is produced is called idiophase.

(ii) Secondary metabolites may be a microbial conversion of a primary metabolite or may be a product of metabolism of the original growth that a microbe produces only after a large number of cells and the primary metabolite have accumulated.

Q.8. What was the ratio of yield of a new strain of *Penicillium* isolated from cantaloupe from a Peoria, Illinois, supermarket, and then treated with X-rays and nitrogen mustard which is a chemical mutagen in comparison to the original culture of *Penicillium notatum?*

Ans. More than 100 times, and improvement in fermentation techniques have nearly tripled even this yield.

Q.9. Name a surface on which the microbial cells are **immobilized** or anchored to make products like high fructose syrups, aspartic acid and some other products of biotechnology.

Ans. Silk fibres or other microscopic spheres or fibres.

Amino Acids

Q.10. Which are the common amino acids which are major industrial products of microorganisms?

Ans. Lysine, glutamic acid, phenylalanine and aspartic acid.

Q.11. Which are the amino acids usually produced by *Corynebacterium glutamicum?*

Ans. Threonine and lysine.

Q.12. Which is the type of mutant used for the maximum production of lysine?

Ans. A mutant of bacterium *Corynebacterium glutamicum* which lacks the enzymes required to produce threonine is used to make lysine on large scale. This mutant is better in comparison to normal bacterium for the prevention of feedback inhibition as both amino acids require a common enzyme in their synthetic pathway.

Q.13. How is glutamic acid produced from the same bacterium that produces glycine?

Ans. The bacterium is induced to produce glutamic acid by feeding it the minimal amount of the vitamin biotin. Consequently the plasma membrane of the bacterium gets weakened and the glutamic acid leaks into the medium.

Q.14. What is the other product for which this technique of inducing the **leaky membrane** is exploited?

Ans. Production of antibiotics.

Citric Acids

Q.15. What are the uses of citric acid?

Ans. It is an antioxidant and pH adjuster in many foods. In dairy products it is used as an emulsifier. Citric acid is also used in industry and pharmaceutical products.

Q.16. How is a major part of citric acid produced commercially?

Ans. It is produced by mould *Aspergillus niger* using molasses as a substrate. This fungus secretes citric acid when given only a limited supply of iron and manganese.

Enzymes

Q.17. Give some common **applications of enzymes.**

Ans. The amylases are used in the production of syrups from corn starch. Glucose isomerase can turn the glucose formed by amylase from starch into fructose which is used in place of sucrose to sweeten many foods. The enzyme proteases adjusts the amount of glutens (that is protein) in wheat. The other proteolytic enzymes are employed as meat tenderizers or in detergents to remove the proteinaceous stains. The rennin produced from fungi is used to form curd from milk.

Vitamins

Q.18. Which microorganisms are used for the production of vitamin B_{12}?

Ans. *Pseudomonas* and *Propionibacterium* spp. As cobalt is a component of B_{12} cobalt chloride is added to the medium.

Q.19. Which is the organism used to produce riboflavin?

Ans. *Ashbya gossypii,* a fungus is used in fermentation to produce vitamin riboflavin.

Q.20. How are antibiotics produced?

Ans. The growth medium is inoculated with spores of appropriate mould or streptomycete and is aerated vigorously. After the antibiotic reaches a satisfactory concentration it is extracted from the solution by precipitation and other industrial procedures.

Beverages

Q.21. What is **beer?** How is it prepared?

Ans. Beer is made using bottom yeast *Saccharomyces carlsbergensis.* The germinated barley releases starches and amylase enzymes which is called malting. The enzymes in malt hydrolyze starch to fermentable sugars called mashing. A liquid called **wort** is sterilized. The flowers of Hop's are added for flavour. Yeast is then added and incubated from 37° - 49°C. In the presence of yeast sugar gets converted into alcohol and CO_2. The alcohol is 4 per cent. The yeast grows on the bottom of the fermenting vessel.

Q.22. What is **ale?**

Ans. It is prepared like beer and is incubated at 50°-70°C. The microorganisms top yeast *S. cerevisiae* converts sugar into alcohol. The alcoholic content is 6 per cent and the yeast grows at the top of the fermentation vessel.

Q.23. What is **sake**?

Ans. The starch of steamed rice is converted into sugar by the fungus *Aspergillus oryzae.* Yeast is added and incubated at 20°C. The yeast converts sugar into alcohol which is 14 - 16 per cent.

Q.24. What is the raw material and which is the organism used to make wine?

Ans. Wine is prepared from grapes using *S. cerevisiae.* and contains 14% or less alcohol.

To prepare wine, grapes are crushed into **must**, SO_2 is added to inhibit wild yeast and then yeast is added to it.

Q.25. What is **red wine**?

Ans. For red wine black grapes are used and pressing does not occur until after fermentation. It is prepared like natural wine incubated at 25°C. Aged in Oak for 3-5 years and in bottle for 5-15 years.

Q.26. What is **white wine?**

Ans. It is almost white to colourless or pale yellow. For white wine white grapes (with green, golden or yellow colour skin) are used, the juice being extracted from the pulp by pressing before fermentation. White wine is incubated at 10 - 15°C and aged for 2-3 years in bottles.

Q.27. What is **sherry**?

Ans. It is like natural wine with additional surface growth floor at 27°C. The alcohol content is 18-21%. The microorganisms used are *S. cerevisiae, S. beticus* and S. *bayanus.* It is S. *beticus* that grows on the surface as a film producing aldehydes from alcohol.

Q.28. What is **sparkling wine** (champagne)?

Ans. It is like natural wine using yeast *S. cervisiae* with secondary fermentation in a bottle. 2.5% sugar and yeast are added to bottled wine. Incubated at 15°C and the bottle is inverted to collect the yeast in the neck of the bottle. In secondary fermentation yeast produces CO_2 and yeast settles quickly.

Q.29. What is **rum** or **jamaica**?

Ans. It is prepared from cane molasses, inoculated from previous fermentations. Wild yeast is used for the purpose. Oak aging adds colour and is finally distilled to concentrate. Alcohol content is 50-95 per cent.

Q.30. What is **brandy?**

Ans. Fruits are crushed and added to *S. cerevisae* and distilled for concentration. It contains 40-43 per cent alcohol.

Q.31. What is **whiskey?**

Ans. Wort is fermented by yeast *S. cerevisae.* It is distilled to concentrate alcohol and aged in charred Oak barrels. Alcohol content ranges from 50-95 per cent.

Q.32. Give the basic steps in making **red wine.**

Ans. (1) Grapes are tested and picked. (2) Crushed and destemmed. (3) Addition of sulphide to kill undesirable microorganisms. (4) Addition of innoculum (yeast). (5) Fermentation is carried out. (6) The resultant is pressed to separate solids from wine. (7) Wine is clarified in settling vats. (8) Filtration of wine. (9) Wine is aged. (10). Bottling of wine.

29 CHAPTER

PHARMACEUTICAL MICROBIOLOGY AND MICROBIAL TECHNOLOGY

Q. 1. Give an introduction to **stoichiometry.**

Ans. Basically, the stoichiometry may be regarded as the branch of chemistry and chemical engineering which deals with the quantities of substances which enter into and are produced by chemical reactions. Stoichiometry gives the quantitative relationship between reactants and products in a chemical reaction. It can be utilized in microbial technology also. Stoichiometry can be classified into two main categories :

(i) Composition Stoichiometry - deals with mass relationships of elements in compounds

(ii) Reaction Stoichiometry - Involves mass relationships between reactants and products in a chemical reaction

In ideal Stoichiometry - All reactants are converted into products.

Common Methods for Solving all Stoichiometry Problems

Mass-Mass Problems:

1. Start with a known mass of reactant or product and find an unknown mass of another reactant or product.
2. All other stoichiometry problems are derivations (shortened versions) of this larger solution:

Find moles of given using molar mass → Use mole ratios to find moles of unknown → Find grams of unknown using molar mass

$$\text{Given (in grams)} \times \left(\frac{\text{1 mole of given}}{\text{given's molar mass in grams}}\right) \times \left(\frac{\text{moles of unknown in balanced equation}}{\text{moles of given in balanced equation}}\right) \times \left(\frac{\text{unknown's molar mass in grams}}{\text{1 mole of unknown}}\right) = \text{grams}$$

Steps for solving problems

1. Start with a correctly balanced chemical equation.
2. Use key words in the problem statement to identify substances as either reactants or products.
3. Determine what units you've been given and what you are being asked to find.
4. Label each step with the correct units! the units from the numerator of the first step become the units in the denominator of the next step, and so forth.
5. Stop when you have an answer with the units that you are searching for.

Limiting Reactant

The reactant that limits the amounts of the other reactants that can combine and the amount of product that can be formed in a chemical reaction.

Excess Reactant

It is the substance that is not used up completely in a reaction.

Identifying the Limiting Reactant:

1. Convert grams of each reactant to moles if the problem has not already been worked out.
2. Use molar ratios from the balance chemical equation to determine which reactant is limiting, and which reactant is in excess.

Stoichiometry with Limiting Reactants

All calculations should start with the amount of the limiting reactant, not the excess reactant.

Calculating Percent Yield

The ratio of the actual yield to the theoretical yield, multiplied by 100

$$\text{percent yield} = \frac{\text{actual yield}}{\text{theoretical yield}} \times 100$$

Q.2. What is a **shake flask** and what do you understand by shake flask volume?

Ans. The flask which is used for shaking particular liquid for particular purpose e.g. for aeration or mixing is called **shake flask.** They are of following types:

(A) Standard shake flask or the Erlenmeyer flask (B) "Flying saucer" shake flask (C) Shake flask with baffles (D) Flat bed "Thompson" or "Roux" bottle as shown in Fig. 29.1.

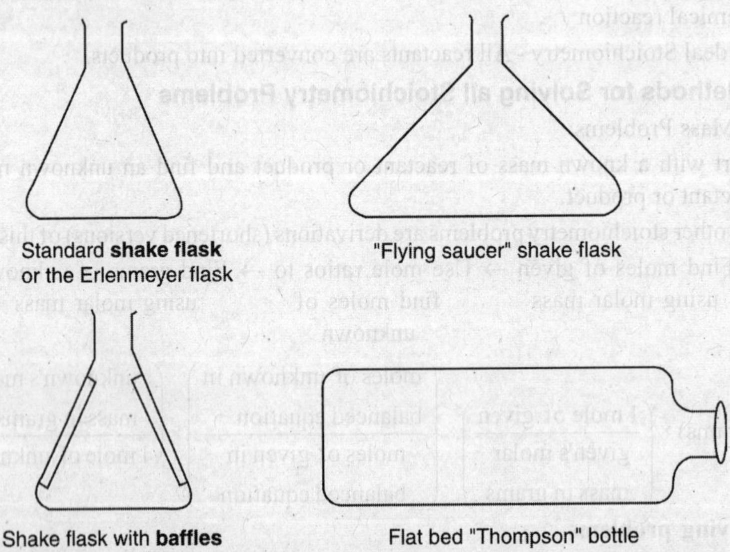

Fig. 29.1. Various types of **shake flask.**

The lower the volume of medium in a shake flask, the better will be the **OTR** (Oxygen Transport Rate). The minimum volume that can be practically obtained (e.g. 50 ml in a 250 ml shake flask) should give the best OTR and hence the best results. This will also be dependent on sample volume. Very low volumes can only be used for short-term fermentations, otherwise the medium will evaporate and the nutrients would become too concentrated for the culture to perform satisfactorily.

Q.3. How does an **airlift bioreactor** or fermentor help you to avoid damaging cell shear?

Ans. Some cell lines are so fragile in culture that any type of mechanical impeller will shear them. **Air lift fermentor** also known as an air tower fermenter or air bioreactor uses the sparged air to gently circulate cells and growth media without damaging them. Due to the action of air lift fermentor many air bubbles are generated at the bottom of vessels which aerate and agitate the culture medium. Thus resulting mixing of culture as well as aeration. The air bubbles lift to the top of the medium and the air passes out through an outlet. The cells and the medium which lift out of the draft tube move down outside the tube and are recirculated. The oxygen supply is quite efficient but sealing up presents certain problems. Fermentors of 2-90 litres are commercially available but 2000 litres fermentors are being used for the production of monoclonal antibodies.

Q.4. What is **fed batch fermentor** ? Explain.

Ans. There are mainly two types of fed batch fermentor: (1). Fixed volume fed batch (2).Variable volume fed batch

In fixed volume fed batch, the limited substrate is fed without diluting the cultures. The culture volume can also be maintained practically constant by feeding the growth limiting factors without affecting the culture volume. Variable volume fed-batch is one in which the volume change with the fermentation time due to the substrate feed. In this way volume change is dependent on the requirement, limitation and objective of the operator.

Q.5. What is **immobilized cell reactor**? Explain.

Ans. Ethanol fermentation is an immobilized cell reactor using *Saccharomyces cerevisiae.* The fermentation set up comprise of a column packed with beads of immobilized cells. The immobilization of *S. cerevisiae* simply performed by enriched cells cultured media harvested at potential growth phase. They are fixed with loaded ICR (Immobilized cell reactor) carried out at initial stage of operation and the cell are entrapped by calcium alginate. The production of ethanol is steady after 24hrs of operation. ICR is a plug flow tubular column constructed with a nominal diameter of 5cm, ID of 4.6cm, Plexiglas of 3mm wall thickness and 85cm length. The medium is sent to ICR column from a feed tank located above the column. A variable speed master flex pump model L/S easy load (Cole-Parmer, Vernon Hills, IL, USA) may be used to transfer feed medium from a 20 litre polypropylene autoclavable nalgene carboy. The carboy serve as a reservoir

Q.6. What are **stirring and-mixing**?

Ans. Diffusion takes place by discontinuous as well as continuous movements. Stirring and mixing are going to trigger diffusion of nutrients. Two rivers can flow together for miles side by side and do not mix properties. Blobs and Mediterranean water can be traced unmixed in the Atlantic for years. Eventually molecular diffusion has very long time scale. Turbulent stirring stretches the two fluids and increase the amount of contact between the fluids which permits mixing to take place much more quickly as the molecular diffusion is due to randomly exchanging molecules in many more places. This is because that the surface where two fluids touch has a larger area. To conclude stirring without diffusion is not mixing and can be reversed.

Q.7. Differentiate between acetic acid and vinegar. Also describe the production of **acetic acid.**

Ans. The word 'vinegar' is derived from the French term 'vinaigre' (vin = Wine+ aigre = sour), meaning "sour wine". It contains about 40% acetic acid by weight and small quantities of alcohol, glycerol, esters, sugars, and salts. **Vinegar** is the product obtained as a result of impartial oxidation of alcohol in a fermenting sugar containing fruit or cane juice, molasses, fermented mash of malted grain, honey, maple skimming, syrups, etc. Since, the vinegar contains considerably large amount of acetic acid it is generally used in the place of acetic acid. But, for the pure acetic acid the vinegar is subjected to purification.

The fermentation of vinegar involves two types of biochemical reactions:

(i) Alcoholic fermentation of carbohydrate. It is the first step in the production of vinegar and takes place under anaerobic condition. This step in which sugar is fermented to alcohol is completed by the action of yeast species.

$$\underset{\text{Glucose}}{C_6H_{12}O_6} \xrightarrow{\text{Yeast}} \underset{\text{Ethyl Alcohol}}{2C_2H_5OH} + 2CO_2 + \text{Energy}$$

(ii) Oxidation of alcohol to acid. This is the second step in the production of vinegar and is aerobic process. This step in which alcohol is oxidised to acetic acid is completed by the action of acetic acid bacteria i.e., *Acetobacter orleansis* and *A. schulzenbachi.*

$$\underset{\text{Ethyl alcohol}}{C_6H_5OH + O_2} \xrightarrow{\text{Acetic Acid bacteria}} \underset{\text{Acetic Acid}}{CH_3COOH} + H_2O + \text{Energy}$$

After the fermentation is over, calcium citrate is precipitated from the fermented broth by adding calcium hydroxide/ $Ca(OH)_2$. It is then filtered, washed, and treated with sulphuric acid (H_2SO_4) to precipitate calcium sulphate and retain the citric acid. The citric acid is purified and finally crystallized. In recent years, citric acid fermentation by yeast is significantly growing because a variety of yeasts, e.g., *Candida, Hansenula*, etc. have been found to ferment citric acid from carbohydrates and hydrocarbons. The yeast, *Candida lipolytica*, appears promising and result in good yield of citric acid from various raw materials. Citric acid is an important chemical used in food and candies, beverages, medicines, flavouring extracts, manufacture of ink, effervescent slats, dyeing, mirror silvering, and detergent industries. It is also increasingly used in the removal of toxic and corrosive gases in air.

Q. 8. Write down the process of industrial **production of alcohol**.

Ans. Ethyl alcohol is known as common solvent and raw material. Much of this alcohol is obtained synthetically from ethylene. However, its production from microbial fermentation using a variety of cheap sugary substrates is still commercially important. It is imperative that the microorganisms used must have a high tolerance for alcohol, must grow vigorously and produce a large quantity of alcohol. Yeasts, particularly *Saccharomyces cerevisiae*, represent the best known microorganisms used in the production of ethyl alcohol. The inexpensive substrates used in alcohol industry are molasses from cane sugar or waste sulphite liquor from paper industries. Starch yielding grams (corn), potatoes, grapes may also be used as substrate if their prices permit. Some countries use sugarbeet for the purpose. The chemical reaction that results in the microbial fermentation of carbohydrate into alcohol can be represented as follows.

$$\underset{\substack{\text{Glucose}\\\text{(A fermentable carbohydrate)}}}{C_6H_{12}O_6} \xrightarrow[\text{Enzyme}]{\text{Yeast}} \underset{\text{Ethyl Alcohol}}{2C_2H_5OH} + 2CO_2 + \text{Energy}$$

Commercial production of alcohol can be done by using molasses. Molasses contains about 50% fermentable carbohydrates (sugars). Big deep tanks of steel or stainless steel are used as containers in the industrial production method. Molasses is diluted to a suitable sugar concentration (15-16%); a small quantity of nitrogen source (e.g., ammonium phosphate, urea. ammonium sulphate) and sulphuric acid (H_2SO_4) are added to it, The pH of this medium is maintained at about 5.0 and an actively growing *Saccharomyces cerevisiae* culture is added. The fermentation starts and is allowed to proceeding for about 24-40 hours at about 25-30°C temperature. The yield of ethyl alcohol ranges about 50% of the fermentable sugar concentration present in the medium. The large amount of CO_2 which is produced during the fermentation process as a result of decarboxylation is recovered and compressed to its solid state. Remaining yeast can be used as an animal feed.

Q.9. What is an enzyme? How can enzymes be produced on commercial scale? Describe.

Ans. **Enzyme** is an organic catalyst formed by a living cell. Enzymes are mostly obtained from plant and animal cells. Now-a-days microorganisms are becoming excellent sources for industrial production of certain enzymes. Many microorganisms are known to excrete enzymes into their growth media during the fermentation processes of various industries such as pharmaceutical, food, textile, etc. High yields and quality of such enzyme is determined to a great extent by suitable strain selection and the cultural conditions. Since these enzymes are used in various industries, their large scale production is essential. Most of the commercial enzymes are obtained from aerobic microorganisms. The following two processes are generally employed: (i) Surface culture method (ii) Semisolid or Solid culture method:

(i) Surface culture method was largely used prior to the World War II. In this method shallow trays were used to grow microorganisms. Hence large scale productions of enzymes were difficult and costly.

(ii) In the semisolid or solid culture method enzyme production and fermentations are carried out in closed, deep tank fermenters varying between 1,000 to 30,000 gallon capacity. A substrate (e.g., wheat bran, sugarbeet and cellulose are taken depending upon the type of microorganisms to be taken to excrete enzyme) moistened with a suitable nutrient solution is sterilized and then inoculated with mould spores (most of the enzymes are of fungal origin). The substrate allows air to penetrate and the mould mycellia grow throughout. When the enzyme level reaches its maximum, the fermented mass is extracted with water and the enzyme is separated out with a suitable solvent. It is now filtered, concentrated and dried.

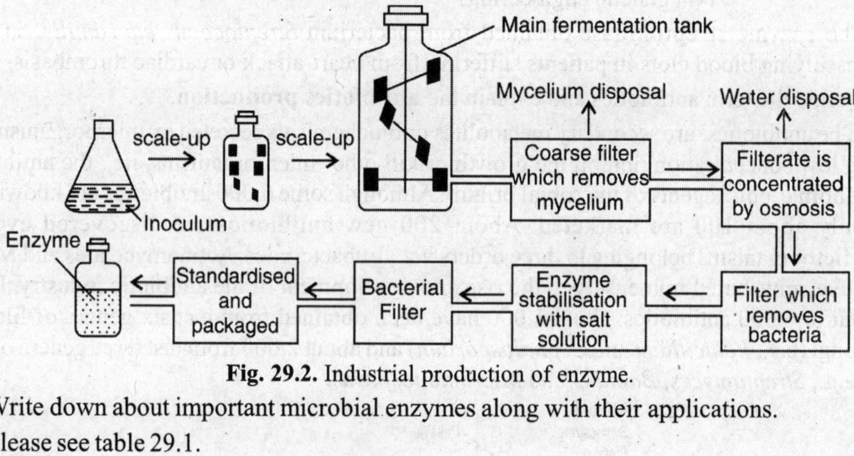

Fig. 29.2. Industrial production of enzyme.

Q.10. Write down about important microbial enzymes along with their applications.

Ans. Please see table 29.1.

Commonly produced enzymes on industrial scale.

Enzyme	**Microorganism**	**Substrate**	**Use**
Amylases	*Aspergillus oryzae*	Wheat bran	Starch hydrolysis to dextrin and sugars mainly in alcoholic fermentation, food preparations and fabric; in preparing adhesives; in designing textiles; in clarifying fruit juices; in manufacturing medicines.
Invertase	*Saccharomyces cerevisiae*	Sugarbeet	In hydrolyzing sucrose to glucose and laevulose; in producing non crystallization syrups from sucrose which is partially hydrolysed by this enzyme; in candy making; in producing artificial honey.

Proteases	*Aspergillus niger, Bacillus* spp.	Wheat bran	In leather processing; ion food processing; in manufacture of liquid glue; in degumming of silk; in clarification of beer protein haze and as an adjunct to soap for cleaning, laundries; in tenderizing meat; in recovery of silver from spent film (photography).
Pectinase	*Aspergillus, Penicillium* sp.	Wheat bran	In clarification, pressing, filtration and concentration of fruit juices and wines; in hydrolyzing pectins in the retting of flex for the manufacture of line in coffee-bean fermentation.
Cellulase	*Trichoderma reesii*	Cellulose	Digestive additive in cellulose hydrolysis to cellobiose; in cheese making.
Lipase	*Rhizopus* sp.	Wheat bran	Digestive additive in hydrolyzing fat into fatty acid and glycerol.
Takadiastase	*Aspergillus spp.*	Wheat bran	Strach dextrinization and designing of textiles; bread supplement; in production of syrups; in digestive aid
Rennin	*Bacillus* spp. (via genetic engineering)		Cheese production

The enzyme **streptokinase** obtained from bacterium *Streptococus equisimilis* is used for dissolving blood clots in patients suffering from heart-attack or cardiac thrombasis.

Q.11. Define the term antibiotics and explain the **antibiotics production**.

Ans. 'The antibiotics' are secondary metabolites or biochemicals secreted by microorganisms which, in low concentration, inhibit the growth or kill other microorganisms, i.e., the antibiotics are 'antimicrobial agents of microbial origin'. Although some 6,000 antibiotics are known to exist, only about 100 are marketed. About 200 new antibiotics are discovered every year. Microorganisms belonging to three orders viz., Eubacteriales, Actinomycetales and Moniliales have contributed immensely to the rise and development of the antibiotic industry. However, out of 6,000 antibiotics, about 1,000 have been obtained from just six genera of filamentous fungi (e.g., *Penicillium* and *Cephalosporium*) and about 2,000 from just three genera of bacteria (e.g., *Streptomyces*, *Bacillus* and *Micromonospora*).

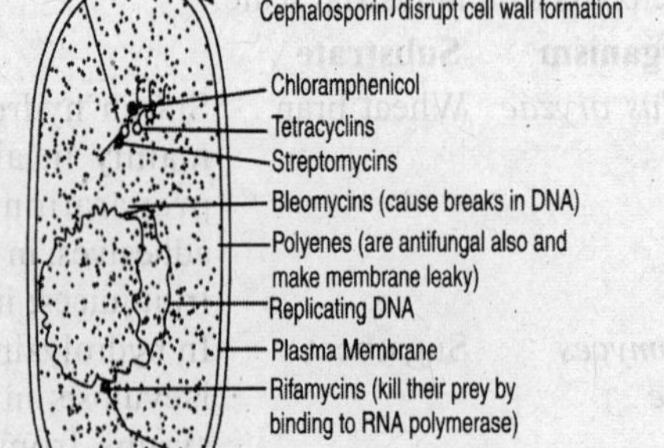

Fig 29.3. Sites of action of different antibiotics, on a bacterial cell.

To be truly useful, an antibiotic must possess following qualities in it :

(i) Antibiotic should be of "broad spectrum", i.e., it should have the ability to inhibit/kill a number of different types of pathogenic microorganisms.

(ii) It should check the development of resistant forms of pathogenic microorganisms.

(iii) It should not result in undesirable side effects in the host.

(iv) It should not disturb the "normal" microbial flora of the host. Such a disturbance may upset the balance of nature.

Penicillin

Alexander Fleming discovered penicillin secretion by the mould *Penicillium notatum* in 1929. While working in a hospital labouratory in London. He saw that *Staphylococus* colonies were lysed due to fungal activity. However, when Chain (1939) purified the active material, called penicillin, it proved remarkably effective in certain infections. Penicillin is not a single chemical compound but a group of substances of related structure and activity. There are six penicillins : Penicillin G (benzyl penicillin V (phenoxymethyl penicillin), penicillin F (A2 – pentenyl penicillin), penicillin X (p-hydroxybenzyl penicillin), penicillin K(n-heptyl penicillin) and penicillion O (allyl-mercaptomethyl penicillin). Penicillin is selective for Gram-positive bacteria, some spirochaetes and the Gram-negative diplococci (*Neisseria*). It was World War II that provided stimulus for commercial production of penicillin; and various strains of *Penicillium chrysogenum* are used for the purpose. At present, almost 50 years later, such strains of the fungus are available that yield 104 times more than the original isolate. Production has also been dramatically increased by improvements in media and to fermenter design. Phenyl ethanoic acid (phenyl acetic acid) is now routinely added to the media since this induces the synthesis of a metabolic precursor of penicillin G (the most active form of penicillin).To produce penicillin commercially, deep fermentation tanks with the capacity of several thousand gallons are filled with fermentation medium consisting of corn steep liquor, nutrient agar salts, lactose, glucose and phenyl ethanoic acid (phenyl acetic acid). This medium is inoculated with conidia-suspension of the fungus and is constantly aerated as pellets. The fermentation is completed in 20 hours. When the fermentation is complete, the masses of fungal growth are separated leaving a clear fluid (the broth). The broth is passed over a filter and washed. The filtrate is when mixed with potassium ions, filtered again and dried. The result is a crystalline potassium salt of penicillin G. Penicillin is assayed to determine its potency before being bottled and sold. G in the name 'Penicillin-G' refers to 'Gold Standard'.

Cephalosporins

These antibiotics are produced from *Cephalosporium acremonium*, a marine fungus. They are effective against Gram-negative bacteria. Cephalosporidine, cephaloglycin and cephalexin are some derivative compounds prepared from cephalosporins. These antibiotics, like penicillin, block the cell-wall synthesis (Fig. 29.3).

Streptomycin

This antibiotic is obtained from *Streptomyces griseus*. Its antibiotic activity was first reported by Waksman in 1944. Streptomycin is a "broad spectrum" antibiotic; inhibits many Gram-negative bacteria, several species of *Mycobacterium* including *M. tuberculosis*. It also inactivates bacteriophages. Streptomycin in commercially produced by the strains of the bacterium in a fermentation medium consisting of soybean meal, glucose and mineral salts. The pH of the medium is maintained at 7.5. The fermentation takes place under submerged conditions at 25 to 30°C for 5-7 days. When the fermentation is complete, the masses of bacterium are separated and the antibiotic is obtained from the fermentation broth using organic solvents.

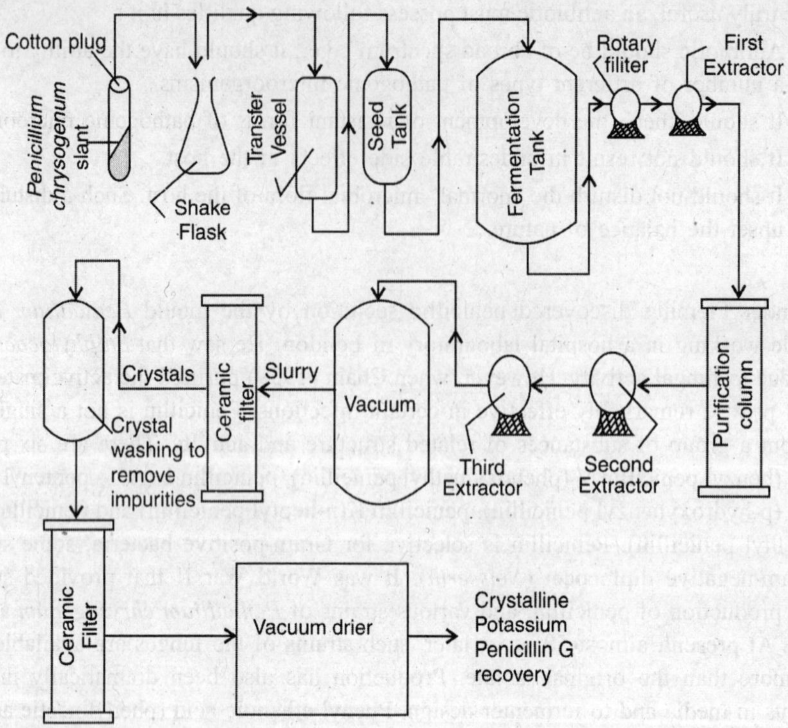

Fig 29.4. Production of antibiotic Penicillin.

Q.12. What is **steroid biotransformation**? Explain.

Ans. Sterol is a polycyclic alcohol such as a cholesterol or ergosterol. Cortisone steroid can be synthesized chemically from deoxycholic acid. This process requires 37 steps many of which must be carried out under extreme condition of temperature and pressure with the resulting product costing over $200 per gram.

$$\text{Deoxycholic acid} \longrightarrow \text{Cortisone}$$

So, cortisone synthesis is an example of steroid biotransformation. The major difficulty in chemically synthesized cortisone is need to introduce an oxygen atom at number 11 position of the steroid ring which can be accompanied by using microorganism. The fungus *Cunninghamella blakesleeana* can hydroxylate the steroid **cortexolone** to form **hydrocortisone** with the introduction of oxygen at the number 11 position. The fungus *Rhizopus arrhizus* for example hydroxylates **progesterone** forming another steroid with the introduction of oxygen at the number 11 position. Steroids are very important in pharmaceutical industry. Steroid hormone regulates various aspect of metabolism. One such hormone **cortisone** has been know to relieve the pain of rheumatoid arthritis. It regulates human sexuality and reproduction cycles. Some of steroids are used as oral contraceptives also. Some cortisone derivatives are also useful in alleviating the symptoms associated with allergic and other undesired inflammatory responses of the human body.

A good example of steroid transformation processes involving microorganisms like *Rhizopus nigricans* which can be by growing the mould in a fermentor using an appropriate growth medium and incubation conditions to obtain high biomass, of course, aeration and agitation are employed to achieve rapid growth. After the growth of microorganism the steroid to be transformed is added. When progesterone is added to a fermentor having *Rhizopus nigricans* which has been growing for about one day, the steroid is hydroxylated at the number 11 position to form | | – α – hydroxyprogesterone. The product is recovered by extraction with methylene chloride or various other solvents followed by purification chromatographically and recovered by crystallization (Fig. 29.5).

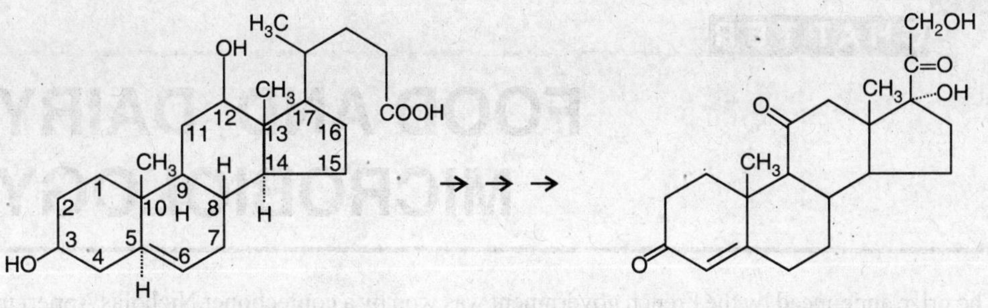

DEOXYCHOLIC ACID   CORTISONE

SYNTHESIS OF CORTISONE FROM DEOXYCHOLIC ACID IN 37 STEPS

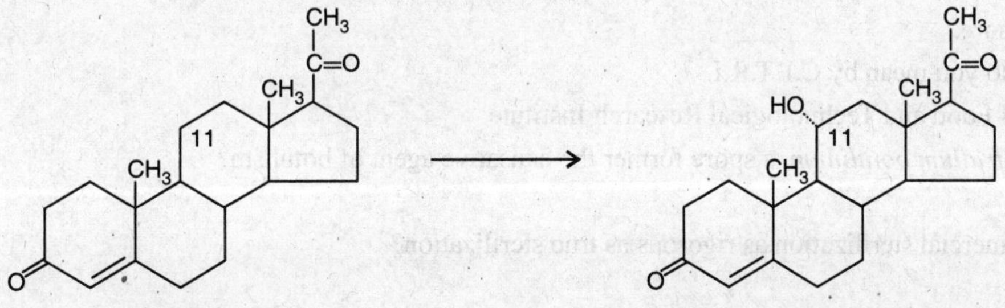

PROGESTERONE   11-OC-HYDROXYPROGESTERONE

PRODUCTION OF 11-OC-HYDROXY PROGESTERONE BY HYDROXYLATING PROGESTERONE THROUGH *RHIZOPUS ARRHIZUS*

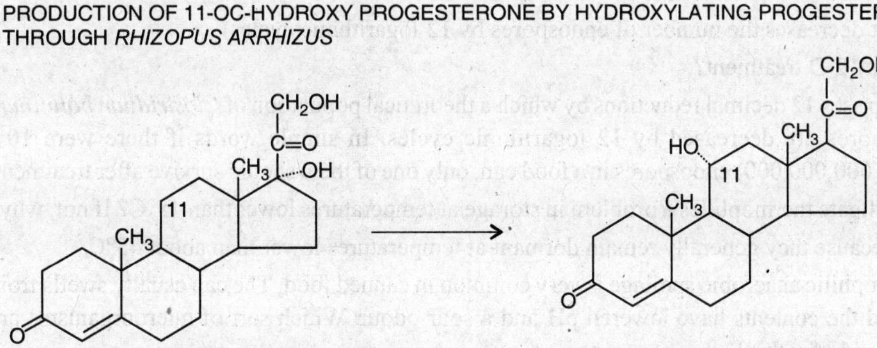

CORTEXOLONE   HYDROCORTISONE

HYDROXYLATION OF STEROID CORTEXOLONE THROUGH *CUNNINGHAMELLA BLAKESLEEANA* OR *CURVULARIA LUNATA* FOR PRODUCTION OF STEROID HYDROCORTISONE

Fig. 29.5. Steroid Biotransformation.

30

CHAPTER

FOOD AND DAIRY MICROBIOLOGY

Q.1. The prize announced by the French government was won by a confectioner Nicholas Appert in 1809. What was it for?

Ans. He showed that food could be preserved if it was sealed in tightly stoppered containers and boiled for a specific duration of time.

Q.2. Where is Central Food and Technological Research Institute situated in India?

Ans. At Mysore.

Q.3. What do you mean by C.F.T.R.I.

Ans. Central Food and Technological Research Institute.

Q.4. Is *Clostridium botulinum,* a spore former the causative agent of botulism?

Ans. Yes.

Q.5. Is commercial sterilization as rigorous as true sterilization?

Ans. No.

Q.6. Do you agree that to ensure complete sterilization of foods enough heat should be applied during the **12 D treatment** to get rid of spores in commercial sterilization?

Ans. Yes. (It decreases the number of endospores by 12 logarithmic cycles).

Q.7. What is 12 D treatment?

Ans. It comprises 12 decimal reductions by which a theoretical population of *Clostridium botulinum* endospores are decreased by 12 logarithmic cycles. In simple words if there were 10^{12} (1,000,000,000,000) endospores in a food can, only one of them should survive after treatment.

Q.8. Are obligate thermophiles a problem in storage at temperatures lower than 45°C? If not, why?

Ans. No. Because they generally remain dormant at temperatures lower than about 45°C.

Q.9. Thermophilic anaerobic spoilage is very common in canned food. The can usually swells from gas and the contents have lowered pH and a sour odour. Which sort of microorganisms are responsible for this?

Ans. A number of thermophilic species of *Clostridium* are responsible for the above.

Q.10. What is flat sour spoilage of canned foods?

Ans. When the thermophilic spoilage occurs but the can is not *swollen by gas* production, it is called **flat sour spoilage,** which is caused by thermophilic bacteria as *Bacillus stearothermophilus* which occur in the starch and sugars used in food preparation.

Q.11. In what conditions do mesophilic bacteria spoil canned foods?

Ans. If the food is unprocessed or if the can leaks.

Q.12. Give the steps involved in **commercial sterilization** process in industrial canning.

Ans. The steps involved in commercial sterilization are:

(i) Washing, sorting and blanching of the food material. In blanching, food is treated with hot water or live steam that softens the product to fill the can better. Blanching also destroys enzymes which may change the colour, flavour or texture of the product and also lowers the microbial count.

(ii) The cans are filled to the extent leaving the least dead space.

(iii) The cans are placed in a steam box to drive out the dissolved air.

(iv) The cans are sealed.

(v) The cans are sterilized in large retorts with steam under pressure.

(vi) The cans are cooled by spraying or submerging in water.

(vii) The cans are labelled, stored and delivered.

Q.13. Which are the important spoilage organisms of acid foods in cans?

Ans. They are moulds, yeasts, and occasionally species of acid tolerant, non-endospore forming bacteria.

Q.14. Which are the organisms that create problems due to being heat resistant and acid tolerant?

Ans. The heat resistant mould *Byssochlamys fulva* that produce heat resistant ascospores and *Aspergillus*. *Bacillus coagulans* can grow at a pH of 4.0.

Q.15. How is aseptic packaging done?

Ans. By feeding the rolls of packaging materials in hot hydrogen peroxide solution and sometimes aided by ultraviolet light. The metal containers can be sterilized with super heated steam or other high temperature methods or high energy electronic beams.

Q.16. Pathogenic bacteria often do not grow under low temperature preservation of food. However, there are exceptions that some bacteria do. List them.

Ans. Clostridia which cause type E botulism, *Yersinia enterocolitica* and *Listeria monocytogenes.*

Q.17. How could pork be made free from parasitic worms which cause trichinosis, without heating?

Ans. By irradiating.

Q.18. What is **irradiation logo**?

Ans. The irradiation logo indicates that the food has received irradiation treatment.

Q.19. High energy electron accelerators are much faster and achieve sterilization within a few seconds, but they have low penetrating power. Therefore, for what sort of food preservation are they suitable?

Ans. Sliced meats, bacon and similar thin products.

Q.20. What are the common additives used to retard the spoilage of food?

Ans. Sodium benzoate, sorbic acid and calcium propionate are the common additives.

Q.21. What are the chemical additives often added to meat products like ham, bacon, hot dogs and sausage?

Ans. These are sodium nitrate and sodium nitrite.

Q.22. What are the two main **functions of nitrites**?

Ans. (i) To preserve the red colour of the meat by reacting with blood components.

(ii) To prevent the germination and growth of any botulism endospores, which may be present.

Q.23. What is the drawback of nitrites?

Ans. Nitrites with amino acids form certain carcinogenic products like **nitrosamines.** Therefore, the amount of nitrites added to food has been reduced these days.

Q.24. Which are the microorganisms responsible for the spoilage of bread?

Ans. *Aspergillus, Neurospora sitophia, Mucor* sp. and *Rhizopus nigricans* cause bread mould.

Q.25. Name the organisms that turn raw milk sour.

Ans. *Streptococcus lactis, S. cremoris* and *Lactobacillus bulgaricus.*

Q.26. Which organism spoils pasteurized milk (turn into sour milk)?

Ans. *Lactobacillus thermophilus.*

Q.27. Which organisms spoil chicken (cause sliminess)?

Ans. *Achromobacter* sp, *Pseudomonas, Alcaligenes* sp. and *Flavobacterium* sp.

Q.28. Which organisms cause spoilage (fishy odour) of fish?

Ans. *Achromobacter* sp, *Flavobacterium, Micrococcus, Pseudomonas* sp. and *Serratia* sp.

Q.29. What are naturally occurring antimicrobial substances in some foods and how do they work?

Ans. Eggs contain lysozyme which causes lysis of cell walls of bacteria as lysozyme breaks down peptidoglycan. Eggs also contain conalbumin and riboflavin both of which chelates iron. The waxy skin of fruits retards the entry of bacteria and fungi. Garlic possesses allyl iothiocyanate that inhibits bacterial growth best at 37°C. Onion possesses lacrimatory factors that inhibit the growth of bacteria and fungi and cranberries have benzoic acid which inhibits the growth of fungi.

Q.30. Which plants were consumed by the Egyptians who built pyramids to prevent and combat infections?

Ans. Garlics and radishes.

Q.31. Who was first to report the antibacterial potential of garlic?

Ans. Louis Pasteur in 1835.

Q.32. What is TDT?

Ans. Thermal death time.

Q.33. Give thermal death times of some spore forming bacteria in a tabulated form.

Ans.

Bacterial spores	*TDT in minutes at 100°C*	*TDT in minutes at 120°C*
Bacillus subtilis	15-20	—
Clostridium botulinum	360	5
Bacillus coagulans	1140	17

Q.34. What are the methods of **preservation of food**?

Ans. (i) High temperature without harming the quality of food.

(ii) Pasteurization.

(iii) Low temperature treatment.

(iv) Desiccation.

(v) By chemical additives.

(vi) Irradiation.

Q.35. Which are the three **methods of pasteurization** in general use?

Ans. These are:

(i) High temperature-short-time method (HTST) need temperatures of 72°C for 15 seconds.

(ii) Low temperature long time (LTLT) need temperatures of 62°C for 30 minutes.

(iii) Ultra high temperature method (UHT) in which temperature of 140°C is maintained for 1.5 seconds and 149°C for 0.5 seconds.

Q.36. Which are the antibiotics used to preserve dairy products?

Ans. The antibiotics nisin and natamycin are used.

Q.37. Match the following.

A.	140°C for 1.5 seconds	(a)	Chemical preservation
B.	Sorbic acid	(b)	Sterilization
C.	72°C for 15 seconds	(c)	Pasteurization HTST
D.	5°C	(d)	Low temperature preservation
E.	121°C For 15 minutes	(e)	Pasteurization UHT

Ans. A-e, B-a, C-c, D-d, E-b.

Q.38. Define 12 D treatment.

Ans. It is a process of sterilization which results in a decrease in the number of *Clostridium botulinum* endospores by 12 logarithmic cycles.

Q.39. Nicholas Appert's discovery may be regarded as a forerunner of hermetically sealed cans. How do hermetically sealed cans prevent the spoilage of food?

Ans. Being completely airtight they do not permit the entry of contaminating organisms.

Q.40. Name three food borne infections.

Ans. Gastroenteritis, typhoid fever and dysentery.

Q.41. Give three examples of microbial intoxication.

Ans. Botulism, staphyloccal food poisoning and mycotoxicoses.

Q.42. What are the public health service standards for Grade A pasteurized milk?

Ans. Standard plate count of fewer than 20,000 bacteria per millilitre of milk and not more than 10 coliforms per millilitre.

Q.43. What are the public health service standards for dry milk products?

Ans. Standard plate count less than 30,000 bacteria per gram or a coliform count of less than 10 per gram.

Q.44. Why does casein, the milk protein curdle?

Ans. Due to action of lactic acid bacteria or the enzyme rennin (or chymosin).

Q.45. What is **cheese**?

Ans. It is the curd separated from the liquid portion of milk called whey.

Q.46. How is hard cheese produced?

Ans. By growing lactic acid bacteria in the interiors of the cheese.

Q.47. What is ripening of cheese?

Ans. Growth of microorganisms in cheese.

Q.48. How are semisoft cheeses ripened?

Ans. By having bacteria grown on the surface.

Q.49. What ripens **soft cheeses**?

Ans. *Penicillium* growing on the surface.

Q.50. How is commercial **butter milk** made ?

Ans. By permitting lactic acid bacteria to grow for 12 hours.

Q.51. How are sour cream, yoghurt, kefir and kumiss produced?

Ans. By growing lactobacilli, streptococci and yeast in low fat milk.

Q.52. How is commercial **yoghurt** produced?

Ans. Commercial yoghurt is produced by evaporating low fat milk in a vacuum pan and the resulting thickened milk is inoculated with a mixed culture of *Streptococcus thermophilus* (for acid production) and *Lactobacillus bulgaricus* (for flavour). The temperature of fermentation is maintained at 45°C for several hours so that S. *thermophillus* outgrows L. *bulgaricus*.

Q.53. What is **water activity** of a food?

Ans. Water activity (A_w) is the amount of moisture available in foods.

Q.54. What is **DEFT** ?

Ans. DEFT is the direct epifluorescent filter technique, which is a microscopic technique. In it the bacteria are concentrated from a significantly large volume of sample by filtering it through a polycarbonate membrane filter. The retained bacteria are then stained on the membrane with acridine orange and counted directly under an epifluorescene microscope. The bacteria are trapped in a single focal plane which is achieved by using a polycarbonate membrane where relatively uniform pores are produced following neutron bombardment of a plastic film rather than cellulose acetate filters which have tortous pores resulting in bacteria to be held at different levels.

Q.55. List the important methods for microbiological examination of food.

Ans. These are:

(i) Direct examination,

(ii) Cultural techniques,

(iii) Enumeration method, e.g., plate counts and most probable number counts (MPN),

(iv) Alternative methods, e.g., dye reduction test, electrical methods, ATP determination,

(v) Rapid methods for the detection of specific organisms and toxins.

31 CHAPTER

SOIL MICROBIOLOGY

Q.1. Name the bacterium that prey on other bacteria.

Ans. *Bdellovibrio*.

Q.2. Give some examples of **mutualism (symbiosis)** in which both organisms are benefitted.

Ans. (i) Association of fungus and alga in lichens.

(ii) Protozoa serving as host to single celled algae which live inside the cytoplasm of protozoan. The protozoa give space and protection while the algae provide additional nutrients for the host.

Q.3. What is **commensalism?**

Ans. In commensalism, one organism benefits from the relationship without affecting the other, e.g., some bacteria can degrade polymers such as cellulose into simpler component glucose which is utilized readily by the neighbouring scavengers.

Q.4. What is cometabolism?

Ans. Some organisms cannot break down the resistant chemicals like pesticides but they can do so if given a primary source of energy such as glucose.

Q.5. Why is soil regarded as "biologic fire"

Ans. The leaf shed by a tree is consumed easily by the soil converting its organic matter into carbon, nitrogen and other chemical compounds.

Q.6. What is the percentage of organic matter in soil?

Ans. It is 2-10% in agricultural soils and upto 95% in peat bog soils.

Q.7. What is **humus?**

Ans. The dark coloured material in soil which consists of partially decomposed organic matter is called humus.

Q.8. Why do soil gases generally have a high proportion of carbon dioxide and a low proportion of oxygen?

Ans. Due to respiration of soil organisms.

Q.9. Where do soil gases exist?

Ans. Soil gases exist in pores and spaces between soil particles or are dissolved in the water contained by it.

Q.10. Name the animals that occur in fertile soil.

Ans. They range from nematodes (minute worms) to large forms as insects, millipedes, centipedes, spiders, slugs, snails, earthworms, mice, moles, gophers (burrowing rodents) and reptiles.

Q.11. At which depth in a garden soil does the maximum number of microorganisms per gram occur?

Ans. At a depth of 3-8 cm.

Q.12. What is **geosmin**?

Ans. The gaseous material produced by the actinomycetes which provides soil with the characteristic of musty odour is called geosmin.

Q.13. What is **biomass**?

Ans. The total mass of living organisms in a given volume is biomass. The biomass of actinomycetes is about equal to the biomass of all bacteria present in the soil.

Q.14. Do moulds outnumber yeasts in soil?

Ans. Yes.

Q.15. The number of protozoa tends to rise or fall with the bacterial populations. Why?

Ans. Because of the predatory habit of many of them.

Q.16. Name the common microbial pathogens of soil.

Ans. The endospore forming bacteria are common survivors in soil, e.g., *Bacillus anthracis* causative agent of anthrax in animals, *Clostridium tetani* which causes tetanus, *Clostridium botulinum* the causal organisms of botulism and *Clostridium perfringens* the causal organisms of gas gangrene.

Q.17. Name the soil bacterium which is a pathogen of insects.

Ans. *Bacillus thuringiensis*.

Q.18. What elements are dependent on **biogeochemical cycles**?

Ans. These are carbon, nitrogen, sulphur and phosphorus.

Q.19. Where from comes the inorganic carbon used in synthesis of organic compounds.

Ans. From CO_2.

Q.20. What is **greenhouse effect?**

Ans. It is the increase of carbon dioxide in the atmosphere due to excessive use of fossil fuels universally resulting in the warming of the earth.

Q.21. How does ammonia released from decomposition of proteins in the soil behave?

Ans. In dry soil the ammonia passes out while in moist soil it solubilizes.

$$NH_3 + H_2O \rightarrow NH_4OH \rightarrow NH_4^+ + OH^-$$

The ammonium ions so formed are used by bacteria and plants for the synthesis of amino acids.

Q.22. What is **nitrification?** Name the two bacterial species associated with nitrification.

Ans. The process of oxidation of ammonium ion to nitrate is called nitrification. The autotrophic nitrifying bacteria living in the soil are *Nitrosomonas* and *Nitrobacter.* Nitrification takes place in the two steps given below:

(i) $$\underset{\text{Ammonium ion}}{NH_4^+} \xrightarrow{\textit{Nitrosomonas}} \underset{\text{Nitrite ion}}{NO_2^-}$$

(ii) $$\underset{\text{Nitrite ion}}{NO_2^-} \xrightarrow{\textit{Nitrobacter}} \underset{\text{Nitrate ion}}{NO_3^-}$$

Q.23. What is **denitrification?** Name the bacterial species causing it.

Ans. The breaking down of nitrates into atmospheric nitrogen is denitrification, e.g.,

$$\underset{\text{Nitration}}{NO_3^-} \rightarrow \underset{\text{Nitrite ion}}{NO_2^-} \rightarrow \underset{\text{Nitrous oxide}}{N_2O} \rightarrow \underset{\text{Nitrogen gas}}{N_2}$$

Common denitrifying bacteria are *Pseudomonas.* Some other genera include *Paracoccus, Thiobacillus* and *Bacillus.*

Q.24. Name the nonsymbiotic nitrogen fixing bacteria.

Ans. They are *Azotobacter* and some cyanobacteria. The other species are *Clostridum, Klebsiella, Enterobacter* (anaerobic) and *Rhodospirillum* and *Chlorobium* (aerobic photoautotrophic).

Q.25. What is **rhizophere**?

Ans. The region where soil and roots make contact particularly in grasslands. This region is rich in microbial population.

Q.26. Name the symbiotic nitrogen fixers.

Ans. *Rhizobium* and *Bradyrhizobium* which form nodules in the roots of leguminous plants.

Q.27. Give an example of nitrogen economics in a forest involving alga and fungus.

Ans. Lichen.

Q.28. Name three **symbionts of cyanobacteria.**

Ans. (i) *Anthoceros* a bryophyte

(ii) *Azolla* a small floating-fern and

(iii) *Cycas* a gymnosperm.

Q.29. Give example of symbiotic nitrogen fixing root nodule formed by a member of actinomycetes.

Ans. *Frankia*.

Q.30. Give steps in the formation of root nodules.

Ans. (i) Attachment of *Rhizobium* on the root hair.

(ii) Formation of infection thread through which bacteria enter.

(iii) The bacterial cells change into pleomorphic bacterioids and the packed root cells enlarge in size.

(iv) The enlarged root cells form a nodule.

Q.31. What is **recalcitrance**? Give an example?

Ans. It is resistance to degradation, e.g., DDT as a recalcitrant.

Q.32. What is one of the major causes of elimination of eagles and other predatory birds that enjoyed abundance of food two decades back in the plains of North India.

Ans. Accumulation of DDT from contaminated food resulting in impairing of their reproductive system.

Q.33. Define **amensalism.**

Ans. It is an interactive association between two populations which is detrimental to one and does not adversely affect the other. In it one microbial population produces a substance which is inhibitory to other populations. Such interpopulation relationship is called amensalism.

The terms antibiosis and allelopathy are also used to describe the cases of chemical inhibition. The first population may not be affected by the inhibitory substance or may gain competitive edge i.e., beneficial.

32

CHAPTER

BIOGEOCHEMICAL CYCLING

Q.1. What is **carbon cycle**?

Ans. It mainly includes the transfer of CO_2 and organic carbon between the atmosphere (the source of inorganic CO_2) and **hydrosphere** (the aqueous envelope of the earth as bodies of water and aqueous vapours) and **lithosphere** (solid part of the earth). The carbon dioxide (CO_2) in lithosphere and hydrosphere reacts with water to form carbonate and bicarbonate which are the main inorganic forms of carbon occurring there.

Q.2. What is **green house effect**?

Ans. Burning of fossil fuels increases carbon dioxide in the atmosphere leading to rise in concentration of carbon dioxide in the **atmosphere** (the mass of air surrounding the earth and a unit of pressure approximately 1×10^6 dynes/cm^2). This rise in concentration of CO_2 and the consequent rise in global temperature is referred to as **green house effect**.

Q.3. How does production of methane by a specialized group of methanogenic archaea represent **a diversion to normal cycling of carbon**?

Ans. It is due to the fact that methane produced by the archaea cannot be used by most of heterotrophic organisms and in this way is lost from biological community to the atmosphere, and hence the methanogens are diversists of the system of normal cycling of CO_2.

Q.4. What are **primary producers** in the ecosystem?

Ans. **Autotrophs**, the organisms whose growth and reproduction are independent of external sources of organic compounds, the required cellular carbon is supplied by reduction of CO_2 and the needed cellular energy is supplied by the conversion of light energy to ATP or the oxidation of inorganic compounds to provide the free energy for the formation of ATP; are the **primary producers.**

Q.5. What is **gross primary production**?

Ans. It is the total amount of organic matter produced by an autotrophic biological community of a **niche** or suitable natural home.

Q.6. What is **net primary production**?

Ans. It is gross primary production minus oxidative metabolism of biological community which removes organic carbon and energy stored in such compounds. It may be regarded as decay of energy stored in a given niche.

Q.7. What is **trophic structure**?

Ans. The feeding relationship between organisms or the routes by which energy and materials are transferred within an **ecosystem** (living and non living components of self supporting systems).

Q.8. What is **food web**?

Ans. The **food web** may be regarded as transfer of energy stored in organic compounds between the organisms in the community. At the base of food web come **primary producers** which manufacture the organic matter for the system. **Grazers** are the organisms which depend for their nutrition on primary producers. However, in a **phytoplankton-based food web** algae and cyanobacteria (blue green algae) are the primary sources of food for grazers. In a **detrital food web,** the microbial biomass produced from growth on dead organic matter, called detritus, is the primary food source for the grazers. The grazers in the chain are eaten by predators and the predators are in turn are preyed on by larger predators. In this way carbon and stored energy are moved to the higher levels of food web. Of course, some of the carbon and energy are lost in each transfer in the process of respiration. (Fig. 32.1).

Q.9. What is **Winogradsky column**?

Ans. The Winogradsky column is a glass or clear plastic cylinder with a core of solid or sediment in it. This model system employed to study the growth of specific microorganisms, has been named after a noted Russian microbiologist Sergei Winogradsky. Specific microbial population grow at different levels due to different environmental conditions. The column is exposed to light so that different photosynthetic populations develop.

Algae and cyanobacteria which are oxygenic photosynthetic microorganisms grow at the surface. At the lower levels are anaerobic green and purple sulphur bacteria while obligately anaerobic heterotrophs appear at the lowest levels of the column.

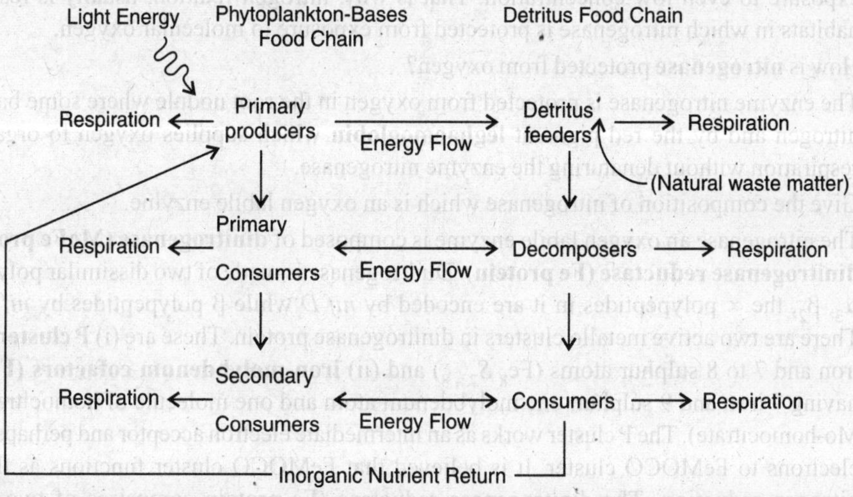

Fig. 32.1. Food web representing integrated feeding structure. Food synthesized by primary **producers** is transferred to higher trophic levels, where **decomposers** turn organic biomass to CO_2 and minerals. In an aquatic (pond, lake, river) ecosystem autotrophs feed consumers. In a terrestrial ecosystem (grassland, forest, desert) also **autotrophs** feed consumers. In a **detrital food web** dead organic matter is used by decomposers. Even in **Tundra** marshy in summer and frozen hard in winter and **antarctica** autotrophs are the primary producers in the food web.

Nitrogen (N) cycle

Q.10. What is **biogeochemical cycling of nitrogen** or **nitrogen cycle**?

Ans. The molecular nitrogen is one of the most abundant substance in the air. The nitrogen moves from atmosphere through biological taxa (**biotas**) living in the soil and aquatic habitats. Thus involves nitrogen fixation in the form of (NH_4^+), nitrification (NO_2^- and NO_3^-) and assimilation into plants as ($R - NH_2$), passed on to animals, leading to ammonification on death of the animal and again fixation of NH_3 to NO_2^- and NO_3^-, NH_3^- and NH_4^+ in soil may be broken down to molecular nitrogen by the microorganisms.

Q.11. Which enzyme system is involved in fixation of atmospheric nitrogen?

Ans. The fixation of atmospheric nitrogen rely on **nitrogenase enzyme system**, which comprise of **nitrogenase** and nitrogenase reductase. The electrons are transferred through ferredoxin or flavodoxin to nitrogenase reductase and then to nitrogenase where they are utilized to reduce N_2 and H^+ to NH_3 and H_2 as shown in the equation below:

$$N_2 + 8e^- + 8H^+ + 16\ \text{Mg ATP} \longrightarrow H_2 + 16\ \text{Mg ADP} + 16\ \text{Pi}$$

Two coproteins are also present in nitrogenase enzyme system. These are a **MoFe protein** containing molybdenum plus iron and a **Fe protein** having iron only. The active site of nitrogenase where reduction of nitrogen takes place is dependent on a iron and molybdenum containing cofactor (Fe MO CO). The production of H_2 which accompanies the reduction of nitrogen adds to ATP requirements of nitrogen fixation. Evolution of H_2 accompanies biochemical nitrogen fixation.

However, only some strains of *Rhizobium* and *Bradyrhizobium* possess hydrogenase and can utilize the hydrogen, and other nitrogen fixing bacteria wastefully evolve H_2 gas.

Nodulated root forming systems which evolve hydrogen often are colonized by hydrogen-oxidizing *Acinetobacter* strains which grow on the hydrogen released by nitrogen fixing bacteria.

It is remarkable that nitrogenase is very sensitive to oxygen and is irreversibly inactivated on exposure to even low concentration. That is why, nitrogen fixation, usually is restricted to habitats in which nitrogenase is protected from exposure to molecular oxygen.

Q.12. How is **nitrogenase** protected from oxygen?

Ans. The enzyme nitrogenase is protected from oxygen in the root nodule where some bacteria fix nitrogen and by the red pigment **leghaemoglobin** which supplies oxygen to organism for respiration without denaturing the enzyme nitrogenase.

Q.13. Give the composition of nitrogenase which is an oxygen labile enzyme.

Ans. The nitrogenase an oxygen labile enzyme is composed of **dinitrogenase (MoFe protein)** and **dinitrogenase reductase (Fe protein)**. Dinitrogenase is made of two dissimilar polypeptides, α_2, β_2, the $\propto$ polypeptides in it are encoded by *nif D* while β polypeptides by *nif K* genes. There are two active metallo clusters in dinitrogenase protein. These are (i) **P cluster** having 8 iron and 7 to 8 sulphur atoms ($Fe_8\ S_{7-8}$) and (ii) **iron-molybdenum cofactors (FeMoCO)** having 7 iron and 9 sulphur, one molybdenum atom and one molecule of homocitrate (Fe_7S_9 Mo-homocitrate). The P cluster works as an intermediate electron acceptor and perhaps transfers electrons to FeMOCO cluster. It is believed that FeMOCO cluster functions as the site of nitrogen reduction. The dinitrogenase reductase (Fe protein comprises of two identical polypeptides γ_2 encoded by the *nif H* gene, and each polypeptide has two iron atoms. The four Fe atoms are organized into an Fe_4S_4 cluster. The major function of Fe protein is to bind and hydrolyse Mg ATP and to transfer electrons from Fe_4S_4 cluster to the P cluster of the MOFe protein.

Both proteins are folded in such a manner as to bring the active centres of each close to each other.

Q.14. Explain the **enzymatic catalysis** for reduction of nitrogen to ammonia.

Ans. In enzymatic catalysis electrons are systematically transferred in a sequence one at a time from the iron centres of dinitrogenase reductase to the iron molybdenum cofactors of dinitrogenase. In this way the electrons are finally transferred to the substrate i.e. nitrogen leading to its reduction. Many many rounds of electron transfer must take place before nitrogen is reduced to ammonia and at least two Mg – ATP molecules are hydrolysed for each electron that is transferred. Subsequently Mg – ATPs bind to the Fe protein, dinitrogenase reductase, but ATP is not hydrolysed until the Fe protein is complexed to MoFe protein or dinitrogenase. Distinctly speaking **dintrogenase** is one of the two proteins that build or comprise nitrogenase; protein of nitrogenase which has attached iron and molybdenum or vanadium cofactor; enzyme composed of two dissimilar polypeptides, α_2, β_2, that function as place or site of nitrogen fixation.

Q.15. Name the common free living Nitrogen – fixing bacteria. Also mention their habitats.

Ans. There are large number of **free living nitrogen-fixing bacteria** depending upon the type of soil or habitat.

These are:

(i) *Azotobacter, Azomonas* and *Derxia* in temperate parts in neutral or alkaline soils and waters.

(ii) *Beijerinckia* are more acid tolerant.

(iii) *Frankia* are actinomycetes which are important symbiotic free living bacteria.

(iv) *Azospirillum lipoterum* and *Azobacter paspali* are nitrogen fixing soil bacteria associate with **rhizosphere** (or the ecological niche making the surface of plant roots) of some tropical grasses.

(v) Cyanobacteria or blue green algae (BGA) like *Anabaena* and *Nostoc* are found in aquatic habitats.

Q.16. Which are the common **symbiotic nitrogen fixing bacteria** ?

Ans. *Rhizobium* and *Bradyrhizobium* make association with leguminous plants. These invade the roots and form root nodules.

Q.17. Which are major steps leading to establishment of symbiotic association between leguminous plants and Rhizobia?

Ans. These are :

(i) Attraction of bacteria to the plant roots by amino acids secreted by roots.

(ii) Binding of bacteria to receptors, **lectins** on plant roots which are receptor specific proteins.

(iii) Activity of plant growth substances causing curling of rootlets.

(iv) Entry of bacteria into root hairs.

(v) Development of infection thread.

(vi) Transformation of root/plant cells to form a tumrous outgrowth.

(vii) Multiplication of bacteria within nodule,

(viii) Transformation of invading bacteria into variable **pleomorphic** forms.

Q.18. Give the genera of nitrifying bacteria.

Ans. These are named below :

Nitrosomonas
Nitrospira
Nitrosococcus
Nitrosolobus
} Convert ammonia to nitrite

Nitrobacter
Nitrospira
Nitrococcus
} Convert nitrite to nitrate

Q.19. Name the bacterium which reduce nitrite to ammonia, called **nitrite ammonification.**

Ans. *Clostridium* species reduce nitrite to ammonium ions and this process is called nitrite ammonification.

Q.20. Name a denitrifying bacteria reducing nitrate to nitrite.

Ans. *Escherichia coli.*

Phosphorus (P) Cycle

Q.21. What is a **phosphate solubilizing microorganisms?** Define.

Ans. Any microorganisms having capability of transforming insoluble organic or inorganic (in mineral form) phosphate into soluble orthophosphate (Pi) in a manner which may be shown to make a pronounced significant contribution to the phosphate nutritional status of a specific plant or plant population within the microorganism's native soil ecosystem.

The soil phosphorus mobilization by microorganisms is of great significance as microorganisms assimilate inorganic phosphates and mineralise organic phosphate compounds and microbial activities are involved in the solubilization and mobilization of phosphate compounds.

Q.22. Give the common **terminologies for microorganisms participating in phosphate solubilization mechanism.**

Ans. A.H. Goldstein and P.U. Krisnaraj in 2007 recognised following common terminologies for microoganisms participating in phosphate solubilization mechanisms :

(i) Phosphate solubilizing microorganisms (PSMs).
(ii) Mineral phosphate solubilizing microorganisms (MPSMs).
(iii) Organic phosphate solubilizing microorganisms. (OPSMs).
(iv) Phosphate mobilizing microorganisms (PMMs).
(v) Mineral phosphate mobilizing microorganisms (MPMMs).
(vi) Organic phosphate mobilizing microorganisms (OPMMs).

Q.23. What is **eutrophication**?

Ans. The eutrophication is the enrichment of natural waters (streams, rivers and ponds) with inorganic materials, specially nitrogen and phosphorus compounds that support the excessive growth of micro and macroorganisms, including photosynthetic organisms like cyanobacteria, algae and hydrophytes belonging to different groups of the plants, water ferns and angiosperms.

Phosphorus is essential to the growth of organisms and can be the nutrient which limits the primary productivity of a water body. The discharge of raw or treated waste water, agricultural drainage, certain industrial wastes to that water may stimulate the growth of photosynthetic aquatic micro and macro organisms in nuisance quantities.

During the subsequent decomposition of accumulated organic matter, water column can be severely depleted of oxygen causing major **fish kills.**

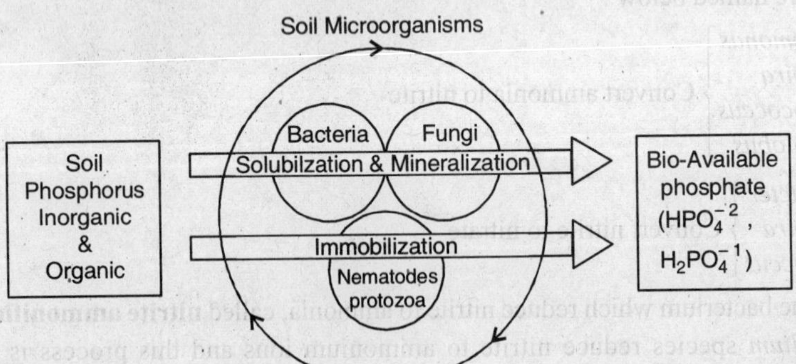

Fig. 32.2. Schematic representation of soil phosphorus mobilization by microorganisms proposed by Richardson (2007) in 1st International Meeting on 'Microbial Phosphate Solubilization'.

Sulphur Cycle

Q.24. How does microbial transformations of sulphur establish the **sulphur cycle** in nature?

Ans. The microorganisms remove sulphur from organic compounds, in the process of desulphurisation by forming sulphate under aerobic conditions, while under anaerobic conditions hydrogen sulphide (H_2S) is normally produced from mineralisation of organic sulphur compounds. Hydrogen sulphide can also be formed by sulphate reducing bacteria which utilize sulphate as a terminal electron acceptor in anaerobic respiration. Hydrogen sulphide can accumulate in toxic concentrations in area of rapid protein decomposition. It can react with metals to form insoluble metallic sulphides. The reduction of sulphates to H_2S is done by sulphate reducing bacteria as *Desulfovibrio*, in **biogeochemical cycling of sulphur.**

Q.25. Which **thermal vent communities** are involved in **sulphur cycling**?

Ans. The **thermal vents** are the deep sea areas of volcanic activity and support the growth of unique biological communities, that depend on chemoautotrophic metabolism for primary production. These include high densities of clams, mussels, vestimentiferan worms (marine tubeworms which lack digestive system), and other invertebrates.

It is interesting to note that energetically, entire vent community is supported by the chemoautotrophic oxidation of reduced sulphur mainly by *Beggiatoa*, *Thiomicrospora* and other sulphide and sulphur oxidisers. The methane produced from reduction of CO_2 with geothermally produced hydrogen by extremely thermophilic *Methanococcus* species has been detected from anoxid hydrothermal fluids, is also oxidised by methanotrophic bacteria which supply additional carbon and energy input for the vent ecosystem.

Q.26. What are **tube worms**?

Ans. The tube worms are *Riftia pachyptila* about a metre in length which grow extensively near deep – sea thermal vents have no guts. The red brown colour of the worms is due to form of haemoglobin which provide oxygen and hydrogen sulphide to chemoautotrophic bacteria which grow inside the tissue of tube worms. *Begiatoa* grow between the strands of tube worms and other chemoautotrophic bacteria on **trophosome** (a spongy tissue) of tube worms.

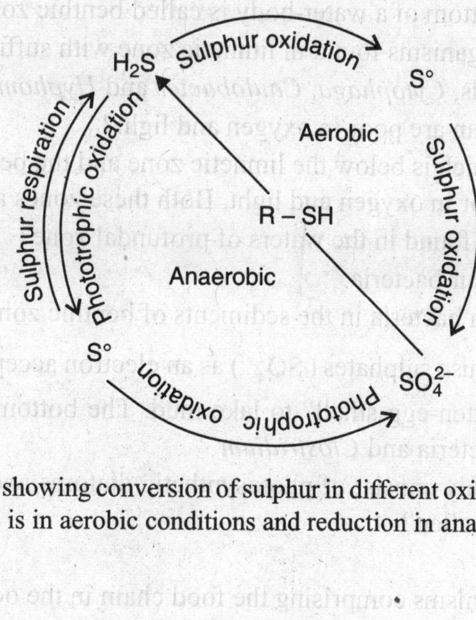

Fig. 32.3. Sulphur cycle showing conversion of sulphur in different oxidation states, in which oxidation of compounds is in aerobic conditions and reduction in anaerobic environment.

33 CHAPTER

WATER AND WASTEWATER MICROBIOLOGY

Q.1. Why do ocean estuaries fed by rivers have higher microbial counts than other shoreline waters?

Ans. Ocean estuaries have a higher nutrient level.

Q.2. Give two main characteristics of aquatic bacteria.

Ans. (i) Bacteria whose main habitat is water usually possess appendages and hold fasts, e.g., *Caulobacter.*

(ii) Some of them have gas vacuoles which they can fill and empty to maintain buoyancy.

Q.3. What is a **littoral zone**?

Ans. This is the region along the shore of an inland body of water where there is considerable vegetation and where light penetrates the vegetation upto the bottom.

Q.4. Define a **limnetic zone**.

Ans. The surface zone of an inland body of water away from the shore.

Q.5. What is **profundal zone**?

Ans. Profundal zone is deeper water in the area of the limnetic zone. It falls below the limnetic zone.

Q.6. What is the **benthic zone**?

Ans. The sediment at the bottom of a water-body is called benthic zone.

Q.7. Which are the microorganisms found in limnetic zone with sufficient oxygen?

Ans. They are pseudomonads, *Cytophaga, Caulobacter* and *Hyphomicrobium.*

Q.8. Which zones in an ocean are poor in oxygen and light?

Ans. The profundal zone which is below the limnetic zone and the benthic zone which is the lower most zone are both poor in oxygen and light. Both these zones are in deeper waters.

Q.9. Which are the bacteria found in the waters of profundal zone?

Ans. Purple and green sulphur bacteria.

Q.10. Which are the common bacteria in the sediments of benthic zone?

Ans. *Desulphovibrio* which use sulphates (SO_4^{2-}) as an electron acceptor and reduce it to hydrogen sulphide, giving a "rotten-egg smell" to lake mud. The bottom sediments may also contain methane producing bacteria and *Clostridium.*

Q.11. The ocean waters mainly consist of photosynthetic diatoms and other algae. What are these suspended organisms called?

Ans. Phytoplankton.

Q.12. Give the types of organisms comprising the food chain in the ocean.

Ans. The protozoa feed on bacteria and the smaller phytoplankton. In turn the krill that are shrimplike crustaceans feed on the phytoplankton, and in turn the krill turn out to be the food of larger animals of sea.

Q.13. What is **microbial luminescence** in deep sea?

Ans. Certain bacteria produce light flashes when they are agitated by the movement of water in wave action.

Q.14. Name the fish that establishes symbiotic relationship with luminescent bacteria at the bottom of the ocean in the benthic zone.

Ans. *Photoblepharon palpebratus,* the flashlight fish uses the flow of its resident bacteria as an aid in attracting and capturing its prey in complete darkness on the floor of the ocean.

Q.15. How do **luminescent bacteria** emit light?

Ans. The luminescent bacteria have an enzyme known as **luciferase** which picks up electrons from flavoproteins in the electron transport chain and then emits some of the electron's energy as a photon of light.

Q.16. Give some of the common diseases spread by water contaminated with human or animal faeces.

Ans. These are typhoid fever, cholera (bacterial) and hepatitis A (caused by a virus). The ingestion of cyst of protozoa *Giardia lamblia* may cause diarrhoea, and that of *Entamoeba histolitica* may cause amoebic dysentery.

Q.17. What harm can be there to water bodies on addition of phosphates?

Ans. 'Phosphates can cause eutrophication (eu = means true or well, troph means nourish) that is an overabundance of nutrients resulting in overgrowth of algae or cyanobacteria, consequently causing death to other organisms.

Q.18. What is bioremediation?

Ans. It is the use of microorganisms to get rid of environmental pollutants, e.g., oil is rich in carbon and sulphur but low in nitrogen and phosphorus which are essential for the growth of microorganisms. Therefore, if an oil soluble fertilizer having these nutrient factors is spread over oil spills mixed with the adapted bacteria the oil is metabolized very fast.

Q.19. How does *Thiobacillus ferroxidans* help to lower the pH of water on receiving coal mining wastes containing iron sulphide (FeS_2).

Ans. The bacterium *Thiobacillus ferroxidans* obtaining energy from oxidation of the ferrous ion (Fe^{2+}) converts the sulphide to sulphate, and consequently sulphate enters the water lowering the pH and causes harm to aquatic life.

Q.20. What do the yellow precipitates in streams polluted with coal-mining wastes indicate?

Ans. They indicate the formation of iron hydroxides which is promoted due to low pH.

Q.21. Give two methods for the analysis of drinking water for coliforms.

Ans. (i) MPN **(Most probable number)** test.

(ii) Membrane filtration test.

(a) For MPN three tests are done. These are presumptive test, confirmed test and completed test. For presumptive test the lactose broth is inoculated and incubated for 24 to 48 hours. Production of gas indicates coliform positive in presumptive test while if the gas is not produced the coliform is absent.

Confirmed Test The samples from the lactose broth are streaked on the EMB lactose plates and are incubated for 24 hours. The typical colonies with dark centres or metallic sheen indicate positive confirmed test. The absence of colonies of coliform is the negative result.

Completed Test Select the typical colonies of coliform (with metallic sheen), inoculate on lactose broth and agar slant and incubate for 24 hours. If gas is produced it shows the presence of coliform group and the completed test is positive.

Also check the slant for Gram-negative rods (which should be present) and for absence of endospores. If both these, i.e., agar slant and lactose broth show the presence of coliform group it is a positive completed test. However, if gas is not produced it is a negative completed test and the original isolates are not coliform.

(b) **Membrane Filtration Test** Some laboratories adopt membrane filtration methods to test for coliforms. The coliform colonies have a distinctive appearance showing dark centres or metallic sheen.

In membrane filter test 100 ml of sample is taken and is filtered through a membrane filter unit. The filter membrane is removed along with the coliform, sieved out on surface. The filter is placed, on a pad of medium similar to the EMB medium. It is incubated for 24 hours. Appearance of four coliform colonies per 100 ml is unsatisfactory while one coliform colony per 100 ml is regarded satisfactory.

Q.22. Name an indicator organism that can be used for testing the purity of water against contamination of faecal (US fecal) matter.

Ans. *Escherichia coli.*

Q.23. What are the diseases caused by some strains of coliforms?

Ans. Diarrhoea and opportunistic urinary infection.

Q.24. What are the criteria for an indicator organism for detection of faecal contamination?

Ans. (i) The organisms should be present in human faeces in substantial numbers.

(ii) The indicator organism should be able to survive in water for sufficient time.

(iii) The indicator organism should be detectable by simple tests.

Q.25. How is it convenient to detect coliforms specifically *Escherichia coli* by using media containing O-nitrophenyl-β-D-galactopyranoside (ONPG) and 4-methylumbelliferyl-β-D-glucuronide (MUG)?

Ans. The coliforms produce enzyme β-galactosidase that acts on ONPG and shows yellow colour indicating their presence in the sample. *E. coli is* quite distinct among coliforms and produces the enzyme, β-glucuronidase which acts on MUG to form a fluorescent compound which glows blue in long wavelength ultraviolet light. These simple tests can be used to determine the presence or absence of coliform or *E. coli* in a sample.

Q.26. How does one get rid of cysts of *Giardia lamblia* and *Cryptosporidium* which are resistant to chlorination?

Ans. By filtration.

Q.27. What is **flocculation**?

Ans. During water treatment, the removal of colloidal substances by addition of a flocculent chemical such as aluminium potassium sulphate (alum), which causes colloidal particles to coalesce resulting in aggregations of finely suspended particles called floc.

Q.28. Give the steps of water treatment in a typical **municipal water purification plant**.

Ans. It comprises the following main steps:

(1) Sedimentation tank or the raw water reservoir to allow particles to settle down.

(2a) Mixing tank in which water is mixed with flocculant.

(2b) Flocculation tank in which aggregates of floc settle down.

(3) Filtration process in which water undergoes filtration.

(4) Disinfection: The water is disinfected by chlorination, ozone treatment or exposure to UV light.

(5) Storage tanks: The water is stored before use by the consumer.

Q.29. Give the steps of a **typical sewage treatment** process.

Ans. The steps adopted in a typical sewage treatment plant are as given below:

(i) Primary treatment which comprises two parts:

(a) Screening, skimming and grinding of sewage.

(b) Settling of solid matter.

(ii) Secondary treatment or biological oxidation: In it the primary effluent undergoes aeration where microorganisms oxidize organic matter. In secondary treatment either a trickling filter is used or an activated sludge (aeration tank) is used but not both in a particular system.

(iii) Disinfection and release: The effluent is disinfected by chlorination and released.

(iv) Sludge digestion: The sludge received from the settling tank of step (1) and the settling tank of secondary treatment of step (2) is digested in anaerobic sludge digesters. The sludge digestion is followed by drying of sludge in drying beds and then the final step is to remove the dried sludge either to a landfill or to agricultural land.

The methane produced in a sludge digester is either burnt or is used for power heaters.

Q.30. What is biochemical oxygen demand (BOD)?

Ans. The BOD is a measure of biologically degradable organic matter in any water. It is measured by the amount of oxygen required by bacteria to metabolize the organic matter present in the water sample.

Q.31. How is BOD measured?

Ans. The BOD bottles with airtight stoppers are filled with test water samples or dilutions of test water samples. The water is initially aerated to provide a relatively high level of dissolved oxygen and is seeded with bacteria if needed. The filled bottles are incubated at 20°C for 5 days and the decrease in dissolved oxygen is measured using a chemical or electronic testing method. The more the amount of oxygen used by the bacteria to degrade the organic matter of the sample, the higher will be the BOD which is expressed in milligrams (mg) of oxygen per litre of water.

Q.32. The trickling filters remove 80% to 85% of the BOD while activated sludge systems remove 75% to 95% of BOD and hence are comparatively less efficient. Why are trickling filters still used?

Ans. They are more convenient to operate and have fewer problems from overloads of toxic sewage.

Q.33. What is a **trickling filter**?

Ans. In a trickling filter the sewage is sprayed over a bed of rocks or moulded plastic honeycomb like structure.

Q.34. Name the common bacterial species which form the flocculent masses (flocs) in activated sludge systems, in which the bacterial cells are embedded in a gelatinous matter.

Ans. *Zoogloea*.

Q.35. What is it that causes bulking of sludge?

Ans. When aeration is stopped in a sludge digester the sludge floats without settling out.

Q.36. What is the permissible number of coliforms for discharge of sewage after treatment, in a small water body and in the ocean, respectively?

Ans. 23 coliforms, per 100 ml for a small water body and 240 coliforms per 100 ml for discharge in the ocean.

Q.37. What is a **lagoon** or an oxidation pond?

Ans. Some small communities and industries use the oxidation ponds or stabilization ponds also called lagoons for the treatment of water from sewage. The system comprises two stages. The first is analogous to primary treatment and the second stage corresponds to secondary treatment.

Q.38. It is difficult to aerate lagoons which are large. What is the alternative to maintain aerobic conditions to lower CO_2 generation during bacterial action of decomposing organic matter?

Ans. By growing algae.

Q.39. What is **tertiary treatment** of sewage?

Ans. When primary and secondary treatments are not able to remove all the biologically degradable material and particularly while the water is discharged in small streams or recreational lakes, it is then that tertiary treatment plants are developed.

Q.40. What is done for the removal of nitrogen and phosphorus in tertiary treatment?

Ans. The nitrogen is converted to ammonia and is released using denitrifying bacteria; and phosphorus is precipitated by combining with chemicals like alum, lime and ferric chloride.

Q.41. What is the constituent of water on which the concentration of bacteria depends?

Ans. Concentration of bacteria is proportional to the amount of organic matter present in the water.

Q.42. What is the significance of waves in a water mass?

Ans. The wave action increases the amount of **dissolved oxygen** (DO).

Q.43. What is sewage?

Ans. It is domestic wastewater.

Q.44. What is the main objective of primary sewage treatment?

Ans. Primary sewage treatment is meant to remove the solid matter known as sludge.

Q.45. What is secondary treatment of sewage?

Ans. It is the biological degradation of organic matter in sewage and follows primary treatment.

Q.46. What is a **septic tank**?

Ans. Septic tanks are designed for primary treatment of sewage in rural areas. They need large leaching fields for the effluent.

Q.47. What is the purpose of oxidation tanks? What is their limitation?

Ans. Small communities can use oxidation tanks also called lagoons for the secondary treatment. Their drawback is that they need large areas for artificial lakes or ponds.

Q.48. What are the main objectives of tertiary treatment?

Ans. The main objectives of tertiary treatment are:

(i) Physical filtration and chemical precipitation to remove all the BOD, nitrogen and phosphorus.

(ii) Tertiary treatment provides drinkable water while secondary treatment provides water which may be used for irrigation only.

Q.49. Why do municipal **landfills** prevent decomposition of solid wastes?

Ans. Because they are dry and anaerobic.

Q.50. What can be done to promote biodegradation of organic matter in a municipal landfill?

Ans. Composting and continuous flow digesters can be used.

Q.51. What is a microstrainer and an **immedium filter**?

Ans. In tertiary sewage treatment the microstrainer is a hollow cylinder of fine mesh stainless steel fabric which rotates on a horizontal axis while an immedium filter is a sand filter.

Q.52. How do heavy metals and their compounds lead to pollution?

Ans. There is no doubt that traces of certain heavy metal ions are essential for the growth of microorganisms. However, a higher concentration may exhibit antimicrobial Activity. Heavy metal ions bind to certain groups, particularly the thiol group and exert antimicrobial activity largely by inactivating proteins, nucleic acids and other microbial constituents . Typically they have little or no activity against bacterial spores.

Antimicrobial activity of mercury ions may be reversed by the presence of substances like cysteine, glutathione, etc. Antimicrobial activity of metals can also be antagonized by other substances, e.g., the inhibition of *Lactobacillus* species by zinc can be reversed by addition of manganese. Some ions appear to have antimicrobial activity only when they are in a particular redox state.

Q.53. How do some bacteria show resistance to heavy metals?

Ans. In bacteria resistance to heavy metals and their compounds may be plasmid borne. Thereby some bacteria can synthesize an inducible plasmid encoded NADPH linked mercuric ions.

Q.54. What is **leaching** and pollution of ground water? Elaborate.

Ans. Leaching is basically the process of water or a liquid slowly filtering through the perforation in soil. Any material on the surface or in the soil may leach into and lead to pollution of ground water, e.g., many wells become contaminated with toxic nitrogen compounds that come from fertilizers used in surrounding agricultural fields. Leaching from municipal landfills, ie., city trash buried in ground often contaminates ground water, particularly while chemical waste materials from industry are disposed off in landfills or deep wells without taking proper precautions. Thus the ground water gets polluted with leached chemical more and more. Mining activity may also add to ground water pollution as surface mining disturbs large amounts of earth in the process of extracting desired materials. In mining raw rock surfaces are exposed to water draining, e.g., the sulphate (SO_4^{--}) present in rock and soil minerals may be leached out as sulphuric acid. Acid drains from coal mines lead to spoilage of surface as well as ground water.

Q.55. How do some bacteria detoxify **heavy metals**?

Ans. Reductase that can reduce Hg_2^+ to relatively nontoxic metallic mercury which is readily lost from the cell. Such cells may also synthesize an organomercural lyase that cleave mercury bond in organomercurals such as phenyl mercuric acetate.

Pseudomonas aeruginosa, a Gram-negative bacteria can reduce Hg^{2+} to HgO which possesses genes for mercury resistance on a plasmid. These genes are known as mer genes that are arranged in an operon and controlled by the regulatory protein Mer R. It is interesting to note that Mer R functions as both repressor and an activator. The mercuric reductase is produced by Mer A genc and Mer D, which is the product of mer D also plays a regulatory role while mer P encodes a periplasmic Hg^{2+} binding protein called Mer P. The Mer P binds Hg^{2+} and transfers it to a membrane protein called Mer T. The Mer T is a product of mer T and Transports Hg^{2+} into the cell for reduction by the enzyme mercuric reductase. In this way Hg^{2+} is reduced to HgO which being volatile gets released from the cell. In *Staphylococcus aureus* plasmid encoded resistance to cadmium may involve *pnf* dependent efflux of cadmium ions via a cadmium proton antiporter which prevents intracellular accumulation of Cd^{2+}.

Q.56. Give some of antimicrobial applications of heavy metals that may also lead to pollution of water.

Ans. Heavy metals and their compounds have been used as agricultural antifungal agents, as antiseptics as chemotherapeutic agents for the treatment of diseases of bacterial and protozoal (causation), as preservatives for textiles and wool and in disinfection of water.

34 CHAPTER

BIOFERTILIZERS

Q.1. Define **biofertilizers.**

Ans. Biofertilizers or microbial inoculants can be defined as preparations containing live or latent cells of efficient strains of nitrogen fixing, phosphate solubilizing or cellulolytic microorganisms used for application with seeds, soil or composting areas with the objective of increasing the number of such cellulolytic microorganisms that accelerate certain microbial processes to increase the extent of the availability of nutrients in a form which can be easily assimilated by plants.

Q.2. Who were the first to propose biological nitrogen fixation?

Ans. J.B. Boussingault and Hellreigel were the first to open up the field of biological nitrogen fixation, in 1886.

Q.3. Who discovered the bacterium now known as *Rhizobium* and its role?

Ans. Beijerinck, a Ducth in 1888 discovered the bacterium now known as *Rhizobium* which is found in the root nodules of legumes and fix up atmospheric nitrogen.

Q.4. Who first cultured rhizobia in the laboratory?

Ans. Nobbe and Hiltner were the first to culture rhizobia in 1895 and called it "Nitragin".

Q.5. Which are the biofertilizers in addition to *Rhizobium?*

Ans. The biofertilizers in addition to *Rhizobium* are: *Azotobacter, Azospirillum,* blue-green algae (BGA) also known as cyanobacteria, *Azolla,* an aquatic fern of freshwater inhabiting *Anabaena azollae* as a symbiont, phosphorus solubilizers *Bacillus* and *Pseudomonas* (bacteria) and *Aspergillus* (a fungus), phosphorus absorber, vesicular arbuscular mycorrhizal (VAM) fungi as *Gigaspora* and *Glomus,* lignocellulolytic organisms *Cellulomonas* and *Arthrobacter* (bacterial species) *Agaricus* and *Trichoderma* (fungal species), *Frankia* and *Wilcoxina (Tricharina)* both actinomycetous and liglinolytic orchid mycorrhizal fungi as *Rhizoctonia.*

Q.6. Which forest trees are benefited by **actinomycetous N-fixing symbiosis** of *Frankia?*

Ans. *Casuarina* and *Alnus.*

Q.7. Can the bacterial *'nif'* gene be transferred to a nonlegume using biotechnological techniques?

Ans. Yes.

Q.8. What are *nif* genes?

Ans. These are genes whose expression is necessary for nitrogen fixation.

Q.9. How many *nif* genes are clustered in *Klebsiella pneumoniae?*

Ans. There are at least 17 *nif* genes in *K. pneumoniae* close to his operon in 8 contiguous transcriptional units *(nif* J, *nif* HDK, *nif,* Y, *nif* ENX, *nif* USVM, *nif* F, *nif* LA and *nif* BQ) spanning Ca. 23 kb on the chromosome.

Q.10. What are the structural genes for nitrogenase?

Ans. *nif*HDK.

Q.11. Name the nonlegume with which Rhizobia can fix up atmospheric nitrogen in symbiotic association.

Ans. *Parasponia*.

Q.12. How can *Rhizobium* be isolated?

Ans. *Rhizobium* can be isolated on yeast extract manitol agar (YEMA) from the clean (surface sterilized by immersing in acidified 0.1% $HgCl_2$) crushed nodules, incubating at 28-30°C for 3 to 10 days.

Q.13. What are the characteristics of the colonies of *Rhizobium?*

Ans. They usually appear as white, translucent, glistening and elevated along the entire margin, and shine against a pink background of congo red dye mixed in the medium.

Q.14. Which are the groups of Rhizobia based on growth habit?

Ans. They comprise two groups based on their growth habits: (1) Fast grower with generation time less than 4 hours (g = 4 hours) producing acid on the YEMA which includes *R. phaseoli, R. trifoli, R. leguminosarum* and R. *meliloti and* (2) Slow growing strains with generation time more than eight hours or equal to eight hours (g = 8 hours) producing alkali on YEMA which includes *B. japonicum,* R. *lupini* and rhizobia of cowpea miscellany groups. Fast growing species form gummy raised colonies 2-4 mm in diameter with 3-4 days while slow growing species form granular colonies not exceeding 1 mm in diameter within 5-7 days.

Q.15. What are the structural parts of a mature nodule of a legume?

Ans. A mature nodule comprises a central bacteroid zone surrounded by many layers of cortical cells.

Q.16. What are **bacteroids**?

Ans. They are bacterium like or modified bacterial cells. The endosymbionts occur within the nodules as modified, membrane enclosed, often pleomorphic cells referred to as bacteroids.

Q.17. On which enzyme nitrogen fixation is dependent?

Ans. Enzyme nitrogenase.

Q.18. How will you differentiate *Agrobacterium* the most common contaminant of *Rhizobium?*

Ans. The *Agrobacterium* takes up the colour of Congo red and appears red in *colour.*

Q.19. Give the presumptive tests carried out to identify an isolate of *Rhizobium.*

Ans. (i) *Rhizobium* grows very poorly on glucose peptone agar medium.

(ii) Doesn't grow on Hofer's alkaline broth at pH 11.0.

(iii) *Rhizobium* does not utilize lactose.

(iv) The best way to identify rhizobia is to conduct plant infectivity tests and check if the nodules are formed.

Q.20. How is a plant infectivity test carried out for rhizobia?

Ans. It is carried out under controlled and sterile conditions either in small tubes or pots. Small test tubes are used for small seeded legumes like clover, berseem, moong and urd while Leonard jar assembly or small pots are used for testing bold seeded legumes as chickpea, arhar and French beans. Agar-agar, vermiculite or sand can be used as the substrate for growing plants. The surface sterilized seeds of the selected crop are sown in appropriate assembly and are inoculated with the bacterial isolate under study for test. The plants are watered at regular intervals with any nitrogen -free nutrient solution. The plants are observed for nodulation, nitrogenase activity and dry biomass after 5 to 6 weeks. Hence an efficient strain needed for a specific crop can be selected. The strain so selected can be judged for its potential in the pots and fields.

It is necessary to lyophilize the selected strain to maintain their efficiency.

Q.21. Why is a selected strain of *Rhizobium* lyophilized?

Ans. A selected strain is lyophilized because subculturing at monthly intervals on artificial medium may lead to deterioration of the activity of the organisms.

Q.22. What is the importance of **carriers** in preparation of *Rhizobium* inoculants?

Ans. The carrier works as a medium which allows the *Rhizobium* to multiply. The shelf life of the product mainly depends on the carrier.

Q.23. What are the suitable carriers used for *Rhizobium* inoculants?

Ans. The suitable carriers for the growth of *Rhizobium* are peat, lignite, Farm Yard Manure (FYM) and charcoal.

Q.24. Where is peat like material found in India?

Ans. The typical peat is not available in our country in large quantities but peat like material is available in Nilgiris (Ootacamund or Udagamandalam) has been found to be a very good carrier.

Q.25. Where is lignite available in India?

Ans. Lignite is available in Neyveli lignite mines of South India.

Q.26. What are the characteristics of a good carrier for *Rhizobium*?

Ans.
(i) It should be cheap and easily available in the country,
(ii) It should be nontoxic to the bacteria used for fertilizer,
(iii) It should have near neutral pH,
(iv) The organic matter should be 40%
(v) Above all it should have high water holding capacity and surface area.

Q.27. Give the method for inoculation of *Rhizobium*.

Ans. A 10 per cent sugar solution or 10 per cent solution of pharmaceutical grade of gum arabic is prepared to help the *Rhizobium* cells to stick on the surface of the seeds. This solution is sprinkled on the seeds required for I hectare and are spread on polythene sheet and mixed uniformly. Then the contents of inoculant of *Rhizobium* are sprinkled on the sticky seeds and mixed thoroughly. The seeds thereafter are spread on gunny bags or on the cement floor in shade to avoid direct sunlight and are sowed on drying. The **pelleting** of inoculated seeds is done by providing a uniform fine coat of calcium carbonate to protect the seeds from adverse effects of soil like acidity, alkalinity and pesticides, etc.

Q.28. What are the substitutes of calcium carbonate for pelleting?

Ans. The substances used for pelleting other than lime are gypsum, bentonite, rock phosphate and charcoal.

Q.29. How much nitrogen may be fixed in *Rhizobium* legume association?

Ans. 50 to 100 kg N/hectare

Q.30. What may be the reasons for poor or no response of the inoculation with *Rhizobium?*

Ans. The reasons for poor or no response of the inoculation with *Rhizobium* are:

(i) Failure of germination of seeds due to poor quality of seeds and lack of moisture.
(ii) Failure of seedlings to produce nodules due to poor quality of inoculants, adverse soil conditions, use of wrong rhizobia and incompatibility of pesticides.
(iii) Ineffective nodulation due to competition created by native population of rhizobia, thereby causing failure of newly introduced legume crops.

(iv) Nutritional deficiencies of elements such as phosphorus, calcium and molybdenum needed in protein synthesis and cell wall development of plants and required by the rhizobia.

(v) Due to competition with other plants for moisture, nutrients and space.

Q.31. List the number of species of *Azotobacter* differentiated on the basis of morphological and physiological features.

Ans. According to 'Bergey's Manual of Determinative Bacteriology' 9th edition (1994) on the basis of motility, water soluble pigments formed for utilization of carbon source by the *Azotobacter* there are 6 species in this genus. They are *A. armeniacus, A. beijerinckii, A. chroococcum, A. nigricans, A. paspali* and *A. vinelandii.*

Q.32. Which are the species of *Azotobacter* that occur commonly in India?

Ans. *A. beijerinckii* and *A. vinelandii* are common in India.

Q.33. Which species of *Azotobacter* occurs in acid soils?

Ans. *A. chroococcum* occurs in acidic soils.

Q.34. Name the species of *Azotobacter* that occurs in neutral and alkaline soils.

Ans. *A. beijerinckii.*

Q.35. Does *Azotobacter* form cysts?

Ans. Yes.

Q.36. Does *Azotobacter* produce spores?

Ans. No.

Q.37. Which is the suitable medium for *Azotobacter?*

Ans. Nitrogen free agar medium, e.g., Ashby's medium.

Q.38. What are the characteristics of the colonies of *Azotobacter* after incubation at 30°C for 3-4 days?

Ans. Soft, milky and mucoid colonies.

Q.39. Does *Azotobacter* show cyst formation maximally in old cultures on nitrogen free medium containing 0.2% butanol?

Ans. Yes.

Q.40. Who was the first to record that *Azospirillum lipoferum* (old name, *Spirillum lipoferum*) fixes up atmospheric nitrogen?

Ans. Beijerinck in 1925.

Q.41. Which species of *Azospirillum* is generally found associated with the roots of C4-plants?

Ans. *A. lipoferum.*

Q.42. Which species of *Azospirillum* is generally found associated with the roots of C3-plants in tropical and subtropical regions?

Ans. *A. brasilense.*

Q.43. Which is the suitable medium for the cultivation of *Azospirillum*?

Ans. Semisolid nitrogen free sodium malate medium.

Q.44. What is the systematic position of *Azospirillum*?

Ans. The genus *Azospirillum* belongs to Group 2 of Aerobic/Microaerophilic, Motile, Helical/Vibroid Gram-negative Bacteria in 9th edition (1994) of 'Bergey's Manual of Determinative Bacteriology'.

Q.45. What is the characteristic of the growth of *Azospirillum* while incubated at 35°C for 48 hours?

Ans. Formation of white pellicle 2-4 mm, below the surface of the medium.

Q.46. Name the species of *Azospirillum*.

Ans. According to 9th edition (1994) of 'Bergey's Manual of Determinative Bacteriology' there are five species in the genus *Azospirillum*. They are *Azospirillum amazonense, A. brasilense, A. halopraeferens, A. irakense,* and *A. lipoferuni.*

Q.47. Which are the major genera with maximum number of nitrogen fixing species?

Ans. *Anabaena, Aulosira, Cylindrospermum, Nostoc, Plectonema, Scytonema* and *Tolypothrix*.

Q.48. Distribution of nitrogen fixing blue green algae (BGA) varies from 4 to 80 per cent in different soils. Which are the genera ubiquitous in all regions?

Ans. *Anabaena* and *Nostoc*.

Q.49. Which is the most common species of *Azolla* in India?

Ans. *Azolla pinnata.*

Q.50. Which species of **BGA** is symbiont of *Azolla*?

Ans. *Anabaena azollae* is found in cavities on dorsal leaves.

Q.51. How much nitrogen generally occur in *Azolla*?

Ans. 0.2 to 0.3% of its fresh weight and 3 to 5% nitrogen of dry weight.

Q.52. What are diazotrophs? Name some of them.

Ans. The procaryotes that can carry on nitrogen fixation are called **diazotrophs**. They are various cyanobacteria, members of Azotobacteriaceae, e.g., *Azotobacter chroococcum,* members of Rhizobiaceae, e.g., *Rhizobium leguminosarum,* some species of *Bacillus* and *Clostridium.*

Bis Norms of Biofertilizers

Q.53. What do you mean by **BIS norms** for common biofertilizers.

Ans. The BIS means Bureau of Indian Standards. The BIS norms are also called 'Standards for Biofertilizers' or 'Standards for Existence of Biofertilizers', which are given below:

1. **Standards for *Rhizobium* (RI)**

IS 8268:2001

(i) Viable Cell Count	10^8 Viable *Rhizobium* Cells/g of the Carrier within 15 days of manufacture and 10^7 Cells/g of Carrier within 15 days before the Expiry at Storage temperature 25-30°C
(ii) pH	6-7.5
(iii) Moisture Percentage	35-40
(iv) Shelf Life	6 Months
(v) **Carrier**	Peat, Lignite, Peat Soil, Humus or Similar Material Neutralized with $CaCO_3$ and then sterilized. Carrier should be in the form of powder capable of passing through 150 to 212 Micron (72-100 Mesh) IS Sieve
(vi) Marking on the Packet	The Inoculatent Packet shall have the following information.

(a) Name of the Product
(b) Name and Address of the Manufacturer
(c) Carrier Used
(d) Batch Number
(e) Date of Manufacture
(f) Date of Expiry
(g) Net Mass
(h) Crop for which intended

	(i) Storage Instructions
	(j) Application
(vii) Pot Culture Test	RI shall show Effective Nodulation Formation. The Total Dry Mass of Inoculated Control and at Least 50% more than the Controlled one

2. Standards for *Azotobacter* (AI)

IS 9138:2002

(i) Viable Cell Count	10^7 Viable *Azotobacter* Cells/g of the Carrier within 15 Days of Manufacture and 10^6 Cells/g of Carrier within 15 Days before the Expiry at Storage Temperature 25-30°C
(ii) pH	6.5-7
(iii) Moisture Percentage	35-40
(iv) Shelf Life	6 Months
(v) **Carrier**	Peat, Lignite, Peat Soil, Humus or similar material neutralized with $CaCO_3$ and then Sterilized. Carrier Should be in the form of powder capable of passing through 150 to 212 Micron (72-100 Mesh) IS Sieve.
(vi) Nitrogen Fixation	The Minimum Nitrogen Fixed Shall not be Less than 100 mg Per g of Sucrose used
(vii) Germination Test	AI Shall have a good effect on seed germination
(viii) Marking on the Packet	The Inoculant Packet shall have the following information:
	(a) Name of the Product
	(b) Name and address of the Manufacturer
	(c) Carrier used
	(d) Batch number
	(e) Date of manufacture
	(f) Date of expiry
	(g) Net Mass
	(h) Crop for which Intended
	(i) Storage Instructions
	(j) Application Istructions

3. Inoculant (AI)

IS: 14806:2000

Azospirillum (ASP)

(i) Viable Cell Count	10^7 Viable *Azospirillum* Cells/g of the Carrier within 15 days of manufacture and 10^6 Cells/g of Carrier within 15 days before the Expiry at Storage Temperature 25-30°C
(ii) pH	7-8
(iii) Moisture Percentage	35-40
(iv) Shelf Life	6 Months
(v) Carrier	Peat, Lignite, Peat Soil, Humus or Similar Material Neutralized with $CaCO_3$ and then Sterilized. Carrier should be in the form of Powder Capable of Passing through 150 to 212 Micron (72-100 Mesh) IS Sieve.

(vi)	Plant Test	ASP Shall Show Effective Root Development on all Crops against which the Inoculant is Intended to be used. There should be 10% increase in Plants over Dry Mass or Uninoculated Control
(vii)	Marking on the Packet	The Inoculant Packet shall have the following information: (a) Name of the Product (b) Name and address of the Manufacture (c) Carrier Used (d) Batch Number (e) Date of Manufacture (f) Date of Expiry (g) Net Mass (h) Crop for which Intended (i) Storage Instructions (j) Application Instructions.

4. Phosphate Solubilizing Bacterial Inoculant (PSB)

IS-14807:2000

(i)	Viable Cell Count	10^7 Viable Bacterial Cells/g of the Carrier within 15 days of Manufacture and 10^6 Cells/g of carrier within 15 Days before the Expiry at Storage Temperature 25-30°C
(ii)	pH	6.5-7.5
(iii)	Moisture Percentage	35-40
(iv)	Shelf Life	6 Months
(v)	Carrier	Peat, Lignite, Peat Soil, Humus or Similar Material Neutralized with $CaCO_3$ and then Sterilized. Carrier should be in the form of Powder Capable of Passing through 150 to 212 Micro (72-100 Mesh) IS Sieve

35 CHAPTER

MYCORRHIZAE

Q.1. Who was the first to carry out mycorrhizal study?

Ans. Hartig in 1851 is believed to be the first person to study the intracellular net of hyphae in roots without knowing its fungal nature. Its structure, later on was named **"Hartig net"**.

Q.2. Who coined the term mycorrhiza?

Ans. A.B. Frank, a German Forest Pathologist coined the term mycorrhiza in 1885.

Q.3. Who started work on mycorrhiza in India?

Ans. B.K. Bakshi, in early sixties.

Q.4. Which are the main types of mycorrhizae?

Ans. The main types of mycorrhizae are:

(i) Ectomycorrhiza
(ii) Endomycorrhiza
(iii) Ectendotrophic
(iv) Host specific

Q.5. What is VAM?

Ans. It is vesicular arbuscular mycorrhiza. It has also been regarded as general mycorrhiza because of its common occurrence. This comes under endomycorrhiza.

Q.6. Which are the main two types of structures in vesicular arbuscular type of mycorrhiza?

Ans. These are smooth oval bodies called **vesicle** and bushy intracellular structures called **arbuscules**. The vesicles are considered to be storage structures while the arbuscules break down and then release the nutrients to the plant.

Q.7. From where does the nutrients released by the arbuscules come?

Ans. These nutrients come from the soil through fungal hyphae to the arbuscules.

Q.8. How can the large spores of VAM be isolated from the soil?

Ans. By sieving.

Q.9. What is **ectomycorrhiza**? Where are they formed?

Ans. In ectomycorrhiza the fungus forms a mycelial mantle over the smaller roots of the trees, e.g., in all species of Pinaceae and some families of angiosperms as Fagaceae, Betulaceae, Juglandaceae and Myrtaceae. The *Pinus* and Oak are typical examples.

Q.10. Do ectomycorrhizae form vesicles or arbuscules?

Ans. No.

Q.11. What sort of mycorrhizae constitute the **truffles** known as culinary delicacy?

Ans. Ectomycorrhizae.

Q.12. Why are pigs trained and used in France to find truffles and ectomycorrhizae and to root them up'?

Ans. To the female pigs the odour is like that of a male mate.

Q.13. How do animals help in dispersal of ectomycorrhizal fungi?

Ans. Animals ingest the fungus and distribute the undigested spores in new locations.

Q.14. What are endotrophic or **endomycorrhizae**?

Ans. They are characterized by intracellular fungus. However, a portion of fungus remains outside in the form of loose hyphae.

Q.15. What are the two subtypes of endomycorrhizae?

Ans. (1) Endomycorrhiza with aseptate fungi or the VAM.

(2) Endomycorrhiza with septate fungi, e.g., ericoid type and **orchid mycorrhiza**.

Q.16. What are ectendotrophic mycorrhizae?

Ans. They combine the characteristics of both ecto and endotrophic mycorrhizae. The fungus besides forming a hyphal mantle and the 'hartig net' as found in ectotrophic or ectomycorrhizae develops haustoria and hyphal coils in the epidermal and cortical cells similar to endomycorrhizae are endotrophic mycorrhizae.

Q.17. Give an example of ectendotrophic mycorrhiza.

Ans. *Monotrapa indica,* the Indian pipe, a nonchlorophyllous plant of Ericaceae grows on the forest floors under *Pinus, Fagus, Quercus, Salix* and other tree species but there is no connection between the tissues of the roots of two plants. Nonchlorophyllous *Monotrapa indica* obtains nutrients from the green trees through a fungal bridge. The fungus *Boletus* forms ectendotrophic mycorrhiza with *Monotrapa* and endotrophic mycorrhiza with the trees.

Q.18. What is the function of mycorrhizae?

Ans. They extend the surface of the root performing the function of root hairs and absorb nutrients particularly phosphorus which is not easily available, being not very mobile in the soil. Ectomycorrhizae also help in conservation of water.

Q.19. Can mycorrhizal fungi be used as biofertilizers?

Ans. Yes.

Q.20. Can orchid mycorrhizal fungus germinate the minute seeds of the orchid plant?

Ans. Yes.

Q.21. Why do orchid seeds depend on mycorrhizal fungus for germination in nature?

Ans. Because they are microscopic in size and are devoid of reserved food material (endosperm).

Q.22. Can orchid mycorrhizal fungus be cultured on synthetic medium *in vitro?*

Ans. Yes.

Q.23. Have VAM fungus or spores been cultured on artificial synthetic medium?

Ans. No.

Q.24. Who cultured VAM mycorrhizal fungus *Glomus deserticola* in association with root organ culture of *Ziziphus nummularia* under *in vitro* conditions?

Ans. Nishi Mathur and Dr. Anil Vyas of J.N.V. University, Jodhpur, in 1995.

Q.25. Which are the common genera of VAM fungus?

Ans. *Glomus, Gigaspora, Sclerocystis, Scutellospora, Acaulospora, Entrophospora.*

Q.26. How does the ectomycorrhiza differ from the ericoid mycorrhiza?

Ans. In ectomycorrhiza the fungal mass forms a sheath called mantle around the outside of the root with only marginal penetration inside the root tissue, while in ericoid mycorrhiza the fungal mycelium gets embedded in the root tissue.

Q.27. Do you agree that mycorrhizae are symbiotic associations between plant roots and fungi?

Ans. Yes.

Q.28. The beneficial effect of mycorrhizal fungus is best observed in soils poor in nutrients. Why?

Ans. In poor soils the trees which are mycorrhizal thrive but nonmycorrhizal ones do not.

Q.29. When trees are planted in **prairie soils** that often lack suitable fungal inoculum the trees which are inoculated artificially at the time of planting show better growth than uninoculated trees. Why?

Ans. The mycorrhizal fungus makes the nutrients available.

Q.30. Do mycorrhizae also protect plants against root diseases?

Ans. Yes.

Q.31. How many scientists in India worked on mycorrhiza in 1989?

Ans. About 200.

Q.32. What is the ratio of scientists working on mycorrhiza in India to that of all countries of Asia?

Ans. 1:1, India : All countries of Asia.

Q.33. Who compiled the Directory of Asian Mycorrhizologists in 1989?

Ans. Mycorrhiza Information Centre, Tata Energy Research Institute, 102 Jor Bagh, New Delhi.

Q.34. What is a **diazotroph**?

Ans. Any organism capable of nitrogen fixation is called diazotroph.

36 CHAPTER

BIOSENSORS: BACTERIA THAT DETECT TOXIC POLLUTANTS

Q.1. What are the **biosensors**?

Ans. The organisms/bacteria which can be used to locate biologically active pollutants are regarded as biosensors.

Q.2. In the USA, 265 million metric tonnes of **hazardous waste** is generated by industrial plants. Where does the major part (80 per cent) of it go?

Ans. It is buried into landfills.

Q.3. What is required to monitor the working of a **bioreceptor**?

Ans. There are two requirements: (1) A receptor which is activated in the presence of pollutants. (2) A reporter to make the change apparent.

Q.4. Name the bacterial species that can be used or engineered to be used as biosensors.

Ans. *Vibrio fischeri, Photobacterium, Xanthomonas, Bradyrhizobium, Lactococcus* and *Escherichia coli,* etc.

Q.5. Which characteristics of the bacterium *Vibrio* or that of *Photobacterium* works as a reporter?

Ans. The lux operon of the bacterium that contains inducers or structural genes for the enzyme luciferase.

Q.6. How do biosensors use the lux operon from *Vibrio* or *Photobacterium?*

Ans. The biosensors employ the lux operon from *Vibrio* or *Photobacterium* as a receptor. The operon has structural and inducing genes for the enzyme luciferase. In the presence of a coenzyme known as $FMNH_2$, the luciferase reacts with the molecule to form an enzyme substrate complex. The enzyme substrate complex emits blue green light which then oxidizes the $FMNH_2$ to form FMN. Therefore, the bacterium containing the lux gene emits visible light when the receptor is activated.

Q.7. Give two examples of the bacteria to which lux operon has been transferred. Why is it done?

Ans. (i) It has been readily transferred to the plant pathogen *Xanthomonas* to monitor progression of infection due to it in plants. As the altered bacteria emit light, their presence can be detected on a photographic film.

(ii) The *Bradyrhizobium* too has been engineered to contain the lux gene.

It has been done to understand the root nodule formation in soybean.

Q.8. What is the use of *Lactococcus* containing lux operon as a biosensor in dairy microbiology?

Ans. The *Lactococcus* containing lux operon can be used to detect the presence of antibiotic in milk. This application is useful because if milk contains antibiotics, the cheese starter culture does not grow.

Q.9. How do some Microtox systems developed recently use marine bacterium *Photobacterium* directly to detect pollutants?

Ans. The photobacteria loose their characteristics of emitting light if they are killed by pollutants.

Q.10. What is an operon?

Ans. It comprises the operator site and the structural genes that it controls.

Q.11. What is **luciferase**?

Ans. Luciferase is an enzyme that accepts electrons from flavoproteins and emits a photon of light.

37 CHAPTER

BIOLOGICAL CONTROL

Q.1. What is biological control?

Ans. The use of one organism to control the growth of another organism is a process known as biological control.

Q.2. Give a simple example of biological control that may be of interest to a microbiologist.

Ans. Some insects causing serious diseases of crop plants can be controlled by the use of bacteria that are pathogenic to those insects.

Q.3. Name some bacterial species which can be used as insect **larvicides**.

Ans. *Bacillus larvae, B. propillae* and *B. thuringiensis*.

Q.4. How do the bacilli work as insect pathogens so as to serve as an insect larvicidal?

Ans. The insect pathogenic bacilli function as biological control agents forming crystalline protein during sporulation referred to as **parasporal body** which is deposited within the sporangium but outside the spore proper, i.e., on the periphery of spores.

Q.5. *Pseudomonas syringae* which are leaf inhabiting bacteria bring about the formation of ice crystals on leaves at temperatures higher than that required for frost. Why?

Ans. *Pseudomonas syringae* produce a protein which causes ice formation.

Q.6. How would you get rid of the frost damage that could be caused by the normal **(ice-plus)** *Pseudomanas syringae?*

Ans. By spraying ice-minus strain of *P. syringae* from which the gene for ice nucleation is deleted. The ice-minus strain brings about the competitive exclusion of ice-plus strain.

Q.7. Do you agree that ice nucleation power of *Pseudomonas syringae* has also been put to positive use to enhance ice nucleation in snow making processes in ski resorts by mixing killed bacteria still retaining their ice-nucleating power and sprayed into the atmosphere by the snow making equipment.

Ans. Yes.

Biological Leaching of Uranium and Copper

Q.8. Which bacterium is used in the recovery of otherwise unprofitable grades of uranium and copper ores?

Ans. *Thiobacillus ferroxidans*.

Q.9. How is copper recovered using *Thiobacillus ferroxidans?*

Ans. Solutions containing ferric ion (Fe^{3+}) are passed through deposits of insoluble copper compounds. The copper is oxidized into soluble compounds. The Fe^{3+} ions in this process are reduced to Fe^{2+}. The Fe^{2+} can be oxidized by *T. ferroxidans.* The soluble copper is then separated from the ore and can be reclaimed.

Q.10. Give steps used in biological oxidation of ores.

Ans. This takes place in the following steps. (1) *Thiobacillus ferroxidans* oxidize Fe^{2+} to Fe^{3+}, which in turn oxidizes Cu^{+} to Cu^{2+} and the Cu^{2+} forms $CuSO_4$. (2) $CuSO_4$ precipitates as (Cu°), while $FeSO_4$ remains. (3) In an oxidation pond *T. ferroxidans* oxidizes $FeSO_4$ to $Fe^{3+} + H_2SO_4$.

Microorganisms as Industrial Products

Q.11. Give the names of microorganisms that constitute industrial products.

Ans. (i) Yeast cultures for making wines and baker's yeast. The cells of yeast are harvested by continuous centrifuagation and are pressed into familiar yeast cakes.

(ii) Symbiotic nitrogen fixers *Rhizobium* and *Bradyrhizobium* mixed with peat moss to preserve moisture which helps mix their inoculum with leguminous seeds.

(iii) For many years the insect pathogen *Bacillus thuringiensis* (Bt) has been used with gardens to control leaf eating insects.

(iv) *Bacillus thuringiensis* strain *israelensis* acts against the larvae of mosquitoes. It is widely used by municipal control programmes. Their commercial preparations contain toxic crystals and endospores.

Microbes as Source of Energy

Q.12. How can wastes of landfill sites be used for the production of energy?

Ans. By producing methane.

Q.13. How do antibiotics fed to cattle inhibit derive methane production from the waste excreted by them?

Ans. The antibiotics interfere with the bacterial activity responsible for the production of methane.

Q.14. What is **gasohol**?

Ans. Gasoline containing ethanol (90% gasoline + 10% ethanol) is available in many parts of USA, is known as gasohol.

Q.15. Name one of the safe landfills that could be used for the production of methane in future.

Ans. At Bhaleswa, in Delhi.

Q.16. What is a **landfill** ?

Ans. The landfill is a site to dump solid waste and is left for decomposition, solid waste containing both organic and inorganic material is deposited and covered with soil to carry on the decomposition.

Q.17. How many landfills were available in Delhi by 1998?

Ans. The three landfills that Delhi had by 1998 are at (i) Ghazipur spread over 70 acres, (ii) Bhaleswa spread over 40 acres and (iii) Okhla 18 acres.

Q.18. What is a **predator** ?

Ans. The organism which preys upon other smaller organism.

Q.19. Define predation.

Ans. It is based on interaction among organisms in which one is benefited and one is harmed, in which process the smaller organism called prey is ingested by larger organism called a predator.

38 CHAPTER

BIOPESTICIDES

Q.1. What are the **biopesticides?**

Ans. The biopesticides have been defined as either microbials themselves or products derived from microbials, plants and other biological entities.

Q.2. Name 5 categories of biopesticides.

Ans. These are microbial pesticides, entomopathogenic nematodes, baculoviruses, plant derived pesticides and insect pheromones.

Q. 3. How do **insect pheromones** work as biopesticides?

Ans. The insect pheromones can be used as biopesticides by using as mating disruption agents in orchards and vineyards.

Q.4. What are **Bt-engineered crops?**

Ans. These are transgenic crops expressing ***Bacillus thuringiensis*** (Bt) insect protein toxins, for the control of lepidopteran pests.

Q.5. Give an example of a recently introduced commercial **biofungicide.**

Ans. The AQ 10® which has been developed from the fungus ***Ampelomyces quisqualis,*** is a recent introduction as a control against PMD (powdery mildew) of fruit primarily grapes.

Q.6. Give an example of **joint action of biofungicides.**

Ans. The joint action or possibly synergy increases the utility of biofungicides as was demonstrated by combining a nonpathogenic strain of fungus ***Fusarium oxysporum*** with a strain of bacterium ***Pseudomonas fluorescens*** to control the ***Fusarium*** wilts.

History

Q.7. Fill in the blanks in the following:

(i) The most common biopesticides used by Indian masses since antiquity that were patented by some foreign companies were ______ and ______.

(ii) The ______ was used by British army in India as insect repellent as quoted by T. Ramachandra Rao, Formerly Director, Virus Research Centre, Pune and Deputy Director of Public Health (Malaria) Maharashtra state in the **"Anophelines of India"** published by **I.C.M.R. New Delhi,** in 1981.

(iii) The villagers in India have used ______ of wheat in storage of wheat grains.

(iv) The most common gum of the tree is called ______ the smoke of which is being used as insect repellent and barries to pathogens in Yajna in India.

(v) Extracts of ______ (______ ______) flowers have been used for centuries to kill insects particularly in Japan and Iran. There has been a reference to it in Arabian Nights.

International Scenario

(vi) Aogostino Bassi disovered in 1834 that the fungus ***Beauveria bassiana*** caused an infectious disease of the ______ ______.

(vii) Metchnikoff in Russia worked with insect population of ***Metarhizium anisopliae*** to control the ______ attacking wheat, round about in 1874.

(viii) Ishiwata in 1901 discovered a ***Bacillus*** sp which attacked the ______ ______.

(ix) Berliner discovered the same species (that was discovered by Ishiwata in 1901) in 1915 which infected flour moths and named the bacterium ***Bacillus thuringiensis*** (Bt) after the German ______ of Thueringen hence came the **nomenclature** of this bacterium.

(x) Mueller in 1939 discovered the insecticidal properties of ______ a low cost petroleum based compound.

(xi) First biopesticide product was commercialized in USA in 1940s when a bacterial product based on ***Bacillus popillae*** was introduced for control of ______ ______.

(xii) Bt products were commercialized in United States in 1950s for control of a variety of ______ and ______ attacking crops.

(xiii) Identification of Bt variety ______ as a potent mosquito larvicide suggested that Bt toxins could be active on a number of targets.

(xiv) The revolution in the field of molecular biology and genetic engineering has enabled us to manipulate highly potent δ–endotoxins produced by ______ in a number of ways to enhance their utility and performance.

(xv) The concept of integrated pest management (IPM) was developed in ______ and ______.

(xvi) Bt subsp **kurstaki** insecticide formulation Dipel ESNT, the active ingredient is encapsulated and then suspended in an oil base for the pupose to ______ ______ ______ ______.

(xvii) The high temperature treatment caused delay in germination as well as death of conidia of ***Metarhizium flavoviride***. But drying conidia by adding ______ ______ to oil formulations highly increased tolerance to temperature.

(xviii) Only two species of bacteria have been developed for control of insects Bt and ______ ______.

(xix) ***B. sphaericus*** has been registered in ______ to work against mosquito larvae.

(xx) Baculoviruses produce polyhedra containing one or more infective virions. When the insect larvae consume contaminated foliage, the polyhedra ______ in the midgut relcasing the virons which attack midgut epithelial cells and spread to hemocoel, on death insect liquefies and releases large quantities of polyhedra on the leaf surface.

Ans. (i) Neem and turmeric, (ii) Pyrethrum, (iii) Straw dust,
(iv) Guggal (gum of ***Comiphora wightii* syn C. mukul),**
(v) Pyrethrum (***Chrysanthemum cinerariaefolium*),** (vi) Silk worm
(vii) Beetle, (viii) Silk worm, (ix) province,
(x) DDT, (xi) Japanese beetle, (xii) Caterpillar, pests,
(xiii) israelensis (xiv) Bt (***Bacillus thuringiensis*),**
(xv) 1950s, 1960s, (xvi) Exclude oxygen from organism,
(xvii) Silica gel, (xviii) ***Bacillus sphaericus*,**
(xix) France, (xx) dissolve.

Q. 8. Match the following

(A) *Streptomyces griseoviridis* registered in Finland, Hungary, Norway and Switzerland.	(a) Works against Dutch elm disease.
(B) *Verticillium dahliae* a fungus registered in Netherlands.	(b) Target pathogens are *Fusarium* spp, *Alternaria, Pythium. Botrytis* and other soil pathogens.
(C) The fungus *Trichoderma harzianum*, registered in Sweden and France.	(c) Is antagonist to silver leaf fungus (*Stereum purpureum*).
(D) The fungus *Trichoderma viride* registered in U.K.	(d) Targets soil borne fungal diseses.
(E) *Phlebiopsis* (= Peniophora) *gigantea* registered in Finland, Sweden and Norway.	(e) Is effective against *Fomes annosus.*

Ans. [Ab], (Ba], [Cd], [Dc], [Ee].

Q.9. How and in which country was the fungal insecticide *Beauveria bassisana* produced?

Ans. This fungal insecticide was developed in China more than 10 years ago by harvesting a vast amount of vigrous mycelium from fermentation (liquid stage) and transferring into semisolid media (solid stage) consisting of bran, sugar, agar and minerals. The final product contained 1.8×10^{11} spores/g. Many village manufacturers adopted this procedure to produce *Beauveria bassiana* for control of pine caterpillar and corn borer.

Q.10. What are **mycoinsecticides?**

Ans. The fungal pathogens developed for the control of insect pests are called mycoinsecticides. Most of these belong to division Deuteromycotina class Hyphomycetes. The most common species are *Metarhizium anisopliae, Beauveria bassiana, Paecilomyces fumosoroseus* and *Verticillium lecanii.* These can be readily isolated from soils from most parts of the world and are known to have a wide host range. However, the strains can differ in their specificity and virulence.

The *Verticillium lecanii* has been developed as a bioinsecticide for the control of aphids, whiteflies, thripes and red spider mites. The fungal spores are dried to produce a wettable powder formulation with a shelf of 6 months at 5°C. It provides protection to glasshouse cuttings by dipping before potting. This product prepared by Christian Hansen's Biosystems was given trade name 'MicroGermin', and was purchased by the Dutch Company, Koppert in late 1995, which already had their product of similar type called **'Mycotal'** with a view to prepare a single product with extensive range effective against pests.

The fungal species *Beauveria bassiana* and *B. brongniartii* are known to produce a cyclodepsipeptide called *beauvericin* which is toxic to insects. *Metarhizium anisopliae* and *M. flavoviride* produce a related insecticidal metabolite known as **destruxins.** In Switzerland pathogenic fungus *Beauveria brongniartii* was developed in 1980s to control subterranean pasture pest (*Melolontha melolontha*) and was registered in 1990. Natural Plant Protection (NPP), France, developed an another strain of the same fungus to control the sugar cane white grub (*Hoplochelus marginalis* of Melolonthinae). The NPP also developed a strain of *Beauveria bassiana* to control European corn borer (*Ostrinia nubilalis*). Both of these pathogens were produced using solid fermentation technology. The pathogen is formulated in clay granules and is applied at 25 kg/ha, and are stable for one month at 35°C but the shelf life is increased at low temperature. A German company Bayer AG developed B101020 which is a strain of *Metarhizium anisopliae* to control black vine weevil (*Otiorhynchus sulcatus*), a major problem in Ornamental crops in glasshouses and nursery stock. Recently the use of *M. anisopliae* has been reported to be effective for the control of western flower thrips, *Frankliniella occidentalis*. The strains of *M. anisopliae* have been noted to be highly pathogenic to several crucifer pests and yet harmless to honey bees.

Q.11. Give the mode of action of bacterium *Burkholderia cepacia* (= *Pseudomonas cepacia*) as a biological control agent.

Ans. It produces antifungal compounds such as **siderophores** and chlorinated phenylpyrrole antibiotics as well as hydrolytic enzyme β–1, 3 – glucanase.

Q.12. Write for the following whether 'True' or 'False'

(i) Either a fluorescent strain of *Pseudomonas* sp isolated from a suppresive soil or its **siderophore,** can result a conducive soil to become suppressive to Fusarium wilts.

(ii) Addition of strong iron (Fe) – chelator (EDDHA) made a soil conducive to Fusarium wilt, suppresive, is indicative that competition for Fe is the chief mode of action of the siderophore producing pseudomonads.

(iii) The growth of germ tubes arising from chlamydospores gets reduced in the presence of siderophore producing pseudomonads due to competition for Fe.

(iv) Some bacteria mostly pseudomonads consistently promote mycorrhizal development. These bacteria are known as **mycorrhizal helper bacteria (MHBs).**

(v) Joint action of microbials for disease control is not possible.

Ans. (i), (ii), (iii), (iv) True; (v) False.

Neem and Government Regulations

Q.13. Fill in the blanks in the following:

(i) In many countries of the world new legislations, Food Quality Protection Act, Worker Protection Standard and Pesticide Regeneration are limiting the availability of traditional _______ pesticides.

(ii) Farmers in developed and developing countries are looking toward the use of natural material as _______ agents.

(iii) The insect growth regulator (IGR), azadirachtin affects over 300 species of _______ including important pests like armyworms, leafminers, aphids, whiteflies, psyllids and many other pests.

(iv) Azadirachtin-based insecticides have negligible effect on natural beneficial _______ and low environmental _______.

(v) Indian neem tree *Azadirachta indica* is a member of family Meliaecae also known as _______ family.

(vi) The first commercial use of azadirachtin based pesticide for nonfood use was approved by the U.S. Environmental Protection Agency (EPA) in _______.

(vii) Vikwood botanicals owned by Robert Larsen introduced _______ for use on trees and shrubs to control leaf miners and gypsy moths.

(viii) In _______ W.R. Grace of New York purchased the patent, registration and technology for **Margosan-O** from Larson

(ix) In _______ W.R. Grace changed the formulation of Margosan O reducing the active content to 0.25% and expanded the registration to include several important insects including whitefies, aphids and armyworms and expanded its use to include the green house and interior scape environments.

(x) In 1993 the EPA granted an exemption of tolerance for using azadirachtin on all food crops at < 20g azadiractin/_______.

(xi) Some Indian companies have brought out products including spic (Neem Gold), E.I.D. Parries (Neemazal) and Margo Biocontrols (Econeem). Most of the azadirachtin based pesticides produced in India are used on three main crops of _______ and _______.

(xii) Registration of some neem based pesticides manufactured by Thermo Trilogy or Biosys have been approved in _______ _______, Taiwan, Israel, Spain, Chile, Mexico, Nicaragua, Costa Rica and Ecuador.

(xiii) A German Company, _______, has received registration of an azadirachtin based insecticide (Neemazal) in _______ and has applied for registration in Germany.

(xiv) It has been reported that the _______ of leafminer *Liromyria trifolia* and the fruit fly *Ceratitis capitata* treated with azadirachtin die before they reach the adult stage.

(xv) The adjuvants are used to improve _______ of a pesticide. The adjuvants containing vegetable or mineral oil gave the most significant improvement to the effect of Neemix.

(xvi) Neemix has been found to bear little or no impact on _______ organisms.

(xvii) Honeybee worker adults (*Apis mellifera*), spiders (*Lycosa pseudoannolata*, *Chiroanthium mildii*), Nematodes (*Steinernema, Heterobditus*), lady beetles (*Coccinellidae, Hippodamia convergens*), earth worms (*Eisema foetida*), nororibated springtails (*Collembola*), ground beetles (*Carabidae*) and Rove beetles (*Staphylinidae*) are beneficial organisms and insect _______.

Ans. (i) Chemical, (ii) pest-control, (iii) insects, (iv) insects, impact, (v) mahogany, (vi) 1985, (vii) Margosan-O, (viii) 1988, (ix) 1990, (x) acre, (xi) tea, cotton, vegetables, (xii) Saudi Arabia, (xiii) Trifolio, Switzerland, (xiv) pupae, (xv) effectiveness, (xvi) beneficial, (xvii) predators.

Fermentation Derived Insect Control Agents

Q.14. What are **spinosyns?**

Ans. During the course of fermentation screening programme extracts from fermentation broth of soil sample collected in 1982 on a Caribbean island were found to be active on mosquito larvae. The microorganism involved was identified as an actinomycete belonging to a less common genus *Saccharopolyspora.* The microorganism *Saccharopolyspora spinosa* was identified as a new species. It produced a new unique macrolides (molecules containing macrocyclic lactone, originally referred to as A 83543 factors) but now called **spinosyns.**

Transgenic Plants Expressing Toxins From *Bacillus thuringiensis* (Bt)

Q.15. What are crystal proteins of Bt?

Ans. The crystal proteins are protoxins which require proteolytical conversion into smaller toxic peptides in the migut of insects. The insecticidal protein crystals are also called δ-endotoxins. It is believed that toxins cause osmotic imbalance by creating pores in the cell membrane of the midgut epithelium of susceptible insects.

Q.16. Which are the transgenic crops other than cotton furnished with Bt gene?

Ans. The corn, potato, egg plant, rice, tomato and soybean are the transgenic crops which express the genes of Bt toxins. Using microprojectile bombardment of immature embryos scientists at CIBA Biotechnology succcessfully placed a synthetic gene encoding a truncated version of the Cry IA (b) protein derived from Bt into corn plants.

A modified version of δ-endotoxin gene, Cry 3A from Bt var *tenebrionis* was used to transform potato plants *Solanum tuberosum* L. This conferred resistance to colardo potato beetle

Leptinotarsa decemlineata under high levels of natural field infestation. The egg plant, *Solanum melongena* L. has also been transformed with a synthetic cry 3A gene from Bt via *A. tumefaciens* – mediated transformation.

A modified cry 1A (b) gene from Bt has been inverted into a rice *Oryza sativa* L. japonica cultivar that confers resistance to two major rice insects rice leaf folder (*Cnaphalocrosis medinalis*) and striped stem borer (*Chilo suppressalis*)

Insect resistant transgenic tomato (*Lycopercon* spp) developed in 1987 is tolerant to tobacco hornworm, tobacco bud worm and bollworm. field tests of genetically engineered tomatoes expressing a gene from Bt were conducted in 1987 in Illinois.

Somatic embryos of the soybean cultivar Jack were transformed using microprojectile bombardment with a synthetic Bt cry 1A (C) gene with the 35 S promoter gene linked to the HPH gene.

Q.17. Name a transgenic tree with δ (delta) endotoxin of Bt.

Ans. *Populus deltoides* plants were transformed using *A. tumefaciens* LBA 4404 strains containing a gene from Bt and two of the three plants regenerated successfully integrated the Bt gene.

Q.18. Which are the Baculovirus types? Give their general properties.

Ans. The Baculoviruses compose a single family Baculoviridae which possesses two genera of occluded viruses *Nucleopolyhedronvirus* called as nuclear polyhedrosis viruses (NPVs) and *Granulovirus* called as granulosis virus (GV). The occluded viruses have been so named because after formation in infected cells the mature virus particles or virions are occluded within a protein matrix forming paracrystalline bodies which are generically referred to as either inclusion or occlusion bodies. The occlusion bodies of NPVs are referred to commonly as polyhedrad because these are typically polyhedral in shape.

39

CHAPTER

APPLICATIONS OF MICROBIOLOGY

Q.1. Match the following:

(A)	San Francisco's 'Sourdough bread' is known for its different taste due to	(a)	*Saccharomyces exiguus* that survives in acidic environment and *Lactobacillus sanfrancisco*
(B)	A. Fleming, E. Chain and H. Florey (1928)	(b)	Discovery of Penicillin and Nobel Prize in 1945.
(C)	Mc Clintock (1983)	(c)	Without breaking down the contents of the host and works as a predator.
(D)	*Bdellovibrio* kills *Escherichia Coli*	(d)	Transposons.
(E)	Can be used as a microbial magnetic tape.	(e)	*Aquaspirillum magnetotacticum.*
(F)	Cabbage fermented with bacterium *Lactobacillus plantarum*	(f)	Citric acid production.
(G)	Molasses fermented with fungus *Aspergillus*	(g)	Commercial production of Sauerkraut.
(H)	$Vent_R$ and Deep $Vent_R$ add bases to DNA at a rate of 1000 bases per minute. Therefore, automatic thermocyclers can be used to repeat the heating and cooling cycles	(h)	TDP (Thermal death point).
(I)	The term for the lowest temperature at which all the microorganisms present in a liquid will be killed in 10 minutes	(i)	*Thermococcus litoralis* and *Pyrococcus* are not denatured at 98°C and therefore, can be used in Polymerase chain reactions (PCRs) that allow production of many copies of DNA.
(J)	The term used for the minimum length of time in which all bacteria present in a liquid will be killed at a given temperature	(j)	DRT (Decimal reduction time) also called D value.

(K)	The degree of heat resistance for bacteria, which is calculated in minutes in which 90 per cent of the population of bacteria get killed at a given temperature.	(k)	TDT (Thermal death time)
(L)	The Xanthan a sticky gum-like substance can be used as a thickness in foods like dairy products, and salad dressing besides in cosmetics, cold creams and shampoos; is produced by	(l)	The bacterium *Xanthomonas campestris*
(M)	A toxin produced by the bacterium if taken in sublethal doses can be used to treat painful muscle spasms, e.g., cerebral palsy, stroke or spinal cord injury or involuntary tremors.	(m)	*Clostridium botulinum* produced toxin by paralyzing muscles involved in unwanted movements
(N)	The monoclonal antibodies can be used	(n)	In trickling filters microorganisms grow on for enormous surface areas and aerobically metabolizes the organic matter in the sewage trickling down.
(O)	The sewage from rotating spray arm is sprayed on a bed of rocks or plastic honeycomb designed to have a maximum surface area and allow oxygen to penetrate into the bed	(o)	Pregnancy test
(P)	Formation of root nodules in legumes is the result of a mutualistic association beneficial to plant and bacteria	(p)	BCG vaccine (Bacillus Calmette-Guerin)
(Q)	A live weakened strain of *Afycobacterium bovis* is used to provide immunity	(q)	*Rhizobium* and *Bradyrhizobium*

Ans. [A a], [B b], [C d], [D c], [E e], [F g], [G f], [H i], [I h], [J k], [K j], [L l], [M m], [N o], [O n], [P q], [Q p].

Q.2. Give major applications of Microbiology.

Ans. Microbiology is one of the most applied branches of science. Its outstanding applications in the field of food microbiology, medical microbiology, industrial microbiology, soil microbiology, water and wastewater microbiology, microbial technology (biotechnology), extraction of metals and environmental microbiology including the use of microorganisms as biosensors is as given below.

(i) It provides us with information about different types of microorganisms enabling us to understand their structure and functions; identifications and differentiations; their classifications; nomenclatures (naming), requirements regarding their nutrition; their isolation and purification; as plant and human pathogens; to derive phylogentic relationships (relationships according to developmental stages in the evolution of an organism) and to understand the origin of life itself.

(ii) Microorganisms as food: Besides **comestible fungi** like mushrooms, microorganisms are also being used as **single cell protein** in the form of yeasts, bacteria, cyanobacteria, fungi as human food or animal feed. The production of the algal microbes as *Chlorella* (green alga and *Spirulina* (cyanobacterium) are being produced in Japan, Taiwan, Mexico, Israel, Thailand and America.

Production of cellulose or lignocellulose utilizing microorganisms serve as human food as such or in the form of their products. Microbial products are also used as animal feed.

(iii) Microorganisms are used in production of a large number of, fermented foods such as leavened bread, sourdough bread, fermented milk products and flavours. The fermented milk products are yoghurt, cheese and several other products.

(iv) The important fermented vegetables are sauerkraut (from cabbage) and Kimchi (from other fermented vegetables in Korea).

(v) Fermented meats and fermented fish are used in different parts of the world due to their increased retentivity, otherwise the meats and fish are highly perishable.

(vi) Beer, vinegar, tempeh, soya sauce, rice wine too are fermented products.

(vii) Microbiology has been very useful in preservation of food by heat processing, by pasteurization and **appertization** (commercially sterile food), by calculating thermal death values, prevention of spoilage of canned foods, aspectic packaging, irradiation, UV radiation, ionizing radiation, high pressure processing, i.e., pascalization, low temperature storage (chill storage and freezing), chemical preservatives (organic acids, esters, nitrite, and sulphur dioxide). In food microbiology one learns about bacterial and nonbacterial agents of food borne illness. Among the helminthes and nematodes are: Platyhelminthus (i.e. liver flukes and tapeworms) and roundworms (e.g., *Trichinella spiralis).*

The protozoa that cause food borne diseases are *Giardia lamblia* and *Entamoeba histolytica.*

(viii) Microbial diseases: Microorganisms are the causative agents of a large number of diseases which have been described under a separate chapter.

(ix) Industrial Microbiology: A large number of products of microbial metabolism after microbial processing of raw materials are produced on industrial scale. A separate chapter has been given on 'Industrial Microbiology'.

(x) Energy from microbial sources: A number of substrates can be used as a source of energy as biogas from methanogenic microorganisms. The microbes like *Methanobacterium* and *Methanococcus* can utilize CO_2 as an electron acceptor finally producing methane. A new species of *Methanobacterium,* i.e., *M. cadomensis* strain 23 has been evolved in Japan for faster production of methane.

Ethanol can also be used for the production of gasohol by mixing 80 per cent gasoline and 20 per cent ethanol.

(xi) Degradation of cellulose and lignin:
Trichoderina reesei can be used to degrade cellulose since it produces extracellular cellulase. The white rot fungus *Sporotrichum pulverulentum* is a cellulase negative organism but a mutant of it has been prepared which can degrade kraft and wood lignocellulase actively. It has been possible to produce biological pulp without any chemical treatment for delignification.

(xii) Mining and extraction of metals:
Thiobacillus ferrooxidans and combination of *Leptospirillum ferroxidans* and *Thiobacillus organoparpus* can be used to degrads pyrite (FeS_2) and chalcopyrite ($CuFeS_2$). The archaeal species *Sulfolobus acidocaldarius* and *S. brierlevi* are capable of oxidizing sulphur and iron for energy depending on CO_2 or other simple organic compounds for carbon. The pyrite and chalcopyrite are also degraded by these archaeobacterial species.

(xiii) Recombinant DNA and genetic recombination:
Recombinant DNA is a wonderful product of genetic engineering, i.e., manufacturing and manipulating genetic material *in vitro*. The process of joining DNA from different sources is genetic recombination. A large number of restriction enzymes/restriction endonucleases have been obtained from various microorganisms that can cut or cleave double stranded DNA leaving staggered ends.

(xiv) Hybridoma and preparation of monoclonal antibodies: Hybridoma is a cell made by fusing an antibody-producing B-cell with a cancer cell. The resulting hybrid myeloma or hybridoma cells have properties of both parent cells immortality and the ability to secrete large amounts of a single specific type of antibody. This was discovered by Kohler.

(xv) Harvesting DNA biotechnology for public health engineering programmes: Such programmes include production of interferon which is an antiviral protein produced by certain animal cells in response to a viral infection, production of human insulin production of somatotropin a human growth hormone and production of a large number of other hormones and vaccines.
The vaccines for cholera, diphtheria, tetanus, pertussis, viral hepatitis type A, type B, influenza, mumps, measles (rubella) plague, poliomyelitis, rabies, rubbela, typhoid, typhus and yellow fever have been developed so far.

(xvi) Microbial technology of nitrogen fixation exploiting symbiotic microorganisms in association with lower or higher plants and asymbiotic or nonsymbiotic (by nitrogen fixing microorganisms independently). Detailed information is covered under a separate chapter on 'biofertilizers'. In nature, in legume root nodules a red pigment containing protein called leghaemoglobin is involved in the process of nitrogen fixation. The key enzyme responsible for biological conversion of molecular nitrogen to ammonia is nitrogenase.

(xvii) Making faster and smarter computers: The Archaeobacterium *Halobacterium halobium* grows in nature in solar evaporation ponds having high concentration of salts. Such salty ponds are found around San Francisco Bay located on the Western coast of USA.
It has been found that the plasma membrane of *Halobacterium halobium* fragments into two fractions, when the cell is broken down. These two fractions are red and

purple. The purple fraction is important in making computer parts (chips). The purple colour is due to a protein which is 75% of purple membrane and has been referred to as bacteriorhodopsin.

Robert Birge at Syracus University's Centre of Molecular Electronics has grown *Halobacerium halobium* in 5-litre batches and has extracted the protein bacteriorhodopsin from the cells and developed the computer chips which are made up of a thin layer of bacteriorhodopsin. The chips so made from the bacterial source can store more information than the conventional silicon chips and process the information faster more like a human brain. The only drawback is that one needs to store the protein chips at –4°C. But Birge believes that this problem will be overcome soon.

40 CHAPTER

MOLECULAR BIOLOGY

Q.1. Match the following by making pairs of capital and small letters written within parentheses (brackets).

(A) Okazaki fragments are later joined as a result of action of	(a) Luciferase operon (lux).
(B) The replication of circular DNA is resulted with a loop of DNA obvious in planar projection	(b) Genetic mapping.
(C) Type 1 and Type II DNA topoisomerases.	(c) The technique used in separation and identification of specific RNA sequences.
(D) *In vitro* method for replication of DNA in which a target nucleotide sequence is copied repeatedly.	(d) Technique to separate DNA fragments by gel electrophoresis.
(E) RNA polymerase moves along a DNA template adding ribonucleotides to synthesize RNA.	(e) Plasmids
(F) RNA forms a 'hairpin loop'.	(f) Ames test.
(G) Exon-intron boundaries.	(g) The lac operon.
(H) Nonsense codons	(h) Shine-Dalgarno sequence.
(I) Occur in bacterial and archaeal cells but not in eucaryotic cells.	(i) Do not code for any amino acid.
(J) Regulate the metabolism of lactose.	(j) Splicing junctions.
(K) Screening for mutagens and carcinogens.	(k) In termination of transcription.
(L) Extrachromosomal genetic elements of the bacterial cells.	(l) In transcription.
(M) E.M. Southern developed Southern blotting	(m) PCR

(N)	Northern blotting.	(n)	Co-exist in bacterial cells.
(O)	Interrupted mating of bacteria is useful in.	(o)	Theta structure.
(P)	The marine bacterium *Photobacterium fischeri* shows bioluminescence due to.	(p)	DNA ligase
(Q)	"Molecular Biology of Human Hepatitis Viruses" published by Imperial College Press, UK, was authored by	(q)	C.L. Mandhar of Panjab University, Chandigarh
(R)	"Molecular Biology of Plant Viruses" published by Kluwer Academic, was authored by	(r)	J. Monjardino of School of Medicine, Imperial College, St Mary's Campus.
(S)	Fourth edition of "Molecular Biology of the Gene" published by The Benjamin/Cummings was authored by	(s)	J.D. Watson, N.H. Hopkins, J.W. Roberts J.A. Steitz and A.M. Weiner jointly.

Ans. [A – p], [B – o], [C – n], [D – m], [E – l], [F – k], [G – j], [H – i], (I – h], [J – g], [K – f], [L – el], [M – d], [N – c], [O – b], [P – a], [Q – r], [R – q], [S – s]

Q.2. What is molecular biology?

Ans. It is the science regarding DNA, RNA and protein synthesis of living organisms, and other related molecules.

Q.3. What is a **molecule**?

Ans. A combination of atoms forming specific chemical compounds.

Q.4. What is a **moiety**?

Ans. A part of a molecule having characteristic chemical property.

Q.5. Define mole % G + C.

Ans. It is proportion of guanine and cytosine in a DNA macromolecule.

Q.6. Who did discover that messenger RNA is involved in protein synthesis?

Ans. Francois Jacob and Jacques Monod discovered it in 1961.

Structure of DNA (Deoxyribonucleic Acid)

Q.7. Give structure of DNA.

Ans. The macromolecule of DNA stores all hereditary information of the cell. DNA is composed of subunits referred to as **nucleotides.** DNA is also known as mastermolecule. At the time of division of a cell its hereditary information is passed on to next generation because of DNA's unique structure. DNA contains four nucleotides, each having one of four **nucleic acid bases,** adenine (A), guanine (G), cytosine (C) and thymine (T). The arrangement or the order of nucleotides determines the hereditary information contained in the DNA. The nucleotides are held together by phosphodiester bonds between deoxyribose, the sugar found in the backbone of DNA. There are two strands of DNA held together by hydrogen bonds forming a coiled molecule known as **double helix.**

Structure of RNA (Ribonucleic Acid)

Q.8. Give structure of RNA

Ans. RNA differs from DNA in several respects, some of which are:

(i) The five-carbon sugar in the nucleotide of RNA is ribose, which has one more oxygen atom in comparison to the deoxyribose found in DNA.

(ii) One of RNA's bases is uracil (U) instead of thymine.

(iii) RNA is normally single stranded.

Other Nucleotides

Q.9. Name some other nucleotide

Ans. The ATP (adenosine triphosphate) is also a nucleotide which consists of adenine, ribose and three phosphate groups. The ATP is known as a high energy molecule because it releases a large amount of usable energy after hydrolysis of a phosphate group resulting the formation of ADP (adenosine diphosphate) production and utilization of ATP are needed for the **bioenergetics** which is transfer of energy through living systems/ cells. When adenosine triphosphate (ATP) is converted to ADP, a high energy phosphate bond is cleaved releasing about 7.5 Kcal/mole which can be used to carry on other chemical reactions.

Some nucleotides work as coenzyme which is a temporary carrier of substances such as electrons. In the process of metabolism coenzymes transport hydrogen atoms and electrons. NAD^+ (nicotinamide adenine dinucleotide and FAD (flavin adenine dinucleotide) are two of such coenzymes. These coenzymes accept electrons and hydrogen during some chemical reactions and release them during other chemical reactions.

Flow of Information

Q.10. What do you mean by **'Central Dogma'**? Discuss.

Ans. **Central Dogma** is one of the major generalizations of molecular biology which shows that only certain transfers are possible. According to it information can flow from DNA to new DNA as in **DNA replication**. However, in the replication of certain RNA viruses we see the flow of RNA to DNA. Within a nondividing cell the major flow of information is from DNA to RNA and then from RNA to protein. Nobel Laureate James D Waston used the term 'Central Dogma' for these flows alone but such a usage is undesirable.

Under special circumstances we may find flow of RNA to DNA and even from single standed DNA to protein, though this latter process can be made to occur under artificial conditions is both inefficient and inaccurate. What we never find are flows from protein to protein or from protein to either DNA or RNA. Correctly stated the **'Central Dogma'** says that when (sequence) has got into protein it cannot go out (Fig. 40.1).

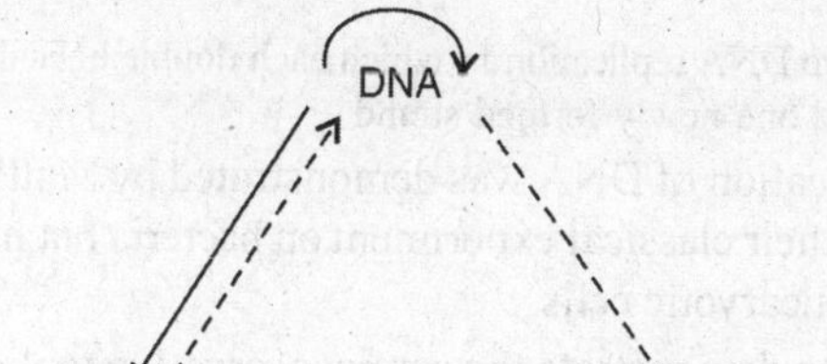

Fig. 40.1. A simplified diagram to explain the **Central Dogma**. The arrows represent the possible flows of detailed sequence information. The solid lines in the diagram represent the most common transfer while the dotted lines are for that found less frequently. Here, absence of some arrows (at protein corner) embodies the central dogma, that information of this type never flows or moves out of protein.

Replication of DNA

Q.11. What is DNA replication? Why is it needed?

Ans. Replication is synthesis of DNA from DNA. As a result of replication of DNA, new molecules of DNA are formed which have same nucleotide sequence as in the genome of the parent organism. It is needed because the genome of the progeny must have appropriate information for the survival and growth of the organism. The changes in the sequence of nucleotides can bring a change in characteristics of the organism considerably. The DNA replication has been designed to ensure that progeny receive accurate copy of the genetic material contained in the parent cell.

Q.12. What is semiconservative process of replication of DNA? Who demonstrated it?

Ans. The process of replication of DNA has been regarded as **semiconservative replication** of DNA because when double stranded DNA replicates each of the two new daughter DNA double helices possesses one intact original conserved strand from the parental double helical DNA and one newly synthesized complimentary strand.

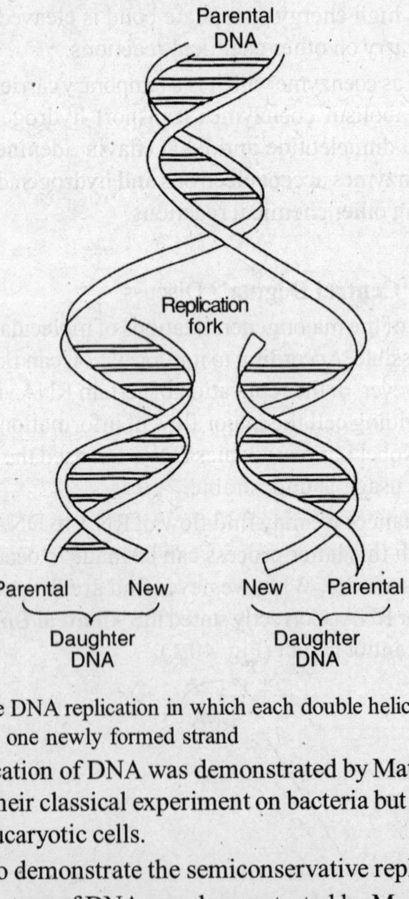

Fig. 40.2. Semiconservative DNA replication in which each double helical daughter DNA molecule has one parental and one newly formed strand

Semiconservative replication of DNA was demonstrated by Matthew Meselson and Franklin Stahl in 1958 based on their classical experiment on bacteria but now it has also been shown to occur in archaeal and eucaryotic cells.

Q.13. Give the methodology to demonstrate the semiconservative replication of DNA?

Ans. The semiconservative nature of DNA was demonstrated by Meselson and Stahl in a series of experiments conducted using the bacterium *Escherichia coli.* It was done by labelling DNA by incorporating heavy nitrogen (^{15}N) and studying this tagged DNA from one generation to the next using density gradient ultracentrifugation. The location of bands was obtained by ultracentrifugation, in other words the distance that was moved by DNA which is a function of molecular weight of DNA.

When the cells were grown in presence of **normal** (^{14}N) which is light in weight, the banding pattern obtained proved that replication of DNA is of semiconservative nature.

To conduct the series of experiments Meselson and Stahl fed the bacteria with isotope of nitrogen (^{15}N). This isotope is nonradioactive but certainly heavier than the normal ^{14}N. When the bacterium was grown on a medium of heavy nitrogen (^{15}N) as the only source of nitrogen, the bacterial cells incorporated heavy nitrogen into their nucleic acids. After it the bacterial culture was transferred to a medium containing "N as a source of nitrogen in the medium. On incubating the cultures the bacterial DNA replicated and the new bacterial cells were reproduced.

The cells were collected for analysis after growing them for different generation times and the DNA present in the cells was analyzed for the presence of ^{15}N and ^{14}N using buoyant density gradient ultracentrifugation. In doing so for analysis of DNA heavy molecules move to a denser part of the gradient in comparison to light molecules. DNA with ^{15}N moves greater distance than the DNA with ^{14}N. They noted that,

(i) In the beginning of the experiment initial DNA formed a single sedimentation band corresponding to heavy DNA containing ^{15}N.

(ii) After one generation a single sedimentation bond was formed corresponding to a hybrid DNA molecule containing a mixture of ^{14}N and ^{15}N.

(iii) If the bacteria were permitted to grow for further replication of DNA so that the parent cell retained the original bacterial chromosome and the progeny received a completely newly synthesized DNA there were two bands after two generations. One band corresponded to light DNA having only ^{14}N and the other corresponding to a combination of ^{14}N and ^{15}N hybrid DNA.

These experiments clearly explain the semiconservative nature of DNA replication.

Q.14. What is replication fork?

Ans. In semiconservative mode of replication of DNA, the region of DNA helix where it unwinds (separation of strands) to localize the DNA synthesis is called **replication fork.**

Q.15. What is theta structure of DNA? Also give the mode of DNA replication in bacterial cell.

Ans. It is a structure formed in replication of circular DNA. Since this structure resembles the Greek letter θ it is known as the **theta structure.**

Semiconservative mode of DNA replication in bacterial chromosome begins at the origin of replication called *ori* C, a region of DNA where specific initiation protein gets attached. The *ori* C in *Escherichia coli* cell-chromosome consists of about 245 bp and contains three 13-bp repeat sequences and four 9-bp repeat sequences where DNA protein initially binds. Therefore, DNA polymerases move bidirectionally from origin of termination of DNA replication abbreviated as *ter* (termination of replication) resulting two replication forks that move in opposite direction around circular chromosome. The bidirectional replication forks move at identical speed after initiation, therefore, both replication forks meet at the termination site. The terminus in *E. coli* and in some other bacteria is not exactly opposite to the origin in the circular chromosome occurring in bacterial cells.

The terminus possesses four *ter* sequences. These are:

(i) The *ter* A and the *ter* D which replicate termination of one strand of DNA.

(ii) The *ter* B and the *ter* C terminate the second strand of the DNA.

All the *ter* sequences contain a 23-bp consensus sequence which is easily recognized by a *turs* protein which prevents the replication of fork further.

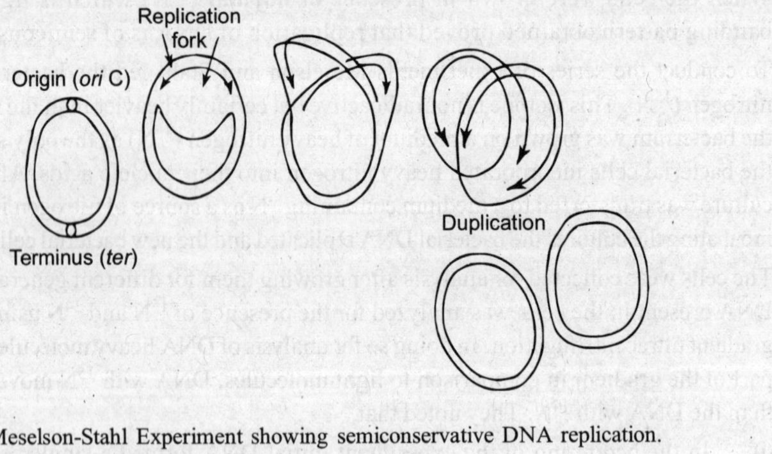

Fig. 40.3. Meselson-Stahl Experiment showing semiconservative DNA replication.

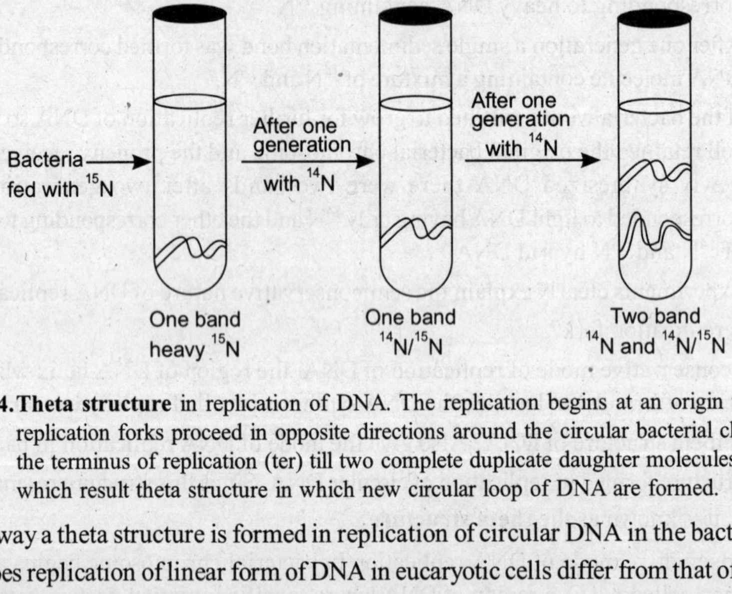

Fig. 40.4. Theta structure in replication of DNA. The replication begins at an origin thereafter two replication forks proceed in opposite directions around the circular bacterial chromosome to the terminus of replication (ter) till two complete duplicate daughter molecules are produced which result theta structure in which new circular loop of DNA are formed.

In this way a theta structure is formed in replication of circular DNA in the bacteria.

Q.16. How does replication of linear form of DNA in eucaryotic cells differ from that of circular DNA of bacteria?

Ans. Normally, the replication of eucaryotic DNA begins at multiple points of origin in each chromosome, of course, proceeds bidirectionally. The DNA of eucaryotes possesses **multiple replicons** which are the segments of a DNA molecule with their own origin and terminal, while there is a single replicon in chromosome of bacterial cells. The eucaryotic genome can be replicated much faster in comparison to bacterial genome, e.g., it is 25 to 30 minutes in the yeast cell and 40 minutes in *Escherichia* coli. The single and multiple origin of DNA synthesis is the basic difference between bacterial and eucaryotic cells.

Proteins Involved in Replication of DNA

Q.17. Which protein-class is responsible for separation or to unwind the double helix of DNA in replication?

Ans. A class of specific proteins known as **helicases** facilitate the separation of strands of DNA double helix in advance of a replication fork. The rep protein and DNA helicases are collectively known as unwinding proteins. **(Fig. 40.5)**

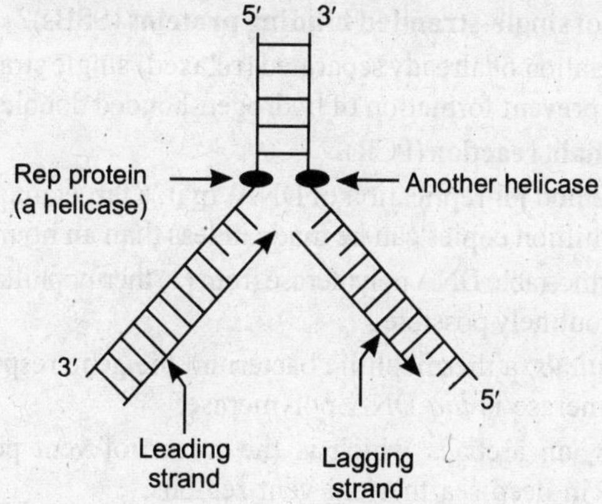

Fig. 40.5. The specific protein called helicases unwind the DNA double helix in advance of a replication fork. The rep protein (a helicase) moves on one strand in the $3' - 5'$ direction while another helicase (e.g., helicase II or III) moves on the complimentary strand in the $5' \rightarrow 3'$ direction.

Q.18. What are **topoisomerases**? Give their types.

Ans. The topoisomerases are the enzymes which can alter super coiling of DNA by bringing a change in the number of topological links between two strands of double helix by introducing transient breaks in phosphodiester backbone of the DNA and catalyze DNA strand passages through these breaks. In this way the topoisomerases can either relax or supercoil DNA macromolecules.

These are the nicking-closing enzymes which are of two types:

(i) Type I topoisomerases: These are enzymes that change negative supercoiling of DNA, hence causing a localized uncoiling effect.

(ii) Type II topoisomerases: These bring about negative supercoiling into the relaxed DNA.

Q.19. Give bacterial topoisomerases. Which one of them is involved in production of negatively supercoiled DNA.

Ans. Several type I and type II DNA topoisomerases are found together in bacterial cells. There are two types I isomerases in *Escherichia coli* known as protein ω and Topo III. The Topo III was the 3rd one to be discovered as name itself indicates; and two type II DNA topoisomerases, called DNA gyrase and Topo IV.

DNA gyrase is the only enzyme which is involved in production of negatively supercoiled DNA.

Q.20. What is the function of *gyr A* gene?

Ans. DNA gyrase of *Escherichia coli is* a type II topoisomerase which comprises four protein subunits. Of these, two are A-subunits while the two are B-subunits. Type A subunit is coded for the gene *gyr A.* The gyrase nicks hydrolyzes' both strands of DNA and then passes the strand around another part of the double helix thus introduces a negative supercoil.

Q.21. Which are the archaeal topoisomerases? What is special about them?

Ans. In the archaeal cells also type I and type II topoisomerases are like that of bacteria. The speciality of the archaeal cells lay in having a novel topoisomerase known as reverse gyrase. The reverse gyrase activity has been observed in extremely thermophilic archaea, but is absent in moderately thermophilic archaea. The reverse gyrase protects archaea from melting or strand separation at the very high environmental temperature of their *niche* or natural places where they grow.

Q.22. What is the function of **single-stranded binding proteins** (SSBs)?

Ans. They prevent reassociation of already separated (relaxed) single stranded regions of the DNA. In other words, they prevent formation of hydrogen-bonded double helix.

Q.23. Define **polymerase chain reaction** (PCR).

Ans. This is an *in vitro* method for replication of DNA. In it, a target nucleotide sequence is copied repeatedly so that a million copies can be made in less than an hour.

Q.24. Give examples of thermostable DNA polymerase from (1) thermophilic bacterium and (2) archaea that has made PCR routinely possible.

Ans. (i) *Thermus aquaticus,* a thermophilic bacterium the gene responsible for the stability of its DNA polymerase is *taq* DNA polymerase.

(ii) *Thermococcus,* an archaea which is the source of vent polymerase. This archaeal species grows in deep sea thermal vent regions.

Q.25. Give three major steps taken in the methodology for polymerase chain reaction.

Ans. (i) The DNA is denatured at high temperature 94° to 95°C.

(ii) Primar annealing by lowering the temperature 37° to 70°C.

(iii) Chain elongation using thermostable DNA polymerase at 60 to 72°C to permit the action of thermostable DNA polymerase.

Q.26. What is **proofreading** by DNA polymerases?

Ans. This is the $3'OH \rightarrow 5'P$ exonuclease activity of DNA polymerases. In this way it is excision of improperly inserted nucleotides by DNA polymerases in DNA replication.

Q.27. What may be the **post-replication modifications** of DNA? Explain.

Ans. In the post replication modifications DNA is liable to addition of methyl groups to some adenine and cytosine residues. The enzymes that bring about methylation are known as DNA methylases. The methylation occurs after the nucleotides have been incorporated by DNA polymerases. The methyl group protects DNA against action of being digested by specific enzymes called restriction endonucleases. But foreign DNA within the cell is digested by endonucleases. The methylation pattern protects the DNA from digestion by cells own endonuclease. However, methylation does not provide DNA protection against being digested by the restriction enzymes produced by the cells of other species. In this way natural exchange of DNA among cells of different species is prevented.

Q.28. How does methylation of DNA at specific site may lead to localized conversion of B-DNA to Z-DNA?

Ans. The hydrophobic methyl group of B-DNA may protrude into the hydrophilic environment of the major groove and cause a destabilizing arrangement.

Transcription or Transferring Information from DNA to RNA

Q.29. What is transcription?

Ans. The **transcription** is the process in which the genetic information stored in the DNA is used to code the synthesis of RNA. The strand of DNA responsible for doing it or working as a template for the synthesis of RNA is known as sense strand.

Q.30. What is mRNA or messenger RNA?

Ans. The **mRNA** has the code which is transcribed from DNA genetic information and is used to specify a sequence of aminoacids; in protein synthesis.

Q.31. How does the longevity of bacterial and eucaryotic mRNA differ?

Ans. The mRNA of bacterial cells generally last for only a few minutes while in eucaryotic cells it remains functional for hours and days.

Q.32. What is tRNA or transfer RNA?

Ans. The tRNA decodes the mRNA sequence or translates it into a correct amino acid sequence. The tRNA carries a specific amino acid to the ribosome for the process of protein synthesis.

Q.33. What is rRNA or ribosomal RNA?

Ans. The rRNA molecules constitute the major part of the ribosome while the remaining part of the ribosome is the ribosomal protein.

Q.34. How does the synthesis of RNA take place in the process of transcription?

Ans. It takes place in the following stages.

(i) Unwinding the double helical DNA molecule for a short sequence of nucleotide.

(ii) Alignment of complimentary ribonucleotides by base pairing opposite the nucleotides of the DNA strand being transcribed and linkage of these nucleotides with phosphodiester bonds by DNA dependent RNA polymerase.

(iii) In transcription RNA polymerase moves along a DNA template adding ribonucleotides to synthesize RNA.

(iv) The process moves from an upstream region where RNA polymerase binds till termination occurs.

(v) The downstream movement is towards the 3′-OH end of the synthesized RNA molecule.

Q.35. Which are the enzymes to synthesize RNA from ribonucleotides?

Ans. These are DNA dependent RNA polymerases.

Q.36. What is open reading frame?

Ans. The **ORF** or **open reading frame** is a region of nucleotide sequence coding for a polypeptide. The ORF is equivalent to a gene and contains codons which code for amino acids.

Q.37. What is **promoter region**? Give its function.

Ans. It is a specific region of nucleotides in DNA. Its function is to initiate the process of transcription.

Q.38. What is meant by "up stream" in DNA undergoing transcription task?

Ans. It indicates that it is transcribed before downstream nucleotides.

Q.39. What is **Pribnow sequence**?

Ans. It is a conserved DNA nucleotide sequence in bacterial genes that take part in initiation of transcription. It is easily recognized by RNA polymerase and is same or almost same as TATAAT for many of the bacterial promoters.

Q.40. What is a consensus sequence?

Ans. It means region of general agreement that is high nucleotide sequence homology. It is a conserved sequence of nucleotides serving same function in diverse cells.

Q.41. What do you mean by TATA box?

Ans. It is a conserved A-T rich DNA sequence in eucaryotic cells that is recognized by TFIID and helps RNA polymerase II in initiating transcription. It is represented as 5′-TATA (T/A) A (T/A).

Q.42. What is sigma (σ) factor?

Ans. It is the factor which ensures that RNA synthesis begins at a correct site. A sigma unit is, however, a subunit of RNA polymerase which helps to recognize the promoter site.

Q.43. What are strong and weak promoters in the process of transcription?

Ans. A promoter with nucleotide sequence which favours transcription by RNA efficiently, is a strong promoter while the one which favours inefficient transcription, is called a weak promoter.

Q.44. How does **elongation of RNA strand** take place in the process of transcription?

Ans. On establishment of initiation of transcription the enzyme releases its sigma factor. A ternary complex of RNA polymerase-DNA-newly synthesized RNA strand results. The RNA polymerase after it moves along the DNA template and elongates the growing RNA strands. In this way new ribonucleotides are added at the rate of about 40 nucleotides per second at 37°C which is very slow in comparison to the rate of replication of DNA in which 1000 deoxyribonucleotides per second are added.

Q.45. How does termination of transcription occur? Give its types in bacteria.

Ans. There are certain sites that are called **termination sites** which act as a signal to stop the transcription. The termination sites may be located either in the RNA sequences which have already been transcribed or in the DNA template.

In bacteria the termination of transcription is of two types, which are:

(i) Simple or rho (ρ) independent termination that do not need any additional factor.

(ii) Rho (ρ) dependent termination which requires an additional protein called rho (ρ) factor.

Q.46. Which are the major post-transcriptional modifications of RNA?

Ans. It involves removal of specific nucleotide sequences known as **introns** which are intervening sequences and splicing (joining) together of remaining sequences known as exons.

Q.47. What do you mean by cap in post-transcriptional RNA modification?

Ans. The cap in it is of 7-methyl guanosine (mG), in eucaryotic: RNA molecules. It also has a poly A tail, and a leader sequence besides a cap.

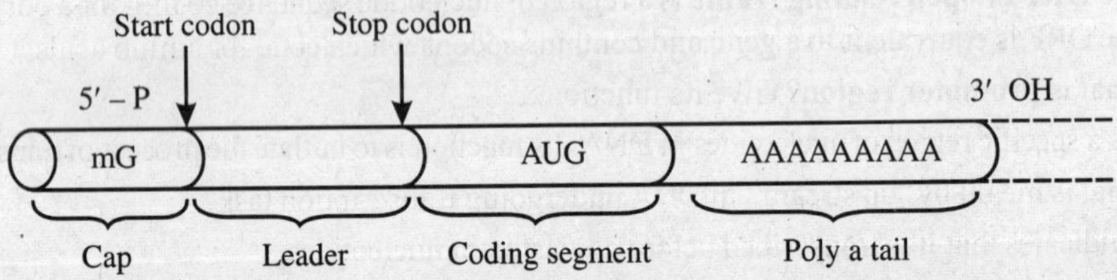

Fig. 40.6. Post-transcriptional modification of RNA showing cap, leader, coding segment and poly A tail sequences.

Translation of Genetic Code or the Protein Synthesis

Q.48. What do you mean by translation of genetic code?

Ans. The assembly of polypeptide chains in which mRNA serves as a template, occurring in ribosomes is the translation of genetic code. In simple words it may be regarded as the process in which information in RNA obtained from DNA in the process of transcription, is directed to the sequence of amino acids in a protein.

Q.49. Define the genetic code.

Ans. The genetic code may be defined as the code for specific amino acids formed by three sequential nucleotides in mRNA. The genetic code is also known as triplet code. The 64 codons (each triplet nucleotide sequence) formed by sequences of 3 nucleotides specify the genetic information of all organisms.

Q.50. Why are **termination codons** or **nonsense codons** named so?

Ans. As they do not code for any amino acid.

Q.51. What is the function of termination codons?

Ans. They work as punctuators that signal the termination of the synthesis of a polypeptide chain.

Q.52. Which is the first amino acid of polypeptide sequence in Archaeal and eucaryotic microorganisms?

Ans. It is methionine coded by the codon AUG.

Q.53. What is the role of transfer RNA in protein synthesis? How does it work with its unique structure?

Ans. The transfer RNA brings the amino acids to ribosomes and properly aligns them during the process of translation. The attachment of an amino acid to its specific tRNA molecule is known as charging and the tRNA molecule which is attached to its amino acid is referred to as charged and the process as **charging of tRNA.** There are at least 20 different types of amino acids and each of the 20 amino acids that occurs in proteins binds to a different RNA.

The unique structure of tRNA (as **clover-leaf model)** plays a very important role in establishing proper alignment of molecules during translation. One of the four lobes is attached to amino acid, contains a nucleotide sequence which interacts with rRNA for proper orientation to ribosomes, third reacts with aminoacyl synthesis and fourth has a region called anticodons which interacts with mRNA.

Q.54. What is **wobble hypothesis**? What is the importance of wobble and degeneracy?

Ans. This hypothesis accounts for the observed pattern of degeneracy in the third base of a codon. According to this hypothesis the third base can undergo with the corresponding first base in the anticodon.

The importance of wobble and degeneracy of the genetic code is that cell does not have to synthesize a different tRNA for each of the 61 sense codons. A simple example is that only two different tRNA anticodons are needed to recognize four different glycine codons.

Q.55. What is **Shine-Dalgarno sequence**? In which groups of microorganisms it is found?

Ans. The Shine-Dalgarno sequence is located before the start sequence of mRNA. This nucleotide sequence allows the mRNA to align with the 30S ribosomal subunit of the bacterial cell. It is about 7 bases upstream earlier) towards the 5′-P end of the AUG start codon on the mRNA and is a polypurine consensus sequence AGGAGG and is referred to as Shine-Dalgarno sequence (Fig. 40.7).

It is found in bacterial and archaeal cells.

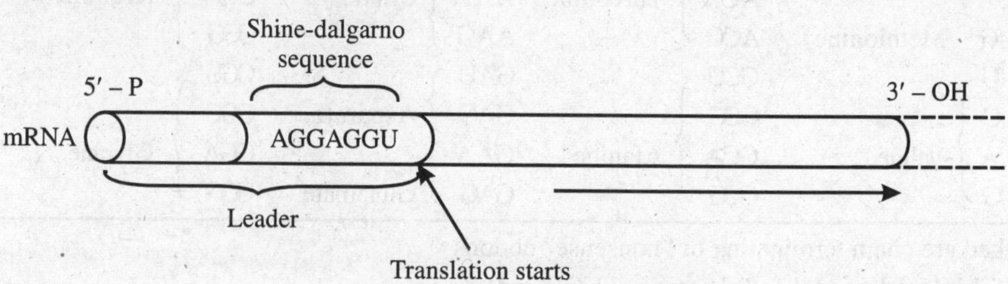

Fig. 40.7. Shine-Dalgarno sequence.

Q.56. What do the codons UGA, UAA and UAG mean in normal translation?

Ans. In normal translation they mean for **"Stop" codons.**

Q.57. Why is genetic code said to be degenerate?

Ans. It is because more than one codon can code for the same amino acid.

Q.58. How many termination codons or nonsense codons are there?

Ans. There are only three termination codons also, called "nonsense codons" as they do not code for any amino acid.

Q.59. The codon AGG normally codes for argine but in altered translation it codes for stop. Where does it occur?

Ans. It occurs in human mitochondria.

Q.60. What is odd in the studies made in the mitochondrial DNA (double-stranded circular DNA genome) as one proceeds from lower to higher eucaryotes?

Ans. While one proceeds from lower to higher eucaryotes, the odd thing in the mitochondrial genome, is that it gets smaller, e.g., the yeast mitochondrial DNA is five times larger than that of the human mitochondria.

Q.61. What has enabled a given tRNA that sometimes it specifically recognizes several codons?

Ans. The wobble in the base pair at the 5′ end on the anticodon enables the given tRNA to recognize several codons.

Q.62. Give the genetic code in the form of a table indicating the amino acid synthesized by each codon. In all how many codons are in it?

Ans. See table 40.1. In all there are 64 codons in the genetic cod. First nucleic acid (or First position) 5′-P end

Table 40.1 The genetic code

	Second nucleic acid or second position								
	U		*C*		*A*		*G*		
	UUU		UCU		UAU		UGU		U
U	UUC	Phenylalanine	UCC	Serine	UAC	Tyrosine	UGC	Cysteine	C
	UUA		UCA		UAA*		UGA*	Stop	A
	UUG	Leucine	UCG		UAG*	Stop	UGG	Tryptophan	G
	CUU		CCU		CAU		CGU		U
C	CUC	Leucine	CCC	Proline	CAC	Histidine	CGC		C
	CUA		CCA		CAA		CGA	Arginine	A
	CUG		CCG		CAG	Glutamine	CGG		G
	AUU		ACU		AAU		CGU		U
	AUC	Isoleucine	ACC		AAC	Asparagine	CGC	Serine	C
A	AUA		ACA	Threonine	AAA	Lysine	CGA	Arginine	A
	AUG	Methionine	ACG		AAG		CGG		G
	GUU		GCU		GAU		GGU		U
	GUC		GCC		GAC	Asparate	GGC		C
G	GUA	Valine	GCA	Alanine	GAA		GGA	Glycine	A
	GUG		GCG		GAG	Glutamate	GGG		G

* Marked are chain terminating or "nonsense" codons

Third nucleic acid or 3rd position (3′OH end).

Q.63. What is **post-translational processing**?

Ans. The newly synthesized polypeptide chains undergo additional enzymatic changes which modify the protein after translation is complete. These changes usually affect the folding of the polypeptide to convert it to its biologically active form. These reactions collectively have been called post-translational modifications or post-translational processing.

Amino acids added to a developing polypeptide may be individually modified, e.g. serine, tyrosine or threonine residues may be phosphorylated by ATP dependent **protein kinases.** Extra carboxyl groups may be added on to asparate or glutamate residues and lysine residue may be methylated and proline residues may be converted to hydroxyproline. Many proteins are folded called **protein folding** into tertiary and quaternary structures. Many of the eucaryotic polypeptides are post-translationally glycosylated. Another feature of protein modification in microorganisms is the splicing of amino acid sequences out of the proteins after their formation. The protein splicing involves the removal of inteins and leaving exteins in the processed proteins. The exteins are exons of proteins which are amino acid sequences which are retained and joined together during protein processing while the **inteins** are introns of proteins that are amino acid sequences which are removed during protein processing.

Q.64. All newly synthesized bacterial proteins start (initiate) with formylmethionine (may be abbreviated as fmet). How is formyl group removed from the fmet polypeptide in bacteria?

Ans. The formyl group is often removed from formyl-methionine by the enzyme deformylase leaving behind the methionine as the first amino acid in the **polypeptide** chain.

Q.65. What is **signal hypothesis**? Describe.

Ans. This is the hypothesis that explains that translocation of a protein from its site of synthesis into or through a membrane which requires that the protein be synthesised in a precursor form that may be called a pre-protein having a specific N-terminal sequence of amino acid residues called signal peptide, signal sequence and leader peptide which is essential for the initiation of translocation but which is excised during translocation.

In both procaryotes and eucaryotes many of the membrane and secretory proteins are synthesized as pre-proteins. Some proteins of cytoplasmic membrane in *Escherichia* appear to lack an obvious N-terminal signal peptide. Some proteins which lack a terminal signal peptide may have an internal sequence of amino acid residues which serves an analogous role while others may require **post-translational processing/modification** before they can be translocated.

The current explanation for the formation of rough endoplasmic recticulum and the mechanism by which newly synthesized proteins pass through the membrane is referred to as signal hypothesis.

The signal for attachment of the ribosome to the membrane is a sequence of very hydrophobic amino acids near the amino terminus of the growing polypeptide chain. At the time of beginning of protein synthesis the ribosome is free. The hydrophobic amino-terminal sequence interacts with a cytoplasmic ribonucleoprotein particle which has been called **signal recognition particle (SRP).** The SRP then binds to protein that is in or on the endoplasmic reticulum. This protein is known as **SRP receptor** as well as **docking protein.** In an unknown way the ribosome and the signal sequence detach from the SRP and the ribosome binds to ribosome receptor, a protein in the endoplasmic reticulum and perhaps adjacent to the pore of the membrane. It has been opined that translocation may stop while the signal sequence is bound to SRP and then resumes while the ribosome binds to the **ribosome receptor.** In either case protein synthesis proceeds. The protein moves through the membrane to the cisternal side of the endoplasmic reticulum. A specific protease called signal peptidase then cleaves through the amino-terminal signal sequence.

The experimental support in favour of signal hypothesis include:

(i) While *in vitro* synthesis of some secretory proteins is carried out with free ribosomes the resulting proteins contain NH_2-terminal sequences about 20 amino acid residues long that are not present in the proteins isolated from intact cells.

(ii) The amino acids in the aforesaid NH_2-terminal extensions are very rich in hydrophobic amino acids.

Inhibitors of Protein Synthesis

Q.66. Why does antibiotic puromycin affect protein synthesis in procaryotes as well as eucaryotes?

Ans. Chemical puromycin resemble the aminoacyl part of **charged tRNA** molecule, therefore, it competes effectively with charged tRNA molecules for the site A. The puromycin, in addition to it also possesses an α amino group that can form a peptide bond with the carboxyl group of a growing peptide chain in a reaction catalyzed by peptidyl transferase. When this bond forms the peptide is cleaved from the tRNA in the P site and the antibiotic puromycin leaves the A site. Puromycin does not bind to P site. Therefore, the polypeptide which possesses a terminal puromycin falls away from the ribosome. In this way polypeptide chain elongation gets blocked in presence of the antibiotic puromycin.

Q.67. Name some antibiotic inhibitors of protein synthesis in procaryotes. Also give the reason for each.

Ans. Some important antibiotics which inhibit the protein synthesis in procaryotes along with the reason are given below.

(i) **Streptomycin:** Because it binds to the S12 protein of the 30 S ribosomal subunit and therefore, inhibits binding of tRNA to the P site. It also causes misreading in a system that is in the 'act of synthesis.

(ii) **Neomycin, Kanamycin:** Both produce same action as streptomycin.

(iii) **Chloramphenicol (or chloromycetin):** It inhibits peptidyl transferase of 70 S ribosome.

(iv) **Tetracycline:** It inhibits binding of charged tRNA to 30 S particle.

(v) **Erythromycin:** The erythromycin binds to 50 S particle and prevents formation of the 70 S ribosome.

(vi) **Puromycin:** Its inhibition is due to premature chain termination by acting as an analogue of charged tRNA.

(vii) **Fusidic acid:** It binds to elongation factor EF-G, consequently GTP is still hydrolyzed and translocation occurs. But EF-G and GDP are not released from ribosome, therefore, the ribosome cannot bind another aminoacyl-tRNA.

(viii) **Kasugamycin:** It inhibits binding of tRNA.

(ix) **Lincomycin:** It inhibits peptidyl transferase complex.

(x) **Kirromycin:** It binds to the elongation factor, written as EF-Tu and stimulates formation of (EF - Tu) - GTP and binding of ternary complex to ribosome therefore, inhibits release of EF - Tu.

Out of the aforesaid antibiotics chloramphenicol, Puromycin, fusic acid also inhibit protein synthesis in eucaryotes.

Gene Regulation or Regulation of Gene Expression

Q.68. What are genes? Define.

Ans. A gene may be defined as a sequence of nucleotides which specifies a particular polypeptide chain or RNA sequence or that regulates the expression of other genes.

The genes which code for proteins are referred to as **structural genes** or **cistrons** while the other genes bearing regulatory function are called **regulatory genes.** The regulatory genes work to control the expression of structural genes. The structural and regulatory genes collectively constitute the genotype which determine the **phenotype,** i.e., observable structural and functional characteristics.

Q.69. Define an **operon.**

Ans. A DNA sequence which codes for one or more structural genes (polypeptides) of related function and the DNA sequence which regulates the expression.

Q.70. What controls induction and repression?

Ans. The **regulatory genes** that produce a regulator protein controls **induction** (i.e., causing increase in rate of synthesis of an enzyme), and **repression** (i.e., blockage of gene expression).

Q.71. What is the lac operon?

Ans. The lac operon stands for lactose operon, an operon which contains genes specifying proteins involved in the utilization of β -galactosides such as lactose. The lac operon occurs in *Escherichia coli* at ca. (= about) 8 minutes on the chromosome map. It has the **structural: promoter-operator lac** Z—*lac* Y—*lac* A. The *lac Z* gene encodes β -galactosidase, lac Y encodes β -galactoside permease and *lac* A encodes thiogalactoside transacetylase (Fig. 40.8).

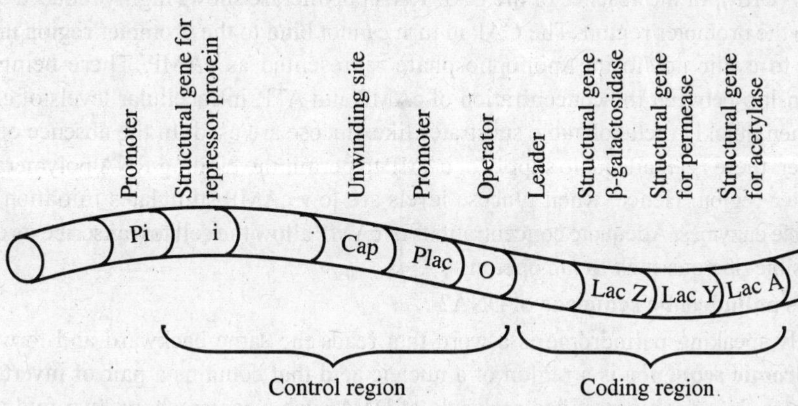

Fig. 40.8. The lac operon which controls the utilization of lactose. There are three structural genes under the control of lac promoter represented as Plac which code for synthesis of enzymes needed for lactose utilization. These enzymes are formed only when lactose is present.

The regulator gene of *lac* operon *lac* I encodes a **repressor protein** that means that operon is under negative promoter control. The repressors (repressor genes) prevent RNA polymerase binding.

Q.72. What is catabolite repression?

Ans. The **catabolite repression** is the repression of transcription of genes coding for certain inducible enzyme systems by glucose or other readily utilizable carbon sources.

Q.73. What are **positive regulators** (activators) and **negative regulators** (repressors)? Describe.

Ans. Bacteria possess many enzymes whose rate of synthesis depends on the availability of external food molecules. These external molecules called **inducers** and corepressors usually determine the rate of synthesis of enzymes by controlling the synthesis of their mRNA templates. Inducers and corepressors act by binding to regulatory proteins referred to as **activators and repressors.** Activators are **positive regulators** because their presence is required for the regulated enzyme to be made while repressors act as **negative regulators** because their regulatory activity is to prevent the synthesis of proteins. Thus, when lactose is absent the *lac* operon (lactose operon) repressor prevents synthesis of enzymes that metabolize lactose. However, upon binding an inducer (a molecule related to lactose) the repressor loses this Ability and permits the production of enzymes. The arabinose operon C protein which is an activator causes making of arabinose enzymes on binding to the inducer arabinose.

Q.74. What is catabolic control of *lac* operon?

Ans. If *Escherichia coli* is grown in a medium that contains both glucose and lactose, the bacterium does not consume both sugars simultaneously rather it shows preference to utilize glucose first until the whole glucose is consumed; then it switches to utilization of lactose as the carbon source. Consequently a biphasic pattern of growth called **diauxic** or **diauxic growth** is established.

When glucose is available to *Escherichia* coli for catabolism in the glycolytic pathway other monosaccharides and dissaccharides are not used by the cell to generate ATP and by blocking the metabolism of these other carbohydrates. Hence, the cell conserves its metabolic resources. The catabolite repression is an example of regulation by positive control. It acts via promoter region of DNA and by doing so it complements the control exerted by the operator region. The efficient binding of RNA polymerase to promoter regions subject to catabolic repression requires the presence of a catabolite activator protein (CAP) also referred to as cyclic AMP receptor protein (CRP). In the absence of the CAP, RNA polymerase shows highly reduced affinity to bind to the promoter region. The CAP in turn cannot bind to the promoter region unless it is bound to cyclic adenosine monophosphate represented as cAMP. There being inverse relationship between the concentration of cAMP and ATP, intracellular levels of cAMP are low when quickly metabolizable substrates like glucose are used. In the absence of glucose, however, there is an adequate supply of cAMP to permit the binding RNA polymerase to the promoter region. Hence, when glucose levels are low cAMP stimulates initiation of many inducible enzymes. Adequate concentrations of cAMP allow the cell to transcribe its catabolite repressible operons such as *lac* operon.

Q.75. What is **palindromic sequence** of DNA?

Ans. Literally speaking palindrome is a word that reads the same backward and forward. The **palindromic sequence** is a region of a nucleic acid that contains a pair of inverted repeat sequences. In a double stranded molecule of DNA such a region shows two-fold rotational (dyad) symmetry or hypernated dyad symmetry if the two IR sequences are separated by another sequence. A double-stranded palindromic sequence can adopt either of two possible formations:

(i) A linear structure with interstrand hydrogen bonding, e.g.,

5′ ... TCCACATGTGGA ... 3′
3′ ... AGGTGTACACCT ... 5′

(ii) A cruciform structure in which of two strands each forms hairpins by intrastrand hydrogen bonding.

Gene Concept

Q.76. Who discovered that X-rays induce mutations?

Ans. Hermann Muller and L.J. Stadler discovered independently in 1927 that X-rays induce mutations.

Q.77. What is cistron?

Ans. A **cistron** is a gene as defined in terms of the CIS-TRANS TEST, i.e., in a diploid cell or merozygote either of two homologous sequences in a genetic nucleic acid in which two mutations in *trans* fail to exhibit complete complementation.

A cistron may also be defined as the functional unit of genetic inheritance a segment of genetic nucleic acid which codes for a specific polypeptide chain. The term cistron has also been used as a synonym for gene.

Q.78. What is **recon**?

Ans. The term **recon** was coined by S. Benzer. According to him it is a unit of genetic subdivision beyond which recombination does not occur.

Q.79. Define a mutator or mutator gene.

Ans. A **mutator gene** or **mutator** is designated as must within which certain mutations cause an increase in the spontaneous mutation rate in other genes, e.g., *Escherichia coli* mutations in the gene encoding the ε subunit of DNA polymerase III (dna Q and *mut D* alleles) can result in extremely high levels of spontaneous mutations. The mutant alleles may differ from the wild type, e.g., by only one or two amino acid changes in the ε subunit. The mutation may lead to reduced accuracy in the polymerising (nucleotide selection) activity or in the proof-reading activity of the enzyme. Some other mutator genes in *E. coli* include in the mismatch repair system.

Q.80. What are split genes? Describe.

Ans. The **split genes** may be defined as the genes that are coded for, by noncontiguous segments of the DNA so that the mRNA and the DNA for the protein product of that gene are not colinear. These are the genes with intervening nucleotide sequences not involved in coding for the gene product.

The split genes have also been regarded as interrupted genes. A structural gene encoding a protein, rRNA or tRNA that contain one to many intervening sequences (introns) that although represented in primary RNA transcript of the gene are absent from the mature RNA molecule (mRNA, tRNA), therefore, do not contribute to the structure of the gene product. Thus mutation of the RNA transcript of a split gene must involve a process of splicing to delete the intron and join together the remaining sequences called exons. In the case of mRNA the sequence of exons includes the coding sequence of the gene as well as noncoding leader and/ or trailor sequences. Introns themselves are usually noncoding.

Most of the nuclear structural genes in higher eucaryotes are split genes.

Q.81. What are overlapping genes?

Ans. The **overlapping genes** are two or more genes in which part or complete gene is co-extensive with part of another. The genes may be translated in different reading frames or in the same reading frame with different start and/or stop points or different splicing patterns. The phenomenon of overlapping genes maximizes the coding capacity of a genome and can also provide a means for the regulation of expression of genes.

Q.82. The history of DNA world is written in gene sequences. Justify this statement.

Ans. The evolution of the organisms from a common ancestor is represented by a branched pathway known as geological tree (also known as **phylogenetic or evolutionary tree).** The branching pattern of tree is calculated using the **principle of parsimony** (based on economy) to determine the minimum number of genetic changes required to derive the sequence of the gene in each organism from a common ancestor. It is sometimes reasonable to assume that highly conserved protein like globin and cytochorme can be used as a **molecular clock** to measure how long the species have been diverging from each other.

41 CHAPTER

TOOLS AND TECHNIQUES OF MOLECULAR BIOLOGY

For Nucleic Acids

Q.1. Name a technique used to separate DNA and RNA molecules according to their size.

Ans. Gel electrophoresis (Fig. 41.1).

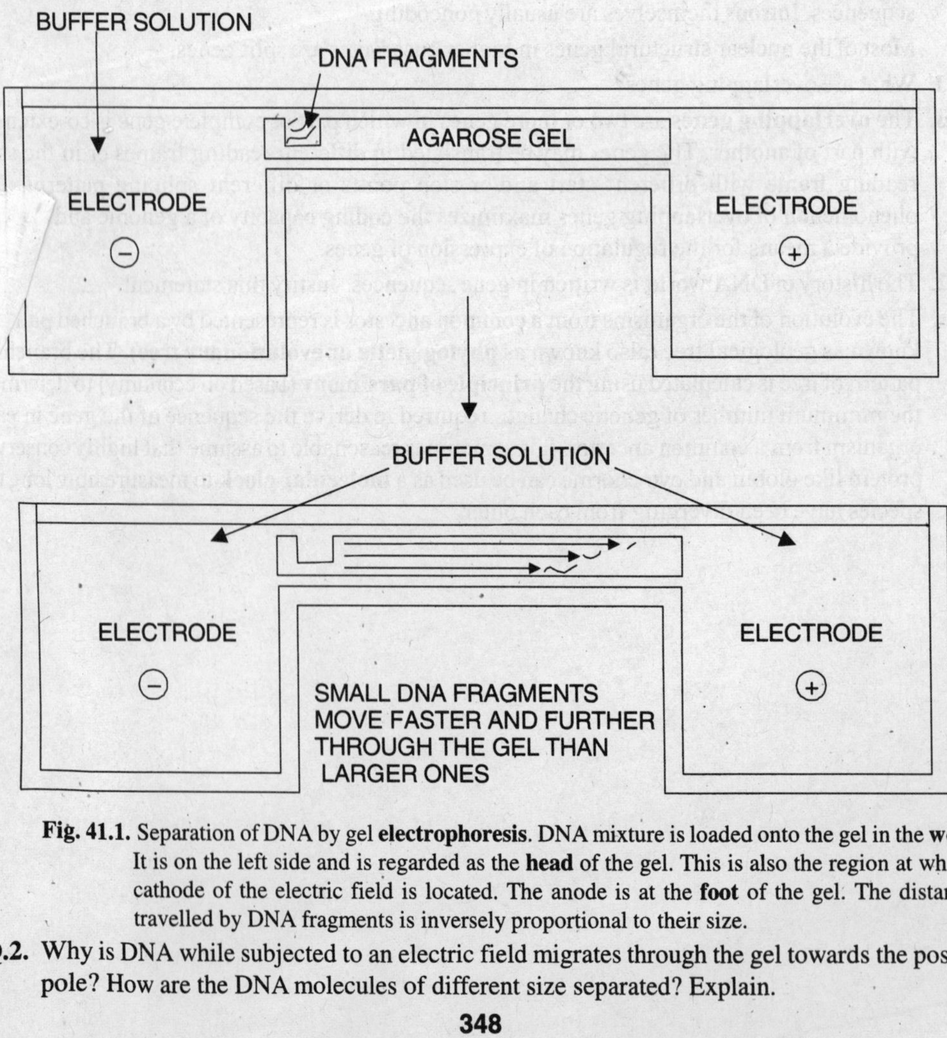

Fig. 41.1. Separation of DNA by gel **electrophoresis**. DNA mixture is loaded onto the gel in the **well.** It is on the left side and is regarded as the **head** of the gel. This is also the region at which cathode of the electric field is located. The anode is at the **foot** of the gel. The distance travelled by DNA fragments is inversely proportional to their size.

Q.2. Why is DNA while subjected to an electric field migrates through the gel towards the positive pole? How are the DNA molecules of different size separated? Explain.

Ans. Because DNA is negatively charged. DNA molecules are flexible. The pores in the gel matrix give the DNA molecules according to their size (volume). Large molecules migrate more slowly through the gel because they have large effective volume in comparison to small DNA molecules. Therefore, the large molecules face more interstices of the gel. Once the gel has been run for a given time the molecules of different sizes can be separated because they travel different distances according to the size of the DNA molecules.

On completion of electrophoresis the DNA molecules can be visualized by staining the gel with fluorescent dyes like ethidium which binds to DNA and intercalates between the stacked bases. Each band shows the presence of a population of DNA molecules according to their size.

Q.3. Name the two alternative types of gel matrices. Also give their characteristics.

Ans. These are : (i) Polyacrylamide and (2) Agarose. The polyacrylamide has got high resolving capability but can separate DNAs over a narrow size range, only. Thus, electrophoresis through polyacrylamide can resolve DNAs which differ from each other in size by as little as single base pair but only with molecules upto several hundred base pairs.

On the other hand agarose has less resolving power than polyacrylamide but can separate from one another DNA molecules of upto tens and even hundreds, of kilobases. Very long DNAs are unable to penetrate the pores even in agarose. However, such type of very long DNAs can be resolved from one another if the electric field is applied in pulses which are oriented orthogonally to each other. This technique is called **pulse-field gel electrophoresis**.

Q. 4. Give a brief account of the techniques used to study the molecular biology of nucleic acids.

Ans. These techniques are described below :

(i) **Gel electrophoresis :** In which electrophoresis through a gel separates DNA and RNA molecules according to size.

(ii) **Use of restriction endonucleases :** Which cleaves *i.e.*, splits or cuts DNA molecules at particular sites.

(iii) **DNA hybridization :** Which can be used to identify specific DNA molecules.

(iv) **Hybridization probes :** Which can identify electrophoretically separated DNAs and RNAs.

(v) **Isolation of specific segments of DNA :** Which is used for separation of specific segments of DNA from much larger DNA molecules followed by their selective amplification. Hence the DNA can be sequenced.

(vi) **DNA Cloning :** This is the ability to construct recombinant DNA molecules and maintain them in cells. This process, however, involves a vector.

(vii) **Cloning DNA in plasmid vectors :** In it for example plasmid vector has a unique recognition site for EcoRI. Treatment with the appropriate restriction enzyme would linesize the plasmid. A target DNA is cleaved with a restriction enzyme to generate potential insert DNAs. Vector DNA which has been cut with the enzyme is mixed with these **insert DNAs** and **DNA ligase** is used to link or join the compatible ends of two DNAs.

(viii) **Vector DNA can be introduced into host organisms by transformation :** Transformation is a process in which host organism can take up DNA from its environment. However, transformation is a relatively inefficient process.

(ix) **Libraries of DNA molecules can be created by cloning :** A DNA library is a population of identical vectors that each contain a different DNA insert. Different types of libraries are made using insert DNA from different sources. To enrich for coding sequences in the library a cDNA library is used. Instead of starting with genomic DNA, mRNA is converted into DNA sequence the process which permits it is called reverse transcription, and is done with a special DNA polymerase (reverse transcriptase) which can make DNA from RNA template. When treated with reverse transcriptase mRNA sequences can be converted into double stranded DNA copies which are known as cDNAs (for copy DNAs or complimentary DNA). These fragments can be ligated into the vector.

(x) **Hybridization can be used to identify a specific clone in a DNA library :** The process by which a labelled DNA probe is used to screen a library is known as **colony hybridization**.

(xi) Use of **Chemically synthesized oligonucleotides :** An oligonucleote is a short, synthetic, single-stranded DNA molecule one of a type which is used as a primer in DNA sequencing or PCR.

(xii) **Polymerase Chain Reaction** with a PCR machine or Thermocycler : Which amplifies DNAs by repeated rounds of DNA replication ***in vitro***. PCR employs the enzyme DNA polymerase which directs the synthesis of DNA from deoxynucleotide substrates on a single stranded DNA template. DNA polymerase synthesize DNA in a 5′ to 3′ end direction and can add nucleotides to the 3′ end of a custom designed oligonucleotide. DNA polymerase can use the oligonucleotide as a primer and elongate it in a 5′ to 3′ direction to generate an extended region of double stranded DNA. It should be noted that to begin with in the first step of the PCR the DNA template is denatured by heating and annealed or attached with synthetic oligonucleotide primers. DNA polymerase is added. It is accompanied heat and repeat. Finally DNA will increase in abundance with each subsequent cycle of the chain reaction.

(xiii) **Nested sets of DNA fragments of** several nucleotide sequences : Nested sets of DNA molecules are created by two ways. In first procedure DNA molecules are radioactively labelled at their 5′ terminal and are then subjected to break preferentially at Gs, Cs, Ts, or As. However, this chemical procedure is no longer in wide use.

The second procedure **chain-terminating nucleotides** are used. The modern automatic sequencing machines called sequenators are based on this technology.

(xiv) **Shotgun sequencing a bacterial genome :** The bacterial species ***Hemophilus influenzae*** was the first organism to have a complete genome sequence and assembly. The genome of this bacterium is composed of 1.8 megabase pairs (mb) of DNA. The ***H. influenzae*** genome was randomly sheared into many random fragments with an average size of 1 kb. These pieces of genomic DNA were cloned into a plasmid recombinant DNA vector. DNA was prepared from individual recombinant DNA colonies and sequenced separately on **Sequenators** employing dideoxy method. This method is known as **shotgun sequencing**.

(xv) Use of shotgun strategy for partial assembly of large genome sequences : Sophisticated computer programmes have been prepared which assemble the short sequences from random shotgun DNAs into larger contiguous sequences known as **contigs**. Short contigs are assembled into larger scaffolds which are typically 1 - 2 Mb long.

(xvi) **Use of paired-end strategy for the assembly of large genome scaffolds :** Special cloning vector known as **(BAC)** or **bacterial artificial chromosome.** The use of BACs is used for the assignment of multiple contigs into a single **scaffold** of several megabases.

(xvii) **Genome - wide analysis :** A variety of **bioinformatics tools** are required to identify genes and know the genetic composition of complex genomes. Now a days computer programmes have been developed which identify protein coding genes through a variety of sequence criteria including the occurrence of extended **open reading frames or ORF** (a series of codons that is or could be a gene), flanked by appropriate 5′ and 3′ splice sites. A limitation of **current gene finder programme** is the failure to identify promoters.

Therefore, most important method for validating predicted protein coding gives and identifying those missed by current gene finder programme is the use of cDNA sequence data cDNAs are produced by reverse transcription from mature mRNAs and thus represent bonafide or real exon sequences. Thus cDNA is used to generate **EST** or **expressed sequence tag,** data, which is a short sequence read from a larger cDNA.

(xviii) **Comparative genome analysis :** A noted finding of comparative genome analysis is the high degree of **syntery** i.e. conservation of genetic linkage between distantly related animals. The most common used genome toôl is **BLAST** or **basic local alignment search tool**. Searching a genome or many genomes, for all the predicted protein sequences which are related to are called **query sequence**.

Q.5. Give the important techniques used for the molecular biological studies of proteins.

Ans. The important techniques involved in the study of molecular biology of proteins are given below :

(i) **Purification from cell extracts :** Purification of proteins is a major part of understanding their function. Purification of protein is designed to exploit its unique characteristic, including size, charge, shape and in many instances, function. The purification of protein requires two major aspects ; First a **specific assay** e.g. **interpretation** assay second, preparations of cell extracts containing active proteins.

(ii) **Using column chromatography :** This includes, (1) techniques of **ion exchange chromatography** on the basis of surface ionic charge using beads. (2) **Gel filtration chromatography** which separates proteins on the basis of size and shape.

(iii) **Affinity chromatography :** This exploits the specific nature of proteins to bind e.g. ATP binding proteins or DNA binding proteins. One of the most common form of protein affinity chromatography is **immunoaffinity chromatography** in which an antibody which is specific for target protein is attached to beads. The bound protein hence can be eluted from the column using salt or in some cases mild detergent. In this some other purposes include recognition and characterization of specific **epitopes** i.e., a sequence of 7 to 10 amino acids which can be attached to any protein. This method permits the modified protein to be purified using the immunoaffinity purification and a heterologous antibody which is specific for the added epitope. Some specific proteins can be precipitated rapidly by attaching the antibody to the same type of bead used in column chromatography, this process is known as **immuno precipitation**.

(iv) **Separation of proteins on polyacrylamide gels :** It is used by using strong ionic detergent **sodium dodecyl sulphate (SDS)** and a reducing agent like mercaptoethanol. After electrophoresis the proteins can be visualized with a stain like **coomassie brilliant blue** which binds to protein.

(v) **Antibodies can be visualized electrophoretically :** Separated proteins : The proteins are altogether different from DNA and RNA but the process has been called **immunoblotting,** by which an individual protein can be seen squintingly amongst thousands of other proteins which is analogous incomplete to southern and northern blot hybridization. Southern, northern and immunoblotting have in common the use of selective reagents to visualize particular molecule in complex mixtures.

(vi) **Sequencing protein molecules :** Sequencing of protein molecules is more complex than sequencing of nucleic acids which can be done using ; (a) **Edman degradation** and (b) **Tandem mass spectrometry (MS/MS)**. Edman degradation is a chemical reaction in which the amino acid residues are sequentially released for the N-terminus of a polypeptide chain. The N-terminal of most amino acids in chain can be modified specifically using a chemical reagent called phenylisothiocyanate (PITC) which possess the potential to modify free a - amino group. The amino acid derivoltized in this way can be easily determined by its elution profile using a column chromatography method called **High Performance Liquid Chromatography** or just HPLC.

Mass spectrometry is a method in which the mass of a very small sample of material can be known on the principle that material travels through an instrument (in a vacuum) in a manner which is sensitive to its mass/charge ratio. The Tandem mass spectrometry or MS/MS has played a wonderful role in protein sequencing and identification.

42
CHAPTER

XENOBIOTICS: THE FOREIGN CHEMICALS IN THE BODY

The term xenobiotic has been derived from Greek word Xenos = foreigner, stranger; and bios, viros = life, plus suffix for adjective (i.e. tic)

A **xenobiotic** is a chemical which is found in an organism but which is normally not produced or expected to be present in the organism. It also covers substances which are present in much higher concentrations than as usual specially drugs like antibiotics, in humans because human body does not produce them itself nor they are part of normal diet.

The term xenobiotic is very often used in context of pollutants such as dioxins and polychlorinated biphenyls and their effect on **biota** because xenobiotics are considered as substances foreign to an entire biological system i.e., artificial substances which did not exist in nature before their synthesis by humans.

Some xenobiotics are resistant to degradation, e.g. they may be synthetic organochlorides as plastics and pesticides.

Due to increased populations and industrial development no only conventional solid and liquid waste pollutants have been produced to alarming levels rather industrial development has produced a range of previously unknown pollution problems for which society was not prepared. It is unfortunate that ingenious industry manufactures have produced many **xenobiotic compounds** which greatly differ in their structure from natural organics.

It has been seen that xenobiotics which are halosubstituted (substituted with chlorine, fluorine and iodine) and nitrosubstituted organics, e.g. propellants, refrigerators solvents, poly-chlorinated biphenyls abbreviated as PCBs, plastics, detergents, explosives and pesticides are completely resistant to biodegradation. Such substances have also been called **recalcitrants.**

There may be some other substance that degrade very slowly in a cometabolic or gratuitous manner in which metabolic transformation of a substance by a microorganisms growing on another substrate take place rather than in a substrate utilization mode. The cometabolised substance is not incorporated into an organism's biomass and the organisms does not derive energy from the transformation of that substance, and such a metabolic transformation of a substance is called **cometabolism.** Moreover, some of the more recalcitrant (totally resistant to microbial attack) xenobiotics are subject to **biomagnification** in the food web which begins with the microorganisms leading to serious damage on the most vulnerable top level carnivores.

As xenobiotics may be viewed as substituted hydrocarbons, the common biodegradation pathways of aliphatic, alicyclic and aromatic hydrocarbons invite the attention of chemists and biochemists that how these biodegradation pathways may explain the structural reasons for recalcitrance of some xenobiotics.

Q.1. What is a **xenobiotic?**

Ans. The scientists recently have coined a new term for foreign chemicals found in the body, which are called **xenobiotics.** The term comes from the words *xenos,* meaning foreign, and *bios,* meaning life. Xenobiotics are found when the body absorbs chemicals that are not nutrients, or when normally occurring substances (nutrients or endogenously produced biochemicals) become denatured and get modified from original molecular structure or are in excess.

In other words, **xenobiotics** are those chemicals, which are normally not produced or expected to be present in an organism, but still they are found in it. It also covers substances, which are present in much higher concentrations than are usual.

It will be worthwhile to add that a **xenobiotic** is a foreign compound to which the body is exposed. The compounds referred to as xenobiotics include, medicinal agents (drugs), industrial chemicals, and environmental chemicals.

Q.2. Discuss implications of xenobiotics.

Ans. Natural compounds can also become xenobiotics if they are taken up by another organism (e.g., uptake of natural hormones excreted from humans by fish downstream of sewage treatment plant outfalls).

The term xenobiotic is also used to refer to organs transplanted from one species to another. For example, some researchers hope that hearts and other organs could be transplanted from animals to humans.

Some authors suggest that we are burdened by as many as 100,000 man-made xenobiotics—and that doesn't take into account the natural ones.

Q.3. What are the major sources of xenobiotics?

Ans. The major sources of xenobiotics are food additives such as colourings, flavours, preservatives, etc. The other sources are fungicides, pesticides, herbicides, antibiotic or hormone residues, various other drugs and environmental pollutants such as smoke produced by automobiles, brick-kilns, industries, crackers and fire works used on **Diwali** and other festive occasions.

Q.4. Mention the diseases that are associated with xenobiotics.

Ans. A considerable number of clinical studies have revealed that xenobiotic chemicals can cause precise and direct toxic effect on human body.

Diseases directly associated with xenobiotics are as under:

(i) **Physical:** Cardiac arrhythmia, eczema, edema, epilepsy, fatigue, headache, hypertension, multiple sclerosis, tinnitus, rheumatism, pain, psoriasis, vasculitis.

(ii) **Psychological:** Autism, aggressive behavior, anxiety, fatigue, insomnia, and organic mental disorders.

Diseases that may be triggered or worsened by food allergies are:

(i) Physical: AIDS, alcoholism, bronchial asthma, constipation, Crohn's disease, dermatitis, diarrhoea, gallbladder disease, glaucoma, hypoglycaemia, irritable bowel syndrome, lupus, obesity, osteoarthritis, Raynaud's syndrome, rheumatiod arthritis, ulcerative colitis, ulcers (duodenal and gastric), and urticaria.

(ii) Psychological: attention deficit disorder, bipolar disorder (manic depression), learning disabilities, schizophrenia.

Q.5. How can we get rid of xenobiotics?

Ans. The body gets rid of xenobiotics by xenobiotic metabolism. This consists of the deactivation and the secretion of xenobiotics, and happens mostly at the liver. Secretion routes are urine, faeces, breath and sweat. An example of a group of enzymes involved in xenobiotic metabolism is the hepatic microsomal cytochrome P450s.

Because of the significant overload of xenobiotics in our modern environment, we should undertake a light detoxification programme.

However, learning the sources of xenobiotics and how to avoid them is even more important. Here are some useful suggestions:

(i) Avoid all artificial colours, flavours, preservatives and sweeteners.

(ii) Eat low on the food chain; the closer a food is to its natural state, the less the toxin risk.

(iii) Eat organically grown food as far as possible. Therefore, **organic farming** should be encouraged.

(iv) Carefully select body care products, avoiding petroleum-based cosmetics and body care products.

(v) Avoid using chemical-based household cleaners, laundry products and air fresheners. Buy toxin-free, biodegradable products made with natural ingredients.

(vi) Drink spring or filtered water.

(vii) Fill your home and work envirnoments with houseplants that absorb toxic gases from the air. Spider plants, Boston ferns, English ivy and Dracaena are easy to grow and provide natural air pollution control.

Biomagnification or Biological Magnification

Q.6. What is **biomagnification** ? Discuss.

Ans. It may be defined as an increase in the concentration of a recalcitrant chemical substance like a pesticide because it is passed to higher members of a food chain.

The organochlorines and some other recalcitrant substances are widely distributed and usually indirectly exposed to the organisms are in very low concentrations, i.e. parts per billion (ppb) range. But these low concentrations even are a matter of concern due to a phenomenon known as **biological magnification** or just **biomagnification** as abbreviation.

It is really very fascinating that aforesaid small amounts in the dissolved form apart from the surrounding water, and being of lipophilic nature become part of lipids of procaryotes and eucaryotes, both.

The concentration of these recalcitrants in the cells of the procaryotic and eucaryotic organisms becomes one to three times more in comparison to surrounding medium. It happens that members of the next higher trophic level ingest these microorganisms. About 10 to 15 per cent of biomass is transferred to higher trophic level and the rest is dispersed in the process of respiration but the persistent lipophilic pollutants are neither degraded nor excreted to a countable amount and thus remain the part of biomass of second trophic level. In this way the magnitude of the recalcitrant goes on increasing in the higher trophic levels represented by birds of prey like kites, vultures, eagles, mammalian carnivores, and large predatory fish by a concentration factor of $10^4 - 10^6$ more than that of the environment (Fig. 42.1).

A biologically active substance like, pesticide may result into serious weakness or even death of the organism. Chlorinated hydrocarbons and DDT have been regarded involved in death and reproductive failure of birds of prey, have reached in the antarctica even.

The chain may be simply represented as pesicide sprayed vegetation → earthworm → robin breast. Certainly it will be destructive to the bird. Some pesticides have been detected even in the milk of the mother. Some countries have, therefore, banned recalcitrants like DDT.

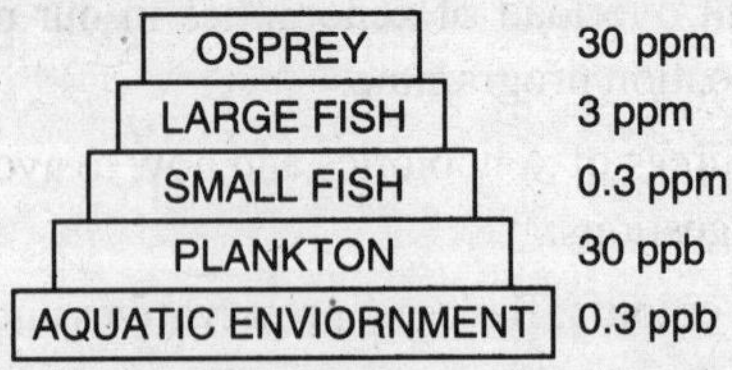

Fig. 42.1. Diagrammatic representation of **biomagnification** of DDT from aquatic environment at the bottom to osprey a large kind of hawk which preys on large fish, at the top of the trophic level.

Q.7. What is cause of xenobiotic organic compound proving recalcitrant?

Ans. The reason is, the unusual substitution of with chlorine, halogen and heavy metals, unusual bonds or bond resuences at tertiary and quarternary carbon atoms, high condensation of aromatic rings and excessive molecular size as in case of polythene and other plastics.

Q.8. Define **biodegradation.**

Ans. This is the process of chemical breakdown or degradation of a substance to smaller units, mediated by microorganisms or their enzymes.

Q.9. What is **mineralization** ? Why was this term coined in place of biodegradation?

Ans. The mineralization is the term used for microbial breakdown of organic materials into inorganic materials mainly due to microorganism.

The term mineralization was required because some times there were reports that DDT an insecticide is degradable as that is converted in parts to DDD, DDE and other closely related compounds but others would argue with the plea that DDT is persistent because the basic carbon skeleton remains unchanged.

Q.10. Which are the main categories of chemical compounds that come under **genobiotic pollutants**?

Ans. The broad categories of **genobiotic pollutants** are:

(i) Recalcitrant hydrocarbons.
(ii) Haloalkyl propellants and solvents.
(iii) Recalcitrant nitroaromatic compounds.
(iv) Polychlorinated biphenyls (PCBs) and dioxins.
(v) Synthetic polymers.
(vi) Alkyl benzyl sulphonates.
(vii) Petroleum hydrocarbons
(viii) Acid mine drainage.

Q.11. What are **recalcitrant halocarbons**?

Ans. The carbon – halogen bond in them is very stable. Cleavage of this bond is not exothermic and requires a substantial energy input as it is an endothermic reaction. There seems to be little prospect for microorganisms ever to evolve a capability for utilization of extensively halogenated carbon compounds as grow material none the less some cometabolism and photodegradation may occur in them. Prominent groups in them are $C_1 - C_2$ haloalkyl propellants, solvents and refrigerants, haloaromatics like chlorobenzenes, chlorophenols and chlorobenzoates, polychlorinated or polybrominated biphenyls and triphenyls, chlorodibenzodioxins and chlorodibenzofurans. Besides them organochlorine insecticides have also been found to be recalcitrants.

Q.12. What are haloalkyl propellants and solvents?

Ans. They are $C_1 - C_2$ alkanes in which all or nearly all hydrogen atoms are replaced by fluorine – chlorine combinations. They are used as solvents and aerosol propellants in spray cans, in cosmetics, paints and insecticides and as the working fluid in the condensor units of airconditioners and refrigerators. They include CCl_3F and CCl_2F_2 freons, designated by the codes F – 11 and F – 12 respectively. The others that may be added to the list are dichloromethane, chloroform, carbon tetrachloride, dischloro-trichloro and tetrachloroethenes (or ethylenes) which are important industrial cleaning solvents. Some of these have also been used in packing syringes and surgical instruments.

Spilled halocarbon propellants and solvents have greatly contaminated ground water. Being inert and quite volatile some of these rise to the atmosphere's part called **stratosphere** or the air between about 10 and 60 km above the earth's surface, on release, and resulting depletion of earth's protective ozone layer through photochemical interactions.

Depletion of the ozone layer will result in increased UV – radiations on the surface of the earth resulting skin cancer and mutagenesis.

It has been revealed that sulphate reducing bacteria transform tetrachloroethane to trichloroe thene and cis – 1, 2-dichloroethene by anaerobic dehalogenation. In an another example aerobically dichloromethane was shown to serve as the only carbon source for pseudomonads and dichloroethane is known to be acted upon as a substrate by a microbial consortium. *Methylococcus capsulatus* is known to act on Chloro – and bromomethane to convert it to formaldehyde, and convert dichloromethane to CO and trichloromethane to CO_2. To conclude, it can be said that methanotrophic bacteria bear hope for **bioremediation** (a biological treatment to biodegradable pollutants) of halocarbon polluted aquifers, the geological formations like subsurface water bodies which supply water for wells and springs.

Q.13. Give an introduction to halobenzenes, halophenols and halobenzoates.

Ans. Chlorobenzenes comprise major industrial solvents. Their aerobic bidegrability declines with the number of halosubstituents but the same halogenated aromatics are dechlorinated anaerobically.

On the other hand *Pseudomonas* and *Alcaligenes* use the dioxygenase to produce chlorocatechols from aerobic biodegradation of trichloro and tetrachlorobenzenes.

The highly chlorinated chlorophenols as pentachlorophenol (PCP) are used as preservatives for wood and canvas. Some chlorophenols are used to make pesticides. A hope has been generated that PCP is converted aerobically by a monooxygenase to tetrachlorohydro-quinone through the oxidative elimination of the chlorine para to phenolic hydroxyl, through a species of *Flavobacterium* and *Mycobacterium chlorophenoticum*.

Dechlorination step by step, to 2, 5 dichlorohydroquinone is followed by ring opening, further, through anaerobic conditions PCP is reductively dechlorinated step by step to phenol, and by further anaerobic treatment of phenol to methane and carbon dioxide.

Q.14. How can halobenzoates be degraded?

Ans. Chlorophenols and chlorobenzoates can be degraded anaerobically by complex methanogenic or sulphidogenic microbial consortia in the presence of additional carbon sources, however, dechlorinations are quite substrate specific.

Desulfomonile tiedjei which is strictly anaerobic rod-shaped sulfidogen removes chlorosubstituents from the meta position of chlorobenzoates. It also dehalogenates highly chlorinated phenols and PCE.

Q.15. Can polychlorinated biphenyls be degraded (PCBs)?

Ans. The PCBs are mixtures of biphenyls with 1 to 10 chlorine atoms per molecule. Their structure is almost like DDT and are also known to accumulate in higher trophic levels.

Although PCBs are relatively resistant to biodegradation yet some microorganisms have be isolated which can transform them. PCBs can be biodegraded aerobically using white rot fungus

Phanerochaete, by *Acinetobacter* and by Alkaligenes and anaerobically by reductive dehalogenation. The degradation of PCBs is by cometabolism and can be increased by adding less chlorinated analogs like dichlorophenyl.

Q.16. Can synthetic polymers be degraded?

Ans. Synthetic polymers are xenobiotics without photodamage in them only the accessible starch granules are degraded to some porosity and brittleness no degradation beyond that has been possible so far. However, efforts are in progress by producing truly biodegradable poly-β-hydroxyalkanoates produced by *Alkaligenes eutrophus, Bacillus cereus* and some pseudomonads. The intracellular storage products bear thermoplastic properties, therefore, can be moulded. But the price of such a product will be 5 to 7 times more than the petrochemical based polyethylene.

Q.17. What are alkyl benzyl sulphonates (ABS)?

Ans. They are main components of anionic detergents. They are surface active with polar sulphate and nonpolar alkyl end in their molecule. Due to their characteristic of emulsification of fatty substances and thereby cleaning occurs while these molecules make a monolayer around lipophilic droplets or particles. Their molecules orient with their nonpolar end towards the lipophilic substance and the sulphonate end towards water. **Non linear alkylbenzyl sulphonates** are recalcitrants and resistant to biodegradation and cause foaming in the rivers in plenty.

Though ABS are easier to manufacture and bear superior detergent properties but the methyl branching of alkyl chain in them interferes with biodegradation of ABS. It is specifically because tertiary carbon atom block the normal β oxidation sequence in the molecule. That is why detergent industry turned from ABS to **LAS (linear alkylbenzyl sulphonates),** which are biodegradable.

Petroleum Hydrocarbons

Q.18. Are petroleum hydrocarbons pollutants?

Ans. Most xenobiotic pollutants are substituted or modified hydrocarbons. Their short chains are toxic to organisms but they evaporate quickly from the oil slicks. However, very long chain alkanes are resistant to biodegradation. Increasing chain length makes alkanes to exceed molecular weight of 500 and consequently alkanes stop to serve as carbon source.

Moreover, branching often reduces the rate of biodegradation because tertiary and quaternary carbon atoms interfere with degradation or at all do not degrade. Aromatic compounds particularly of condensed polymer type are degraded at a much slow rate in comparison to alkanes.

The alicyclic compound frequently do not serve as the sole carbon source for microbial growth until and unless they are provided with long aliphatic side chain but they cannot be degraded through cometabolism using 2 or more strains that exhibit cooperative mechanisms because of their complementary metabolic ability.

Biodegradation of Petroleum Hydrocarbons

Q.19. What are **hydrocarbons**?

Ans. The hydrocarbons are the substances formed of hydrogen and carbon or the compounds containing carbon and hydrogen only, e.g. benzene, paraffin and coal gas. They occur in petroleum. Therefore, hydrocarbons have also been called petroleum hydrocarbons.

Q.20. Can some types of hydrocarbons be used as substrates for growth by various microorganisms?

Ans. Yes, some type of hydrocarbons can serve as growth substrates. This type of organisms are ecologically important, e.g in the degradation of petroleum pollutants and some may be utilized in commercial production of **single cell protein (SCP)** from hydrocarbons. However, their draw back lies in the fact that certain hydrocarbon utilizing organisms can cause spoilage of hydrocarbon products such as fuels. Hydrocarbon metabolism is strictly aerobic and appears always to involve the introduction of oxygen into the molecule in a process requiring a monooxygenase (= hydroxylase) or dioxygenase.

Aliphatic hydrocarbons which are straight chain paraffins (n-alkanes) can be utilized by bacteria like strains of *Acinetobacter, Corynebacterium, Mycobacterium, Nocardia, Pseudomonas,* by yeasts as species of *Candida* and the mycelial fungi *Aspergillus, Botrytis, Fusarium, Helminthosporium, Hormoconis,* and *Penicillium*. Some organisms can use only short-chain alkanes, some can use only long-chain alkanes. However, in most cases n-alkane metabolism appears to occur by co-oxidation (is read as omega oxidation), i.e. terminal methyl group of the alkane is oxidized by a monooxygenase to form a primary alcohol which is in turn oxidised via the aldehyde to the corresponding fatty acid by alcohol dehydrogenase and aldehyde dehydrogenase activities; the fatty acids then can be degraded by conventional β-oxidation.

In *Pseudomonas* strains the alkane-oxidising enzyme system is complex, involving ϖ-monooxygenase a rubredoxin, and an NADH-rubredoxin oxidoreductase. In eucaryotes and probably in some bacteria like *Acinetobacter* strains H_2 alkane monooxygenase is linked to the cytochrome P–450 electron carrier systems. **Branched-Chain alkanes** (alkylalkanes) and unsaturated hydrocarbons (alkenes, olefins) are generally somewhat less susceptible to microbial degradation in comparison to n-alkanes. They may be oxidised in the same way as n-alkanes but alkenes may not be oxidised at the double bond resulting in the formation of a diol.

Since alkane metabolism occurs intracellularly, the hydrocarbon must be taken into the cell. Uptake may take place by different mechanism in different organisms and at least in some cases may need prior emulsification of the hydrocarbon by an extracellular **biosurfactant** or **bioemulsifier.**

Alicyclic hydrocarbons (Cycloparaffins and cycloalkanes) are cyclic, non-aromatic hydrocarbons. They are cyclic, non-aromatic hydrocarbons. They are generally less susceptible to microbial attack in comparison to either aliphatic or aromatic compounds. Degradation of cyclohexane by strains of *Nocardia* or *Pseudomonas* involves oxidation of the cyclohexane to cyclohexanol by a cyclohexane monoxygenase. Cyclohexanol is oxidised to cyclohexane and then an oxygen atom is introduced into the ring (forming a lactone) by cyclohexanone monoxygenase. The lactone, then can be hydrolysed to form a non cyclic dicarboxylic acid.

Aromatic hydrocarbons (benzene, naphthalene, anthracene etc.) are present in the petroleum and are formed by incomplete combustion of almost any organic material. They are thus common pollutants and many are recognised carcinogens. *Pseudomonas* spp of bacteria metabolise aromalic hydrocarbons by initially incorporating two atoms of oxygen into the substrate to form a *cis*-dihydrodiol; and this reaction catalysed by a multicomponent enzyme system comprising of dioxygenase, a flavoprotein and iron sulphur proteins. The *cis*-dihydrodiol is oxidized to a catechol which is in turn a substrate for another dioxygenase system that breaks open aromatic ring. The fungi, in contrast, oxidize aromatic hydrocarbons using a cytochrome P – 450 dependent monoxygenase to form a reactive arene oxide. This is turn can either undergo isomerization to form a monohydric phenol, or may be hydrolysed enzymatically to result into a trans-dihydrodiol.

43 CHAPTER

BIOCHEMISTRY OF THE MOLECULES OF LIFE

Q.1. What are the four elements that make up over 98% of all living matter? Also name the other two elements which are next to them in importance.

Ans. The four major elements are carbon, oxygen, hydrogen and nitrogen. The other two elements next to them in importance are: phosphorus and sulphur.

Q.2. What is the importance of weak bonds?

Ans. The weak bonds make important biological systems and determine the very important properties of the molecules and are responsible for their proper functioning,

Q.3. What are macromolecules?

Ans. The macromolecules consist of many repeating subunits. Each subunit consists of a small simple molecule.

Q.4. What is an element? Name their three components.

Ans. An element is a pure substance comprising a single type of atoms. The three components of an atom are electron, proton and neutron.

Q.5. What are strong bonds?

Ans. The strong bonds are often covalent bonds formed when atoms share electrons to fill in their outer shell achieving thereby maximum stability. The stronger the bond, the more energy is needed to break it.

Q.6. Discuss whether the hydrogen bonds are weak or strong.

Ans. The hydrogen bonds are weak but biologically they are very important. The two strands of DNA are held together by hydrogen bonds. The hydrogen bonds also determine the shape of proteins.

Q.7. What are monomers? Give three examples of macromolecules and their monomers.

Ans. The small repeating subunits which make large or macromolecules are called monomers. The three examples of macromolecules and their monomers are as follows:

(i) The proteins, which are chains of aminoacids that form polypeptides.

(ii) The polysaccharides, which are the chains of monosaccharides which form branching structures.

(iii) The nucleic acids DNA and RNA each of which is made up of nucleotides.

Q.8. Which are two major groups of amino acids?

Ans. The carboxyl group and the amino group are two groups bonded to the same carbon atom.

Q.9. How many aminoacids are there in a protein?

Ans. There are twenty aminoacids each differing in its side chains.

Q.10. How are **peptide bonds** synthesized?

Ans. The peptide bonds are formed by the formation of a covalent bond between the amino group of one aminoacid and the carboxyl group of another aminoacid, accompanied by removal of HOH which is called dehydration synthesis.

Q.11. Give two examples of **substitute proteins.**

Ans. They are **glycoproteins and lipoproteins** which contain covalently bonded molecules, sugars and lipids other than aminoacids.

Q.12. What are polysaccharides? Give three categories of carbohydrates.

Ans. The polysaccharides are polymers of monosaccharides. The three categories of carbohydrates are monosaccharides (pentoses and hexoses), disaccharides, and polysaccharides.

Q.13. Name the two nucleic acids. What are their monomers (subunits)?

Ans. They are:

(i) Deoxyribonucleic acid (DNA)

(ii) Ribonucleic acid (RNA).

Their monomers (subunits) are nucleotides which are unbranched.

Q.14. Give composition of nucleotides of DNA.

Ans. The nucleotides of DNA are made up of three units: These are (1) a **nitrogen base** called **purine** made up of adenine or guanine or **pyrimidine** (thymine or cytosine), (2) **deoxyribose** (bonded covalently), and (3) a **phosphate molecule.**

Q.15. How are double helix strands of DNA occurring in a cell joined?

Ans. The two strands in a helix are joined together by hydrogen bonding between adenine and thymine and between guanine and cytosine.

Q.16. How does RNA differ from DNA?

Ans. The nucleotides in ribonucleic acid are same as in DNA except that ribose replaces deoxyribose and there is uracil in place of thymine. There are three different types of RNA which are found in the cell, e.g., rRNA, mRNA and tRNA.

Q.17. Are lipids macromolecules? Are they biologically important?

Ans. No, the lipids are too small and heterogeneous that cannot be considered as macromolecules. They are of great-biological significance.

Q.18. Give general characteristics of lipids.

Ans. Important general characteristics of lipids are as follows:

(i) The lipids are insoluble in water but soluble inorganic solvents. This difference in their nature for solubility is due to nonpolar hydrophobic nature.

(ii) The lipids are not made up of similar subunits.

(iii) Some of the simple lipids which are made up of C, H and O include fats and steroids. The fat is made up of glycerol covalently bonded to fatty acids. The steroids consist of four-membered ring structure.

(iv) The compound lipids consist of fatty acids and glycerol and usually have elements other than C, H and O, e.g., phospholipids, lipoproteins and lipopolysaccharides. They are of great importance in the cell envelope of bacteria.

(v) The phospholipids are of two parts. The polar end is soluble in water while the nonpolar end containing C and H is insoluble in water but soluble in organic solvents.

Q.19. Fill in the blanks with appropriate terms.

(i) Protein molecules are in three-dimensional shape due to hydrogen bonds ———— and bridges.

(ii) A hydrogen atom covalently bonding to an oxygen or nitrogen atom is a weak bond known as ———.

(iii) A strong bond, as a result of sharing of electrons between atoms is called ———.

(iv) ——— are the lipids which are generally liquid at room temperature.

(v) The process by which polymers are made from monomers is called ———.

(vi) The negatively charged particles which take part in bonding of atoms are ———.

(vii) The membrane-molecules bearing hydrophilic and hydrophobic parts are called ———.

(viii) The example of a disaccharide is ———.

(ix) The monomers or building blocks of nucleic acids are ———.

(x) The particles found in the nucleus of an atom that do not have a charge are known as ———.

(xi) The proteins, polysaccharides and nucleic acids are the large molecules referred to as ———.

(xii) The bonds that involve the complete transfer of electrons from one atom to another are called ———.

(xiii) ——— are formed as a result of combination of fatty acids and glycerol and are solid at room temperature.

(xiv) The large molecules formed as a result of joining together small molecules of same type or monomers, are referred to as ———.

(xv) ——— is the principal sterol of fungi and yeast and is classed as mycosterol.

Ans. (i) disulphide, (ii) hydrogen bond, (iii) covalent bond, (iv) oils, (v) dehydration synthesis, (vi) electrons, (vii) phospholipids, (viii) sucrose, (ix) nucleotides, (x) neutrons, (xi) macromolecules, (xii) ionic, (xiii) fats, (xiv) polymers, (xv) ergosterol

Q.20. Choose the correct answers for the following.

(i) Which of the following bear electrical charge?

(a) Protons (b) Electrons
(c) Ions (d) Neutrons
(e) Atoms

(ii) There is a single positive charge on the sodium ion (Na) because:

(a) It possesses more protons in its nucleus than the electrons in the orbit.
(b) It possesses more neutrons in the nucleus than the electrons in the orbit.
(c) It possesses more electrons in the nucleus than the neutrons in the orbit.

(iii) Number of protons in the nucleus of an atom determines:

(a) atomic weight (b) valency
(c) molecular weight (d) atomic number
(e) number of electron orbitals.

(iv) Covalently bonding of the molecules of amino acids forms.

(a) nucleic acids (b) lipids
(c) proteins (d) disaccharides
(e) polysaccharides

(v) Two strands of DNA are held together by weak bonds of:

(a) nitrogen bonds (b) ionic bonds
(c) disulphide bonds (d) hydrogen bonds

(vi) The number and the sequence of amino acids reveal:
 (a) Tertiary structure of protein
 (b) Primary structure of protein
 (c) Secondary structure of the protein
 (d) Primary structure of polysaccharide

(vii) The monomers join to form macromolecules by process of:
 (a) Hydrogen bonding (b) dehydration synthesis
 (c) hydrolysis (d) dehydrolysis

(viii) Peptide bonds are formed in:
 (a) proteins (b) nucleic acids
 (c) lipids (d) polysaccharides

(ix) Ester linkages are found in:
 (a) polysaccharides (b) proteins
 (c) nucleic acids (d) lipids

(x) Which of the functional groups are found in aminoacids?
 (a) COOH (b) NH_2
 (c) R group (d) fatty acid

(xi) Lactose is a:
 (a) protein (b) disaccharide
 (c) DNA (d) monosaccharide
 (e) polysaccharide

(xii) In carbohydrates ratio of carbon to hydrogen and oxygen is:
 (a) 1 : 2: 1 (b) 1 : 2 : 2
 (c) 2 : 1 : 4 (d) 1 :3 :3

(xiii) DNA differs from RNA because DNA possesses
 (a) Two strands, sugar deoxyribose and thymine
 (b) One strand, sugar ribose and uracil

(xiv) Lipids with unsaturated fatty acids generally are:
 (a) solid (b) liquid

(xv) There are alternating units of one of the following in the backbone of the molecules of RNA.
 (a) deoxyribose and ribose (b) uracil and ribose
 (c) ribose and phosphate (d) purine and pyrimidine

Ans. (i) a, b, c (ii) a (iii) d (iv) c
(v) d (vi) b (vii) b (viii) a
(ix) d (x) a, b, c (xi) b (xii) a
(xiii) a (xiv) b (xv) c

Q.21. How many pairs of electrons are shared by carbon and oxygen in the structure C = 0?

Ans. Two pair.

Q.22. Is it right that phospholipids have a polar and nonpolar end and occur in the membrane of the cell?

Ans. Yes, it is right.

Q.23. Name the type of bond between the oxygen and the hydrogen atoms in a molecule of water.

Ans. It is a polar covalent bond.

Q.24. Match the following.

(A)	Glycerol is found in	(a)	Applicable to all A to E
(B)	Possess alternate sugar and phosphate bonds	(b)	Proteins
(C)	Consist of CH_2O units	(c)	Nucleic Acids
(D)	Have COOH and NH_2 groups	(d)	Lipids
(E)	Possess C, H, and O	(e)	Carbohydrates

Ans. [A – d], [B – c], [C – e], [D – b], [E – a].

Q.25. How do bacteria detect and move towards nutrients in their environment?

Ans. The movement of bacteria towards some specific substances and away from others is known as **chemotaxis.** In 1960s Julius Alder began to investigate the molecular basis of bacterial chemotaxis. The chemotaxis begins with the detection of chemicals by specific chemoreceptors on the surface of cell. Information from these sensors is transmitted to a processing system which analyzes and integrates many stimuli. This sensory processing system consequently sends signals to the motors that drive the flagella. These signals determine whether a bacterium continues to swim smoothly in a straight line or changes its course all of a sudden. Approximately twenty different chemoreceptors have been detected in *Escherichia coli.* Each of these proteins is located either in the plasma membrane or in the periplasmic space. The galactose binding protein which is a soluble protein in the periplasmic space serves as a recognition protein for positive chemotaxis to galactose in addition to being part of the pump that actively transports it into the cell. The chemoreceptor for glucose is a component of membrane-bound phosphotransferase system for the active uptake of sugars. The surface of the bacterium *E. coli* also contains chemoreceptors for attractants like serine, cysteine, alanine and glycine. Some of the typical examples of chemoreceptors for negative chemotaxis are fatty acids, alcohols, hydrophobic aminoacids, indole, H^+ (pH < 6.5), OH^- (pH > 7.5) and sulphides.

Q.26. Name two neurotransmitters.

Ans. Acetylcholine is a transmitter at motor end plates (neuromuscular junctions) which are junctions between nerve and striated muscle. The γ -aminobutyrate, also known as gamma-aminobutyric acid or GABA is also a neurotransmitter.

44

CHAPTER

CARBOHYDRATES OR "HYDRATES" OF CARBON OR SACCHARIDES

Q.1. Which are the most abundant **biomolecules** on the earth? Elaborate.

Ans. The carbohydrates are the most abundant **biomolecules**, the organic compounds which generally occur as essential component of living organisms.

The **carbohydrates** are polyhydroxy aldehydes or ketones or substances which yield compounds of this types on their hydrolysis. They are said to be hydrates of carbon in which the ratio of C : H : O is 1 : 2 : 1. Many of the carbohydrates are according to empirical formula $(CH_2O)_n$ while others are not. Some carbohydrates also contain, nitrogen, phosphorus and sulphur.

There are three main classes of carbohydrates. These are **monosaccharides, oligosaccharides** and **polysaccharides.** The term saccharide has been derived from the Greek word, *sakkharon* = sugar.

Q.2. Fill in the blanks in the following

(i) **Molisch test** a very sensitive general reaction for carbohydrates, is performed by mixing a few drops of alcoholic solution of α- napthol with the sugar solution and stratifying the mixture over concentrated sulphuric acid. A ______ or ______ coloured ring appears at the junction of liquids.

(ii) A dilute solution of orcinol in 30 percent hydrochloric acid containing a little feric chloride constitutes ______ ______ reagent. Pentoses give a green colour as do aldohaxuronic acids.

(iii) In **phloroglucinol - hydrochloric acid test** a ______ ______ colour appears.

(iv) Naphthoresorcinol test is used for detection of common uronic acids such as glucuronic, galacturonic and mannuronic. The sugar is dissolved in 1N hydrochloric acid having a small amount of naphthoresorcinol and heated for five minutes in boiling water bath. The mixture is cooled and extracted with ether. The ether solution becomes ______, ______ or ______ coloured when the test is positive.

(v) Barfoed's test is used to distinguish monosaccharides from reducing ______.

(vi) Reducing disaccharides reduce ______, ______, and ______ reagents.

(vii) Moore's test consist of heating the carbohydrate with dilute ______ a yellow to brown or reddish brown colour is produced if the carbohydrate contains a free sugar group (hemiacetal group).

(viii) Seliwanoff's test (resorcinol hydrochloric acid test) is given by ______. When heated with reagent (resorcinol in hydrochloric acid) ketohexoses give red colour.

Ans. (i) Red or violet (ii) Bial's test (iii) Cherry red (iv) Purple, (violet or pink)
(v) Disaccharides (vi) Fehling, Benedict and Shaffer – Hartman
(vii) Alkali (viii) Ketohexoses.

Classification of Carbohydrates

Q.3. Give the classification of carbohydrates

Ans. The carbohydrates are generally classified into 3 groups which are as follows:

1. **Monosaccharides**

The monosaccharides are known as simple sugars which possess free aldehyde (– CHO) or ketone (= CO) group and 2 or more hydroxyl (– OH) groups. Their general formula is $C_n (H_2O)_n$ or $C_n H_{2n} O_n$.

Table 44.1. Monosaccharides with 3 to 7 carbon atoms.

Name of Sugar	Formula
Trioses	$C_3H_6O_3$
Tetroses	$C_4H_8O_4$
Pentoses	$C_5H_{10}O_5$
Hexoses	$C_6H_{12}O_6$
Heptoses	$C_7H_{14}O_7$

The **hexoses** which include aldohexose D-glucose and the ketohexose D-fructose are the most common monosaccharides in nature. The aldopentoses D-ribose and 2-doxy D-ribose are components of nucleotides. A nucleotide in the nucleoside phosphosylated at one of its pentose hydroxyl groups.

The monosaccharides are referred to as **aldoses** or **ketoses** on the basis of **aldehyde** and **ketone** group that they have. The examples of two families of monosaccharides are given in table 44.2.

Table 44.2. Common aldoses and ketoses.

Aldoses	Ketoses
Alda sugars	Ketosugars
Glycerose	Dihydroxyacetone
Erythrose	Erythrulose
Ribose	Ribulose
Glucose	Fructose
Glucoheptose	Sodoheptulose

2. **Oligosaccharides** (oligo = few). These are the compound sugars which produce 2 to 10 molecules of the same or different monosaccharides on hydrolysis

Disaccharides – Sucrose, Lactose, maltose, Cellobiose, Trehalose, Gentiobiose, Melibiose

Trisaccharides – Rhamninose, Gentianose, Raffinose (= Melitose), Rabinose, Melezitose

Tetrasaccharides – Stachyose, Scorodose

Pentasaccharide – Verbascose

3. **Polysaccharides** (poly = many). These are also compound sugars which bear more than 10 molecules of monosaccharides on hydrolysis.

Their general formula is $(C_6H_{10}O_5)$. Some common examples are:

Homopolysaccharides – Starch, Glycogen, Inulin, Cellulose, Pectin, Chitin

Heteropolysaccharides – Specific soluble sugars of pneumococcus type III, Hyaluronic acid, Chondrotin.

Optical Property

Q.4. What are optical isomers?

Ans. **Optical isomers** (enantiomers) differ from each other in the disposition of the various atoms or groups of atoms in space round the asymmetric carbon atom. These are infact the mirror images of each other. For example the glyceraldehyde has only one asymmetric carbon atom (numbered as 2) and it can therefore, exist in 2 isomeric forms:

D-glyceraldehyde L- glyceraldehyde

One form in which H atom at carbon 2 is projected to the left side and OH group to the right is designated as **D-form** and the form where H atom is projected to the right side and OH group to the left is called as **L-form**. Similarly, a compound having 2 asymmetric carbon atoms (e.g., a tetrose).

The number of possible isomers of any given compound thus depend, upon the number of asymmetric carbon atom present in the molecule. According to the rule of n (also called as **Le-Bel-Van't H off rule**) the total number of isomers of a compound will be equal to 2^n where n represent the number of asymmetric carbon atoms present in the molecule.

Q. 5. Give salient features of the carbohydrates.

Ans. The outstanding features or characteristics of the carbohydrates are listed below:

1. Three major classes of carbohydrates are monosaccharides, oligosaccharides and polysaccharides.
2. The simplest of carbohydrates are monosaccharides from which disaccharides, olgosaccharides and polysaccharides are formed.
3. **Monosaccharides** are colourless, crystalline solids which are freely soluble in water and insoluble in nonpolar solvents.
4. Two families of monosaccharides are aldoses and ketoses.
5. Monosaccharides except dihydroxyacetone have asymmetric centres since they contain one or more asymmetric or **chiral carbon atom** and hence exist in optically active isomeric forms. A simple example is of the aldose, glyceraldehyde, having a **chiral centre** (the middle carbon atom) hence possess two different optical isomers or **enantiomers** (**Fig. 44.1.**), shown by ball and stick model. It can also be represented by **Fischer projection formulas** and **perspective formulas.**

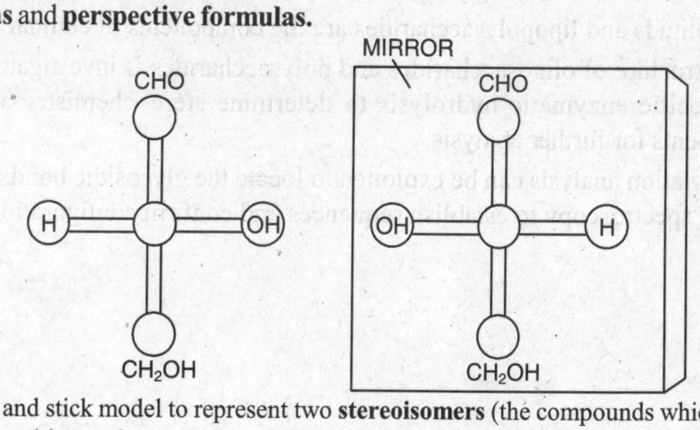

Fig. 44.1. Ball and stick model to represent two **stereoisomers** (the compounds which have the same composition and same order of atomic connections but different molecular arrangements) of glyceraldehyde which are mirror images of each other.

Three dimentional sugar structure on paper can be represented by Fischer projection formula as below.

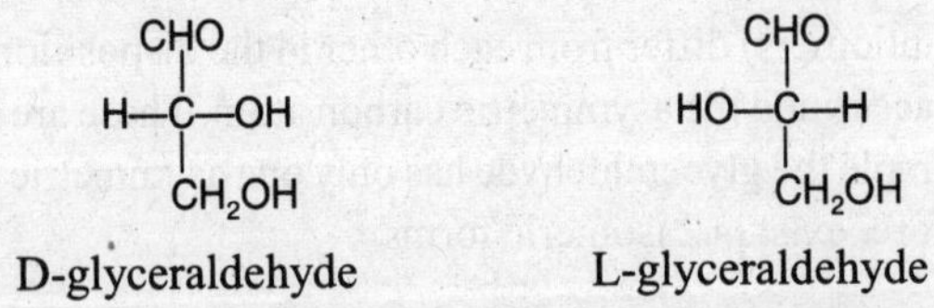

Fig. 44.2. Presentation of stereoisomers of glyceraldehyde by Fischer projection formulas.

6. The common monosaccharides occur in cyclic forms. Infact the monosaccharides with five or more carbon atoms in the backbone usually occur in aqueous solution as **cyclic or ring structures.**
7. **Haworth perspective formulas** are generally used to show the ring forms of monosaccharides. The six membered pyranose ring is not really planar, as Haworth perspectives suggest but tends to assume either the boat or chair conformation. However, two forms of molecules are interconvertible without the breakage of any bonds. But two configurations can be interconverted only by breaking a covalent bond–in the case of α and β configurations the bonds involving the pyranose oxygen atom. Three dimentional conformations of the monosaccharide units are important in determining the biological properties and functions of some polysaccharides.
8. A variety of hexose derivatives occur in organisms.
9. Simple monosaccharides are reducing agents.
10. Disaccharides contain a glycosidic bond.
11. Most of the carbohydrates found in nature occur as polysaccharides or the polymers of high molecular weight. Polysaccharides are also known as **glycans.**
12. The polysaccharides may be **homopolysaccharides** having single type of monomeric units and **heteropolysaccharides** having two or more different kinds of monomeric units.
13. Important storage polysaccharides in nature are **starch** in plant cells, and **glycogen** in animal cells.
14. Cellulose and chitin are structural homopolysaccharides.
15. Bacterial cell wall has heteropolysaccharide.
16. Glycosaminoglycans and proteoglycans are components of extracellular matrix.
17. Most proteins secreted by eucaryotic cells are glycoproteins.
18. The oligosaccharides of glycoproteins have biological functions.
19. Glycolipids and lipopolysaccharides are the components of cellular membranes.
20. The structure of oligosaccharides and polysaccharides is investigated by a combination of specific enzymatic hydrolysis to determine stereochemistry and produce simple fragments for further analysis.
21. Methylation analysis can be exploited to locate the glycosidic bonds and high resolution NMR spectroscopy to establish sequences and confirm configurations.

45 CHAPTER

LIPIDS

Q.1. What are **lipids?**

Ans. The **lipids** are water-insoluble biomolecules generally containing fatty acids, sterols or isoprenoid compounds. They show diverse biological functions. The fats and oils are major stored forms of energy in many organisms and phospholipids and sterols make up about half the mass of biological membranes.

Other lipids are present comparatively in small quantities play a very useful role as cofactors of enzymes, carriers of elections, light absorbing pigments, hydrophobic anchors, emulsifying agents, hormones and intracellular messengers. Thus the lipids play a wonderful role in cell structure and functions.

Q.2. Give the general characteristics of biological lipids.

Ans. The chemical, structural and physical characteristics or properties of the biological lipids are as under:

(i) The fats and oils are stored forms of energy in living organisms and are highly reduced compounds.

(ii) The complete oxidation of fatty acids into CO_2 and H_2O in cells is highly **exergonic** or energy producing in the same way as the explosion of fuels in the internal combustion engines.

(iii) The fatty acids are hydrocarbon derivatives or carboxylic acids with hydrocarbon chains of 4 to 36 carbons.

(iv) The nonpolar hydrocarbon chain accounts for the poor solubility of fatty acids in water.

(v) **Triacylglycerols** also called **triglycerides, fats** or **neutral fats** are fatty acid esters of glycerol.

(vi) Triacylglycerols provide stored energy and insulation.

(vii) Most of the natural foods contain triacylglycerols.

(viii) The ester linkages of triacylglycerols are susceptible to hydrolysis by either acid or alkali Na^+ and K^+ salts of fatty acids are known as soaps. The **practical application of the soaps** is for their ability to solubilize or disperse water insoluble materials by forming microscopic aggregates called **micelles.**

(ix) **Biological waxes** are esters of long chain saturated and unsaturated fatty acids which serve as energy stores and water impermeable coatings.

(x) The lipids form the structure of biological membranes. The biological membranes comprise of a double layer of lipids serving a barrier to the passage of polar molecules and ions.

There are 3 general types of membrane lipids, which are: (1) **glycerophospholipids** in which hydrophobic regions are composed of two fatty acids joined to glycerol, (2) **sphingolipids** in which a single fatty acid is joined to a fatty amine, and (3) **sterols** or the compounds characterized by a rigid system of four fused hydrocarbon rings.

(xi) Some animals and unicellular organisms are rich in ether lipids.

(xii) **Sphingolipids,** the second largest class of membrans lipids, are derivatives of sphingosine.

(xiii) Sphingolipids are the sights of biological recognition.

(xiv) specific **phospholipases** degrade membrane phospholipids.

(xv) **Sterols** have four fused hydrocarbon rings.

(xvi) **Amphipathic lipids** or the lipids containing both polar and nonpolar domains, form aggregates.

(xvii) The steroid hormones carry messages between tissues. Major groups of steroid hormones are the male and female sex hormones and the hormones of the adrenal cortex, cortisol and aldosterone.

(xviii) The hydrolysis of phosphatidylinositol produces intracellular messengers.

(xix) **Ericosanoids** which are fatty acid like devivatives with a variety of extremely potent hormone like actions on various tissues of vertebrate animals, are potent biological effectors.

(xx) Vitamins A, D, E and K are fat soluble

(xxi) Liquid quinones carry electrons.

(xxii) Dolichols form activated hydrophobic sugar derivatives.

(xxiii) Lipid extraction requires organic solvents.

(xxiv) Adsorption chromatography separates lipids of different polarity.

(xxv) Gas-liquid chromatography resolves mixtures of volatile lipid derivatives.

(xxvi) Specific hydrolysis is useful in determination of lipid structure.

Break Down of Lipids

Q.3. How do phospholipases degrade membrane phospholipids?

Ans. There is a specific hydrolytic enzyme for each of the bonds in a glycerophospholipid. The **phospholipases** that have been characterized are phospholipase A_1, phospholipase A_2, phospholipase C and phospholipase D. The pholspholipases of A type remove one of the two fatty acids producing a lysophospholipid. These estrases do not attack the ether link **plasmalogens**, the phospholipids with an alkenyl ether substituent on the C–1 of glycerol. The lysophospholipases remove the remaining fatty acid.

The break down of phospholipid is part of at least two signaling processes in animal cells. The **extracellular signals,** certain hormones for example, activate a phospholipase C which specifically cut phosphatidylinositols, releasing diacylglycerol and inositol phosphates, that work as intracellular signals. The other **extracellular stimuli** activate a phospholipase A, which releases arachidonic acid from membrane lipids. Arachidonate works as a precursor in the synthesis of one of the eicosanoids which act as **intracellular messengers.**

46

CHAPTER

PROTEINS: STRUCTURE AND FUNCTION

Q.1. What are proteins?

Ans. A protein is a macromolecule composed of one or more polypeptide chains, each with a characteristic sequence of aminoacids linked by **peptide bonds.**

The cells of organisms contain thousands of different proteins each with a different function or biological activity. These functions include enzymatic activity, molecular transport, nutrition, cell or organismal motility, structural roles, defense regulation etc. The proteins consists of very long polypeptide chains with 100 to 2000 amino acid residues joined by **peptide linkages.** However, all proteins are made from the same set of 20 amino acids.

The genetic information of a cell is ultimately expressed as protein. There is a segment of DNA (a **gene**) for each protein which encodes information specifying its sequence in amino acids.

Bioluminescence

Q.2. How do fireflies produce light in darkness? Also give some other examples of bioluminescence.

Ans. The production of light by a firefly is as a result of light producing reaction involving **luciferin** a complex carboxylic acid **and** ATP which is catalyzed by enzyme **luciferase.**

Some fungi that include some mushrooms, marine dinoflagellates, jellyfish and crustaceans are also able to generate bioluminescense which needs a considerable amount of energy. In the firefly ATP is used in a set of reactions which converts chemical energy into light energy. McElroy and his colleagues at Johns Hopkins University isolated the major biochemical components involved luciferin, a complex carboxylic acid and luciferase an enzyme. The regeneration of light flashes or **bioluminescense** needs activation of luciferin by an enzymatic reaction with ATP in which a pyrophosphate cleavage of ATP take place to form luciferyl adenylate.

This compound is then acted upon by molecular oxygen and luciferase to bring about the oxidative decarboxylation of the luciferin to produce oxyluciferin. This reaction which has some intermediate steps is accompanied by emission of light as shown in **Fig. 46.1**.

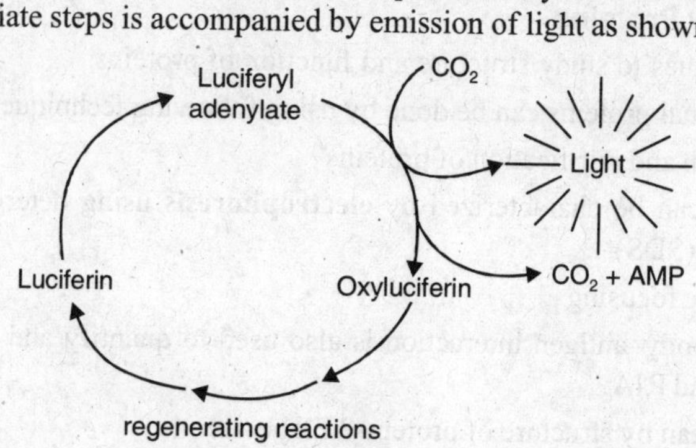

Fig. 46.1. Important steps in **bioluminescence cycle** or luciferin-luciferase reaction with scintillation effect.

Q.3. How can proteins be classified according to their biological roles?

Ans. The proteins according to their biological functions can be classified into the followign types:

(i) **Enzymes** or the proteins with catalytic activity.

(ii) **Transport proteins** e.g., haemoglobin of erythrocytes.

(iii) **Nutrient** and **storage proteins,** e.g., proteins of seeds, ovalbumin the major protein of egg white and casein (the milk protein).

(iv) **Contractile** or **motile proteins**, e.g., actin and myosin in skeletal muscle and protein dynein in flagella and cilia.

(v) **Structural proteins,** e.g., collagen (protein in leather), keratin (in hair, nails and feathers), fibroin (silk fibres and spider webs), and resilin (hinges of some insects).

(vi) **Defense proteins,** e.g., immunoglobulins or antibodies; fibrinogen and thrombin (blood clotting proteins); and snake venoms, bacterial toxins and plant proteins like ricin.

(vii) **Regulatory proteins,** e.g., insulin which regulates sugar metabolism and growth hormones of pituatory.

(viii) **Monellin,** a protein of an African plant which has intensely sweet taste.

(ix) **Antifreeze proteins,** found in antarctic fish.

Q.4. What are conjugate proteins? How are they classified?

Ans. Some proteins also have chemical components in addition to amino acids. These are called **conjugated proteins.** The non-amino acid part of the conjugated protein is usually called its **prosthetic group.**

The conjugate proteins are classified on the basis of the chemical nature of their prosthetic groups. The major classes of proteins with their prosthetic group and examples are given below:

(i) **Lipoproteins,** with lipids as prosthetic groups e.g., b_1– lipoprotein of blood.

(ii) **Glycoproteins,** with carbohydrates (sugars) as prosthetic group. e.g., immunoglobulin G.

(iii) **Phosphoproteins,** with phosphate group, e.g., casein of milk.

(iv) **Haemoproteins,** with haeme or iron porphyrin, prosthetic group, e.g., haemoglobin.

(v) **Flavoproteins,** with flavin nucleotides as prosthetic group, e.g, succinate dehydrogenase.

(vi) **Metalloproteins,** with iron as prosthetic group, is **ferritin,** with zinc as **alcohol dehydrogenase,** with calcium is **calmodulin,** with molybdenum is **dinitrogenase** and with copper is **plastocyanin.**

Techniques to Study Proteins

Q.5. Give the techniques to study structure and function of proteins.

Ans. Study of individual proteins can be done by using following techniques:

(i) Separation and purification of proteins

(ii) Proteins can be characterized by **electrophoresis** using detergent **sodium dodecyl sulphate** (SDS).

(iii) Isoelectric focusing

(iv) The antibody–antigen interaction is also used to quantify and localize proteins, e.g. ELISA and RIA.

Q.6. What do you mean by structure of proteins? Elaborate.

Ans. There are four levels of structure or architecture in proteins, which are known as primary structure, secondary structure, tertiary structure and quaternary structure.

The **primany structure** covers all the covalent bonds between amino acids and is characterized by the sequence of amino acids in the peptide. It also tells about locations of disulphide bonds. The **secondary structure** results from hydrogen bonding between side chains of amino acids to form α-helices and β-sheets. The tertiary structure is because of folding of proteins as a result of interaction between the amino acid side chains usually controlled by chaperone proteins. The **quaternary structures** results when more than one polypeptide makes up the functional protein.

Q.7. Why do woollen clothes shrink when washed in hot water or dried in an electric dryer and the silk on the other hand does not shrink under similar conditions? Explain.

Ans. The major components of silk fibres and spider webs is **fibroin,** a type of structural protein and a member of fibrous proteins called β- keratins. **β-conformation** makes fibroin of silk soft and flexible in comparison to **α-keratin** of hair, feather and nails (which has α helix cross-linked by disulphide bonds) due to which α-keratin is tough, insoluble protective structure of varying hardness and flexibility.

Heating and steaming of wool increases the spacing of structural units. It is because of soft and flexible nature of β-conformation of β-keratins of silk. The **Indian masses** have been obtaining silk by boiling the cocoons in hot water since ancient times.

47

CHAPTER

ENZYMES: BIOLOGICAL CATALYSTS OR BIOCATALYSTS

Q.1. Fill in the blanks in the following:

(i) Catalysts increase the velocity of chemical reactions. Enzymes are biological ____________.

(ii) Most but not all biological catalysts are ____________.

(iii) Some ____________ molecules can act as enzymes which are called **ribozymes.**

(iv) The protein **trypsin** catalyzes hydrolysis of ____________ bonds in proteins and polypeptides.

(v) The **catalase** an enzyme carried in many cells increases the uncatalyzed rate of H_2O_2 ____________ about one billion times.

(vi) The catalysts function by lowering the activation ____________.

(vii) Nonprotein biocatalysts are ____________.

(viii) Some ribonucleic acid molecules function as enzymes which are called ____________.

(ix) The ribonuclease P, an enzyme cleaves the ____________ of tRNA to yield the functional tRNA.

(x) Thomas Cech and coworkers on examining the removal of an intron (intervening sequence or IVS) from preribosomal RNAs of the protist ____________ found that the rRNA itself carried out the excision of its 413 – nucleotide intron and the necessary replicing as well.

(xi) The enzymatic reactions can be described by **Michaelis – Menton equation,** with two parameters, Michaelis constant, K_m and ____________ ____________, K_{cat}.

(xii) Blood clotting involves a cascade of proteolytic activation of specific proteases, culminating in the transition of fibrinogen to ____________.

(xiii) Using **molecular engineering** new and radically modified enzymes can be created by site – directed ____________, protein hybridization or catalytic antibody formation.

(xiv) Louis Pasteur in the 1850s concluded that fermentation of sugar into alcohol by ____________ is catalyzed by **"ferments".**

(xv) James Sumner in 1926 isolated and crystallized ____________ and postulated that all enzymes are proteins.

Ans. (i) Catalysts, (ii) Proteins,
(iii) RNA (iv) Peptide,
(v) Decomposition, (vi) Energy,
(vii) **Ribozymes,** (viii) Ribozymes,

(ix) Precursors (x) **Tetrahymena,**
(xi) Turnover number, (xii) **Fibrin,**
(xiii) Mutagenesis, (xiv) Yeast
(xv) **Urease.**

Q.2. What is a **cofactor** in an enzyme?

Ans. Some enzymes require chemical groups other than their amino acid residues for activity. A **cofactor** may be either one or more inorganic ions like Fe^{2+}, Mg^{2+}, Mn^{2+}, or Zn^{2+} or a complex organic or metal organic molecule called a **coenzyme**. Some enzymes require both a coenzyme and one or more metal ions for activity.

Q.3. What is a **prosthetic** group in an enzyme?

Ans. A coenzyme or metal ion which is covalently bound to an enzyme protein is called a **prosthetic group**. A complete catalytically active enzyme along with its coenzyme and or metal ions is known as **holoenzyme**.

Q.4. What is meant by **apoenzyme** or **apoprotein**?

Ans. The protein part in an holoenzyme is called **apoenzyme** or **apoprotein**.

Classification of Protein Enzymes

Q.5. How are Enzymes classified? Discuss.

Ans. The enzymes are classified by the reactions they catalyze by adding the "ase" to the name of their substrate or to a suffix word or phrase which is meant for their activity, e.g., urease which catalyzes the hydrolysis of urea, and DNA polymerase which catalyzes the synthesis of DNA. But some enzymes, e.g., pepsis and trypsin bear names which do not denote their substrates. Some times same enzyme has two or more names or two different enzymes have same name. To reduce confusion a rational naming and numbering system has been devised by "**Enzyme Commission of International Union of Molecular Biology**. According to it the enzymes are divided into **six major classes** with subgroups and sub-subgroups to define their functions in a better and precise manner. Each enzyme is assigned a **four digit classification number** and a systematic name which identifies the reaction catalyzed, e.g. The formal systematic name of the enzyme which catalyzes the reaction.

$$\text{ATP} + \text{D-glucose} \longrightarrow \text{ADP} + \text{D-glucose} - 6 - \text{Phosphate}$$

Would it be **ATP : glucose phosphotransferase** it catalyzes the transfer of a phosphate group from ATP to glucose. Its **enzyme classification number** or **E.C. number** is 2.7.1.1. Here, The first digit 2 denotes the class name transferase, the second digit 7 stands for the subclass **phosphotransferase**, the third digit 1 strands for **phosphotransferases** with hydroxyl group as acceptor, and the fourth digit I denotes the **D-glucose** as a phosphate group acceptor. In case when the systematic name of an enzyme is long or cumbersome, a trivial name may be used, as in case of hexokinase. It is not possible here to go in more detail. However, the major 6 groups according to international classification of enzymes are :

(i) **Oxidoreductases :** Catalyze oxidation reduction reactions. They catalyze transfer of electrons (hydride ions or hydronium ion).

(ii) **Transferases :** Catalyze transfer of functional groups from one molecule to another.

(iii) **Hydrolases :** Catalyze hydrolytic cleavage that means catalyze hydrolysis reactions (transfer of functional groups to water)

(iv) **Lyases :** Catalyze removal of a group from or addition of a group to a double bond, or other cleavages involving electron rearrangement.

(v) **Isomerases :** Catalyze intra molecular rearrangement. Catalyze transfer of groups within molecules to yield isomeric forms.

(vi) **Ligases :** Catalyze reactions in which two molecules are joined, e.g. formation of C – C, C – S, C – O, and C – N bonds by condensation reactions coupled to ATP cleavage.

Specificity of Enzymes

Q.6. Define **specificity** of enzymes.

Ans. It is the ability of an enzyme or a receptor to discriminate among competing substrates or ligands. The **ligand** is a small molecule which binds specifically to a larger one, e.g. a hormone is a ligand for its specific protein receptor.

Q.7. What do you mean by **catalytic power** of **biocalalysts** or enzymes.

Ans. The rate enhancement produced by enzymes or **biocalalysts** are often in the range of 7 to 14 orders of magnitude.

The rate enhancements produced by some of the enzymes are as under :

Carbonic anhydrase	10^7
Phosphoglucomutase	10^{12}
Succinyl CoA transferase	10^{13}
Urease	10^{14}

Q.8. Where does the energy come from to provide a dramatic lowering of the activation energies for **specific reactions** ?

Ans. The specific groups of an enzyme as specific amino acid side chains, metal ions and coenzymes lower the activation energy and thereby accelerate the reaction by providing a lower energy reaction path.

In the formation of each weak interaction the enzyme substrate or **ES complex** is accompanied by a small release of free energy which gives a degree of stability to the interaction. The energy derived from enzyme – substrate interaction is known as **binding energy**. The binding energy is a major source of free energy used by enzymes to lower the activation energies of reactions. The binding energy, in this way provides **specificity** as well as **catalysis**. Also, the weak interactions are optimised in the reaction transition state. The enzyme active sites are complementary not to the substrates per se rather to transition states of the reactions they catalyze. Lock and key model **(Fig.47.1)** is one of the most common explanation for interaction of substrate and enzyme.

Active Sites In Enzymes

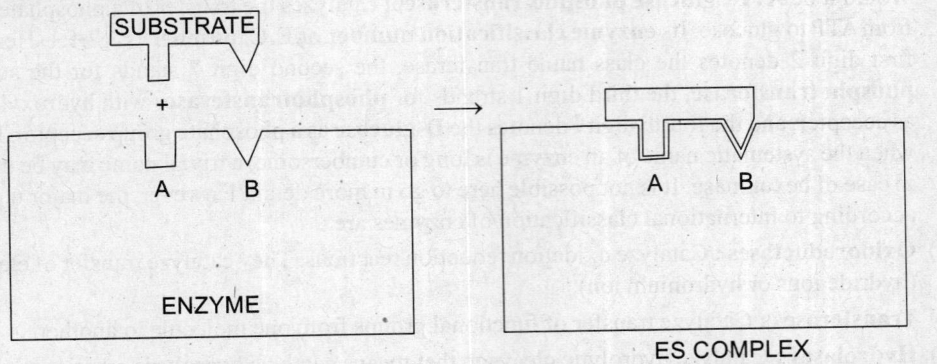

Fig. 47.1. Lock and key model of enzyme and substrate interaction. The **active site** of the enzyme marked by depressions A and B is matchable in shape to that of the substrate.

Q.9. What is an **active site** in an enzyme?

Ans. The region of an enzyme surface which binds the substrate molecule and catalytically transforms it is called **active site** or **catalytic site**.

The **fatty acid synthase complex** is known to have 7 different active sites. The **fatty acid synthase system** in ***Escherichia coli*** consists of seven separate polypeptides which are tightly associated in a single, organized complex. The proteins act together to catalyze the formation of fatty acids from acetyl – CoA and malonyl – CoA. The intermediates in this whole process remain attached covalently to one of the two thiol group of the complex. One of the point of attachment is the – SH group of a Cys residue in one of the seven proteins (β – ketoacyl – ACP synthase) while the other is – SH groups of acyl carrier protein with which the acyl intermediates of fatty acid synthesis form a thioester.

Q.10. Give the requirements of active sites in enzymes.

Ans. The active site of an enzyme generally consists of a **pocket** on the enzyme surface lined with the amino acid side chains necessary to bind the substrate and catalyze its chemical transformation. For example in carboxypeptidase which leads to removal of carboxyl – terminal amino acid residues from its peptide substrates, comprises of a single chain of 307 amino acids. The two essential catalytic groups in the active site are furnished by Arg^{145} and Glu^{270}.

Q.11. How do enzymes work ?

Ans. Most biological molecules are quite stable in the neutral pH, mild-temperature, aqueous environment found inside cells. The reactions required to digest food, send nerve signals or contract muscle simply do not occurs at a useful rate in absence of catalysis.

The enzyme – Catalyzed reaction occurs within the confines of a **pocket on the enzyme** known as **active site**. The molecule which is bound by the active site and is acted upon by the enzyme is known as **substrate**.

The enzyme substrate complex is central to the action of enzymes and it is the starting point for mathematical treatments defining the kinetic behaviour of enzyme – catalyzed reactions and for theoretical descriptions of enzyme mechanisms. A simple enzymatic reaction may be given as below :

$$E + S \rightleftharpoons ES \rightleftharpoons EP \rightleftharpoons E + P$$

Here E, S and P represent enzyme, substrate and the product.

Activity

Q.12. What is **activity** or **effective concentration** or **capacity to function**?

Ans. The **Theory of activity** was developed by G.N. Lewis in order to account for the behaviour of **strong electrolytes**. A very simple example to understand this theory would be that : "when a football player is surrounded closely on all sides by other players his capacity to run and execute the maneuvers of football is different from when no one is near him. In general the activity of the player is conditioned by other players surrounding him." The activity theory as applied to strong electrolytes considers these electrolytes to be 100 per cent ionized. Each ion possesses an **activity** or **capacity to function**.

The position of equilibrium of any chemical reaction is known by its equilibrium constant which can be shown for a generalized reaction.

$$A + B \rightleftharpoons C + D$$

Hence, an equilibrium constant can be defined in terms of the concentrations or reactants (A and B) and products (C and D) present at equilibrium:

$$K_{eq} = \frac{[C][D]}{[A][B]}$$

To be very concise, the concentration terms should be the **activities** or **effective** concentrations in non ideal solutions, of each species. In very accurate work the equilibrium constant may be approximated by measuring the concentration at equilibrium. **Enzyme activity** is affected by pH.

Q.13. When do enzymes typically exhibit maximum **catalytic activity**?

Ans. The enzymes typically show maximum catalytic activity at a characteristic pH known as **optimum pH.**

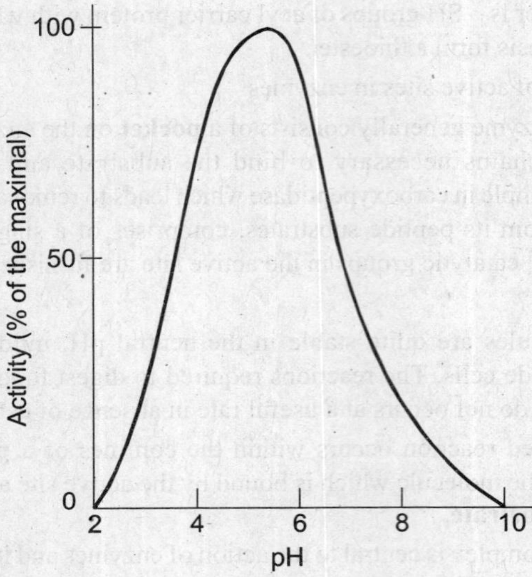

Fig. 47.2. Optimum enzymatic activity of Lysozyme at pH 5.2.

Q.14. How does an **uncontrolled diabetes** of an individual turns fatal.

Ans. The pH of human blood plasma normally is 7.40. The pH regulating mechanisms fails when over production of metabolic acids causes acidosis. The pH of blood, so, may fall to 6.8 or below leading to irreparable cell damage and death. In other diseases the pH may rise to lethal levels. Therefore, biological control of pH of cells and body fluids is a matter of central importance in all aspects of metabolism and **cellular activities.**

Q.15. Can the pH changes be used to measure **acetylcholine levels**?

Ans. Yes, the concentration of **acetylcholine** a **neurotransmitter** can be determined from the pH changes which accompany its **hydrolysis**. When incubated with a catalytic amount of the enzyme **acetylcholinestrase**, acetylcholine is quantitatively converted into choline and acetic acid which dissociates to yield acetate and hydrogen ion.

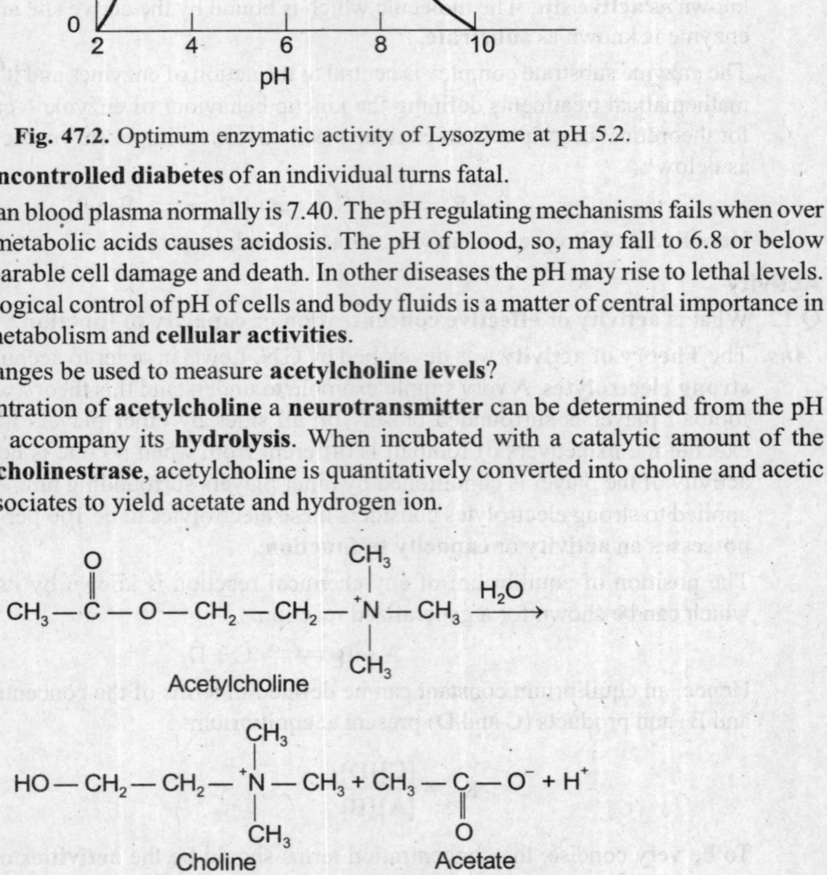

Isozymes or Isoenzymes

Q.16. Define **isozymes**.

Ans. These are multiform of enzymes which catalyze the same reaction but differ from each other in their amino acid sequence, substrate affinity, V_{max}, and/or regulatory properties. These are also known as **isoenzymes**.

Q.17. What is **Cori cycle**? Also discuss the isozymes (isoenzymes) associated with it.

Ans. The liver provides glucose to contracting skeletal muscle which derives ATP from the glycolytic conversion of glucose to lactate. The glucose is then synthesized from lactate by the liver. These conversions comprise the **Cori cycle** as shown in the figure below.

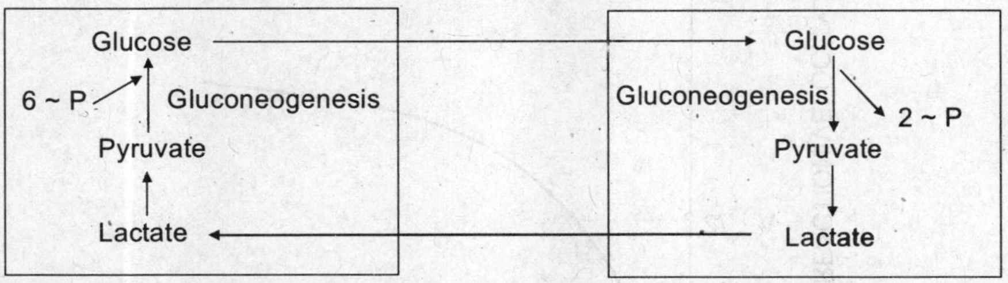

Fig. 47.3. Demonstration of **Cori cycle**. The lactate formed by active muscle is converted into glucose by the liver. Hence, this cycle shifts part of the metabolic burden of active muscle to the liver.

These conversions take place despite of differences in the catalytic properties of **lactate dehydrogenase enzymes** in **skeletal muscle** and **liver**. The lactate dehydrogenase is a tetramer of 35 – kDa subunits. There are two types of polypeptide chains referred to as M and H which can form 5 types of tetramers, which are : M_4, M_3H, M_2H_2, M_1H_3 and H_4. These species of enzymes are known as **isozymes** or **isoenzymes**. The M_4 isozyme possesses a much higher affinity for pyruvate than the H_4 isozyme. The other isozymes have intermediate affinities. The principal isozyme in skeletal muscle and liver is M_4 while the major one in the heart muscle is H_4. Of course, these isozymes have been studied deeply but the reasons for the existence of their multiple forms yet remains unknown.

Enzyme Kinetics

Q.18. Which are two hallmarks of enzyme catalyzed reactions?

Ans. These are enormous activity and discriminating specificity. Crystallographic pictures of polypeptide chains of enzymes show that they are coiled in intricate shapes or conformations which apparently impart **specificity** and **activity**.

Q.19. What is **enzyme kinetics**?

Ans. Much information may be obtained concerning the mechanism of enzyme – catalyzed reaction by kinetic studies, which means studies of reaction rates under various conditions, e.g. the reversible reaction.

$$S \underset{}{\overset{E}{\rightleftharpoons}} P$$

Where S is the substrate, P the product and E the enzyme. Under constant and suitable conditions of temperature and pH. The rate of reaction depends on the concentration of substrate [S], the concentration of product [P] and concentration of enzyme [E]. The reaction velocity in the absence of enzyme is negligible. The velocity of over all reactions may be obtained by determining the changes in [S] or [P] with time.

In the study of kinetics of forward reaction alone one usually determines the velocity over a short time interval at the beginning of the reaction where P is still negligibly small. This has the additional advantage minimizing the effect of enzyme inactivation during the experiment.

Michaelis – Menten Equation for Simple Enzymes

Q.20. What is **Michaelis – Menten equations**?

Ans. A theory explaining the changes which take place in enzymatic reactions was proposed by Leonor Michaelis and Maud Menten in 1913. They postulated that substrate combined reversibly with enzyme to form an enzyme substrate complex [ES] which in turn is decomposed to yield the product and the free enzyme. The latter could then react with more substrate and the cycle is repeated.

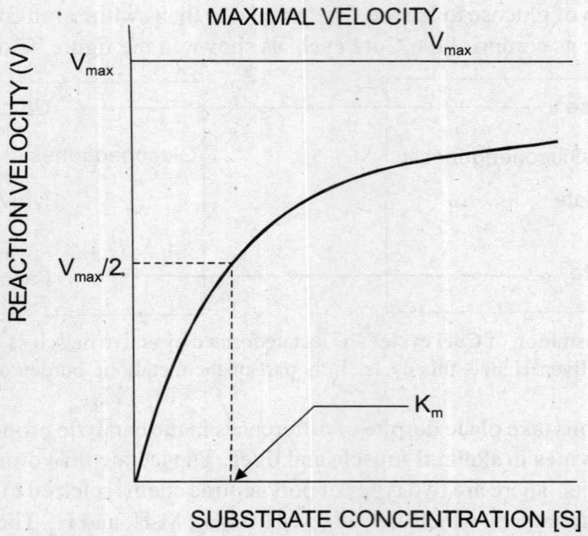

Fig. 47.4. Velocity of an enzyme catalyzed reaction is a function of substrate concentration. In this plot V is for reaction velocity, S represents substrate concentration, V_{max} is the maximal velocity and K_m is the Michaelis constant. This typical plot is for an enzyme which obeys or follows Michaelis – Menten Kinetics.

Hence **Michaelis – Menten equation** can be defined as the equation that describes the hyperbolic (Fig. 47.4) dependence of the initial reaction velocity, V_o, on substrate concentration, [S], limited to early times in the course of the reaction. The biochemists Michaelis and Menten concerned themselves with the steady state rate and this type of analysis is called **steady state kinetics**.

Q.21. Define **Michaelis – Menten constant (K_m)**.

Ans. It is the substrate concentration at which an enzyme – catalyzed reaction proceeds at one half its maximum velocity.

Q.22. Which characteristic of the enzymes makes them to follow **Michaelis – Menten kinetics** ? Which enzymes are exception to it.

Ans. All enzymes which exhibit a hyperbolic dependence of V_o on [S] are said to follow **Michaelis – Menten kinetics**. The regulatory enzymes and allosteric enzymes are exception to Michaelis – Menten kinetics in many enzyme catalyzed reactions :

$$V_o = \frac{V_{max}\,[S]}{K_m + [S]}$$

Q.23. What is **Michaelis – Menten Kinetics**?

Ans. A kinetic pattern in which the initial rate of an enzyme catalyzed reaction exhibits a hyperbolic dependence on substrate concentration.

Q.24. What is **steady – state kinetics**?

Ans. The enzyme while mixed first with a large excess of substrate there is an initial period called the pre-steady state during which the concentration of enzyme substrate complex increases or gets build up. **Pre-steady state** is usually too short to be easily observed. The reaction soon achieves a **steady state** in which the ES and the concentration of any other intermediates is almost constant for some time. The measured V_o often reflects the steady state even though V_o is limited to early times in the course of the reaction.

Determination of Kinetic Parameters

Q.25. Name a best **kinetic parameter** used in comparison of catalytic efficiency.

Ans. The factor K_{cat}/K_m

Q.26. What is the usefulness of kinetic parameters K_{cat} and K_m.

Ans. They are generally useful for the comparison and study of different enzymes to know whether their reaction mechanisms are simple or complex. Each enzyme possesses the optimum values of K_{cat} and K_m which reflect the cellular environment.

Q.27. How can V_{max} and K_m be determined or estimated?

Ans. They can be measured or determined by graphical methods.

Multi-Step Reactions or Reactions Carried by Multienzyme Complexes

Q.28. What are the **multistep reactions**? Discuss.

Ans. These reactions involve more than one step and are carried by **multienzyme complexes**. The intermediates are channelled between glycolytic enzymes, which is a typical example. The enzymes of glycolysis usually described as soluble components of cytosol but there is a growing evidence that within the cell exists the multi enzymes complexes. There exists kinetic evidence for the **channelling** of 1, 3 biphosphoglycerate from glyceraldehyde – 3– phosphate dehydrogenase to phosphoglycerate kinase without entering solution is corroborated by physical evidence that the said two enzymes form stable complexes which are monocovalent complexes. Other examples are pyruvate dehydrogenase complex which requires 5 coenzymes; succinyl– CoA synthetase also called succinic thiokinase reaction; fatty acid synthase from bacteria and plants is a complex of 7 different polypeptides; construction of purine ring of insinuate (IMP) synthase, synthesis of AMP and GMP from IMP, biosynthesis of pyrimidine nucleotides UTP (uridine – 5 – triphosphate) and CTP (cytidine – 5' – triphosphate via orotidylate, and other biological oxidation reduction reactions.

Q.29. Give fundamental **principles of biological oxidation – reduction**.

Ans. The principles of biological oxidation – reduction may be briefed as under :

(i) The enzymes involved in biological oxidation – reduction are dehydrogenases carriers of reducing equivalents and oxidases.

(ii) Many of these enzymes are organized in multienzyme systems that work as terminal respiratory chain.

(iii) The first process in terminal respiratory chain consists in the activation of certain specific hydrogen atoms by specific dehydrogenases.

(iv) Second process involves the transfer of the hydrogen atoms (reducing equivalents) from metabolites to carriers of reducing equivalents.

(v) Final stage taking place in terminal oxidation is the combination of the reducing equivalents from the carriers with oxygen.

Q.30. What do you mean by **rate – limiting step**?

Ans. A reaction consists of a number of steps out of which one or more steps are slowest (*i.e.*, have lowest rate of reaction). These steps have highest activation energy. These steps are called *rate limiting steps*. Considering a simple case the rate limiting step is the highest energy point in the curve or diagrams that are often made for inter conversion of S and P, while S and P reaction (or S $\rightleftharpoons$ P) is catalyzed by an enzyme. The ES and EP complexes are intermediates. These intermediates occupy valleys while reaction coordinate diagrams are drawn.

In the **citric acid cycle** there are 3 strongly exergonic steps in the cycle which are catalyzed by enzymes citrate synthase, isocitrate dehydrogenase and α – ketoglutarate dehydrogenase Each of these enzymes can become a **rate – limiting step** under certain circumstances. The availability of substrates **for** citrate synthase (acetyl – CoA and oxaloacetate) varies with the metabolic circumstances sometimes limits the rate of citrate formation.

To conclude the **rate–limiting** step may be :

(i) Generally the step in an enzymatic reaction with the greatest activation energy or the transition state of highest free energy.

(ii) It is the slowest step in a metabolic pathway.

Enzyme Inhibition

Q.31. What is enzyme inhibition? Describe.

Ans. It is the prevention of enzymatic process as a result of the interaction of some substance with an enzyme so as to decrease the rate of enzymatic reaction. The substance which causes such an effect is termed as **inhibitor**. The enzyme inhibitors are important as **chemotherapeutic agents** as regulators in normal control of enzymatic processes in living organisms and as useful agents in study of biochemistry. Thus enzyme inhibitors are of applied value and are studied in applications of enzymes.

The inhibitors have been classified as **competitive, non-competitive** and **uncompetitive**. The enzymes are subject to **reversible** and **irreversible inhibition**. One **common type of reversible inhibition** is called **competitive inhibitor**.

Competitive inhibitors are generally compounds that the substrate and combine with enzyme to form **E1 complex.** This type of inhibition can be analyzed quantitatively by **steady–state kinetics.** As the inhibitor binds reversibly to the enzyme, the competition can be biased to favour the substrate simply by adding more substrate. When enough substrate is present the probability that an inhibitor molecule will bind is minimised and the reaction thus gives a normal V_{max}. None-the-less, the [S] at which $V_o = \frac{1}{2} V_{max,}$ the K_m will increase in the presence of inhibitor. Thus the effect on apparent K_m and the absence of an effect on V_{max} is indicative of competitive inhibition and is calculated readily in a double reciprocal plot. The **equilibrium constant** for inhibitor binding K_I can be obtained from the same graphic plot.

Therapeutic application of competitive inhibition is to treat patients who have ingested methanol, a solvent found in gas – line antifreeze. The methanol gets converted to fromaldehyde by the action of the enzyme alcohol dehydrogenase. Formaldehyde damages many tissues and result blindness because the eyes are specifically sensitive. Ethanol competes effectively with methanol as a substrate for alcohol dehydrogenase. The therepy for treatment of methanol poisoning is the infusion of ethanol by intravenous mode. It slows the formation of formaldehyde and thus a major part of methanol is excreted harmlessly through using.

It will be worth while to exonerate that three types of **reversible inhibitors** are :

(i) **Competitive inhibitors** that binds to enzymes active site.

(ii) **Noncompetitive inhibitors** generaly bind at a separate site.

(iii) **Uncompetitive inhibitors** also bind at a separate site but they bind only to the ES complex. K_1 is the equilibrium constant for inhibition binding.

The **double-reciprocal plot** is an easy way to determine whether the inhibitor enzyme is competitive or noncompetitive (Fig. 47.5.).

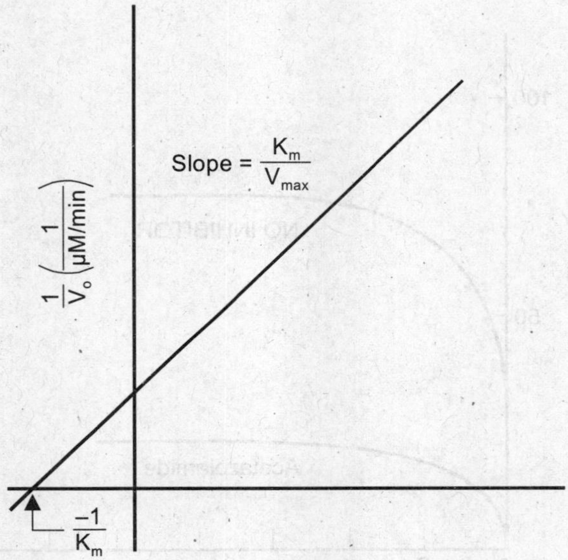

Fig. 47.5. A double - reciprocal plot also called lineweaver – Burk plot, which is a transformation of Michaelis – Menten equation and yields a straight line. Here Michaelis – Menten equation has been transformed algebraically into forms which are more useful in plotting experimental data.

Q.32. What are **suicide inhibitors**?

Ans. The **suicide inhibitors** are a special class of irreversible inhibitors which are comparatively unreactive until they bind to the active site of enzyme. A suicide inhibitor is designed to carry out few chemical steps of the normal enzyme. But it instead of forming a normal product, the inhibitor is converted to a very reactive compound which combines irreversibly with the enzyme. Therefore, these have also been referred as **mechanism based inactivators** since they utilize the normal enzyme reaction mechanism to inactivate the enzyme. This sort of inhibitors find application in therapeutics by finding new pharmaceutical agents, a process known as rational drug design.

Allosterism

Q.33. Write a short note on **allosterism**?

Ans. Many enzymes have sites referred to as **allosteric sites,** which are quiet different from the substrate binding sites. The legends that bind at the allosteric site are called **allosteric effectors** or **modulators**.

Binding of an allosteric effector causes a conformational change of the enzyme, so that the affinity for the substrate or the ligand also changes. These effectors can bind reversibly and non covalently to all allosteric sites and they affect the rate of reaction. Positive (+) allosteric effectors increase the enzyme affinity for the substrate or other ligand. The reverse is true for negative (–) allosteric effector. This is a type of regulation known as **allosterism** and the enzyme regulated this way is referred to as an allosteric enzyme. The allosteric site to which the positive effector bind is referred to as an activator site; the negative effector binds at an inhibitory site.

Allosterism is an effective mechanism by which the enzymatic activities can be controlled to ensure that biological processes remain coordinated all times to meet the immediate metabolic requirement of a cell.

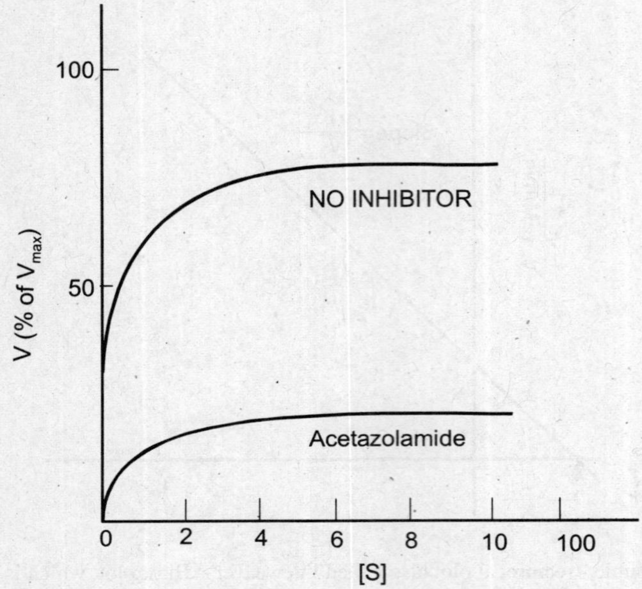

Fig. 47.6. Plot exhibiting inhibition of carbonic anhydrase by acetazolamide. The drug acetazolamide strongly inhibits carbonic anhydrase. Acetazolamide is used as a diuretic to increase production of urine and to reduce high pressure within eye ball to treat glaucoma. Carbonic anhydrase plays a wonderful role in these and other secretory processes as it participates in regulating the pH and bicarbonate content of a number of body fluids. The upper curve is for reaction velocity for the carbonicanhydrase reaction. The lower curve is in the presence of acetazolamide. This explains the **inhibitory nature** of acetazolamide.

Q.34. What is an **allosteric enzyme**? Define.

Ans. It is a regulatory enzyme with catalytic activity modulated by non covalent binding of a specific metabolite at a site other than the active site.

The term **allosteric** has been coined from Greek **allows** = other, and **stereos** = solid or shape. Therefore, allosteric enzymes are those enzymes that possess "other shapes" or conformation induced by the binding of modulators. The activity of regulatory enzymes is modulated by means of various signal molecules which are generally small metabolites or cofactors.

Q.35. How are allosteric enzymes regulated?

Ans. Allosteric enzymes are regulated by non covalent binding of modulators. The modulator or modulators are metabolites which when bound to the allosteric site of an enzyme alters its kinetic characteristics.

Q.36. How are allosteric enzymes exceptions to many general rules?

Ans. It is because that modulator or modulators for allosteric enzymes may be either inhibitor or stimulatory. Generally the activator is a substrate itself and regulatory enzymes for which substrate and modulator are identical are known as **homotropic**. However, while the modulator is a molecule other than the substrate the enzyme is called heterotropic. The modulator may be one, two or more in some enzymes.

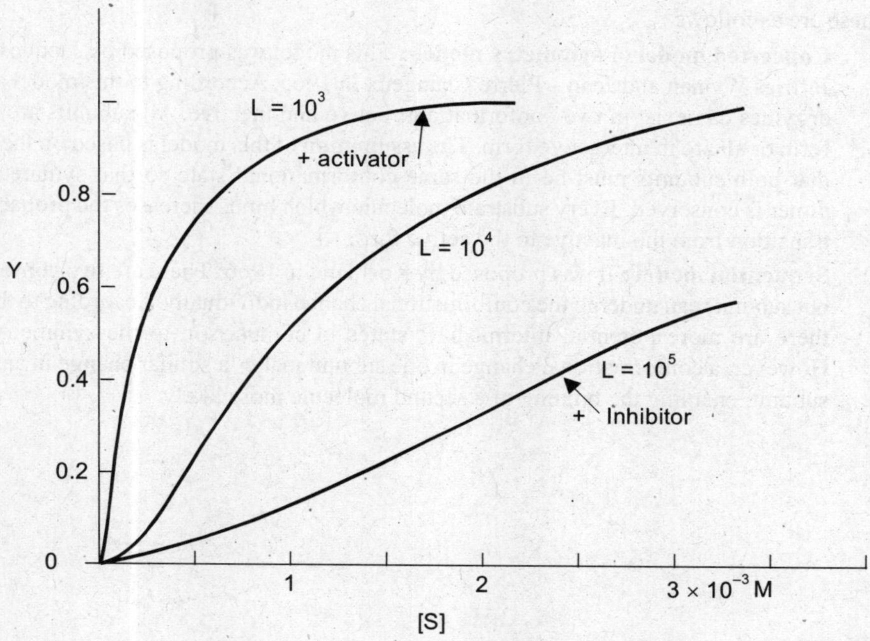

Fig. 47.7. Y plotted versus [S]. Here Y is saturation as a function of substrate concentration [S]. Effects of an allosteric activator and inhibitor are also shown.

Kinetic Analysis of Allosteric Enzymes

There exists some differences between non regulation enzymes and allosteric enzymes which involves the kinetic properties.

The allosteric enzymes exhibit relationship between V_o and [S] that does not obey normal Michaelis – Menten behaviour and rather differ from normal simple enzymes. They do not give saturation with the substrate when [S] is sufficiently high but for some allosteric enzymes if V_o is plotted against [S] a **sigmoid saturation curve** is formed instead of the hyperbolic curve by not regulatory enzymes.

Of course, we can find a value of [S] on the sigmoid saturation curve at which V_o is half maximal but one cannot refer to it with the designation K_m as the enzyme does not obey the hyperbolic Michaelis – Menten relationship, rather the symbol $[S]_{0.5}$ or $[K]_{0.5}$ is generally used to represent the substrate concentration giving half-maximal velocity of the reaction catalyzed by an allosteric enzyme.

Principles of Allosteric Regulation

Q.37. Give the principles of allosteric regulation.

Ans. Allosteric enzymes do not follow Michaelis – Menten kinetics. The allosteric enzymes which do not obey Michaelis – Menten display **sigmoidal plots** of the reaction velocity, V, versus substrate concentration [S] rather than the hyperbolic plots predicted by the Michaelis – Menten equation. Here, an important similarity lies that oxygen binding curve myoglobin is hyperbolic whereas that of haemoglobin is **sigmoidal**. Thus this example is analogous to enzymes.

Two models have been proposed regarding the allosteric interactions. These are:

(i) The concerted model for allosteric interactions.
(ii) Sequential model for allosteric interactions.

Q. 38. Which are the two models that explain the kinetic behaviour of allosteric enzymes? Explain.

Ans. These are as follows :

(i) **Concerted model** or **symmetry model :** This model was proposed by Jacques Monod, Jeffries Wyman and Jean – Pierre Changeux in 1965. According to this model allosteric enzymes can exist in two conformations, active and inactive. All subunits are in active form or all are in interactive form. The assumption of this model is based on the thinking that both subunits must be in the same conformational state so that symmetry of the dimer is conserved. Every substrate molecule which binds increases the probability of a transition from the inactive to the active form.

(ii) **Sequential model :** It was proposed by Koshland in 1966. There are two conformations but subunits can undergo the conformational change individually. According to this model there are more potential intermediate states in comparison to the symmetry model. However, a conformational change in one subunit makes a similar change in an adjacent subunit, enabling the binding of a second molecule more likely.

48

CHAPTER

ENZYME TECHNOLOGY OR APPLICATIONS OF ENZYMES

Q.1. What are the **applications of enzymes**? Discuss.

Ans. The enzymes possess high catalytic activity and lack undesirable side reactions, therefore, have been exploited recently on an increasing scale, in food, pharmaceutical and chemical industries. According to Nicholas C. Price and Lewis Stevens (1999), because of advancement of genetic engineering a larger number of enzymes are available on large scale which has increased and ushered the **scope of enzyme technology.** From 1983 to 1995 the sales of enzymes have doubled and a tremendous rise is expected in enhancement of sales. Although the enzymes present in the bacteria and yeasts have been used by man long back but the use of isolated enzymes in industry reverse to the enzymes found in the cells of organisms, has started in 1970s only. Now over 60 enzymes have found their way in industrial use.

The applications of enzymes include the exploitation of reactions that are catalyzed by the **enzymes in intact organism** *i.e.,* in complete and undamaged organisms and the isolated enzymes, *i.e.,* freed from microorganisms besided **immobilized enzymes** *ie.,* attaching an enzyme to an insoluble support to permit its reuse and continuous use, and the **immolized whole cells.** Distribution of sales of different types of enzymes is shown in **Fig. 48.1** and **Fig. 48.2**.

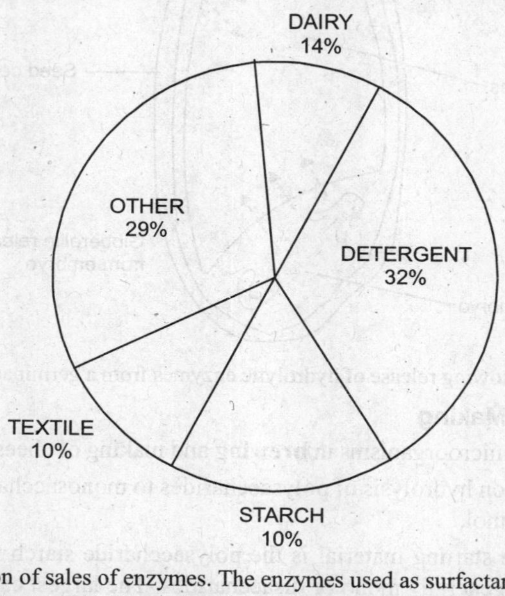

Fig. 48.1. Distribution of sales of enzymes. The enzymes used as surfactants or surface active agents in detergents are 32%. The estimation of above data is for 1994.

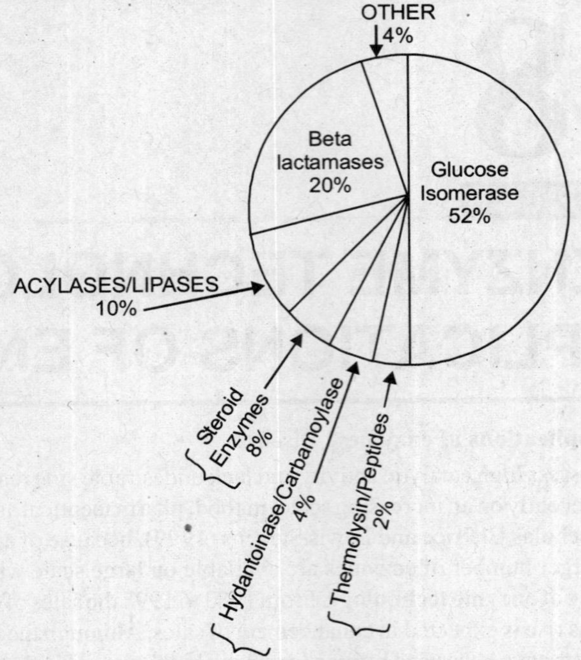

Fig. 48.2. Diagramatic presentation of percentage of total sales of enzymes used in biotechnology or industry.

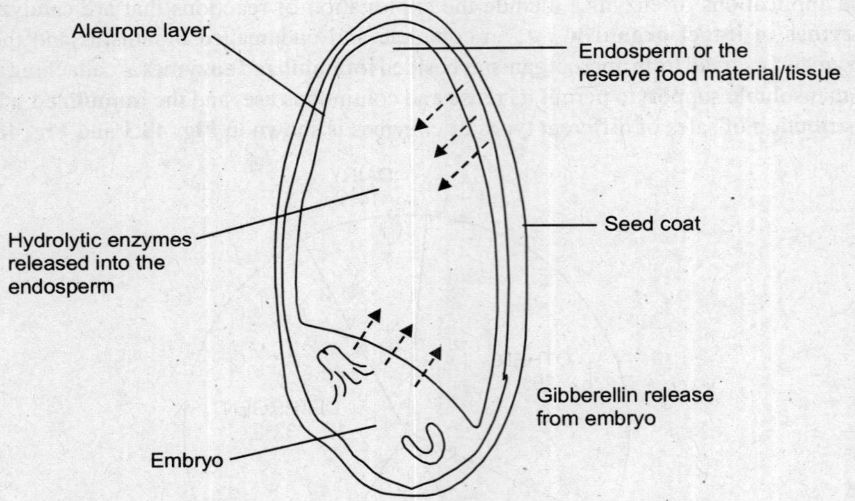

Fig. 48.3. Diagram showing release of hydrolytic enzymes from a germinating seed of barley.

Brewings and Cheese-Making

Q.2. Discuss the role of microorganisms in **brewing** and making of cheese.

Ans. The enzymes carry on hydrolysis of polysaccharides to monosaccharides or in glycolysis lead to formation of ethanol.

In brewing also the starting material is the polysaccharide starch while in wine making the starting material is generally mono or disaccharides. The largest carbohydrates fermented by most of yeasts are trisaccharides.

The main raw material used in brewing is the starch from barley. Therefore, the barley seeds are permitted to germinate which leads to release of **amylase.** The release of hydrolytic enzymes from germinating barley is shown in **Fig. 48.3**. The enzyme amylase leads to break down of starch. The starch grains are present in the cells of endosperm. The walls of the cells are surrounded by a number of polymers like 1, 3 -β -D-glucans. It is needed to degrade these glucans so that the amylase may come in contact with starch granules. The release of some peptidases is also important which catalyze the degradation of endogenous protein releasing peptides and amino acids. The amino acids and peptides are required for the growth of yeast cells which occur at the fermetation stage. The starch granules dispersed in water are maintained above 60°C to prevent gelatinization. Also the enzyme amylase has to be stabilized to remain active at this temperature. The pH is maintained 5 to 6 facilitated by addition of Ca^{2+} and having a high substrate concentration.

The amylases hydrolyse the α-1, 4 links found in amylose and amylopectin. It may be stated that amylopectin is polymer of D-glucose units linked α-1, 6 as well as α-1, 4 and constitute from 17 to 30 per cent of starch, the rest of that is amylose with α-1, 4 links only.

The hydrolysis of amylose and amylopectin leads to production of a mixture of **glucose, maltose, maltotriose** and higher unfermentable sugars called **dextrins.**

α-Amylase is an endoamylase which randomly hydrolysis starch to dextrins while **β-amylase** which is an exoamylase attacks both starch and dextrins, from nonreducing end removing maltose units. However, neither attack α-1, 6 links present in amylopectin.

The second stage consists of anaerobic glycolysis in which **ethanol** is the end product.

Strain Improvement Programme to Make Brewing and Wine-Making More Profitable

Q. 3. How have new strains of yeast been possible by **recombinant DNA technology**?

Ans. The new strains of yeasts which show amylolytic activity have been prepared using recombinant DNA technology. To achieve this objective glucoamylase genes from *Aspergillus,* β-1, 3/1, 4-glucanase genes from *Bacillus subtilis* and **cellulase** (endo-β-1, 4 glucanase) and β-galactosidase genes from *Trichoderma* have been transferred and expressed in brewer's yeast.

Q. 4. Name the organisms and the enzymes involved in **brewing** and **wine making, cheese-making** and production of **vinegar**.

Ans. These are presented in **Table 48.1.**

Table. 48.1. Organisms and enzymes used in food industry. After Price and Stevens (2001).

Process/Industry	Organism used/involved	Enzymatic steps and Enzymes
Brewing and Wine-making	Barley (*Hordeum vulgare*)	α-Amylase and β-amylase, endo - 1, 3 - β D - glucanase
	Saccharomyces spp	Limited proteolysis, oligosaccharides (Oligosaccharides → monosaccharides) Glycolytic enzymes (monosaccharides → ethanol)
Cheese-making	*Streptococcus lactis* *Streptococcus cremoris*	Lactose → lactate + limited lipolysis and proteolysis
	Propionobacteria spp	Conversion of lactate → propionate + acetate + CO_2 + H_2O
	Penicillium camemberti and *P. roqueforti*	Limited lipolysis and proteolysis plus other undefined activities
	Calf stomach	Chymosin (rennin), Caseinogen → Casein
Vinegar production	*Acetobacter*	Ethanol to acetate is formed.

Microorganisms in Production of Organic Acids

Q. 5. Name the mould used to produce **citric acid.** Why Mn^{2+} and Fe^{3+} are used in low concentration.

Ans. The growth conditions of *Aspergillus niger* are arranged to stimulate the overproduction of citric acid. The growth medium used has high concentration of sugars and low concentration of Mn^{2+} Fe^{3+}. The low concentration of Mn^{2+} is kept because Mn^{2+} deficiency represses all the TCA cycle enzymes except **citrate synthase** permitting citric acid accumulation. The citrate is an allosteric inhibitor of phosphofructokinase. Mn^{2+} deficiency also lead to accumulation of NH_4^+ ions which relieves citrate inhibition.

Q. 6. Give the use of microorganisms in production of organic chemicals.

Ans. A large number of microorganisms are involved in production of organic chemicals. Some of these are listed in **Table. 48.2.**

Table. 48.2. Production of **organic acids** and **antibiotics** using bacteria and moulds. (After Price and Stevens, 2001).

Organic compound, Organic acid or Antibiotic	Organism Exploited	Enzymatic Steps	Uses of organic compound
Citric acid	*Aspergillus niger*	Monosaccharides → citrate, involve glycolysis tricarboxylic acid cycle	It is used in soft drinks, jams, jellies, flavouring food, blood transfusion, sequestring agent in electro plating and tanning of leather
Itaconic acid	*Aspergillus* sp	Citrate formation, plus citrate ⇌ aconitate + CO_2	Copolymer in acrylic resins
Gluconic acid	*A. niger*	Glucose → gluconolactone → gluconate	As a food additive and calcium gluconate is used in Ca therapy
Fumaric acid	*Rhizopus* sp	TCA cycle	Paper sizing
Amphotericin B	*Streptomyces* spp		As an antibiotic
Chloramphenicol	*Streptomyces* spp		As an antibiotic
Cycloheximide	*Streptomyces* spp		As an antibiotic
Erythromycin	*Streptomyces* spp		As an antibiotic
Novobiocin	*Streptomyces* spp		As an antibiotic
Streptomycin	*Streptomyces* spp		As an antibiotic
Griseofulvin	*Penicillium* spp		As an antibiotic for dermatophytosis (ring worms)
Penicillin	*Penicillium* spp		As an antibiotic
Cephalosporin	*Acremonium*		Like penicillin but resistant to penicillinases
Cyclosporins	*Trichoderma polysporum*		Antifungal antibiotic and immunosuppressant

Lovastatin	*Aspergillus terreus*		Treatment of hypercholesterolaemia
Bacitracin	*Bacillus subtilis*		Antibiotic for topical application
Gramicidin	*Bacillus brevis*		As an antibiotic
Carotenoid pigmets	*Blakeslea trispora*	Synthesised from acetate	Colouring agent
Riboflavin	*Ashbya gossypi*	Synthesis from glycine, formate & CO_2	A nutrient

Q. 7. Which method is generally used for isolation of intracellular enzymes?

Ans. It is high pressure homogenization method, generally used.

Enzymes Isolated on Industrial Scale and Their Applications

Q. 8. Name the enzymes used in industry to **melabolize carbohydrates.**

Ans. These are amylase, exo-1, 4 -α-D-glucosidase (amyloglucosidase), cellulase, polygalacturonase, β-D-fructofuranosidase, glucose oxidase, xylose isomerase generally called glucose isomerase.

Q. 9. Which enzymes are used to metabolize proteins in the industry?

Ans. These are papain, chymosin (**rennin**), trypsin, chymotrypsin, fungal proteases and bacterial proteases.

Bread Making

Q. 10. On what activity aeration or CO_2 production depend in manufcture of bread?

Ans. It depends on adequate α- and β- amylase activity.

Cheese-Making

Q. 11. Why is **chymosin** and not **trypsin** (a proteolytic enzyme) used in cheese-making?

Ans. Chymosin (rennin) obtained largely from calf stomachs is an aspartic acid rotease leads to clotting of milk which involves the cleavage of **a single** peptide bond in k-casein between Phe 105 and Met 106, releasing the acidic **c-terminal peptide**. The release of c-terminal peptide is followed by Ca^{2+} induced aggregation of **the modified micelles** to form a gel. The **micelle** are an aggregate of amphipathic molecules in water with the nonpolar portions in the interior and the polar portions at the exterior surface exposed to water.

On the other hand proteolytic enzymes like **trypsin** also result clotting of casein (milk protein) but further degrade casein. If it happens in cheese-making it leads to impart undesirable flavours.

Q. 12. What are the substitutes of chymosin obtained from calf stomachs with the ever increasing number of vegetarians?

Ans. The fungal aspartate proteases from *Mucor pucillus, Rhizomucor miehei* and *Cryphonectria parasitica* serve as substitute for chymosin but they do not possess good clotting proteolysis ratio. However, recombinant DNA chymosin in which **the chymosin gene** has been closed into expression systems such as *Aspergillus oryzae* and *Kluyveromyces lactis* have also been used in cheese-making since 1994.

Meat Tenderizing (Softening) and Marination

Q. 13. Which enzymes are used in tenderization of meat?

Ans. The toughness in meat is mainly due to collagen and elastin, and also due to actomysin. Lower-quality cuts of meat are just as nutritious as prime cuts. Therefore, to make maximum use of carcass (dead body of animal for cutting up as meat. The **carcass meat** is contrasted from tinned or corned meat) efforts have been made to tenderize the quality cuts. Proteases are used both for the purpose of tenderization as well as marination. Traditionally, enzyme **papain** is used but bromelain trypsin, chymotrypsin and microbial proteases from the mould *Aspergillus* are also used for the purpose.

As Sweeteners of Processed Foods

Q. 14. Which are the enzymes used as swetening agents?

Ans. Two important enzymes used for the process are **exo-1, 4 - α -D- glucosidase** obtained from the fungus *Aspergillus niger,* and xylose isomerase from *Streptomyces* spp. The **xylose isomerase** is also called **glucose isomerase** since it also isomerize glucose.

In Clarification of Beers, Wines and Fruit Juices

Q. 15. Give the role of enzymes in classification of beverages.

Ans. Turbidity in beers, wines and fruit juices is undesirable because turbit products are not accepted by the customer. The fruit juices are more easily prepared from pulp if the viscosity is not very high.

In case of beer the turbidity called as **haze** may be formed on cooling because of a complex between tannin and protein. It can be removed by adding small amounts of proteases often **papain.** But a very careful balance has to be maintained because proteins because of their foaming properties account for the formation of **head** or the froth on the top. The proteolysis not only remoes the chilled haze but good head also. Therefore, the head needs to be protected.

In case of wines and fruit juices the turbidity is because of starch or pectin which can be removed by addition of **amylases** or **polygalacturonases.**

As Surfactants in Detergent

Q. 16. Discuss the role of enzymes as **surfactants** in making detergents.

Ans. The **surfactants** are the surface-active-agents which decrease the tension between molecules lying on the surface of a liquid. The first patent for use of enzymes in detergent was granted in 1913. It was a crude extract of pancreas (pancreatin) in a solution of sodium carbonate. The large scale production of detergents with enzymes, however, took place in 1960s so that by 1969 half of the detergents sold in the market had enzymes.

Some health hazards, primarily skin allergies were noticed due to such enzymes consequently use of enzymes declined rapidly in 1970s.

However, a method of encapsulation of enzyme particles using nonionic surfactants was developed. The environmental friendly detergents thus became popular as the use of enzymes in detergents also lowered the washing temperature.

The present **'biological' detergents** have a mixture of enzymes amylase and neutral and alkaline proteases which are active in the pH range 6.5 to 10 and at temperatures from 30°C to 60°C. All proteases employed for detergents are obtained from *Bacillus* usually either neutral or alkaline protease. The carbohydrate hydrolysing enzymes are α-amylases from either *Bacillus licheniformis* or *B. amyloliquifaciens* which bear high temperature optima. Since 1988 lipases were also introduced as enzymes for detergents.

The use of **cellulases** for garments has been to act on damaged fibres of the garments. Hence, the microscopic effect of this **microscopic'shaving' of garments** provides softness to the fibre, brightens the colours and enhances the removal of particulate material that adhere to the surface of garments.

Medical Applications of Enzymes

Q. 17. Discuss the medical applications of enzymes in analysis and therapy.

Ans. About 50 different enzymes are being used for clinical diagnoses. Two of the major enzymes are **peroxidases** from horseradish and **alkaline phosphatase** from beef intestinal mucosa. Both of these are required for immunoassays. The enzymes are used in test strips, ELISA, biosensors and autoanalysers.

The usage of enzymes in therapeutical way may be divided into two categories:

(i) Topical application of enzyme as an extracellular agent.

(ii) Intracellular application of enzyme to treat metabolic deficiency and other related diseases.

The examle of extracellular enzymes are the **proteolytic enzymes** in clot removal; **collagenase** to treat skin ulcers, **hyaluronidase** in drug administration, **trypsin** as an anti-inflammatory agent and wound cleanser, **rhodanase** (thiosulphate sulphur transferase) in cyanide poisoning and **uricase** in gout.

Human tissue plasminogen activator has been cloned in *Escherichia coli* a bacterium and is produced in transformed mammalian cells which has been approved for treatment of myocardial infection.

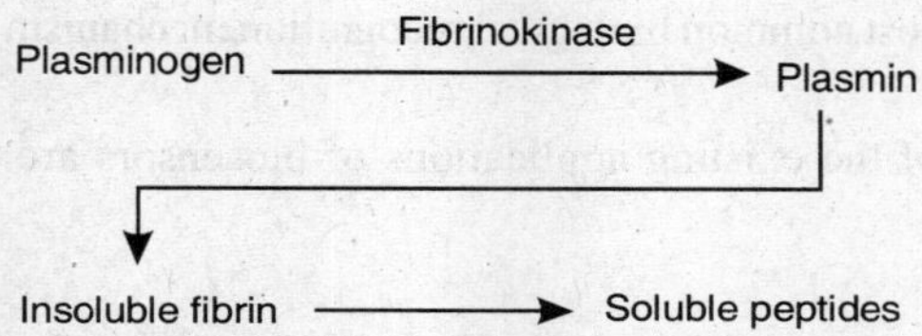

It is interesting that **hyaluronidase** obtained from beef testes is a useful agent to aid diffusion when coinjected with other drugs such as antibiotics, adrenalin, heparin and local anaesthetics. It is because that hyaluronidase causes hydrolysis of polyhyaluronic acid the main polysaccharide constituent of connective tissue, thereby help in diffusion of other substances by reducing the viscosity in the area of injections given.

Q. 18. Give the two main uses of immobilized enzymes.

Ans. Two main uses of immobilized enzymes are:

(i) Preparative – as large scale isomerization of glucose and production of L-aminoacids racemic mixtures

(ii) Analytical – as use of immobilized glucose oxidase.

Q. 19. How enzyme works as sweetener in production of syrups from corn starch?

Ans. It is well known that fructose is more satisfactory than glucose as it is sweeter and does not crystalize so readily in concentrated solutions. The over all process comprises of two enzyme catalyzed steps as below:

$$\text{Starch} \xrightarrow{\text{exo-1, 4 -}\alpha\text{- D- glucosidase}} \text{glucose} \xrightleftharpoons{\text{xylose isomerase}} \text{fructose}$$

Immobilized Whole Cells used in Amino Acid Interconversions

Q. 20. How immobilized whole cells may be used for amino acid interconversions? Discuss.

Ans. Not only single enzymes rather whole cells also can be immobilized by trapping the cells in polymerizing acrylamide and N, N′ – methylenebisacrylamide around the cells. Undamaged or complete cells of *Escherichia coli* and *Pseudomonas putida* having suitable enzymes, have been used for the production of aspartate, phenylalamine tyrptophane, serine, urocanic acid 6 – aminopenicillanic acid by entrapment technique.

Enzymatic Synthesis of Acrylamide

Q. 21. What is acrylamide? Discucss its enzymatic synthesis.

Ans. Acrylamide is used as a starting material for the production of a large number of polymers, as an ingredient in coagulators and as soil conditioner. Acrylamide can be made by reacting acrylonitrile and water in the presence of appropriate inorganic catalysts like copper but this leads to undesirable side reactions.

On the other hand nitriles are found in plants, bacteria and fungi, *e.g.,* in antibiotics like toyocamycin and treponemycin and some microorganisms are capable of using nitrile both as carbon and nitrogen source. Aliphatic nitriles in this way are degraded in two steps :

(i) The first step is catalysed by nitrile hydratase

(ii) The second is catalysed by an amidase

While *Rhodococcus rhodochrous* is grown on the optimum culture medium to induce nitrile hydratase, the latter stands for more than 50% of total soluble protein. The purification of enzyme is not necessary rather simply to entrap the organisms in a polyacrylamide gel. Thus it is one of the major products made using immobilized enzymes.

Enzymatic Biological Recognition Mechanisms or Use as Biosensors

Q. 22. Give one of the most common biological recognition mechanism or use of **biosensors** in clinical diagnosis.

Ans. More than half of the existing applications of biosensors are the devices to measure blood glucose.

49

CHAPTER

METABOLISM: CATABOLISM AND ANABOLISM

Salient Features

- The **metabolism** is a chain of energy releasing and energy consuming reactions to support life. It is the sum of all chemical reactions within a living organism performing the act of energy balancing.
- The **catabolism** is the breakdown of complex organic substances into simpler ones.
- The **anabolism** is the building of complex organic molecules from simple molecules involving **dehydration** through reactions which release water and require energy to form new bonds, as formation of proteins from amino acids, nucleic acids from nucleotides and polysaccharides from simple sugars.
- The catabolic reactions are **hydrolytic reactions** which use water and chemical bonds are broken in them.
- This coupling or linking of energy consuming and energy releasing reactions is made possible through molecules called **ATP** (adenosine triphosphate).
- A unique feature of **chemoautotrophic** bacteria is that they can grow on inorganic substances like carbon dioxide, sulphur, hydrogen gas and ammonia.
- The enzymes are biological catalysts and make chemical reactions faster by lowering the activation energy.
- Many enzymes are **holo enzymes** means complete or whole enzymes consisting of an **apoenzime** which is a protein part and co-factor a nonprotein part. The **co-factor** can be a metal iron, copper, magnesium, zinc, calcium or cobalt or a complex organic molecule called **coenzyme** NAD (Nicotinomide adenine dinucleotide), NADP (Nicotinomide adenine dinucleotide phosphate), FMN (Flavin mononucleotide), FAD (Flavin adanine dinucleotide) and COA (Coenzyme A)
- When the end product of a pathway inhibits the activity of an enzyme it is called **feedback inhibition**.
- **Oxidation** (removal of elections) and **reduction** (gain of elections) are reaction that go side by side.
- In **Photophosphorylation** energy from light as **photons** is trapped by chlorophyll. The electrons are passed through a series of electron acceptors thus releasing energy used for synthesis of ATP..
- There are two main types of **glucose**-catabolism : (1) **Respiration** in which glucose is completely broken down, (2) Fermentation in which glucose is broken down partially.
- **Glycolysis** is the most common pathway for oxidation of glucose leading to formation of **pyruvic acid** (= pyruvate) as an end product; generating two ATP and two NADH molecules from one molecule of glucose.
- The **alternatives to glycolysis** are : 1. **Pentose phosphate pathway** used to metabolize five carbon sugars in which one ATP and 12 NADPH molecules are produced from one molecule

of glucose. 2. **Enter-Doudoroff pathway** found in some gram negative bacteria neither employ glycolysis nor pentose phosphate pathway, and produce one ATP and two NADPH molecules from one molecule of glucose.

- The organic molecules are oxidized in **respiration** and energy is generated from electron transport chain. The oxygen serves as final electron acceptor in aerobic oxidation. The final electron acceptor in **anaerobic respiration** is an inorganic molecule other than oxygen.
- The pyruvic acid cannot enter directly in **KREBS CYCLE**. The decarboxylation of pyruvic acid produces one molecule of CO_2 and one acetyl group. In Krebs cycle two acetyl groups are oxidized and electrons are picked up by NAD^+ and FAD for electron transport-chain. The oxidation of one molecule of glucose produces six molecules of NADH, two molecules of $FADH_2$ and two molecules of ATP. Decarboxylation produces six molecules of CO_2.
- A sequence of carrier molecules form the **electron transport chain** (ETC) also called electron transfer chain. Electrons are brought to the electron transport chain by NADH. The electron transport chain comprises mainly of three classes of carrier molecules, flavoproteins, cytochromes and ubiquinones.
- There exists a **CHEMIOSMOTIC MECHANISM OF ATP GENERATION**, in which the protons pumped across the membrane generate a proton motive force while electrons move through a series of acceptors or carriers. The energy so produced from movement of protons back across the membrane is used by enzyme **ATP synthase** to produce ATP from ADP and inorganic phosphorus. Now it is obvious that in eucaryotes electron carriers are located on the inner mitochondrial membranes while in procaryotes electron carrier are in the plasma membrane.

  AEROBIC RESPIRATION is found in procaryotes as well as eucaryotes. In aerobic procaryotes 38 ATP molecules are produced on complete oxidation of one molecule of glucose in glycolysis, the Krebs cycle and ETC (Electron transport chain). However, in eucaryotes 36 ATP molecules are produced from complete oxidation of one molecule of glucose.

  In **ANAEROBIC RESPIRATION** the oxygen is not the final electron accepter rather may be inorganic ions like nitrate (NO^{-3}), sulphate ($SO^{2-}{}_4$) and carborate ($CO^{2-}{}_3$). The total ATP yield also is less than that of aerobic respiration since only part of Krebs cycle goes under anaerobic conditions.

  The **FERMENTATION** also is an energy yielding metablism, and release energy from sugars or other organic molecules by oxidation. It produces two ATP molecules by substrate level phosphorylation. The electrons removed from the substrate reduce NAD+. The final electron acceptor in fermentation is some organic molecule. The pyruvic acid in lactic acid fermentation is reduced by NADH to lactic acid. In alcohol fermentation acetaldehyde is reduced by NADH to produce ethanol.

- In **LIPID CATABOLISM** the enzymes **lipases** hydrolyze lipids into glycerol and fatty acids. The glycerol is catabolized by conversion to dihydroxyacetonephosphate while fatty acids are catabolized by beta oxidation. The **beta oxidation** produces two carbon units that are linked to COA to make acetyl COA.
- In **PROTEIN CATABOLISM** before amino acids are catabolized they must be converted to substances which are capable to enter the Krebs cycle. Transamination, decarboxylation and dehydrogenation are required to convert amino acids to suitable substances to enter Krebs cycle.
- In **PHOTOSYNTHESIS** the light energy from sun is converted to chemical energy which is further used for fixation of carbon.
- Photosynthesis takes place under two stages: 1 **LIGHT REACTION OR PHOTOPHOSPHORYLATION** and 2. The **DARK REACTIONS** or **CALVIN-BENSON CYCLE**. The photophosphorylation is further of two types which are **CYCLIC PHOTOPHOSPHORYLATION** and **NON CYCLIC PHOTOPHOSPHORYLATION**.

- The **ANABOLISM** or major pathways of energy consumption are biosynthesis of polysacchazsides, that of lipids, amino acids and proteins, and purines and pyrimidines.

Q.1. What is the role of ATP (adenosine triphosphate) in coupling or linking of energy releasing reactions, the **catabolism** and energy requiring reaction, the **anabolism**?

Ans. The adenosine triphosphate (ATP) stores energy produced by catabolic reactions and releases it later to run anabolic reactions and other cellular activities. The ATP molecule consists of an adenine, a ribose and three phosphate groups. When the terminal phosphate group goes out from ATP, adenosine diphosphate (ADP) gets formed and consequently energy is released to sum the anabolic reactions. Representing the phosphate group with P the reaction can be written as:

$$\text{ATP} \longrightarrow \text{ADP} + \text{P} + \text{Energy}$$

The energy released by catabolic reactions is used to combine ADP with P, The phosphate group synthesize the ATP again which can be represented as:

$$\text{ADP} + \text{P} + \text{Energy} \longrightarrow \text{ATP}$$

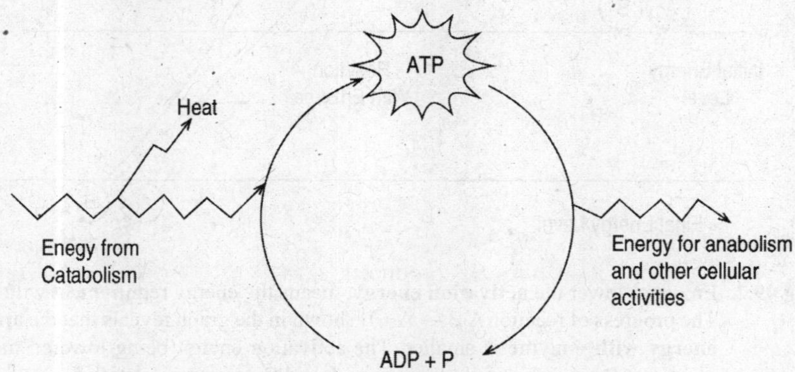

Fig. 49.1. Generalized representation of relationship between catabolism, anabolism and adenosine triphosphate (ATP).

The anabolic reactions are linked to ATP break down while catabolic reactions are linked to synthesis of ATP (Fig. 49.1). The cells metabolic pathways are determined by its enzymes and the enzymes in turn by the genetic makeup of the cell.

Enzymes in Cellular Metabolism

Q.2. What is turnover number of enzyme? Give the characteristics of enzymes which support the cellular metabolism.

Ans. The **turnover number of an enzyme** is the maximum number of substrate molecules an enzyme may convert to product each second, which commonly is between 1 and 10000 and may be higher upto 500,000 for instance the enzyme **DNA polymerase** – I which takes part in synthesis of DNA gives a turnover number of 15 while the enzyme **lactate dehydrogenase** that removes hydrogen atom from lactic acid gives a turnover number of one thousand.

The atoms, ions and molecules must collide for chemical reactions to take place which depends on amount of energy required to begin the reaction called **activation energy**, specific configuration of the particles, physiological temperature and pressure of the cell or the organism which are too low for the chemical reactions to take place. Practically raising the temperature and pressure may damage or kill the organism. The solution to such problems are provided by a class of proteins referred to as **enzymes**, which are also known as **biological catalysts** and this term has also been abbreviated as biocatalysts. Depending on the type of reaction catalyzed by them the enzymes have been grouped under six classes. The enzymes involved in oxidation – reduction reactions have been called **oxidoreductases**, in which oxygen and hydrogen are either gained or lost. The specific examples of oxidoreductases are cytochrome oxidase and Lactase dehydrogenase. The other enzyme classes are **transferases**

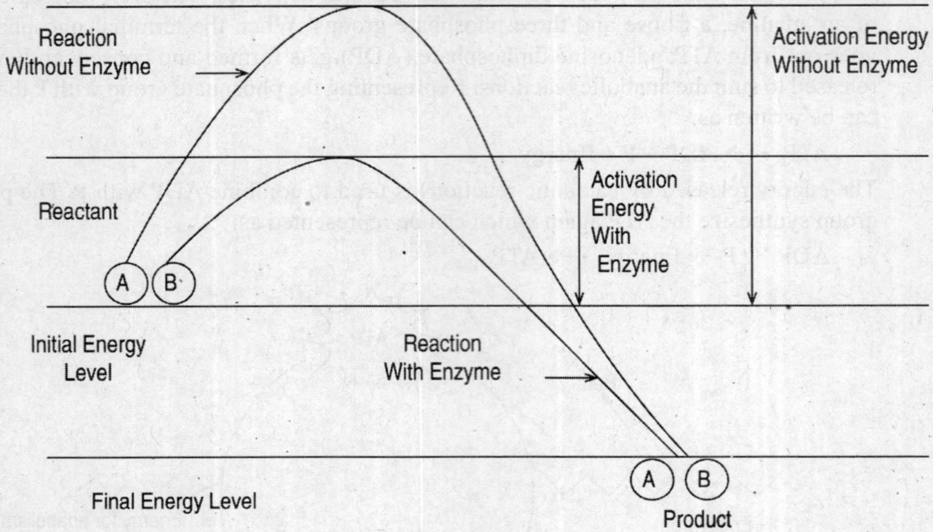

Fig 49.2. Enzymes lower the **activation energy**, means the energy requirement with enzyme is lesser. The progress of reaction A B $\rightarrow$ A + B shown in the graph reveals that the arrow of **activation energy with enzyme** is smaller. The activation energy being lowered more molecules of reactant AB are converted to products A and B as more molecules of reactant AB have the activation energy required for the reaction.

which transfer functional groups as amino group, acetyl group or phosphate group, the specific examples of which are acetate kinase and alanine deaminase; **hydrolases** which carry hydrolysis or addition of water e.g. Lipase and suezase; lyases which remove groups of atoms without hydrolyses e.g. oxalate decarboxylase, and isocitrate **lyase**; **isomerases** which catalyze rearrangement of atom within a molecule; and **ligases** which catalyze joining of two molecules using energy usually obtained from break down of ATP the specific examples being acetyl – COA synthetase and DNA ligase.

Enzyme Components

Some enzymes consist of proteins only but most of them consit of protein part called apoenzyme and nonprotein part called a cofactor. Both apoenzyme and cofactor collectively form **holoenzyme** or the complete enzyme. If the cofactor is removed the **apoenzyme** will not function. The **cofactor** may be a metal ion or some complex organic molecule known as **coenzyme**. These coenzymes can help the enzyme by accepting atoms removed from the substrate or by donating the atoms required by the substrate. NAD^+ (Necotinamide adenine dinucleotide) and $NADP^+$ (Necotinamide adenine dinucleotide Phosphate) are two among the most important coenzymes seen in cell metabolism.

Both of these possess derivatives of the **B vitamin** nicotinic acid or niacin and both work as carriers of electrons. However, NAD^+ is primarily involved in energy yielding **catabolic reactions** and $NADP^+$ is primarily involved in energy requiring **anabolic reactions**. The two flavin coenzymes

FMN (flavin mononucleotide) and **FAD** (flavin adenine dinucleotide) which contain derivatives of B vitamin **riboflavin** (vitamin B2 or lactoflavin) a water soluble, photolabile heat stable vitamin, both of which are electron carrier. **CoA** is also a coenzyme called coenzyme A has a derivative of pantothenic acid another water soluble vitamin B. The CoA can perform synthesis and breakdown of fats in a series of oxidizing reaction in a cycle known as Krebs cycle.

The **metal ions** like iron, copper, magnesium, manganese, zinc calcium and cobalt also work as cofactors. Such cofactors catalyze the reaction by forming a bridge between the enzyme and a substrate as magnesium (Mg^{2+}) is needed by many phosphorylating enzymes which transfer a phosphate group from ATP to another substrate. Thus Mg^{2+} can form a link between the enzyme and the molecule of ATP. This sort of mechanism has been adopted by most of the trace elements to activate the cellular enzymes in the living cells.

Mechanism of Enzymatic Action

The enzymes perhaps work through the following states: 1. The surface of the substrate contracts a specific region on the surface of the enzyme molecule which has been called the **active site**. 2. A temporary intermediate compound called **enzyme substrate complex** gets formed. 3 The substrate molecule is transformed by the rearrangement of rusting atoms, the breakdown of the substrate molecule or combination with some other substrate molecule which may take place. 4. The transformed substrate molecules, and the products of the reaction are released from the enzyme molecule because they no longer fit in the active site of the enzyme . The unchanged enzyme becomes free to react with fresh substrate molecule. In this way an enzyme speeds up a chemical reaction.

The enzymes are specific in their action and act on particular substrate as a specific enzyme may be able to hydrolyze a peptide bond only between two specific amino acids. Some enzymes can hydrolyze starch but not cellulose, even though both starch and cellulose are polysaccharides made of glucose subunits but the orientation of subunits in both of the said polysaccharides is different. The **glucose 6 – phosphate** which is important in cell metabolism can be acted upon by four different enzyme and each reaction will form a different product.

Factors Affecting Enzymatic Activity

The rate of chemical reactions increases with the increase or rise of **temperature** but beyond certain limit the rise of temperature may be harmful because enzymes being proteins get denatured one of the simplest examples being hardening of albumin or egg while on certain state of heating. The proteins get denatured on heating because of breakage of hydrogen bonds and other noncovalent bonds. The enzymes may also get denatured by concentrated acids, bases, heavy metal ions like lead, organic or mercury, alcohol and UV radiations. When the hydrogen ion concentration (H^+) or pH of the medium is changed the amino acid of some enzymes are affected and the three dimentional structure of the protein changes under conditions of high substrate concentration the enzyme is said to be in a state of **saturation.**

Some chemical substances also work as enzyme inhibitor which may be of two types: **competitive inhibitors** and **noncompetitive inhibitors**. A typical example of competitive inhibitors is **sulphanilamide**, a sulpha drug which inhibits the enzyme, the normal substrate of which is **para amino benzoic acid or** PABA.

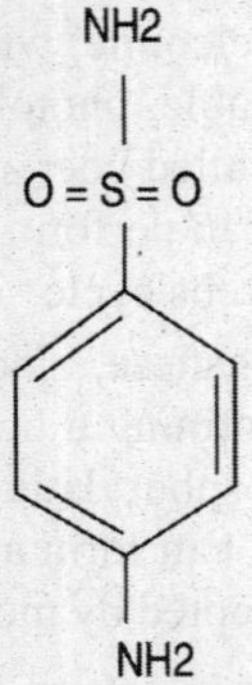

Sulphanilamide

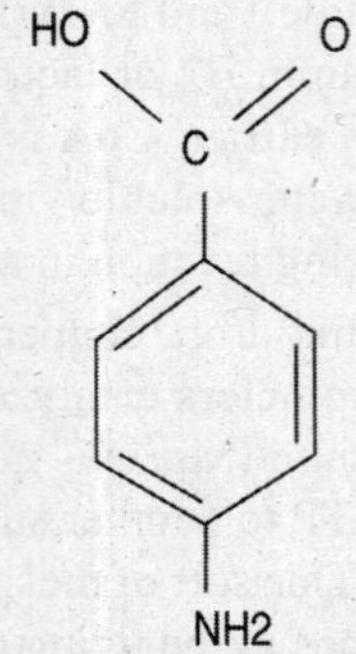

PABA(Para aminobenzoic acid)

PABA is an essential nutrient required by many bacteria in the synthesis of **folic acid** a water soluble photolabile vitamin, which functions as a coenzyme. When sulphanilamide is given to bacteria, the enzyme which generally converts 'para amino benzoic acid' to folic acid combines with the sulphanilamide, consequently the bacteria fail to grow and folic acid is not synthesized.

It is of great interest that human cells do not use PABA (para amino benzoic acid) to prepare their folic acid. That is how the sulphanilamide kills bacteria without producing any harm to human cells.

NONCOMPETITIVE INHIBITORS: These do not compete with the substrate for the active site of the enzyme rather interact with another part of the enzyme. This gave rise to the term allosteric from greek *allows* = other, *stereos* = solid or shape. This process is called **allosteric inhibition** in which the inhibitor binds to site on the enzyme other than the substrate's binding site this binding causes the **active site to change its shape** and turning the active site nonfunctional. Consequently the activity of enzyme is reduced. However, this effect may be either reversible or irreversible depending on the probability that active site of the enzyme can return to its original shape or it cannot return.

Some chemicals may bind the metal ion activators and thus prevent the enzymatic reaction. Some typical examples are that cyanide may bind iron in iron containing enzymes; and fluoride may bind calcium or magnesium. Such substances like cyanide and fluoride therefore, have also been called **enzyme poisons** since they inactivate the enzymes permanently.

FEEDBACK INHIBITION: The allosteric inhibitors defined and discussed under noncompetitive inhibitors produce a type of biochemical control called **feedback inhibition** or **end product inhibition**. This control process steps the cell from wasting chemical resources by producing more of a substance that the cell needs. Many steps are required for the synthesis of a particular chemical compound referred to as **end product** in performance of some metabolic reactions. The process may be compared to an assembly line with each step catalyzed by a separate enzyme.

Generally the **feedback inhibitors** act on the first enzyme in a metabolic pathway just like shutting down an assembly line by stoping the first worker. Since the enzyme is inhibited, the product of the first enzymatic reaction in the pathway is not synthesized or produced. This unsynthesized product would have normally worked as the substrate for the second enzyme in the pathway. Thus the second reaction also stops. Though the first enzyme of the pathway is inhibited but the entire pathway shuts down and no new end product is formed.

The bacterium *Escherichia coli* server as a typical example to demonstrate the feedback inhibition in the synthesis of the amino acid isoleucine which is required for the growth of the all. There are 5 steps in this metabolic pathway to convert the amino acid **threonine** to **isoleucine** enzymatically. But if isoleucine is added to the growth medium for *Escheria coli* it inhibits the first enzyme in the pathway and consequently bacteria stop synthesizing isoleucine. This continues as such until the supply of isoleucine is depleted.

This sort of feedback inhibition is also involved in regulating the cell's production of other amino acids, vitamins and pyrimidines,

Q.3. Describe the process of production of energy and generation of ATP for conveniency of cells.

Ans. In all cells the nutrient molecules like all other molecules possess energy associated with the **electrons** which form **bonds** between their atoms. But this energy being spread throughout the molecule is not convenient for the cells to use. Many reactions going in the catabolic pathways are able to concentrate the energy into the bonds of ATP (Adenosine 5-triphosphate). In both procaryotic and eucaryotic cells ATP actively participate in a range of processes and reactions which involve the conversion and expenditure of energy.

The concept of **"Oxidation - reduction"** and the **"generation of ATP"** are the vital aspects concerning production of energy.

Oxidaction – Reduction

The oxidation is the process of removal of electrons(e^-) from an atom or molecule which is accompanied by production of energy. The molecule on oxidation loses one or more electrons while the molecule which undergo reduction gains one or more elections. The oxidation and reduction reactions are always coupled that means each time one substance is oxidized while other is simultaneously reduced. This type of pairing of reactions has been called **oxidation – reduction.** In cellular oxidations when electrons and protons (**hydrogen ions H^+**) are removed at the same time that is equivalent to removal of hydrogen atoms since a hydrogen atom is one proton and one electron. In most of the biological oxidations there is less of hydrogen atoms therefore, they have also been called **dehydrogenation reactions**. NAD accepts two electrons and one proton. One proton is released into the surrounding medium.

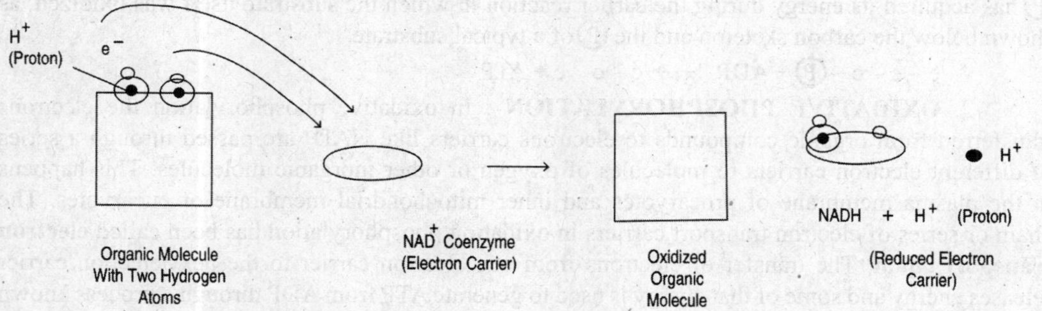

Fig. 49.3. Diagramatic representation of biological oxidation. Two electrons and two protons which are equivalent to **two hydrogen atoms** are transferred from an **organic substrate molecule** to a coenzyme NAD^+. The **coenzyme** NAD^+ receives one hydrogen atom and one electron. One proton is released into the medium. The NAD^+ on getting reduced to NADH becomes a **more energy rich molecule**. NAD also called **coenzyme I** or **cozymase** works as a hydrogen carrier in a wide range of **redox reactions**.

It is obvious that **reduced coenzyme** NADH possesses more energy in comparison to NAD^+ also called **coenzyme I** or **cozymase**. This gained energy can be used to generate ATP. The biological oxidation - reduction reactions have also been called **redox reactions**.

It is interesting to note that cells use nutrients some of which are sources of energy. The cells degrade them from **highly reduced compounds** with many hydrogen atoms to highly oxidized compounds. A simple example is that while a cell oxidises a molecule of glucose, which is $C_6H_{12}O_6$

to carbon dioxide (CO_2) and water (H_2O). The energy stored in the glucose molecule is removed in a stepwise manner and finally is trapped by ATP that is used in the energy needing reactions. The compounds like glucose possess many hydrogen atoms and therefore, are highly reduced compounds and bear a large amount of potential energy.

Generation of ATP

ATP transports chemical energy within a cell for metabolism. ATP was discovered by carl Lohmann in 1929 and then independently by Cyrus Fiske and Yellapragada Subbarow of Horward Medical School. However, its correct structure was not determined until some years later. It was proposed to be the intermediate molecule between energy yielding and energy requiring reaction by Fritz Albert Lipmann in 1941. It was first artificially synthesized by Alexander Todd in 1948.

A phosphate group Ⓟ, is added to ADP (adenosine diphosphate) with the input or consumption of energy to form ATP (adenosine triphosphate).

Adenosine ~ Ⓟ ~ Ⓟ + Energy + Ⓟ ⟶

Adenosine ~ Ⓟ ~ Ⓟ + Ⓟ

Adenosine triphospate (ATP)

The symbol ~ stands for a **high energy bond**. The addition of Ⓟ to a chemical compound is called **phosphorylation**. There are three mechanism of phosphorylation to generate ATP from ADP that are used by the organisms.

1. **SUBSTRATE LEVEL PHOSPHORYLATION** : The ATP is generated in it when a high energy Ⓟ is directly transferred from a phosphorylation compound or a substrate to ADP. In it the Ⓟ has acquired its energy during the earlier reaction in which the substrate itself was oxidized, as shown below the carbon skeleton and the Ⓟ of a typical substrate.

   c – c – c ~ Ⓟ + ADP ⟶ c – c – c + ATP

2. **OXIDATIVE PHOSPHORYLATION** : In oxidative phosphorylation the electrons transferred from organic compounds to electrons carriers like NAD^+ are passed through a series of different electron carriers to molecules of oxygen or other inorganic molecules. This happens in the plasma membrane of procaryotes and inner mitochondrial membrane of eucaryotes. The chain or series of electron transport carriers in oxidation phosphorylation has been called **electron transport chain**. The transfer of electrons from one electron carrier to the next election carrier releases energy and some of that energy is used to generate ATP from ADP through a process known as chemiosmosis. In **chemiosmosis** ATP is formed from ADP by means of enzyme called ATP ase.

3. **PHOTOPHOSPHORYLATION** : This process takes place in photosynthetic cells which bear light trapping pigments like chlorophylls. In photosynthesis sugars are synthesized with the help of energy of light using **energy – poor building blocks** carbon dioxide and water. The **photophosphorylation** begins its work by converting light energy to chemical energy of ATP and NADPH which are finally used to synthesize organic molecules. Like oxidative phosphorylation an electron transport chain is involved. This cyclic electron flow causes the transmembrane pumping of protons means it generates **pmf** or proton motive force. The **pmf** can be used for ATP synthesis.

Biochemical Pathways and Production of Energy

Q.4. Why do organisms store and release energy by a series of controlled reactions instead of releasing energy all of a sudden at once?

Ans. The readily release of energy all at once might have damaged the cell disrupting its biochemical pathways because of sudden burst of energy. That is why the organisms release and store energy from organic molecules in the form of a series of controlled reactions.

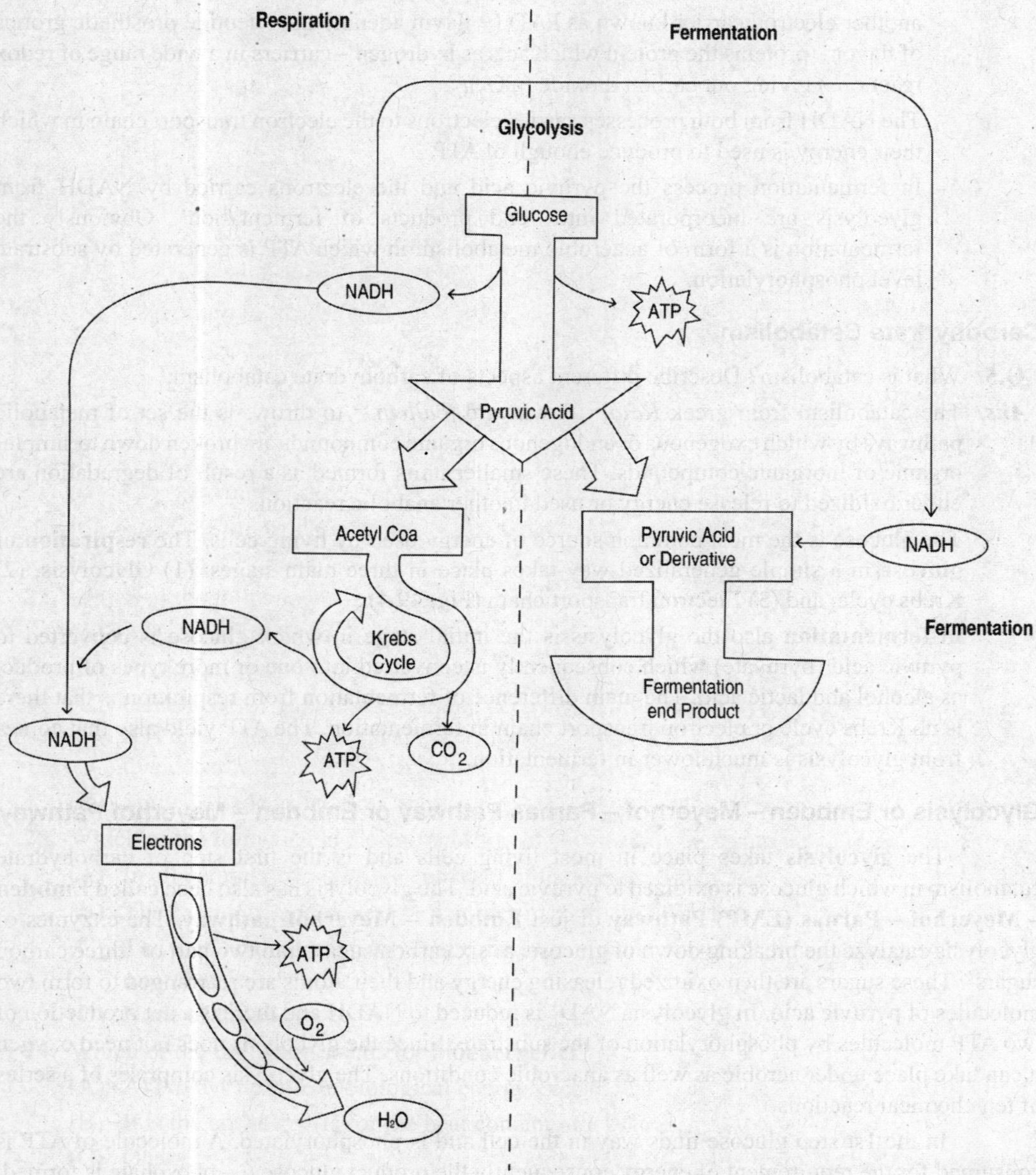

Fig. 49.4. General view of **respiration** and **fermentation**.

A general view of respiration and fermentation makes it clear (Fig. 49.4) that how does the process of **glycolysis** that makes a part of both respiration and fermentation (from **glycose** an older term for glucose + lysis = degradation is the metabolic pathway which converts glucose $C_6H_{12}O_6$ into pyruvate $CH_3COCOO^{+}H^{+}$ via ten enzymatic steps) produces ATP and reduces NAD^+ to NADH in the oxidation of **glucose** to **pyruvic acid.** In respiration the pyruvic acid is converted into the first reactant in the **Krebs cycle** also called tricarboxylic acid (TCA) cycle produces ATP and reduces NAD^+ and another electron carrier known as FAD (= flavin adenine dinucleotide prosthetic groups of flavor – proteins the protein which acts as hydrogen – carriers in a wide range of redox reactions) giving out carbon dioxide (CO_2).

The NADH from both processes carries electrons to the electron transport chain in which their energy is used to produce enough of ATP.

In fermentation process the pyruvic acid and the electrons carried by NADH from glycolysis are incorporated into 'end products of fermentation'. Obviously the fermentation is a form of anaerobic metabolism in which ATP is generated by substrate level phosphorylation.

Carbohydrate Catabolism

Q.5. What is catabolism? Describe different aspects of carbohydrate catabolism.

Ans. The catabolism from greek *Kato* = downward, *ballein* = to throw; is the set of metabolic pathways by which exogenous or endogenous organic compounds are broken down to simpler organic or inorganic compounds. These smaller units formed as a result of degradation are either oxidized to release energy or used in other anabolic reactions.

The glucose is the most common source of energy used by living cells. The **respiration of glucose** in a simple generalized way takes place in three main stages: (1) Glycolysis, (2) Krebs cycle, and (3) Electron transport chain (Fig. 49.4).

In **fermentation** also the glycolysis is the initial stage in which glucose is converted to pyruvic acid (pyruvate) which consequently is converted into one or more types of product as alcohol and lactic acid. The main difference of fermentation from respiration is that there is no Krebs cycle or electron transport chain in fermentation. The ATP yield also that comes from glycolysis is much lower in fermentation.

Glycolysis or Embden – Meyerhof – Parnas Pathway or Embden – Meyerhof Pathway

The **glycolysis** takes place in most living cells and is the first step of carbohydrate catabolism in which glucose is oxidized to pyruvic acid. The glycolysis has also been called **Embden – Meyerhof – Parnas (EMP) Pathway** or just **Embden – Meyerhof pathway.** The enzymes of glycolysis catalyze the breaking down of glucose, a 'six carbon sugar' into two part of 'three carbon sugars'. These sugars are then oxidized releasing energy and their atoms are rearranged to form two molecules of pyruvic acid. In glycolysis NAD^+ is reduced to NADH and there is a net production of two ATP molecules by phosphorylation of the substrate. Since the glycolysis does not need oxygen it can take place under aerobic as well as anaerobic conditions. The glycolysis comprises of a series of ten chemical reactions.

In the 1st step glucose finds way in the cell and is phosphorylated. A molecule of ATP is consumed for the requirement of energy consequently the product glucose 6 – phosphate is formed. In the 2nd step glucose 6 – Phosphate is rearranged to form its isomer fructose – 6 phosphate. In the 3rd step the P from another ATP is used to produce fructose 1, 6 – diphosphate still a 6 – carbon compound. The total number of ATP molecules used upto this step is two. In the 4th step the enzyme splits or breaks the sugar into two 3 – carbon molecules which are dihydroxyacetone phosphate (DHAP) and glyceraldehyde 3 – phosphate (GP). In the 5th step dihydroxyacetone phosphate and glyceraldehyde 3 – phosphate are interconverted since they are reversible or interconvertible. In the

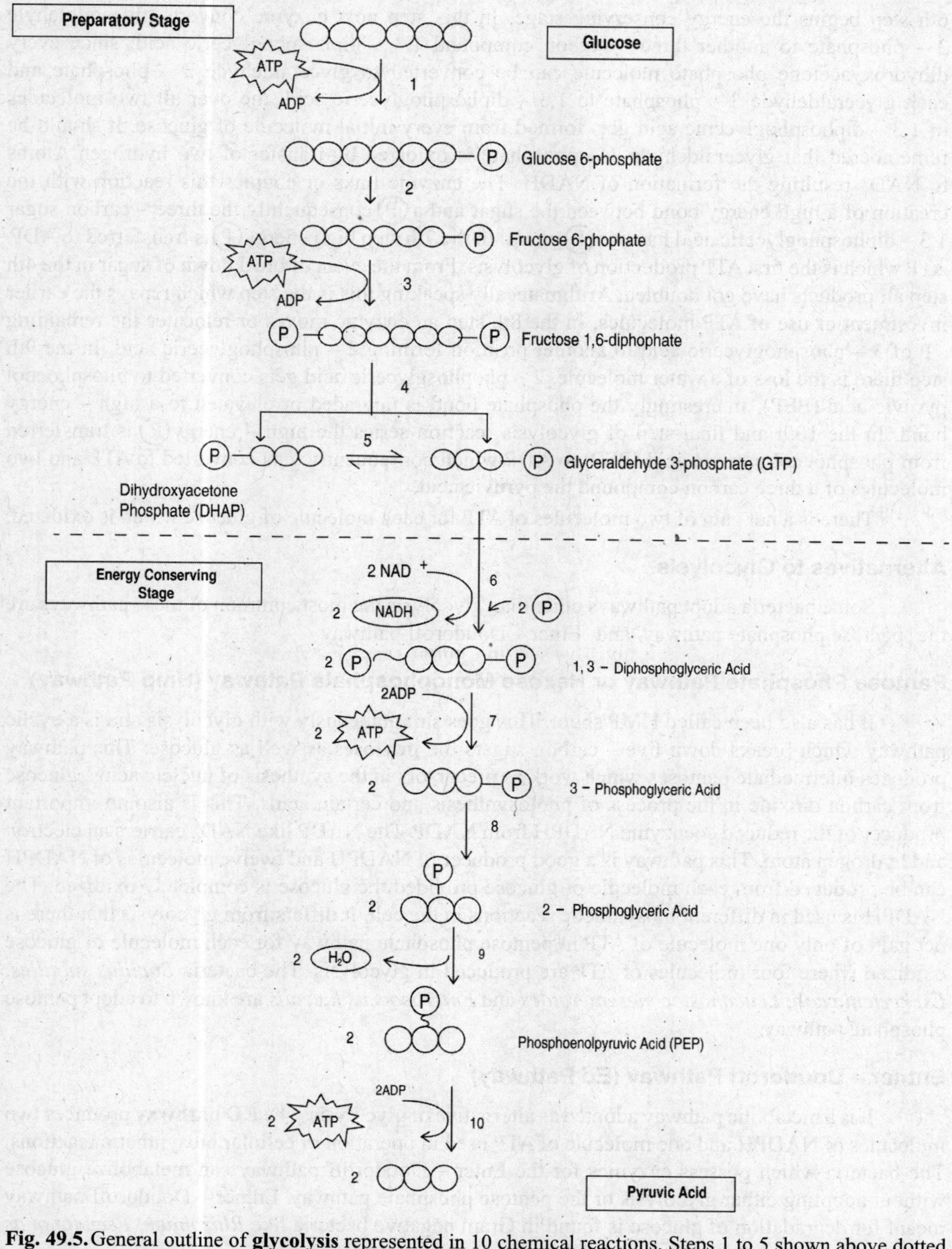

Fig. 49.5. General outline of **glycolysis** represented in 10 chemical reactions. Steps 1 to 5 shown above dotted line represent **preparatory stage** and that from 6 to 10 below dotted line represent **energy conserving stage**.

6th step begins the energy conserving stage. In this step next enzyme converts glyceralolehyle 3 – phosphate to another three – carbon compound 1,3 – diphosphoglyceric acid; since every dihydroxyacetone phosphate molecule can be converted to glyceraldehyde 3 – phosphate and each glyceraldehyde 3 – phosphate to 1,3 – diphosphoglyceric acid; the over all two molecules of 1,3 – diphosphoglycenic acid gets formed from every initial molecule of glucose. It should be remembered that glyceraldehyde 3 – phosphate is oxidized by transfer of two hydrogen Atoms to NAD^+ resulting the formation of NADH. The enzyme links or couples this reaction with the creation of a high energy bond between the sugar and a Ⓟ consequently the three – carbon sugar 1,3 – diphosphoglyceric acid has two Ⓟ group. In the 7th step high energy Ⓟ is transferred to ADP. ATP which is the first ATP production of glycolysis. From the event of breakdown of sugar in the 4th step all products have got doubled. Arithmatecally speaking this is the step which repays the earlier investment or use of ATP molecules. In the 8th step an enzyme moves or relocates the remaining P of 3 – phosphoglyceric acid to another position forming 2 – phosphoglyceric acid. In the 9th step there is the loss of a water molecule, 2 – phophosglyceric acid gets converted to phosphoenol pyruvic acid (PEP). Interestingly the phosphate bond is upgraded or elevated to a high – energy bond. In the 10th and final step of glycolysis reaction series the high – energy Ⓟ is transferred from phosphoenol pyruvic acid (PEP) to ADP which consequently gets converted to ATP and two molecules of a three carbon compound the pyruvic acid.

There is a net gain of two molecules of ATP for each molecule of glucose which is oxidized.

Alternatives to Glycolysis

Some bacteria adopt pathways other than glycolysis the most common of these pathways are the 'pentose phosphate pathway' and 'Etner – Doudoroff pathway'.

Pantose Phosphate Pathway or Hexose Monophosphate Pathway (Hmp Pathway)

It has also been called HMP shunt. This goes simultaneously with glycolysis and is a cyclic pathway which breaks down five – carbon sugars the **pentoses** as well as glucose. This pathway produces intermediate pentoses which work as precursors in the synthesis of nucleic acids, glucose from carbon dioxide in the process of photosynthesis and certain acids. This is also an important producer of the reduced coenzyme NADPH from NADP. The NADP like NAD^+ carriers an electron and hydrogen atom. This pathway is a good producer of NADPH and twelve molecules of NADPH can be produced from each molecule of glucose provided the glucose is completely oxidized. The NADPH is used in different biosynthetic reactions in the cell. It differs from glycolysis that there is net gain of only one molecule of ATP in pentose phosphate pathway for each molecule of glucose oxidized where four molecules of ATP are produced in glycolysis. The bacteria *Bacillus subtilies*, *Escherichia coli*, *Leuconostoc mesenteroides* and *Enterococcus faecalis* are known to adopt pentose phosphate pathway.

Entner – Doudoroff Pathway (Ed Pathway)

It is a metabolic pathway adopted as alternative of glycolysis. The **ED pathway** produces two molecules of NADPH and one molecule of ATP used in operation of cellular biosynthetic reactions. The bacteria which possess enzymes for the Enter – Doudoroff pathway can metabolize glucose without adopting either glycolysis or the pentose phosphate pathway. Entner – Doudoroff pathway meant for degradation of glucose is found in Gram negative bacteria like *Rhizobium, Pseudononas* and *Agrobacterium*. It is generally not found in Gram positive bacteria.

The ability of *Pseudomonas* to oxidize glucose by this pathway is used as a test for identification of *Pseudomonas* in clinical laboratory.

Some bacteria like *Escherichia coli* which metabolize glucose mainly via the EMBDEN – MEYERHOF – PARNAS pathway can metabolize gluconate and other aldonates via reactions of Entrer – Doudoroff pathway.

Respiration

The respiration is an ATP generating energy – yielding metabolism in which the oxidation of an energy substrate involves an **exogenous electron acceptor** i.e. an externally derived oxidizing agent. The glucose an energy yielding carbohydrate on being broken down to pyruvic acid, in the next step either may undergo fermentation or the respiration. In aerobic respiration the final electron acceptor is O_2 while in anaerobic respiration the final electron acceptor is an inorganic molecule other than molecular oxygen or rarely an organic molecule.

Krebs Clycle or Tricarboxylic Acid (TCA) Cycle or Citric Acid Cycle

The **Krebs cycle** was named after German biochemist Hans Krebs who identified its steps in 1930's. It metabolizes two carbon units called acetyl groups to CO_2 and H_2O. It also has been called **tricarboxylic acid cycle** or TCA cycle because some molecules in the cycle have three carboxyl (COOH) groups or **citric acid cycle** because citric acid is an important intermediate compound.

It should be noted that pyruvic acid which is a product of glycolysis cannot enter the Krebs cycle directly. Therefore, pyruvic acid must be converted to **acetyl CoA**, a two carbon compound by removal of CO_2, transfer of electron to NAD to reduce to NADH and addition of coenzyme. These reactions in procarytes take place in cytoplasm while in eucaryotes in the matrix of mitochondria.

It is worth noting that the oxidation of one molecule of glucose produces two molecules of pyruvic acid. Once the pyruvic acid has undergone decarboxylation and its derivative, acetyl group has got attached to CoA, the resulting acetyl CoA is ready to enter the Krebs cycle.

Each reaction in the Krebs cycle is controlled by a specific enzyme and the molecules are passed from one enzyme to next as they go through the cycle the reactions from a cycle because oxaloacetic acid (oxaloacetate), a **first reactant** is generated at the end of the cycle. As one acetyl group is metabolized the oxaloacetate combines with another to form **citric acid**. (Fig. 49.6). For the sake of convenience the whole cycle may be divided into 8 steps. As has been just told that the 1st step or turn of cycle starts as enzymes remove the CoA part from acetyl CoA and combine the remaining two – carbon acetyl group with oxaloacetic acid. The addition of acetyl group to oxaloacetic acid or oxaloacetate a four – carbon molecule result the formation of the six – carbon molecule citric acid. The 2nd step is the rearrangement or formation of an isomer in the form of **isocitric acid**. The 3rd step result the formation of **α - Ketoglutaric acid**. 3rd and 4th steps carry on oxidations and decarboxylations. Two carbon atoms that came from oxaloacetic acid are disposed of through these steps. The carbons are released as CO_2 and the oxidations generate NADH form NAD^+. The CoA is added in the cycle in the second oxidation shown in the 4th step, and result the formation of **succinyl CoA.** In the 5th step ATP is produced by substrate level phosphorylation. CoA is removed from succinyl CoA leaving behind succinic acid, a four – carbon compound. In step 6 to 8 the enzymes rearrange chemical bonds leading to formation of three different molecules. These are **fumaric acid, malic acid** and **oxaloacetic acid**. In the 6th step the oxidation produces $FADH^2$. In the 8th step a final oxidation generates NADH and converts malic acid to oxaloacetic acid which is ready to enter the next round of Krebs cycle, again. Obviously the events of Krebs cycle which are of special significance are oxidation of carbon, transfer of electrons to co-enzymes and substrate level energy transfer.

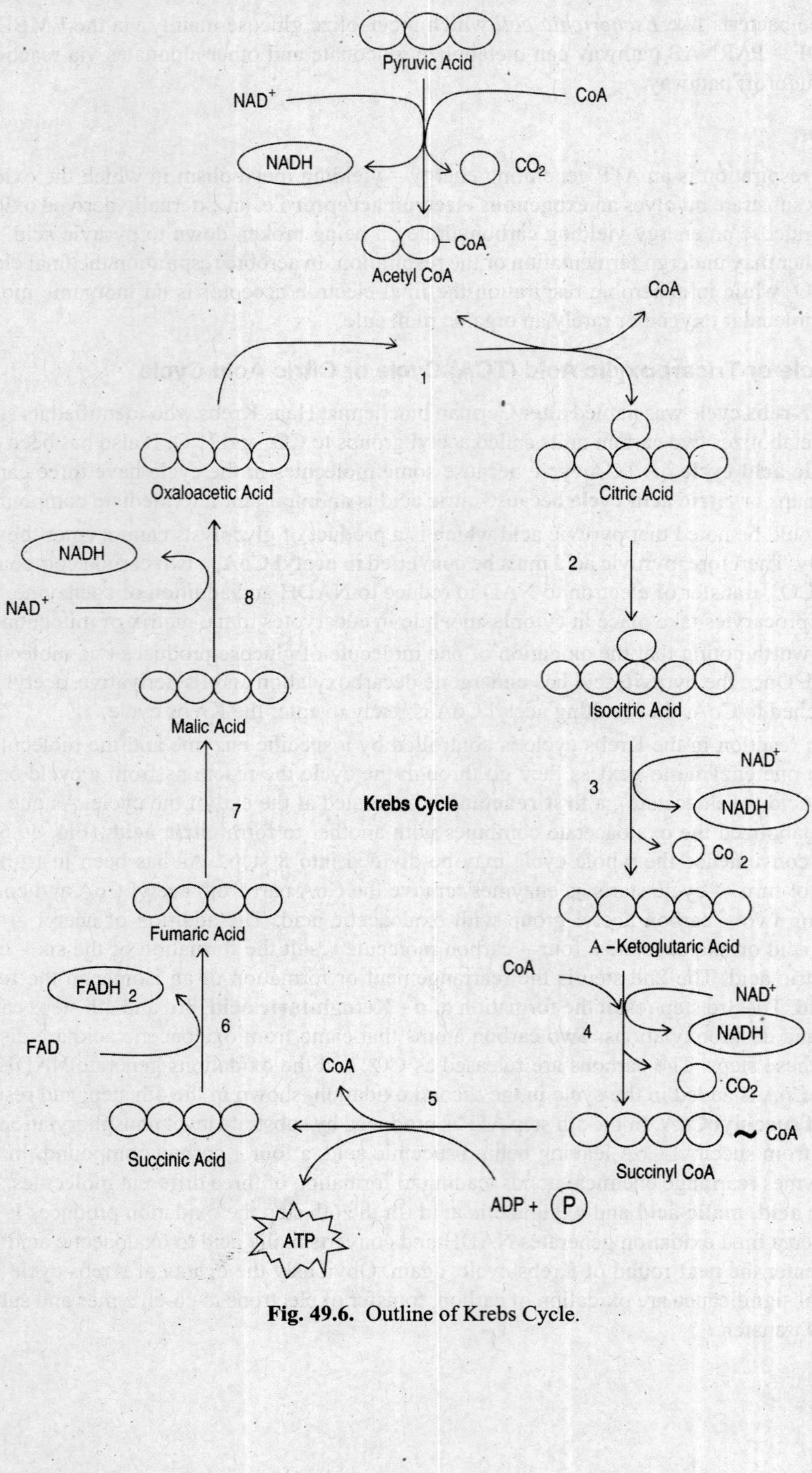

Fig. 49.6. Outline of Krebs Cycle.

Electron Transport Chain

The electron transport chain (Fig. 49.7) comprises of a series of large number of "**carrier molecules**". Each carrier is reduced as it picks up electrons and is oxidized as it passes them on the next carrier in the chain. The final oxidation is irreversible. Eventually the electrons combine with the final electron acceptor oxygen to form water. It may be noted that FADH2 adds its electrons to the electron transport chain at a lower level than NADH. Because of this the electron transport chain produces about one-third less energy for ATP generation when $FADH_2$ donates electrons when NADH is involved.

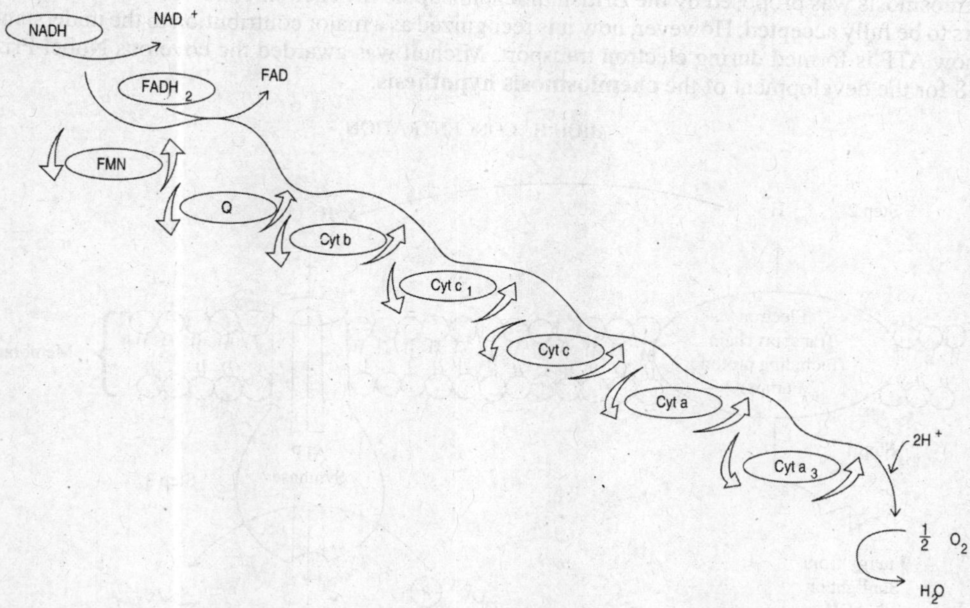

Fig. 49.7. Generalized electron transport chain to demonstrate process of respiration. Say, in the mitochondrial electron transport chain the electron pass in the chain in a gradual and stepwise fashion to relase the energy in manageable quantities. The arrow indicating downwards stands for reduction while upwards for oxidation.

The presence of some carrier like FMN and Q which accept and release protons and electrons and some other carriers like cytochromes which transfer electron only, is the interesting feature of electron transport chain. The electrons flow down the chain is accompanied at several points by the active transport or pumping of protons from the matrix side of the inner mitochondrial membrane to the opposite side of the membrane consequently there is a buildup of protons on one side of the membrane. This buildup of protons provides energy for the generation of ATP through a mechanism which has been called **chemiosmosis**.

The carrier molecules taking part in the electron transport chain have been categorized into three classes. (1). These are **flavoproteins** which contain flavin – a coenzyme obtained from riboflavin (vitamin B2) that carrier on alternating oxidation and reductions. One of the important flavin coenzyme that takes part in the chain is flavin mononucleotide (**FMN**). (2) The second class of carrier molecules comprises of **cytochromes** – a class of haemoproteins, the proteins with an iron containing group that exist in reduced form (Fe^{2+}) and oxidized form (Fe^{3+}). The cytochromes which take part in the electron transport chain include cytochrome b (**cyt b**) cytochrome c1 (cyt c1), cytochrome c (cyt c), cytochrome a (cyt a) and cytochrome a3 (cyt a3). (3) The third class is of **Ubiquinones** or **coenzyme** Q often written as "Q" which are small nonprotein carrier.

Chemiosmotic Mechanism of ATP Generation

The chemiosmosis is the process of energy capture in which a proton gradient is created by means of electron transport which is them used to drive or carry synthesis of ATP.

The electron from hydrogen atom removed from reactions of the krebs cycle are transferred through the electron transport system which generates the high energy bonds of ATP. ADP is converted to ATP by a large ATP synthesizing complex **adenosine triphosphatase** or **ATP synthase** or ATP are in a process known or **chemiosmosis**. This process is the result of series of chemical reactions that take place in and around a membrane. Although the mechanism for this process of chemiosmosis was proposed by the British biochemist pate Mitchell in 1961 but it took a number of years to be fully accepted. However, now it is recognized as a major contribution to the understanding of how ATP is formed during electron transport. Michell was awarded the covetous Nobel Prize in 1978 for the development of the **chemiosmosis hypothesis**.

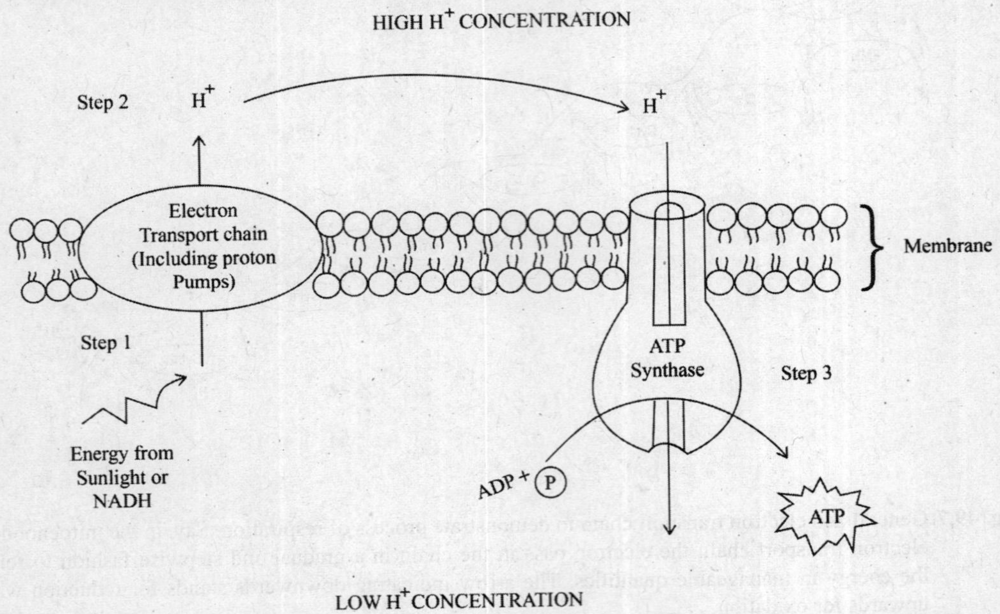

Fig. 49.8. A generalized mechanism of chemiosmosis. In 1st step electron from NADH or sun light are passed on to electron transport chain. Some of the carrier in the chain pump proton across the membrane this type of carrier molecules are known as **proton pumps**. The 2nd step shows that phospholipid membrane is generally impermeable to protons, therefore, unidirectional pumping creates a difference in concentration of protons on the two sides of the membrane. A special protein channel **ATP synthase** carries the diffusion of protons from higher proton concentration side of membrane to low H+ (proton) concentration side.

The chemiosmosis occurs in the cell membrane of procaryotes like *Escherichia coli* **and** the inner mitochondrial membrane in eucaryotes. As the electrons are transferred along **the electron** transport chain the protons (H+) are pumped outside the membrane, hence, the **concentration** of ions is higher outside the membrane than the inside.

This process results the lowering of the proton concentration on the inside and the development of a force that drives the protons back into the cell or mitochondrial matrix to equalize their concentration on both sides of the membrane. Any concentration gradient naturally tends to equalize itself.

In addition to the proton concentration gradient across the membrane there is an electrochemical gradient which makes the membrane a type of biological battery that can provide power for the generation of ATP.

The H^+ (proton) excess on one side of membrane gives that side a positive charge than the other side. The power generated by the gradient is called the **proton motive force**. The protons flow through special channels within the synthase complex. The energy is thus released and used to form ATP from ADP and inorganic phosphorus (Pi).

Aerobic Respiration

In aerobic respiration the final electron acceptor in the electron transport chain is oxygen (O_2). When one molecule of glucose is oxidized the different transfers that take place in electron transport chain generate about 34 molecules of ATP. Out of these 3 molecules of ATP are generated from each of the ten molecules of NADH (i.e. a total of 30); and two from each of two molecules of FADH2 (i.e. a total of 4). To count the total number of ATP molecules generated for each molecule of glucose the 34 from chemiosmosis are added to those generated by oxidation in glycolysis and Krebs cycle. A single molecule of glucose in procaryotes in aerobic respiration generates 38 molecules, (Fig. 49.9) four of which come from substrate level phosphorylation in glycolysis and Krebs cycle.

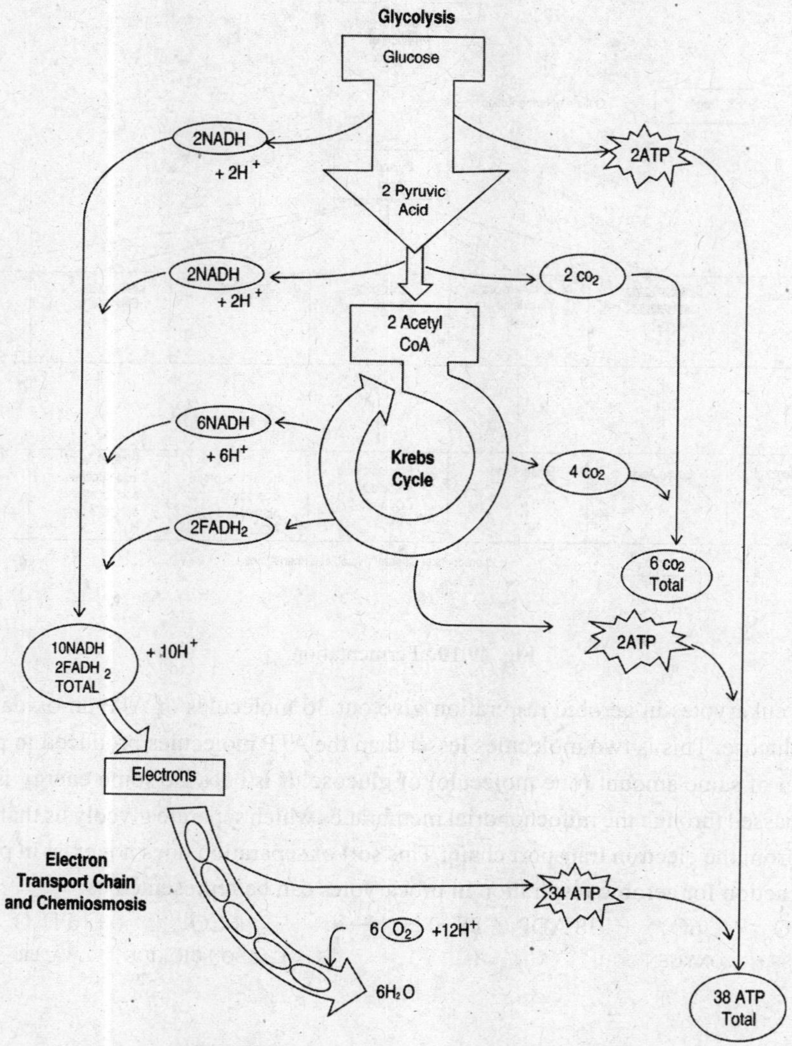

Fig. 49.9. Outline of aerobic respiration in procaryotes.

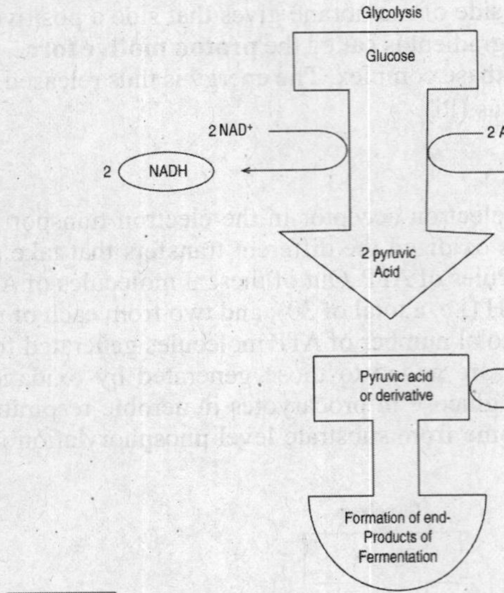

Part-a Outline of fermentation

Pyruvic Acid

Organism	*Streptococcus, Lactobacillus, Bacillus*	Saccharomyces cerevisiae (Yeast)	Propionibacterium	Clostridium	Escherichia, Salmonella	*Enterobacter*
End - products of Fermentation	Lactic Acid	Ethanol and CO_2	Propionic acid, acetic acid CO_2 and H_2	Butyric acid butanol, acetone isopropyl alcohol and CO_2	Ethanol, lactic acid, succinic acid, acetic acid, CO_2 and H_2	Ethanol, lactic acid, formic acid, butanediol, acetoin, CO_2 and H_2

Part-b End products of different microbial fermentations

Fig. 49.10. Fermentation.

However, the eukaryotes in aerobic respiration give out 36 molecules of ATP on oxidation of one molecule of glucose. This is two molecules lesser than the ATP molecules produced in procaryotes from oxidation of same amount (one molecule) of glucose. It is because some energy is lost when electrons are passed through the mitochondrial membranes which separate glycolysis that underwent in cytoplasm from the electron transport chain. This sort of separation does not exist in procaryotes. The overall reaction for aerobic respiration in procaryotes can be represented as:

$$\underset{\text{GLUCOSE}}{C_6H_{12}O_6} + \underset{\text{OXYGEN}}{6O_2} + 38\,ADP + 38\,P \longrightarrow \underset{\text{CARBON DIOXIDE}}{6CO_2} + \underset{\text{WATER}}{6\,H_2O} + 38\,ATP$$

Anaerobic Respiration

The microorganism which adopt anaerobic respiration are called **anaerobes** which do not use free oxygen as their final electron acceptor, rather their final electron acceptor is some inorganic oxygen containing molecule as some bacteria like *Pseudomonas* and *Bacillus* can take nitrate ion (NO^{3-}) as a final electron acceptor. They reduce the nitrate ion to nitrite ion (NO^{2-}) nitrous oxide (N_2O) or nitrogen gas (N_2). *Desulpho vibrio* use oxygen containing inorganic sulphate ($SO^{2-}{}_4$) as final electron acceptor to form hydrogen sulphide (H_2S). Some other bacteria take carbonate ($CO^{2-}{}_3$) to form methane (CH_4). Because the anaerobes use lesser part of Krebs cycle of the metabolic pathways they produce fewer ATP molecules than do aerobic organisms.

Fermentation

The **fermentation** is an energy yielding metabolism in which energy substrate is metabolized without involvement of an **exogenous electron acceptor**. The energy substrates may be sugars or other organic molecules like amino acids, organic acids, purines and pyrimidines. It does not require Krebs cycle or an electron transport chain. It uses an organic molecule as the final electron acceptor. It produces small amounts of ATP (only one or two molecules of ATP) for each molecule of starting substrate molecule since major part of original substrate like glucose remains in the chemical bonds of the organic end products like lactic acid or ethanol.

The electrons along with protons from reduced enzymes as NADH, NADPH are transferred to pyruvic acid and the derivatives of pyruvic acid (Fig. 49.10A). The final electron acceptors are reduced to end products of a variable range depending upon the microorganism. The NAD^+ and $NADP^+$ are regenerated and can enter the next round of glycolysis. Thus the second stage of fermentation is to ensure the uninterrupted or continuous supply of NAD^+ and $NADP^+$ so that the glycolysis may continue.

Lactic Acid Fermentation

The lactic acid fermentation proceeds through glycolytic pathway. The glycolysis is the first phase of lactic acid fermentation in which a single molecule of glucose is oxidized to two molecules of **pyruvic acid (pyruvate)**. The oxidation of glucose produces energy which is used to form two molecules of ATP (Fig. 49.11). Thereafter , two molecules of pyruvic acid are reduced to two molecules of lactic acid which is the end product of reaction. There is no further oxidation and most of the energy produced by the reaction remain stored in the lactic acid. Obviously this fermentative reaction produces fewer energy. The reaction may be represented as:

Glucose +2 ADP + 2 P $\rightarrow$ 2 lactic acid + 2 ATP.

The lactic acid bacteria *Streptococcus* and *Lactobacillus* which produce lactic acid and the glucose is converted entirely or almost entirely to lactic acid which has been called **homolactic fermentation**, and the bacteria since they produce lactic acid are called **homolactic** or **homofermentative** . The lactic acid bacteria can cause the spoilage of food. However, these can also be exploited to form yogurt and butter milk from milk, sauerkraut from fresh cabbage and pickles from cucumbers. Interestingly when deprived of oxygen muscle tissue of animals including humans also carries out lactic acid fermentation by same pathway.

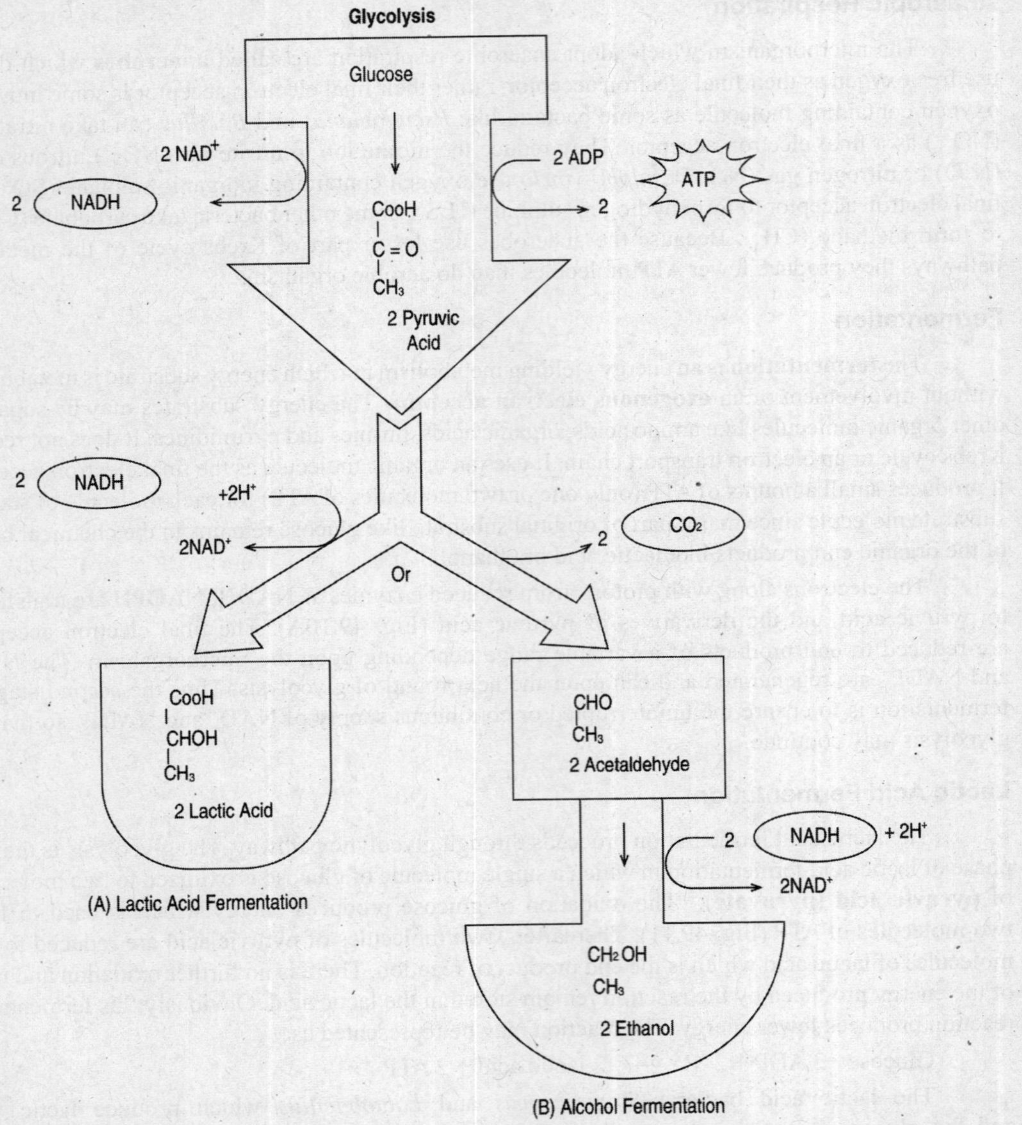

Fig. 49.11. The fermentation. (A) Lactic acid fermentation. (B) Alcohol fermentation.

Alcohol Fermentation

The alcohol fermentation also like lactic acid fermentation starts with the **glycolysis** of a molecule of glucose to yield two molecules of **pyruvic acid** (structural ionic form is pyruvate) and two molecules of ATP. This is followed by conversion of two molecules of pyruvic acid to two molecules of **acetaldehyde** and two molecules of CO_2. The two molecules of acetaldehyde are in the next step reduced by two molecules of NADH to from two molecules of **ethanol** this also is a low energy yielding process since most of the energy stored in the initial glucose molecule remains in the ethanol. The alcohol fermentation is performed by some bacteria and yeasts.

The microorganisms which produce lactic acid as well as other acids or alcohols are called **heterolactic** or **heterofermentative**. Such microorganisms use pentose phosphate pathway. The heterolactic fermentation may be represented with the reaction:

Glucose + ADP + P $\rightarrow$ Lactic acid + Ethanol + CO_2 + ATP

Mixed Acid Fermentation

In the mixed acid fermentation which is the characteristic of *Escherichia coli* and related bacteria the pyruvic acid (structural ionic form pyruvate) is converted to at least six different end products. Production of a large number of fermentation end products using various microorganisms is shown in Fig. 49.10.

Lipid Catabolism

The fats are also called **lipids** which consist of fatty acids and glycerol. Some microorganisms produce extra cellular enzymes referred to as lipases which are capable to split fats into **fatty acids** and **glycerol**, and each of these components can be metabolized separately. Some microbes bear capability to convert glycerol into dihydroxyacetone phosphate which is one of the three carbon intermediates produced in the process of glycolysis. The dihydroxyacetone phosphate (Fig. 49.12) subsequently is metabolized by glycolysis and the Krebs cycle. However, the fatty acids are catabolized in somewhat different manner. The mechanism by which the fatty acid are oxidized has been called **β (beta) oxidation**. In this process of β(beta) oxidation the fragments of long chain of fatty acid are removed two at a time so as to form acetyl CoA from them. As the molecules of acetyl CoA are formed they enter the Krebs cycle in the same way or the 'acetyl CoA' molecules farmed as a result of oxidation of pyruvic acid or pyruvate. Hence, the Krebs cycle function in oxidation of glycerol fatty acids and glucose.

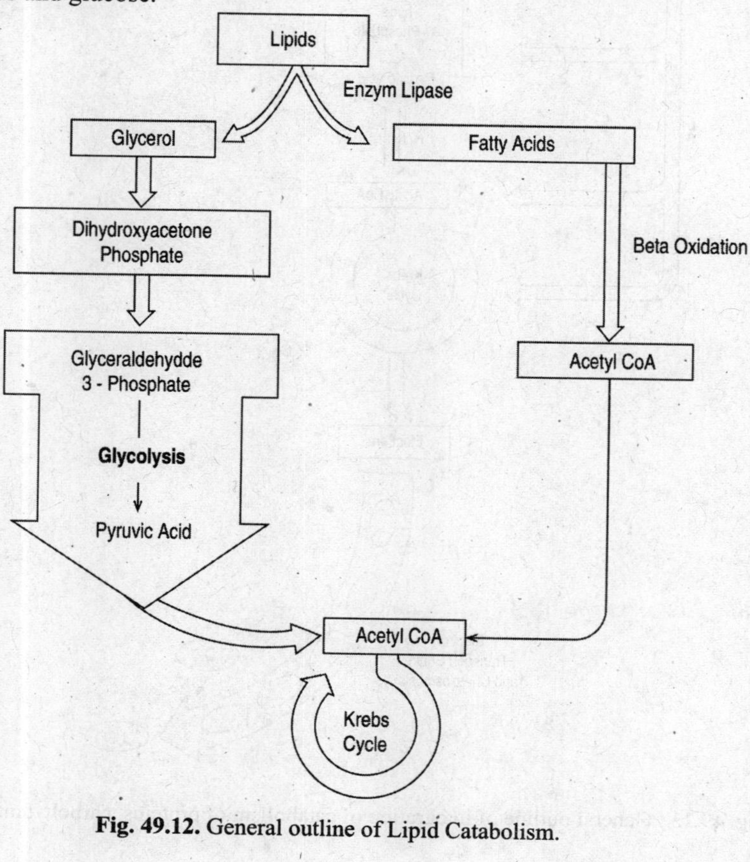

Fig. 49.12. General outline of Lipid Catabolism.

Protein Catabolism

The proteins are so large in Size that they cannot pass unaided through a plasma membrane. The microorganisms produce extra cellular enzymes called **proteases** and **peptidases** which are capable to cleave or break down proteins into their component amino acids which however, can cross the membrane. It should be remembered that before the amino acids are catabolized they need to be converted or transformed to substances which can enter the Kreb's cycle one of the process of conversion of amino acid in which the amino group is removed and changed into ammonium ions

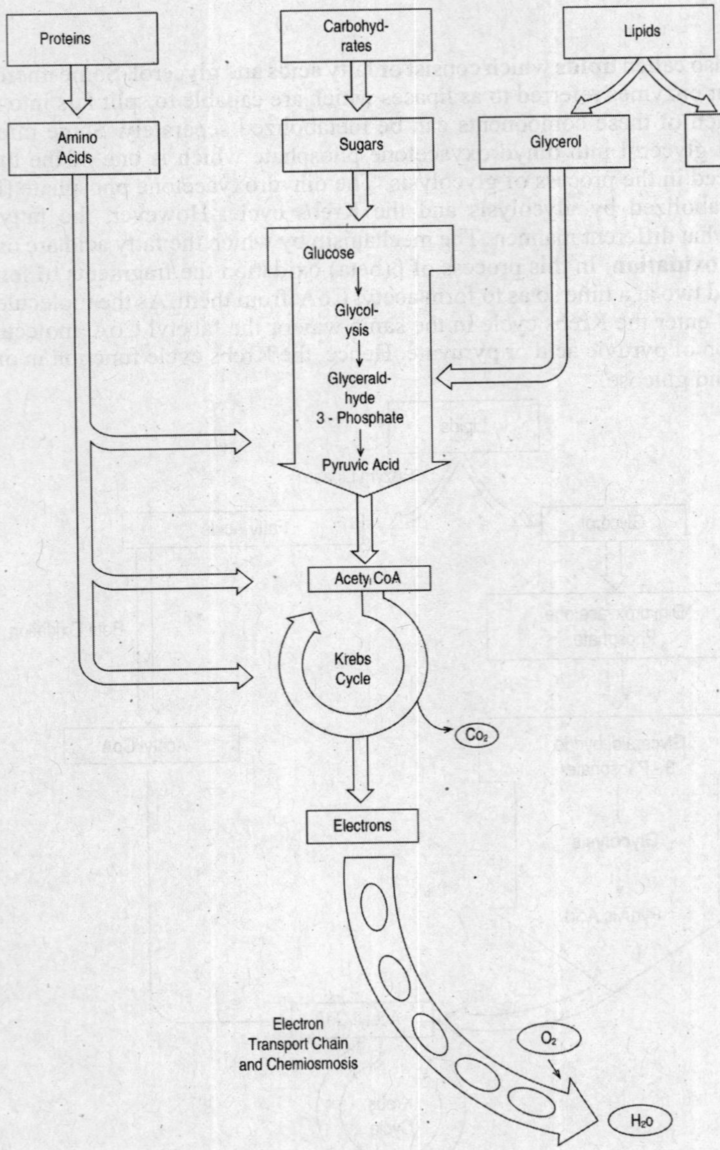

Fig. 49.13. General outline of integration of catabolism of proteins, carbohydrates and lipids.

(NH^{4+}) that in turn is released out of the cell, which process is known as **deamination**. The another process of conversion of amino acids is by removal of –COOH group known as **decarboxylation** and removal of hydrogen atoms called **dehydrogenation**. Although these processes are very complex nonetheless there exists interrelationship (Fig. 49.13) in the processes of catabolism of carbohydrate, lipids and proteins, all of which are sources of electrons and proton for respiration. Their molecules make entry in glycolysis or the krebs cycle at different levels. Glycolysis and the Krebs cycle have been regarded the catabolic channels which provide route through which high energy electrons from all types of organic molecules flow on their energy – generating pathways.

Photosynthesis

Q. 6. What is **photosynthesis**? What are light and dark reactions? Describe.

Ans. The term photosynthesis has been coined from *photon* = light; synthesis = assembly; i.e. synthesis of organic compounds from water and carbon dioxide using energy absorbed from sun. Thus **photosynthesis** is a process in which the light energy of sun is converted into chemical energy. The chemical energy is consumed to convert carbon dioxide (CO_2) of atmosphere to more reduced carbon compound mainly sugars. The synthesis of sugars by using carbon from CO_2 gas is also called **carbon fixation** which is an important aspect of **recycling of carbon**. The cyanobacteria (blue green algae), algal and green plants are major contributors of this process of recycling of carbon through the photosynthesis process. The photosynthesis may be summarized by the equation:

$$6CO_2 + 12\,H_2O + \text{Light energy} \longrightarrow C_6H_{12}O_6 + 6O_2 + 6\,H_2O$$

To carry on this process the electrons are taken from water which is an **energy poor molecule** as a result of decomposition or **photolysis** of water using light energy absorbed by chlorophyll pigments. These electrons are transferred to sugar which is an **energy rich molecule**, of course, the process is also pushed or enhanced indirectly by the light energy. The photosynthesis occurs in two stages:

1. Light reaction in which light energy is used to convert ADP and Pi (inorganic phosphorus) to ATP. Besides the electron carrier NADP is reduced to NADPH. The coenzyme NADPH like NADH is an energy rich carrier of electrons.
2. Dark reactions, which are light independent, and in which electrons along with energy generated by ATP are used to reduce CO_2 to sugars.

Light Reactions: Photophosphorylation

Though there are three ways of formation of ATP from ADP but **photophosphorylation** takes place in photosynthetic cells only. The light energy is absorbed by chlorophyll pigment molecules in the photosynthetic cells. Some of the electrons of the chlorophyll molecules get excited. The 'chlorophyll – a' is used by green plants, algae and cyanobacteria, which is located in the membranous thylakoids of chloroplasts in green plants while in thylakoids in cyanobacteria. The other bacteria use bacteriochlorophylls. The excited electrons of the molecules of chlorophyll jump to the first of the series of carrier molecules serving as **electron transport chain**. The electrons are passed through the series of carriers, protons are pumped across the membrane by proton pumps and the ADP is transformed to ATP. Two types of photophosphorylation have been named on basis of the 'path' of electron transport. These are: (1) Cyclic photophosphorylation and (2) Noncyclic photophosphorylation. In **cyclic photophosphorylation** the electrons return to chlorophyll forming a cycle. In cyclic photophosphorylation the electrons which got excited due to light return back to chlorophyll after completing their journey along the electron transport chain. The energy from the electron transport chain is used in generation of ATP. In the **non cyclic photophosphorylation** which is considered to be more common the electrons released from chlorophyll do not return to chlorophyll

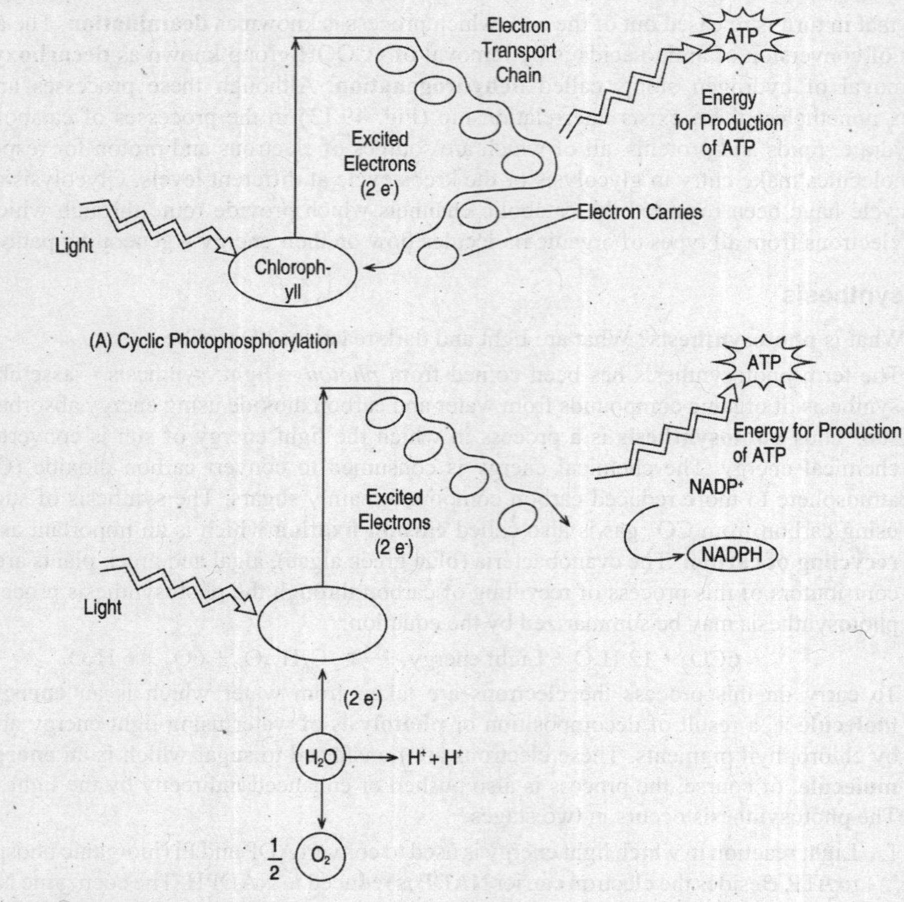

Fig. 49.14. General outline for representation of photosynthesis.

rather get incorporated into NADPH. The excited electrons released or jumped out of chlorophyll are replaced by electrons from water (Fig. 49.14B) or another oxidizable compound like hydrogen sulfide (H_2S).

The noncyclic photophosphorylation lead to production of ATP by chemiosmosis using energy released in electron transport chain.. The chlorophyll electrons are passed along the electron transport chain to the electron acceptor NADP. The NADP combined with electrons, and with hydrogen ions from water resulting the formation of NADPH.

Dark Reactions: Calvin Benson Cycle

Do you know why Calvin cycle has been given the name a **dark reaction cycle**? It has been named so because this cycle does not involve any reactions directly accompanied wih input of light energy. But it certainly requires ATP and NADPH.

It is also called reductive pentose phosphate cycle or **C_3 cycle**. Calvin Benson cycle is a cyclic pathway used for the fixation of carbon dioxide (CO_2) by a wide range of autotrophs (Fig. 49.15).

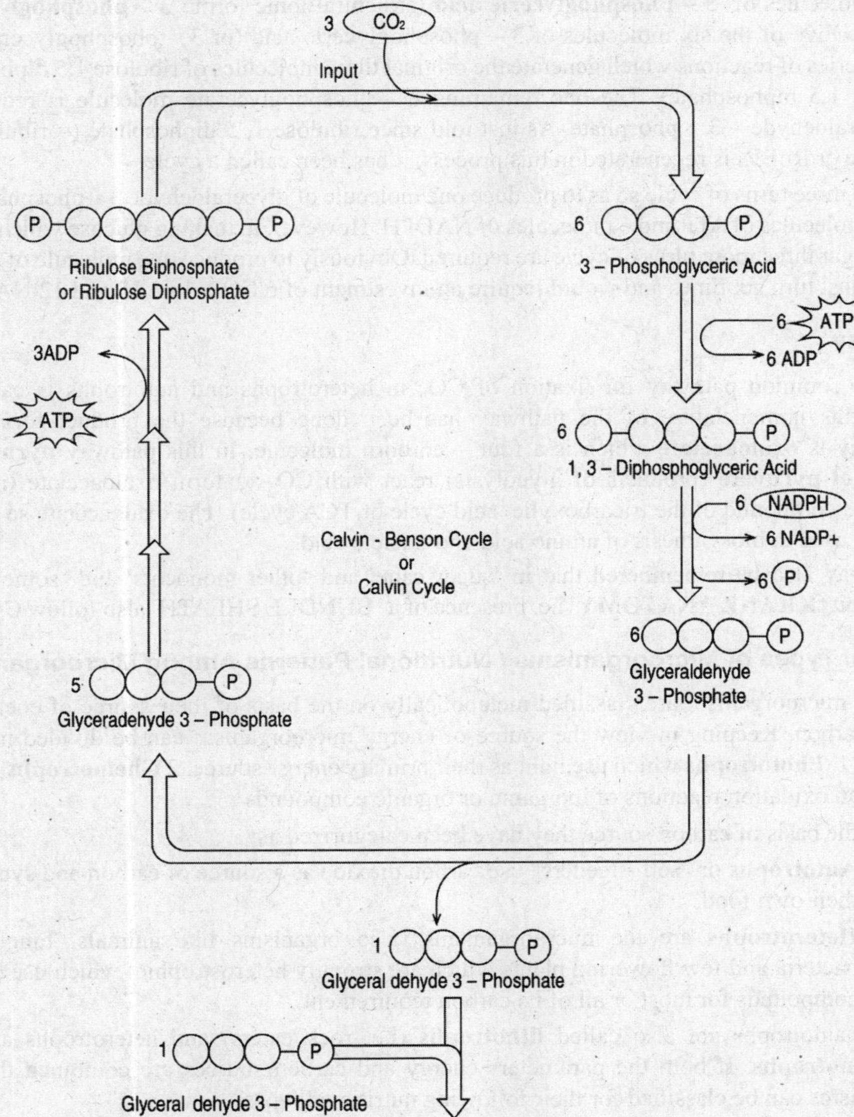

Fig. 49.15. Outline of Calvin-Benson Cycle.

The key enzymes of Calvin cycle are Ribulose 1, 5 bisphosphate carboxylase – Oxygenase (RuBisC) responsible for CO_2 fixing reaction and phosphoribulokinase (ribulose 5 – phosphate kinase) responsible for regenerating the **CO_2 acceptor**: **ribulase 1, 5 bisphosphate (RUBP)** (= ribulose 1,5 diphosphate or **Ribulose diphosphate**) To bring about synthesis of one molecule of a three – carbon sugar known as **glyceraldehyde 3 – phosphate** (GP), the cycle of reactions must take place three times to fix up three molecules of carbon dioxide (CO_2). This is to be kept in mind that glyceraldehyde 3 – phosphate is also formed in preparatory stage of glycolysis and constitute an important intermediate compound. It may further be explained that for every three molecules of CO_2

fixed six molecules of **3 – phosphoglyceric acid** (structural ionic form, **3 – phosphoglycerate**) are formed. Five of the six molecules of 3 – phosphoglyceric acid (or 3 – phosphoglycerate), go through a series of reactions which generate the original three molecules of ribulose 1,5 diphosphate (= ribulose 1,5 biphosphate). The one remaining 3 – phosphoglycerate molecule is reduced to form glyceraldehyde – 3 – phosphate. As just told since ribulose 1, 5 diphosphate (= ribulose 1,5 biphosphate or RuBP) is regenerated in this process, it has been called a cycle.

The three turns of cycle so as to produce one molecule of glyceraldehyde 3 – phosphate need input of 9 molecules of ATP and 6 molecules of NADPH. However, in making glucose which is a six – carbon sugar three more turns of cycle are required. Obviously to produce one molecule of glucose the cycle must turn six times and would require an investment of 6 CO2, 18 ATP and 12 NADPH.

C_4 Pathway

The common pathway for fixation of CO_2 in heterotrophs and autotrophs is called **C_4 pathway.** this nomenclature of the pathway has been done because the product formed via this pathway is **oxaloacetate** which is a four – corborn molecule. In this pathway **pyruvate** or **phosphoenol pyruvate** (products of glycolysis) react with CO_2 to form oxaloacetate (also, an intermediate compound of the tricarboxylic acid cycle or TCA cycle). The oxaloacetate so formed then can be used in biosynthesis of amino acid and nucleic acid.

It may also be remembered that in 'sugar cane' and 'other monocots' and 'some dicots' which exhibit **'KRANZ ANATOMY'** i.e. presence of a 'BUNDLE SHEATH' also follow C4 cycle.

Nutritional Types of Microorganisms / Nutritional Patterns Among Microorganisms

The microorganism are classified metabolically on the basis of their source of energy and source of carbon. Keeping in view the source of energy microorganism can be divided into two categories : 1. **Phototrophs** which use light as their primary energy source. 2 **Chemotrophs** depend for energy on oxidation reactions of inorganic or organic compounds

On the basis of carbon source they have been categorized as:

1. **Autotrophs** or "self - feeders" use carbon dioxide as a source of carbon and synthesize their own food.
2. **Heterotrophs** are the microorganisms (also organisms like animals, fungi, most bacteria and few flowering plants which are strongly heterostrophic) which use organic compounds for most or all of its carbon requirement.

The autotrophs are also called **lithotrophs** (i.e. rock eaters) and heterotrophs are also called **organotrophs**. If both the parameters energy and carbon sources are combined then the microorganisms can be classified for their following nutritional types:

1. PHOTOAUTOTROPHS

The photoautotrophs use light as source of energy and the carbon dioxide as the main source of carbon, e.g. photoautotrophic bacteria, **'green sulphur'** and **'purple sulphur'** bacteria and cyanobacteria, algae and green plants. The hydrogen atoms of water are used to bring about reduction of carbon dioxide and the oxygen is released. Since this phototrophic process gives out oxygen it has also been called **oxygenic**.

Besides cyanobacteria (blue green algae) there are also some other families of photosynthetic procaryotes. Their classification is done on the basis of style how do they reduce CO_2. This category of bacteria cannot use H_2O to reduce CO_2 and they cannot carry on photosynthesis in presence of oxygen. They do it under anaerobic conditions. Eventually, they do not produce oxygen in process of photosynthesis and are called anoxygenic. The green sulphur and purple sulphur bacteria come under the category of anoxygenic. The *Chlorobium* an example of green sulphur bacteria use sulphur (S), sulphur compounds as H_2S or hydrogen gas (H_2) to reduce carbon dioxide from organic compounds. These bacteria using light – energy and suitable enzymes oxidize sulphide (S^{2-}) or sulphur (S) to sulphate ($SO^4{}_{2-}$) or the hydrogen gas to water (H_2O). The *Chromatium* the purple sulphur bacterium also use sulphur, sulphur compounds or hydrogen gas to reduce carbon dioxide, Their specific chlorophyll and location of the sulphur stored are the criteria to differentiate them from green sulphur bacteria.

The types of chlorophyll used by photosynthetic bacteria are known as **bacteriachlorophylls** and they absorb light at longer wavelengths in comparison to that absorbed by chlorophyll 'a'. The green sulphur bacteria possess bacteriochlorophylls in the vesicles known as **chlorosomes** or **chlorobium vesicles** which occur beneath the plasma membrane and are attached to the plasma membrane. The bacteriochlorophylls in the purple sulphur bacteria (which occur in sulphide rich aquatic and terrestrial habitats) are present in the invaginations of plasma membrane.

The arehaeobacterium *Halobacterium halobium* found in salt lakes, and high salt concentration of solar evaporation ponds used to manufacture salt around 'San Francisco Bay' of USA, uses **bacteriorhodopsin** instead of chlorophyll in rare type of photosynthetic system. The plasma membrane of the *Halobacterium* can be differentiated into 'red' and 'purple' fractions. The red fraction which comprise the most of the membrane contains cytochromes, flavoproteins and other parts of the electron transport chain which carries out oxidative phosphorylation.

Photoheterotrophs

They use sunlight as source of energy but cannot transform carbon dioxide to sugar, instead they use alcohols, fatty acids, other organic acids and carbohydrates as source of carbon. The examples are *Chloroflexus* (green non sulphur bacteria) and *Rhodopseudomonas* (purple nonsulphur bacteria).

Chemoautotrophs

They use the electrons obtained from reduced inorganic compounds and CO_2 is their major source of carbon. They use inorganic sources energy like hydrogen sulphide (H_2S) as in case of *Beggiatoa*, elemental sulphur in *Thiobacillus theoxidans*, ammonia (NH_3) for *Nitrosomonas*, nitrite ions (NO^{2-}) in *Nitrobacter*, hydrogen gas (H_2) in *Hydrogenomonas* and ferrous ions (Fe^{2+}) in *Thiobacillus ferroxidans*.

Chemoheterotrophs

They do not have any demarcation between their source of energy and source of carbon, rather they can use the same organic substance for both the purposes, e.g. glucose. They specifically take electrons from hydrogen atoms in organic compounds as a source of energy. Depending upon the source of organic molecules the chemo-heterotrophs have also been classified as **saprophytes** (grow on dead organic matter) and **parasites** (derive nutrition from living host).

Anabolism: The Biochemical Pathways Requiring Energy

Q 7. What are the main types of pathways of anabolism? Describe.

Ans. The main types of pathways of anabolism are as under:

Gluconeogenesis

The **gluconeogenesis** is the process of biosynthesis of glucose from noncarbohydrate substrates. Do you know how a cell can biosynthesize glucose from proteins, using proteins as substrate? To achieve the objective of conversion of proteins to glucose, the amino acids obtained or derived from proteins can be converted into pyruvate or phosphoenol pyruvate that are intermediate metabolites of gluconeogenic pathway. A parallel example is the breakdown of lipids into 3-carbon intermediates of the gluconeogenic pathway.

Biosynthesis of Polysacchrides or Glycan

The polysacchasides are macromolecules consisting of monosaccharide or glycose (= glucose) residues joined by glycosidic linkages. The polysaccharides containing only one kind of monosacchaside are called **homosacchasides** (= **homoglycans**) those having two or more different kinds are called **heteropolysaccharides** (= **heteroglycans**). The microorganism first of all synthesize polysacchasides. The carbon atoms needed to synthesize glucose are derived from

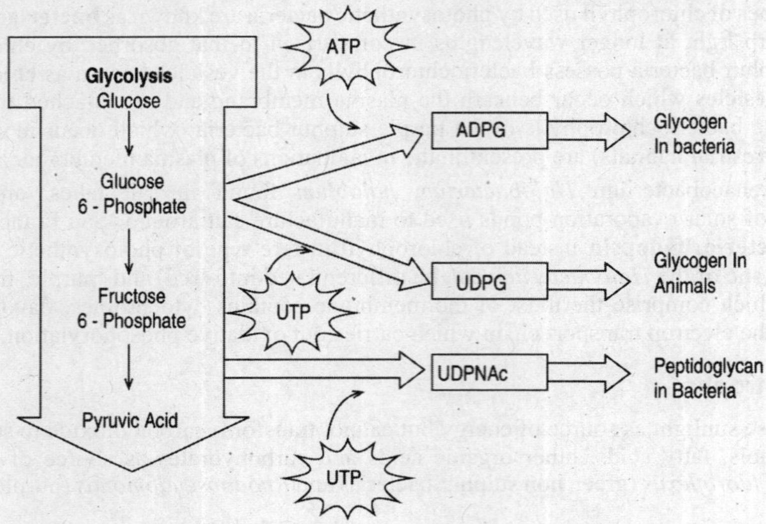

Fig. 49.16. General outline of biosynthesis of polysaccharides. The polysaccharide **glycogen** to be stored by bacterial cell, also by animal cell and the peptidoglycan the structural molecule of eubacterial cell wall are end products.

intermediates produced in the process of glycolysis (Fig. 49.16) and Krebs cycle or from lipids or amino acids. The bacteria can assemble the molecules of sugars like glucose into more complex polysaccharides. One of typical examples is that of glycogen. To prepare glycogen from glucose in the bacteria the units of glucose are phosphorylated and linked. The product of phosphorylation of glucose is glucose 6 – phosphate. This sort of process require expenditure of energy that is provided often by ATP and dehydration or loss of water. The major intermediate in the synthesis of glycogen in bacteria is known as **adenosine diphosphoglucose** (ADPG) that is formed when a molecule of ATP is added to glucose – 6 – phosphate. This may be followed by synthesis of next building block which in turn is linked with similar units through dehydration synthesis to form glycogen. Similarly a carbohydrate **uridine diphosphoglucose** (UDPG) is used and a nucleotide known as uridine triphosphate (UTP) works as source of energy in animals for synthesis of **glycogen** (Fig. 49.16). The bacteria also need to synthesize peptidoglycan a tough polysaccharide that comprise the wall of bacterial cell for which they use a compound related to UDPG called **UDP – N – acetylglucosamine** (UDPNAc). The UDPNAc which works as a starting material in biosynthesis of peptidoglycan is formed from fructose – 6 – phosphate and this reaction also is perfomed with UTP as source of energy.

Biosynthesis of Lipids

The lipids serve as structural components of biological membrane and most of the membrane lipids are **phospholipids**. The lipids are synthesized by many routes. The cell can synthesize fats by joining **glycerol** and **fatty acids**. The glycerol part of the fat comes from dihydroxyacetone phosphate. An intermediate formed in the process of glycolysis. The long chain hydrocarbons called

fatty acids are synthesized when two carbon fragments of acetyl CoA are combined in biochemical reaction (Fig. 49.17). Like the synthesis of polysaccharides the building units or blocks of fats and other lipids are linked via dehydration reactions which need energy not in the form of ATP always, e.g. in the synthesis of phospholipids the energy is obtained from hydrolysis of a nucleotide called **cytidine triphosphate** (CTP). The lipids also function in storage of energy and on break down the products

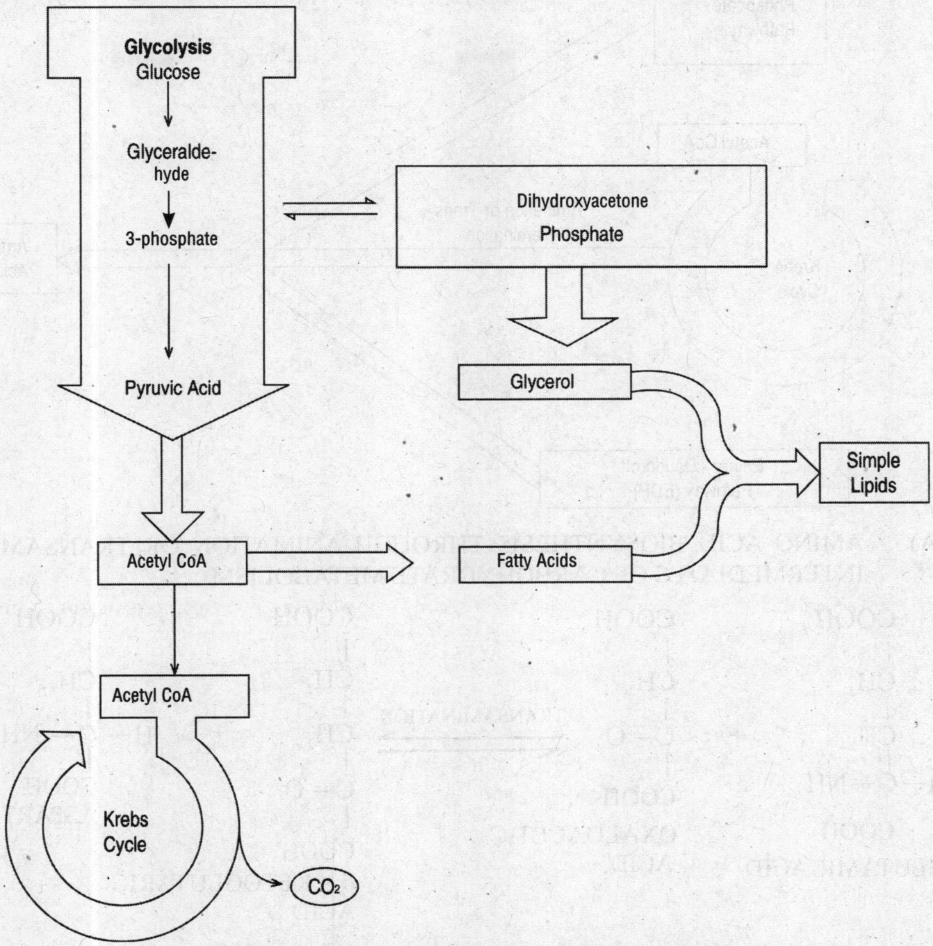

Fig. 49.17. Outline of pathway of biosynthesis of simple lipids.

of lipids after biological oxidation feed into the Krebs cycle. The cholesterol a lipid of a different structure is also found in plasma membrane of eucaryyotic cell. The waxes are important component of cell wall of acid fast bacteria also are lipids. The carotenoids that make red, orange and yellow pigments of some microorganisms are also a category of lipids.

Biosynthesis of Amino Acids

The amino acids serve as the building blocks of protein synthesis. The proteins play an instrumental role in the cell as enzymes, structural components, toxins and some receptor specific proteins. There are twenty or more amino acids which bear a three letter as well as one letter standard abbreviation for each amino acid, e.g. alanine = Ala = A, arganine = Arg = R, asparagine = Asn = N, aspartic acid = Asp = D and glutamine = Gln = Q. The colon bacterium *Escherichia coli* has the necessary enzymes to produce amino acids from starting material like glucose and inorganic salts to carry on synthesis of all the amino acids required by the bacterium. The microorganisms equipped with required enzymes can carry on the synthesis of all amino acids directly or indirectly

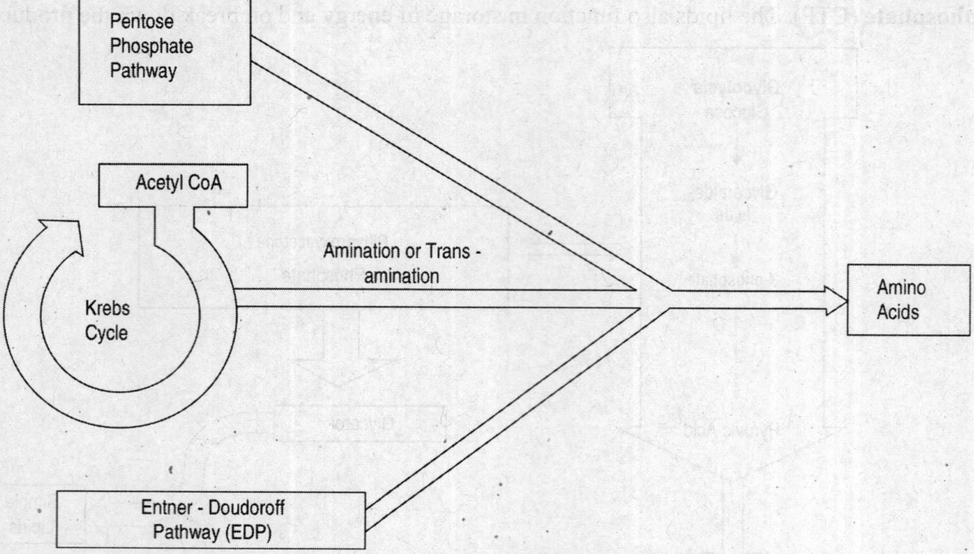

(A) AMINO ACID BIOSYNTHESIS THROUGH ANIMATION OR TRANSAMINATION INTERMEDIATES OF CARBOHYDRATE METABOLISM

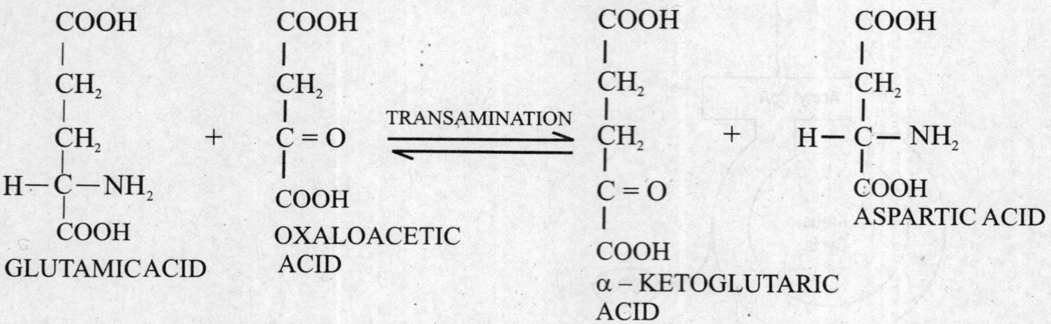

(B) REPLACEMENT OF AMINO GROUP IN TRANSMINATION PROCESS. NEW AMINO ACID **ASPARTIC ACID** HAS BEEN MADE USING AMINE GROUP FROM OLD AMINO ACID **GLUTAMIC ACID**.

Fig. 49.19. General outline of pathway of biosynthesis of amino acids.

formed in the metabolism process of carbohydrates. The Krebs cycle (Fig. 49.19) is one of the important servers for intermediates or precursors. The addition of an amino group to **pyruvic acid** a product of **glycolysis** to an appropriate organic acid of the Krebs cycle converts that organic acid into amino acid which process has been called **amination**. The amino acid if comes from a preexisting amino acid then the process is termed as **transamination**. Some precursors for amino acid biosynthesis are derived from the pentose phosphate pathway and Entner – Doudoroff pathway. The joining of amino acids to make proteins is done by dehydration synthesis and requires energy in the form of ATP. The protein synthesis, of course, is carried out at the direction of genes.

Biosynthesis of Purines NAD Pyrimidines

The DNA (**Deoxyribonuclic acid**) and RNA are known as macromolecules of information which comprise of repeating units of **nucleotides** that is turn are made of nitrogenous bases, larger **purines** adenine and guanine and smaller **pyrimidines** cytosine and thymine; and (uracil in RNA), a five carbon **pentose** sugar, and a phosphate group. The five carbon sugar in RNA nucleotide is **ribose** which possesses one more oxygen atom than deoxyriboxe sugar occurring in RNA (**ribonucleic acid**) a class of nucleic acids which comprises messenger RNA, ribosomal RNA and transfer RNA. The five – carbon sugars of the nucleotides are derived either from **pentose phosphate pathway** (which provides an alternative route for the oxidation of glucose and plays a role in synthesis of biological molecules required by the cell) or the **Entner – Doudoroff pathway** (also an alternative pathway for oxidation of glucose to pyruvic acid. Some amino acids like **aspartic acid** (=Asp=D), **glycine** (=Gly=G) and **glutamine** (=Gln=Q) derived from intermediates produced during glycolysis and in Krebs cycle take part in the biosynthesis of puriner and pyrimidines. The carbon and nitrogen atoms derived from these amino acids result into the formation of **purine** and **pyrimidine rings** (Fig.49.20).

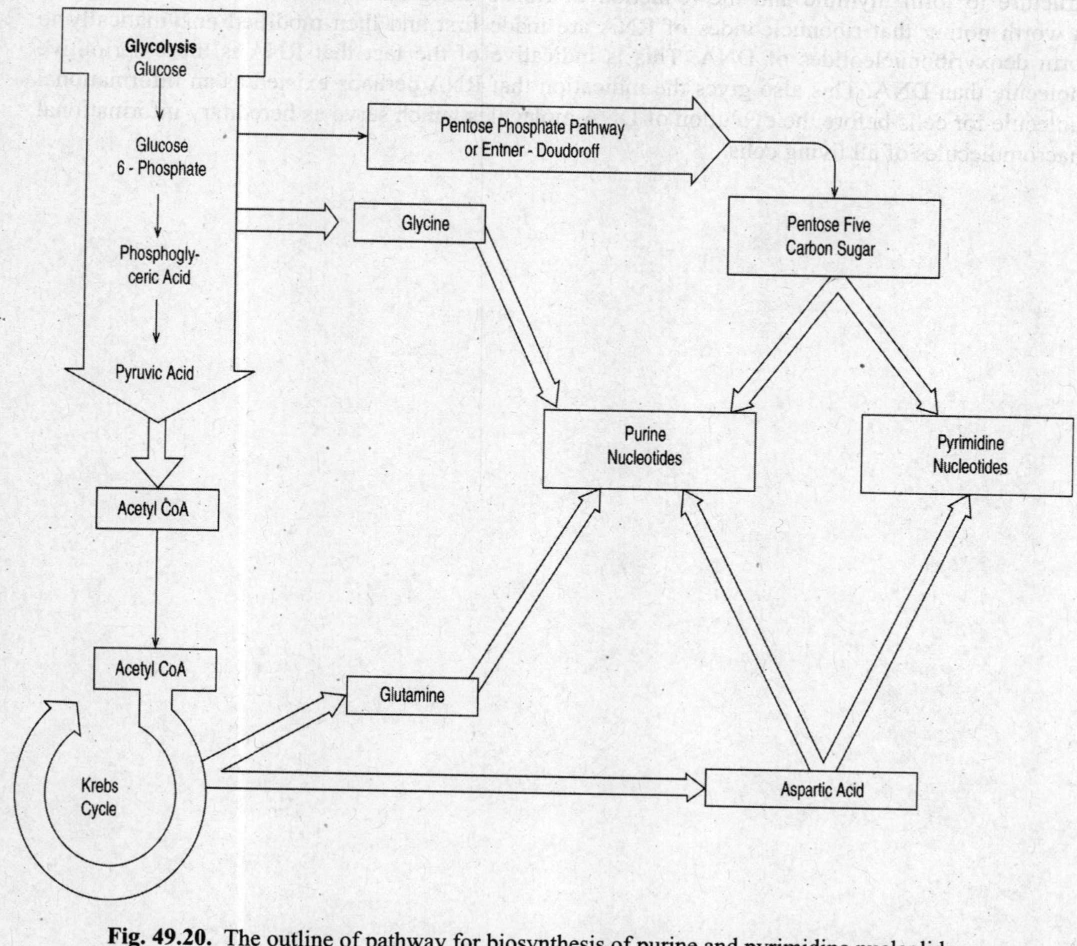

Fig. 49.20. The outline of pathway for biosynthesis of purine and pyrimidine nucleolides.

The energy for this biosynthesis is given by ATP. Moreover, the nucleotides ATP, NAD^+ and $NADP^+$ also play the role in activating and inhibiting the rate of cellular metabolism. To elaborate, the pyrimidine ring begins with the formation of carbamoyl aspartate from the reaction of carbamoyl phosphate and catalyzed by the enzyme aspartate transcarbamoylase. This is the key regulatory step in the synthesis of the pyrimidine ring. The enzyme aspartate transcarbamoylase as stated above is subject to allosteric feedback inhibition by the products of the reaction and to allosteric activation by ATP. The onward steps in the formation of pyrimidine ring involve dehydration, cyclic ring formation and oxidation to form **orotate** (uracil 4 - carboxylate). After the formation of pyrimidine ring, ribose and phosphate are added to the molecule involving 5 – phosphoribosyl 1 – pyrophosphate (PRPP). A carboxyl group is thereafter removed as CO_2, forming uridine monophosphate (UMP) which is then phosphorylated to form UTP one of the nucleotides of ribonucleic acid (RNA).

The nucleotide UTP can be modified to form cytidine triphosphate by the replacement of a Keto group with an amino group in the pyrimidine ring. Cytidine triphosphate CTP is a nucleotide in RNA and DNA. The ribose sugar is reduced to the deoxyribose form in DNA involving the reducing power of NADPH and the enzyme ribonucleoside reductase. The UTP is also the precursor of thymine triphosphate (TTP) found in DNA. The formation of thymine for incorporation into DNA involves the addition of a methyl group derived from serene (=Ser= S) to the uridine ring structure to form thymine and the reduction of ribose using NADPH to make deoxyribose. It is worth noting that ribonucleotides of RNA are made first and then modified enzymatically to form deoxyribonucleotides of DNA. This is indicative of the fact that RNA is more primitive molecule than DNA. This also gives the indication that RNA perhaps existed as an informational molecule for cells before the evolution of DNA molecules which serve as hereditary informational macromolecules of all living cells.

50
CHAPTER

MILESTONES OR LAND MARKS IN MICROBIOLOGY

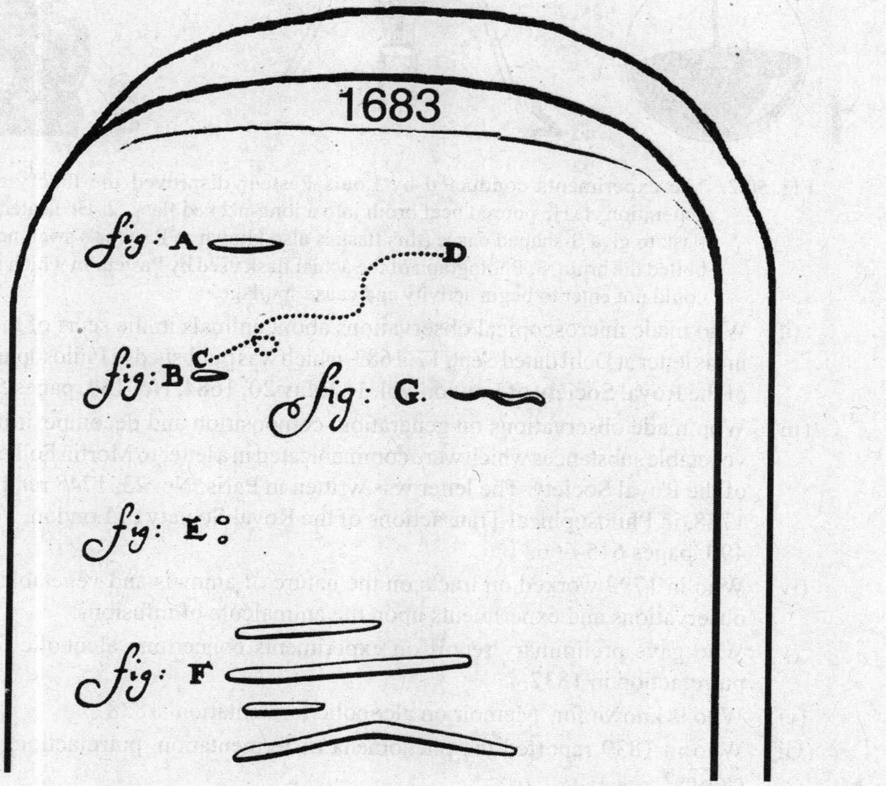

Fig. 50.1. A drawing by Antony van Leeuwenhoek which he made in 1683 served as a mile stone of microbiology. He was the first to observe animalcules which are now known as bacteria and protozoa.

Q.1. Name the scientist concerned with the following contributions.

(i) Who made observations concerning "animalcules" (little animals) in rain, well, sea and snow-water and recorded in a Dutch letter of 9th October 1676, which was published in Philosophical Transactions of Royal Society of London, Vol. II, 1677, No. 133, pages 821-831.

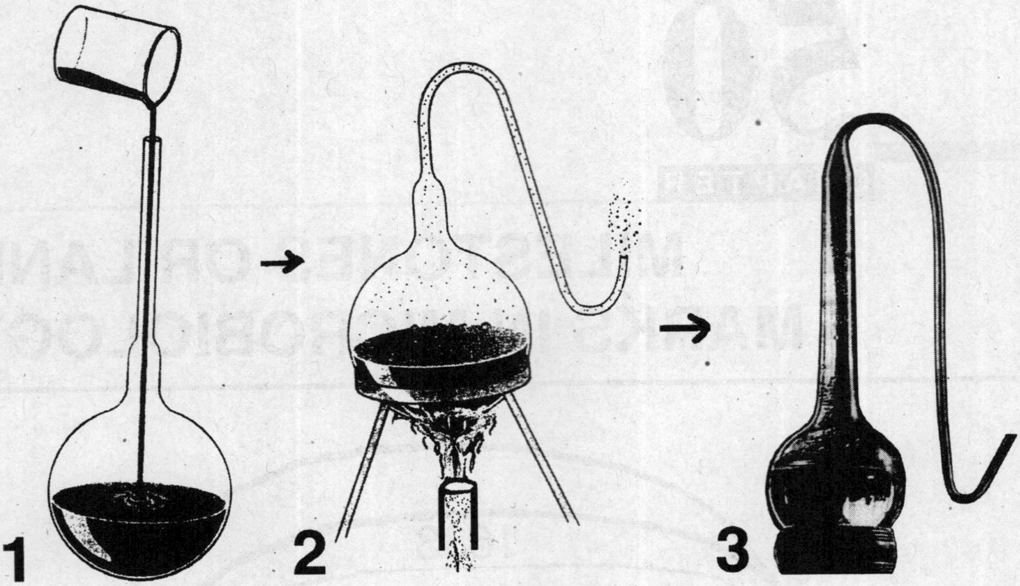

Fig. 50.2. The experiments conducted by Louis Pasteur disproved the theory of spontaneous generation. 1. He poured beef broth into a long-necked flask. 2. He heated the neck of the flask to give S-shaped curve (this flask is also known as Pasteur's swan necked flask) and boiled the broth. 3. Photograph of the actual flask used by Pasteur in which microorganisms could not enter to begin activity and cause spoilage.

(ii) Who made microscopical observations about animals in the scurf of the teeth recorded in his letter at Delft dated Sept. 17, 1683, which was published in Philosophical Transactions of the Royal Society of London Vol. 14, May 20, 1684, No. 159, pages 568-574, 1 plate.

(iii) Who made observations on generation, composition and decomposition of animal and vegetable substances which were communicated in a letter to Mortin Folkes Esq.; President of the Royal Society. The letter was written in Paris, No. 23, 1748 read Dec. 15 and 22, 1748, in Philosophical Transactions of the Royal Society of London, Vol. 45, 1748 No. 490, pages 615-666.

(iv) Who in 1799 worked on tracts on the nature of animals and vegetables and also made observations and experiments upon the animalcule of infusions.

(v) Who gave preliminary report on experiments concerning alcoholic fermentation and putrefaction in 1837.

(vi) Who is known for 'Memoir on alcoholic fermentation' 1838?

(vii) Who in 1839 reported the phenomena of fermentation, putrefaction, decay, and their causes?

(viii) Who reported in 1857 'The lactic acid fermentation'?

(ix) Who is known for 'Memoir on alcoholic fermentation' published in 1860 in French in Annales de Chimie et de Physique, Vol. 58, 3rd Series, pages 323-426.

(x) Who contributed the French article on "Animal infusoria living in the absence of free oxygen, and the fermentations they bring about" published in 1861 in Comptes rendus de l'Académie des sciences, Vol. 52 pages 344-347.

(xi) Who contributed on 'Influence of oxygen on the development of yeast and on the alcoholic fermentation' in 1861?

(xii) Who contributed in 1861 "On the organised bodies which exist in the atmosphere; examination of the doctrine of spontaneous generation"?

(xiii) Who made studies in 1876 on "The biology of bacilli"?

(xiv) Who made "Further researches on the deportment and vital persistence of putrefactive and infective organisms from a physical point of view" in 1877?

(xv) Who contributed on "The lactic fermentation and its bearing on pathology" in 1878?

(xvi) Who reported "alcoholic fermentation without yeast cells", in 1897?

(xvii) Who contributed on "Contagion, contagious diseases and their treatment" in 1546?

(xviii) "Contagion that infects by fomes (now called fomites)" in 1546?

(xix) Who worked on "Miasmatic, contagious and miasmatic-contagious diseases", in 1840?

(xx) "Lecture on the genesis of puerperal fever (childbed fever)" in 1850.

(xxi) "On a new method of treating compound fracture, abscess, and so forth" with observations and conditions of suppuration.

(xxii) Who put forward "on the antiseptic principle in the practice of surgery", in 1867?

(xxiii) Who demonstrated "the etiology of anthrax, based on the life history of *Bacillus anthracis*" in 1876?

(xxiv) Who discovered "Method for study of pathogenic organisms" in 1881?

(xxv) Who discovered "the etiology of tuberculosis", in 1882?

(xxvi) Who is associated with "the etiology of tuberculosis (Koch's postulates)" in 1884?

(xxvii) Who gave "A method for staining the tubercle bacillus", in 1882?

(xxviii) Who made "An enquiry into the causes and effects of the variolae vaccinae, a disease discovered in some of the western countries of England, particularly, Gloucestershire, and known by the name of the cowpox", in 1798?

(xxix) Who was to make an attempt to publish in Transactions of the Royal Society, the results of his first case? When the work was rejected he collected 23 cases and published his book.

(xxx) Attention to the causal agent of fowl cholera was invited by ______.

(xxxi) Who contributed "On a vaccine for fowl cholera and anthrax", in 1881.

(xxxii) Who in 1884 discovered the "disease of Daphnia (common water flea) caused by a yeast".

(xxxiii) "The mechanism of immunity to diphtheria in animals to diphtheria and tetanus" in 1890, was reported by ______.

(xxxiv) Studied "mechanism of immunity to diphtheria in animals", in 1890.

(xxxv) Who discovered that mice and rats are immune to diphtheria while equal volumes of the inoculum (cultures) were fatal to guinea pigs, rabbits and sheep.

(xxxvi) "A broth culture of the bacteria isolated from the membrane of a child which had died in January 1890 from diphtheria can kill a guinea pig in 3-4 days of injecting only 0.05 cc". The experiment was done by.

(xxxvii) "A *Contagium virum fluidum* (living germ that is soluble) as the cause of mosaic disease of tobacco leaves" in 1899.

(xxxviii) "Invisible microbe that is antagonistic to the dysentery bacillus" was discovered in 1917.

(xxxix) Isolated a crystalline protein possessing the properties of tobacco-mosaic virus in 1935.

Chemotherapy

(xxxx) The chemical foundation of the study of disinfection and the action of poisons, in 1897.

(xxxxi) Delivered a lecture on "Modern chemotherapy" in 1908.

(xxxxii) Showed antibacterial action of a *Penicillium* with special reference to their use in isolation of *Bacillus influenzae* (Hemophilus), in 1929.

(xxxxiii) "A contribution to chemotherapy of bacterial infections" in 1935.

(xxxxiv) Studies on bacteria (systematics, organisation and development of bacteria, and relationships of bacteria), in 1875.

(xxxxv) Worked out "differential staining of Schizomycetes in tissue sections and in dried preparations" in 1884.

(xxxxvi) Made a "minor modification of the plating technique of Koch" in 1887.

(xxxxvii) Discovered "bacterioids" in the root nodules, in 1888, and called the bacterial species *Bacillus radicicola*.

(xxxxviii) Gave information on Auxanography, a method useful in microbiological research involving diffusion in gelatin, in 1889.

(xxxxix) Worked on physiological studies of bacteria, in 1889.

(xxxxx) Worked on nitrifying bacteria, in 1890, and showed that nitrifying organisms obtained their energy from the oxidation of ammonia.

(Li) Made enrichment culture studies with urea bacteria in 1901.

(Lii) Worked on oligonitrophilic microorganisms.

(Liii) Reported new substances (now known as vitamins) for the growth of the yeast, in 1901.

(Liv) Reported fixation of free atmospheric nitrogen by *Azotobacter* in pure culture, in 1908.

(Lv) Reported on "Unity and diversity in the metabolism of microorganisms" in 1924.

Ans. (i) Antony van Leeuwenhoek
(ii) Antony van Leeuwenhoek
(iii) Turbevill Needham
(iv) Lazaro Spallanzani
(v) Theodore Schwann
(vi) Charles Cagniard-Latour
(vii) Justis Liebig
(viii) Louis Pasteur
(ix) Louis Pasteur
(x) Louis Pasteur
(xi) Louis Pasteur
(xii) Louis Pasteur
(xiii) Ferdinand Cohn
(xiv) John Tyndall
(xv) Joseph Lister
(xvi) Eduard Buchner
(xvii) Girolamo Fracastoro
(xviii) Girolamo Fracastoro
(xix) J. Henle
(xx) Ignaz Semmelweis
(xxi) Joseph Lister
(xxii) Joseph Lister
(xxiii) Robert Koch
(xxiv) Robert Koch
(xxv) Robert Koch
(xxvi) Robert Koch
(xxvii) Paul Ehrlich
(xxviii) Edward Jenner
(xxix) Edward Jenner
(xxx) Louis Pasteur
(xxxi) Louis Pasteur
(xxxii) Elias Metschnikoff
(xxxiii) Emil von Behring and Shibasaburo Kitasato
(xxxiv) Emil von Behring
(xxxv) Emil von Behring
(xxxvi) Emil von Behring
(xxxvii) M.B. Beijerinck
(xxxviii) F. d'Herelle
(xxxix) Wendell M. Stanley
(xxxx) B. Krönig and Th. Paul
(xxxxi) Paul Ehrlich
(xxxxii) Alexander Fleming
(xxxxiii) Gerhard Domagk
(xxxxiv) Ferdinand Cohn
(xxxxv) Christian Gram
(xxxxvi) R.J. Petri
(xxxxvii) Martinus W. Beijerinck
(xxxxviii) Martinus W. Beijerinck
(xxxxix) S. Winogradsky
(xxxxx) S. Winogradsky
(Li) Martinus W. Beijerinck
(Lii) Martinus W. Beijerinck
(Liii) E. Wildiers
(Liv) M. Beijerinck
(Lv) Albert J. Kluyver

Inoculation by Brahmins of India with Preceding Year's Pokes of Smallpox to Immunize People

Fig. 50.3. During East India Company days a Brahmin of Bengal, inoculating a boy with a Barber's nail cutter using crusts of preceding year's pokes of **smallpox** and the Ganges water to which local people called teeka, much before Edward Jenner inoculated 8 years old boy James Phipps in 1796" (Published in 1798).

Dharampal, a well known Gandhian published a book entitled "Indian Science and Technology in the eighteenth century" in which he reported some contemporary European accounts describing the status of science in India in different branches. Two accounts one written by Rev. Coult and sent to Dr. Coult on diseases in Bengal dated 10 February 1731 and the other by Dr. J.L. Holwell, F.R.S., sent to the President and Members of College of Physicians in London in 1767 AD on "Inoculating for smallpox in the east Indians." Both of these accounts provide evidence that the Indians knew how to immunize people against smallpox. (Fig. 50.3)

"The operation of inoculation called *teeka* by natives has been known to the Kingdom of Bengal as near as I can learn about 150 years i.e. approximately 1581 A.D and according to brahmiman records was first performed by one **Dhanwantary** a physician of champanagar a small town by the side of the Ganges about half way to cossimbazar whose memory is now held in great esteem as being thought the author of this operation which secret, they say, he had immediately of God in a dream." Their method of performing this operation is by taking a little of pus (when the smallpox come to maturity and are of good kind) and dipping these in the point of pretty large sharp needle. Therewith make several punctures in the hollow under deltoid muscle or sometimes in the forehead after which they cover the part with a little paste made of boiled rice.

Rev Coult was skeptical about the procedure, since Rev. Coult had depended upon descriptions of procedures by Brahmins.

J.Z. Holwell in 1767 wrote the following:

"On perusing lately some tracts upon the subject of inoculation, I determine to put together a few notes relative to the manner of inoculation, practiced time out by the Brahmins of Hindustan to this I was chicfly instigated by considering the great benefit that may arise to mankind from a knowledge of this foreign method which so remarkably tends to support the practice now generally followed with such marvelous success."

The cotton which he preserves in a double calico rag is saturated with matter from the pustules of the preceding year for they never inoculate with fresh matter nor with matter from the disease caught in the natural way

He extends prohibition of fish, milk and ghee for one month from the day of inoculation.

51 CHAPTER

BIOINFORMATICS

Q.1. When was HGP (Human genome project) started?

Ans. Human genome project was started in 1987.

Q.2. What is Bioinformatics?

Ans. Bioinformatics is an integrated field, which combines computational, mathematical and statical methods to manage and analyze the biological data or information.

Or

Bioinformatics is the management and analysis of biological informations contained in biological databases.

Or

Bioinformatics is a new subject of genetic information collection, analysis, and dissemination to the research community.

Q.3. Which are the main sub-disciplines of bioinformatics?

Ans. There are three main sub-disciplines within bioinformatics:

a. To develop, new algorithms and statical methods which may be used to access the relationships among informations present in large databases.

b. To analyze and interpret various types of information, which include nucleic acid and amino acid sequences with protein structures and protein domains.

c. To develop new computer tools that enable efficient access and management of different types of biological information.

Q.4. What is **medicalinformatics**?

Ans. **Medicalinformatics** is a new field, which is used by physicians to manage and analyze clinical information contained in hospital database.

Q.5. Which types of issues or problems related to biological data are dealt with the bioinformatics?

Ans. Mainly, bioinformatics deals with development of data analysis tools, molecular modelling of various biological macromolecules in two dimensional and three dimensional structures, metabolic pathways, in pharmaceutical industries to develop new drug molecules, peptide vaccines, proteins etc.

Q.6. What is the main role of a **bioinformatician** in present biological research and development area?

Ans. The main role a **bioinformatician** is to create a framework that can support the needs of information based R&D for biological research.

Q.7. Why do we move towards the use of bioinformatics ?

Ans. There are many reasons, that provide support to the use of bioinformatics such as,

a. For handling the biological data and manage it in such a way to give more meaningful information with the help of highly sophisticated computers.

b. To communicate among people, projects, and institutions engaged in biological R&D and applications.

c. To organize, give access and retrieving facility of biological information, documents and literature.

d. To analyze and interpret biological data through the computational facilities.

Q.8. Which type of skills are required to be a good bioinformatician ?

Ans. To be a good bioinformatician, it is necessary to have a sound knowledge of molecular biology, biochemistry, molecular biophysics, molecular modelling, computer, information technology, and biostatistics. Beside, a bioinformatics professional must understand the central dogma of molecular biology and bioinformatics software packages.

Q.9. Why **dry lab** term is used for defining bioinformatics?

Ans. The meaning of **dry lab** is related to work without any type of chemicals, solutions in laboratory conditions. Bioinformatics mainly relate to work with computer and internet to analyze the biological information that is present on internet servers in different databases, due to these specific requirements dry lab term is used for defining the bioinformatics.

Q.10. What is **data mining**?

Ans. Discovery and retrieving of required data from different databases is known as **data mining**.

Q.11. What is KDD?

Ans. Knowledge Discovery in Databases (KDD) is an emerging field combining the theoretical and practical issues of extracting high level information from volume of low level data.

Q.12. How the term **bioinformaticist** is different from **bioinformatician**?

Ans. A **bioinformaticist** is an expert who knows how to use bioinformatics tools as well as to write interface for effective use of the tools. While a **bioinformatician** is a trained individual who only knows to use bioinformatics tools without having the knowledge of its construction.

Q.13. What is **integrative bioinformatics**?

Ans. The study of informative processes in biotic systems is known as **integrative bioinformatics**.In other words the high throughput interpretation of biological pathways is known as **integrative bioinformatics**.

Q.14. Define term **database**.

Ans. Database is the combination of same type of information or files that are collectively called as database.

Q.15. What is OMIM?

Ans. OMIM is a database that provides all the information about the human inheritance. The complete meaning of OMIM is Online Mendelian Inheritance in Man.

Q.16. Which type of databases are used in bioinformatics?

Ans. There are more than 200 databases which are used in bioinformatics but the main categories of database relate to annoted database, curated database, federated databases, integrated databases, interoperability databases, non-redundant databases, proprietary databases, redundant databases, relational databases, indepth flat files and indexed flat files.

Q.17. What is **federated database**?

Ans. Federated database is an integrated repository information of multiple data sources presented with consistent and coherent semantics.

Q.18. What is **interoperability** ?

Ans. The ability of different platforms of computers, networks, operating systems and applications to work effectively, without prior communication in order to exchange the information in a useful and meaningful manner is known as interoperability.

Q.19. What is **non-redundant database**?

Ans. A database that have redundant entries present in merged condition. These entries are typically those that are 100% character by character identical.

Q.20. What is **complementarity**?

Ans. The sequence specific or shape specific recognition that occurs when two or more molecules binds together.

Q.21. What is consensus sequence?

Ans. The most commonly occurring amino acid or nucleotide at each position of an aligned series of proteins or polynucleotide is known as consensus sequence.

Q.22. Define the **consensus map**.

Ans. The location of all consensus sequences in a series of multiple aligned proteins or polynucleotide is called the consensus map.

Q.23. What do you understand by the term-**conserved sequence**?

Ans. Conserved sequence is a sequence present in DNA or Protein that is consistent across species or has remained unchanged within the species over its evolutionary changes.

Q.24. What are **proprietary databases**?

Ans. These are a charge subscription type database, which require fees to access those on commercial level, such as LifeSeq and Gene Logic's.

Q.25. Define **redundant databases**.

Ans. In general, the primary type database is known as redundant databases or when sequences were first created in the database is called as redundant database.

Q.26. Define the following terms.

Ans. **a. Allele :** Different forms of a gene, which occupy the same position on the chromosome.

b. Amplification : The process of repeatedly making copies of the same piece of DNA.

c. BAC (Bacterial artificial chromosome) : A long sequencing vector, which is created from a bacterial chromosome by splicing a DNA fragment of 100kb from another species.

d. Base pair : The complementary bases on opposite strands of DNA which are held together by hydrogen bonding.

e. BEAUTY : BEAUTY or BLAST enhanced alignment utility, a tool developed by Baylor College of Medicine which uses BLAST to search several custom databases and incorporates sequence family information, location of conserved domains and information about any annotated sited or domains directly into the BLAST query results.

f. BLOCKS : a database of ungapped multiple alignments for protein or peptide families in PROSITE.

Q.27. What is **search engine**?

Ans. A search engine is a type of utility or tools, which provide facility to retrieve information from different databases. In general life we use many search engines such as Goggle, Rediff and Yahoo but for bioinformatics there are mainly two search engines BLAST and FASTA.

Q.28. What is **BLAST**?

Ans. Basic local alignment searching tool, used to find out the queried sequence from different databases of protein, DNA, RNA etc.

Q.29. Where is the **EMBL** situated?

Ans. The European Molecular Biology Laboratory is situated in Heidelberg, Germany.

Q.30. What is **FASTA**?

Ans. An alignment tool for protein sequences created by Pearson and Lipman in 1988.

Q.31. What is **FISH**?

Ans. FISH (Fluorescence *in situ* hybridization) is a method, which is used to pinpoint the location of a DNA sequence on a chromosome.

Q.32. How is **HMM** useful in bioinformatics?

Ans. HMM (Hidden Markov Model) is a computer algorithm which locates the essential, unique features which can distinguish a protein or gene family by analyzing a range of known sequences from the family.

Biological Database

Q.33. What is a biological database?

Ans. A **biological database** is a large, organized body of persistent data, usually associated with computerized softwares designed to update queries and retrieve components of the data stored within the system.

Q.34. Which types of biological data are stored in biological databases?

Ans. In general, biological database is a complex structure of many individual types of databases, collectively known as biological database and comprises nucleotide sequences, nucleotide structures, gene expressions data, protein structures, protein functions, metabolic pathways, cell type, cell signaling, tissues type, organs and organism type.

Q.35. What are the main properties of a **good database**?

Ans. A good database has following qualities :

a. Broad coverage of the chosen topic,
b. Up-to-date information gathering,
c. Support staff,
d. Commitment to the future, and
e. Having good query interface.

Q.36. What is the main function of biological databases?

Ans. There are two main functions of biological databases,

a. Provide biological data for research purpose to scientists,
b. To make biological data available in computer readable form.

Q.37. In which year, first biological sequence database was published?

Ans. It was Atlas of Protein sequences and structure by Margaret Dayhoff and colleagues, published in the year 1965.

Q.38. How many sub databases are contained in **nucleotide databases** ?

Ans. Nucleotide databases contain a lot of sub databases but there are mainly 13 databases which are used as DNA/RNA database such as Genebank, EMBL, DDBJ, Genomic database, Yeast database, *E. coli* database, Specialized Database, ESTs database, STSs database, Kabat database, EPD database and REPBASE database etc.

Q.39. Name a few protein databases?

Ans. These are : Swiss-Prot, PIR, TREMBL, GenPept, Kabatpro, Alu, PDB etc.

Q.40 What is **PROSITE**?

Ans. It is a database of protein motifs. This database is very useful, which provides a list of distinct structural motifs in proteins.

Q.41. Which type of information is present in GDB ?

Ans. GDB is a genome database, which contains mapping information of the human genome project. It also contains a huge information about locations of genes, DNA segments, expressed sequences tags, clinical phenotypes, alleles information, and probes information.

Q.42. Define the following.

Ans. **a. ALA mode :** It is an environment for building models of ligand and monomer molecular components.

b. NDB : It is the nucleic acid database.

c. HotMolecBase : This is a database having extensive information about biological factors, proteins, and genes.

Q.43. What is the **microbiological database**?

Ans. **Microbiological database** is a listing of all published microbial genomes and chromosomes such as **LakeLadoga.**

Q.44. What is the **virological database**?

Ans. It is a database which comprises all type of viruses which infect all types of living beings such as vertebrates, invertebrates, protozoa, plants, bacteria, fungi, archaea and algae.

Q.45. Write the main properties, which are used to classify viruses in virological database?

Ans. There are many properties of viruses, which are used to classify viruses in virological database such as, virion properties, physiochemical and physical properties, genome type, proteins, G+C ratio, lipids type and biological properties.

Q.46. What is the **primary database**?

Ans. A database consisting of data derived experimentally such as nucleotide sequences and three-dimensional structures are known as primary databases.

Q.47. What are **secondary databases**?

Ans. Those data that are derived from the analysis or treatment of primary data such as secondary structures, hydrophobicity plots, and domains are stored in secondary database.

Q.48. Name the few websites important for bioinformatics?

Ans. a. **NCBI:** http://www.ncbi.nlm.nih.gov

b. **EBI:** http://www.ebi.ac.uk

c. **DISC:** http://www.dna.affrc.go.jp

d. **ExPASy-SwissProt and TrEMBL**: http://www.expasy.ch

e. **SwissProt, TrEMBL and PIR:** http://www.ebi.ac.uk

f. **NCBI-Entrez** (for searching sequence database): http://www.ncbi.nlm.nih.gov/entrez/**g. PDB:** http://www.rcsb.org/pdb/

h. **BLAST:** http://www.ncbi.nlm.nih.gov/BLAST/

Q.49. What is **EMBL**?

Ans. The **EMBL** is the European Molecular Biology Laboratory, nucleotide sequence database, which is maintained by the European Bioinformatics Institute in Hinxton, Cambridge.

Q.50. Which institute maintain the **Gene Bank** ?

Ans. The Gene Bank is a nucleotide database, which is maintained by the National Center for Biotechnology Information (NCBI).

Q.51. Define HTTP://WWW.NCBI.NLM.NIH.GOV/.

Ans. The complete expanded form of the HTTP://WWW.NCBI.NLM.NIH.GOV is,**HTTP:** Hyper text transfer protocol,**WWW:** World wide web,**NCBI:** National Center for Biotechnology

Information**NLM:** National Laboratory for Medicine,**NIH:** National Institute of Health,**GOV:** Government site.

Q.52. What is **DDBJ**?

Ans. It is the DNA Databank of Japan, which is run by the National Institute of Genetics of Japan.

Q.53. What is **SGD**?

A.53. The *Saccharomyces* genome database is a scientific database of the molecular biology and genetics of the yeast ***Saccharomyces cerevisiae***.

Q.54. What is **SWISSPROT** and how is it useful?

Ans. **SWISSPROT** is a protein sequence database, which strives to provide a high level of annotations such as description of the function of a protein, its domain structure, protein, and nucleic acid data, post-translational modifications etc.

Q.55. What is **PIR**?

Ans. The protein information resources (PIR) is a division of the National Biomedical Research Foundation (NBRF) in the US.

Q.56. Give a list of databases often used in bioinformatics.

Ans. The LIST OF DATABASES USED IN BIOINFORMATICS is as under :

Database type	**Its meaning**
ACEDB	A *Caenorhabditis elegans* database
AFDB	*Archaeoglobus fulgidus* genome database
AMSDb	Antimicrobial sequence database
AnoDB	An anopheles database
AtDB	*Arabidopsis thaliana* database
BBDB	*Borrelia burgdorferi* genome database
BioImage	Biological image database
CCSD	Complex carbohydrate structure database
CSD	Cambridge structural database
CSNDB	Cell signalling networks database
DDBJ	DNA data bank of Japan
DPD	DNA patent database
ECDC	*E. coli* database collection
EPD	Eukaryotic promoter database
EpoDB	Erythropoiesis database
GBASE	Genome database of mouse
GCRDb	G-protein coupled receptor database
GDB	Genome database
GRBase	Growth regulation database
GSDB	Genomic sequence database
HEXAdb	Hexosaminidase A locus database
HGMD	Human gene mutation database
HIDB	*Haemophilus influenzae* Rd genome database
HOVERGEN	Homologues vertebrates gene database
HPDB	*Helicobacter pylori* genome databasae
IGD	Integrated Genomic database

IMGT	ImMunoGeneTics database
JIPID	Japan international protein information database
MAD	Malaria antigen database
MDB	Metalloprotein site database and browser
MGDB	*Mycoplasma genitalium* genome database
MHCPEP	MHC-binding peptide database
MMDB	Molecular modelling database
MPDB	Molecular probe database
MT	Metallothioneins database
MycDB	*Mycobacterium* database
NDB	Nucleic acid database
OPD	Oligonucliecotide probe database
ORDB	Olfactory receptor database
PDB	Protein data bank
PMD	Protein mutation database
Pfam	Protein families database
ProDom	Protein domain database
SCPD	*Saccharomyces cerevisiae* promoter database
SGD	*Saccharomyces* genome database
SRPDB	Signal recognition particle database
SSUrRNAdb	Small ribosomal subunit RNA database
WormPD	Worm proteome database
YPD	Yeast protein database
DbCFC	Cytokine family cDNA database
DbEST	Database of expressed sequence tags
DbGSS	Database of genome survey sequences
DbSNP	Database of single nucleotide polymorphism
DbSTS	Database of sequences tagged sites

Q.57. Fill in the blanks in the following :

(i) J.D. Watson and Francis Crick proposed the double helix model of DNA based on X-ray diffraction pattern obtained by ________ and ________.

(ii) In 1954, ________ group develop heavy atom method to solve the phase problem in protein crystallography.

(iii) The first sequenced protein was ________.

(iv) The ________ was created by linking computers at Stanford and UCLA.

(v) The detail of Needleman-Wunsch algorithm for sequence comparison was published in ________.

(vi) In 1972, ________ created the first recombinant DNA molecule.

(vii) The Brookhaven Protein Data Bank was announced in the year ________.

(viii) ________ and ________ found Microsoft Corporation in year ________.

(ix) Two-dimensional electrophoresis was announced by ________.

(x) The National Center for Biotechnology Information (NCBI) was established at the ________.

(xi) Pearson and Lupman published the algorithm for sequence comparison by using ________ tool.

(xii) Use of BLast was started from the year ________.

(xiii) The research institute in Geneva (CERN) announced the creation of the protocols which make up the ________.

(xiv) ________ database was formed in Maryland.

(xv) The PRINT database of ________ was published by Attwood and Beck.

(xvi) The *Haemophilus influenza* genome (1.8 M) was sequenced in the year ________.

(xvii) Baker's yeast genome comprising of approximately 6000 genes was sequenced in the year ________ having ________ size.

(xviii) ________ produced the first commercial DNA chip in the year ________.

(xix) In 1997, bacterium ________ genome was published having ________ size.

(xx) Genome of *P. aeruginosa* was published having size ________.

(xxi) *Drosophila melanogaster* genome was sequenced in the year ________.

(xxii) The first genome maps were produced in Drosophila. In 1910 the Morgan lab identified a spontaneous mutant male fly that had ________ coloured eyes rather than the brilliant red coloured seen in normal strains.

Ans.

(i) Rosalind Franklin, Maurice Wilkins.

(ii) Perutz's.

(iii) Bovine insulin.

(iv) ARPANET.

(v) 1970.

(vi) Paul Berg and his group.

(vii) 1973.

(viii) Bill Gates, Paul Allen.

(ix) P.H.O'Farrell.

(x) National Cancer Institute.

(xi) FASTA.

(xii) 1990.

(xiii) World Wide Web.

(xiv) Gene Logic.

(xv) Protein motifs.

(xvi) 1995.

(xvii) 1996, 12.1Mbp.

(xviii) Affymetrix, 1996.

(xix) *E. coli*, 4.7Mbp.

(xx) 6.3Mbp in year 2000.

(xxi) 2000 with size 180Mbp.

(xxii) White.

52
CHAPTER

OBJECTIVE ASSESSMENT

Choose the correct out of the following.

1. Which of the following microorganism was used in industrial production of acrylamide at Nitto Chemical Industry, Japan, in 1985?

(A) *Aspergillus*

(B) Using nitrite hydratase of *Rhodococcus* sp. N-774

(C) *Acetobacter*

(D) None of the above.

2. In glycolysis, a molecule of glucose is degraded in a series of enzyme catalyzed reactions to yield.

(A) Three molecules of pyruvate (B) Two molecules of pyruvate

(C) One molecule of pyruvate (D) No pyruvate

3. Which of the following before he lost his head in French Revolution recognized that animals somehow transform chemical fuels (foods) into heat and that this process of respiration is essential to life?

(A) Arthur Kornberg (B) George Hitchings

(C) B. McClintock (D) Antoine Lavoisier

4. Which of the following procaryotes lack the peptidoglycan layer ?

(A) Gram negative bacteria (B) Gram positive bacteria

(C) Archaeobacteria (D) Cyanobacteria

5. What are the spherical vesicles in animal cells, containing enzymes capable of digesting proteins, polysaccharides, nucleic acids and lipids, called?

(A) Lysosomes (B) Ribosomes (C) Peroxisome (D) Plasmodesmata

6. The histones and DNA associate to form complexes 10 nm in diameter, are called :

(A) Myosin

(B) Nucleosomes

(C) Thylakoids

(D) Actin – myosin complexes or contractile systems

7. Which of the following in bacterial cell serve as the points adhesion to surface of animal cells?

(A) Pili (B) Flagella (C) Mesosome (D) All of the above

8. In which of the following animal viruses genomic material is RNA?

(A) Adenoviruses (B) SV 40

(C) Poliomyelitis, Influenza and Sindbis (D) Herpes, Vaccinia, Parvoviruses and HIV

9. Which of the following gave correct explanation for the phenomenon of **optical activity** by separating two types of crystals identical in shape but mirror images of each other, from crystalline material, paratartaric acid also called racemic acid that got accumulated in wine casks ?
   (A) James Sumner, in 1926 (B) Leonor Michaelis, in 1913
   (C) Vietor Henri, in 1903 (D) Louis Pasteur, in 1843
10. Which of the following equation relates p^H, p^Ka and buffer concentration?
   (A) Lineweaver – Burk equation (B) Michaelis – Menten equation
   (C) Henderson – Hasselbalch equation (D) None of the above
11. Which of the following worked out the sequence of amino acids in the polypeptide chains the hormone insulin?
   (A) James D. Watson and Francis Crick (B) Georges Köhler
   (C) Frederick Sanger (D) Cesar Milstein
12. Electrophoretic method generally used for estimation of purity and moleculer weight of proteins, makes use of :
   (A) Sodium dodecyl sulphate (SDS)
   (B) Agar
   (C) Gum arabic
   (D) Powdered thalli of algae *Gelidium* and *Gracilaria*
13. The study of reaction rates and how they change in response to changes in experimental parameters is called :
   (A) Kinetics (B) Allosterism (C) Zymogen (D) Lysozyme
14. The lipids are :
   (A) Water soluble components which cannot be extracted by nonpolar solvents.
   (B) Not found in the membrane of chloroplast of photosynthetic alga *Euglena*.
   (C) Not found in the membrane of mitochondria of *Euglena*.
   (D) Water – insoluble components of cells which can be extracted by nonpolar solvents.
15. The biological membranes are composed of :
   (A) Lipids and proteins in varying combinations which are specific to each species, cell type and organelle.
   (B) Lipids only
   (C) Proteins only
   (D) None of the above.
16. The membrane protein bacteriorhodopsin is a characteristic of :
   (A) *Halobacterium* (B) *Escherichia coli*
   (C) Mitochondrion of bovine heart (D) None of the above
17. Which of the following is found in porous rigid walls of bacterial cells :
   (A) Cellulose the structural polysaccharide with D-glucose units in β 1 – 4 linkage
   (B) Chitin a linear polymer of N–acetyl glucosamine joined in β 1 – 4 linkages
   (C) Peptidoglycan linear polysaccharides of alternating N – acetyl muramic acid and N – acetylglucosamine units, cross – linked by short peptide chains.
   (D) Heteropolysaccharides (glycosaminoglycans) and proteins collectively called proteoglycans

18. Which of the following characteristic is concerned to unusual structure of H – DNA ?

(A) It is the most stable structure and is considered standard point of reference in any study of the properties of DNA.

(B) It is arranged in right handed double helix favoured in many solutions devoid of water.

(C) It shows the pairing and inter winding of three strands of DNA to form a triple helix.

(D) Left handed helical rotation and DNA backbone takes a zig-zag appearance.

19. Which of the following are **purines**?

(A) Thymine and Cytosine
(B) Adenine and guanine
(C) Cytosine and Uracil
(D) Thymine and Uracil

20. Which of the following is a **nucleoside**?

(A) A nitrogenous base, pentose and a phosphate

(B) Guanine isolated from guano (bird manure)

(C) Thymine isolated from thymus tissue

(D) Characteristic components are a nitrogenous base and pentose.

21. Which of the following are the end products of anabolism ?

(A) Energy poor products : CO_2, H_2O and NH_3

(B) Cell macromolecules : proteins, polysaccharides, lipids and nucleic acids

(C) Precursor molecules : amino acids, sugars, fatty acids and nitrogenous bases

(D) None of the above.

22. Which of the following is true for firefly flashes ?

(A) ATP is not involved

(B) Radio active material glows

(C) Generation of light flashes requires activation of luciferin by an enzymatic reaction with ATP in which pyrophosphate cleavage of ATP occurs, to form luciferyl adenylate. Oxidative decarboxylation of luciferin is accompanied by emission of light.

(D) Some volatile substances burn.

23. Which of the following is the correct form of **second law of thermodynamics**?

(A) In all natural processes the entrophy of the universe increases.

(B) The extent of randomness or disorder in a system decreases.

(C) The total amount of energy in the unverse remains constant although the form of energy may change.

(D) None of the above.

24. Which of the following stands for **bioenergetics**?

(A) ATP plays no role in biological energy exchanges.

(B) It is the **enthalpy** (H) or the heat content of a system.

(C) Quantitative study of energy transductions which take place in living cells of nature and function of the chemical processes underlying these transductions.

(D) None of the above.

25. Glycolysis is the break down of 6 – carbon glucose into :

(A) Three molecules of three-carbon pyruvate which takes place in 10 steps.

(B) Four molecules of three-carbon pyruvate.

(C) Two molecules of three-carbon pyruvate which take place in 10 steps.

(D) None of the above.

26. Hans Kreb is known for :
(A) Preparatory phase or conversion of glucose to glyceraldehyde-3-phosphate.
(B) Citric acid cycle or TCA (tricarboxylic acid cycle).
(C) Pay off phase or conversion of glyceraldehyde 3-phosphate to pyruvate
(D) A linear sequence of enzymatic catalyzed steps for breakdown of glucose.

27. β oxidation is :
(A) Oxidative degradation of fatty acids into acetyl-CoA.
(B) Formation of a chain of fatty acids by successive oxidations at β-carbon atom.
(C) An extended zig-zag arrangement of polypeptide chain.
(D) Synthesis of fatty acids by peroxisomes of plants and animals.

28. Bacteria and free-living protozoa release ammonia :
(A) To their aqueous environment.
(B) As uric acid that is part of purine catabolic pathway.
(C) As guano which is a valuable fertilizer because of its nitrogen content.
(D) By making urea enzymatically.

29. Halophilic (salt loving) bactericum ***Halobacterium halobium*** use light energy to make ATP:
(A) Through water of Salt lake and the dead Sea.
(B) Through membranous pigment containing structures called **chromatophores**.
(C) Through patches of light absorbing pigment of the protein **bacteriorhodopsin**.
(D) Through water of the brine ponds.

30. Photosystem I passes electrons from its excited reaction centre :
(A) P 700 through a series of carriers to ferredoxin which then reduces $NADP^+$ to NADPH
(B) P 680 to plastoquinone
(C) P 800 of "Z-scheme"
(D) None of the above.

31. Gluconeogenesis is :
(A) The biosynthesis of carbohydrate glucose from simpler non-carbohydrate non -hexose precursors like oxaloacetate or pyruvate.
(B) Conversion of glucose to pyruvate.
(C) Inhibition of enzymes of Calvin cycle.
(D) Photorespiration waste.

32. Malonyl–CoA, a three carbon intermediate participates in :
(A) Break down of fatty acids.
(B) Break down as well as biosynthesis of fatty acids.
(C) Biosynthesis of fatty acids.
(D) None of the above.

33. *Mycoplasma laidlawii* is a common :
(A) Saprophytic species (B) Halophilic species
(C) Thermophilic species (D) Parasitic species.

34. Who discovered the species of mycoplasma first of all in 1898?
(A) Thomas and Krisnaswami (B) Nocard and Roux
(C) John G. Holt (D) S.M. Paul Khurana

35. Who named *Rickettsia* in honour of Howard Taylor Ricketts?
(A) Robert Koch (B) H. da Rocha-Lima
(C) Vinson (D) Louis Pasteur

36. Where do *Chlamydia* reproduce ?
(A) Inside the host cell (B) Outside the host cell
(C) In dead organic matter (D) In milk

37. Which of the following discovered in 1892 that TMV is filterable ?
(A) Edward Jenner (B) Carlton Gajdusek
(C) Dimitri Ivanowski (D) Moris and Smith

38. Where is sex pilus found in/amongs the following?
(A) Donor bacterial cell (B) Recepient bacterial cell
(C) Donor and Recepient both (D) None of these

39. The zygospores of Mucor when present are with :
(A) Unequal suspensors (B) No suspensors
(C) Equal suspensors (D) Suspensor on one side only

40. Which of the following is true regarding a cleistothecium?
(A) A closed fructification (B) An open fructification
(C) Closed at the top, open at the bottom (D) Closed at the bottom and open at the top

41. Which of the following possess heterocyst?
(A) *Vaucheria* (B) *Chlorella* (C) *Gymnodinium* (D) *Nostoc*

42. The cells of cyanobacteria possess flattened membranaceous sacs. What are they called?
(A) Golgi bodies (B) Thylakoids (C) Mesosome (D) Ribosome

43. Which of the following is correct for the Hyphochytridiomycetes?
(A) They grow on sand (B) They grow in oil
(C) They are aquatic (D) They grow in mineral oil

44. Which of the following can be used to prepare the biochips of computer ?
(A) *Bacillus* (B) *Halobacterium* (C) *Chlamydia* (D) *Acetobacter*

45. On which of the following *Pleurotus sajor-caju* can be grown?
(A) Plastic sheets (B) Used polybags (C) Straw in polybags(D) Wet soil

46. Which of the following bacterium is used as biofertilizer?
(A) *Mycobacterium* (B) *Salmonella*
(C) *Escherichia coli* (D) *Rhizobium*

47. What for predacious fungi can be used in view of the following?
(A) In biological control (B) As biofertilizers
(C) As food (D) None of these

48. What is the wine which has lost carbon dioxide in the final steps of its preparation called?
(A) Sparkling wine (B) Gasoline (C) Gasohol (D) Still wine

49. Which of the following is a comestible fungus?
(A) *Puccinia* (B) *Ustilago* (C) *Agaricus bisporus* (D) *Cystopus*

50. Who authored "Bacterial Fertilizers" in IÇAR Handbook of Manures and Fertilizers in 1965.
(A) A.C. Gaur (B) W.V.B. Sundara Rao
(C) P. Tauro (D) J.B. Boussingault

51. At which place the national facilityfor Blue-Green Algae (BGA) is situated?
(A) N.B.R.I., Lucknow (B) C.D.R.I., Lucknow
(C) F.R.I., Dehradoon (D) I.A.R.I., New Delhi

52. Which of the following is used as a carrier for *Rhizobium* inoculants.
(A) Sterile sand (B) Bee hive (C) Peat powder (D) Wheat straw

53. Which of the following is causative agent of botulism?
(A) *Escherichia coli* (B) *Clostridium botulinum*
(C) *Klebsiella* (D) *Bacillus*

54. Which of the following is indicator of pollution due to faecal matter?
(A) *Caulobacter* (B) *Escherichia coli*
(C) *Staphylococcus aureus* (D) *Streptococcus pyogenes*

55. The study of which of the following is called limnology?
(A) Sea water (B) Fresh water (C) Lime (D) Lichens

56. Which of the following is a source of agar a polysaccharide?
(A) A fungus (B) An alga (C) A bacterium (D) *Agaricus*

57. Which of the following is the potable water?
(A) Water in the pot of copper (B) Water of an earthen pot
(C) Water of a river (D) Water fit for drinking.

58. Who amongst the following discovered 'little animalcules'?
(A) Robert Koch (B) Leeuwenhoek (C) Louis Pasteur (D) Lister

59. Which of the following is true for sterilization?
(A) To make free from all kind of organisms.
(B) Selective killing of microorganisms.
(C) Heating to boil.
(D) Heating to 72°C.

60. Which of the following is true for a bacterial smear?
(A) A loopful of the bacterial culture. (B) A thin film of bacteria on slide.
(C) A thick layer of bacteria. (D) A drop of bacterial suspension.

61. Which of the following is not a member of Enterobacteriaceae ?
(A) *Salmonella* (B) *Enterobacter* (C) *Proteus* (D) *Rhizobium*

62. Which of the following possesses a set of degradation plasmids to use toxic compounds e.g. camphor, toluene, and octane?
(A) *Escherichia coli* (B) *Pseudomonas*
(C) *Streptococcus faecalis* (D) *Rhizobium*

63. Who was the first to have a patent of oil eating bacterial strain from the U.S. Supreme Court, in 1980.
(A) Edward S. Ayensu (B) Michael Smith
(C) Anand Mohan Chakraborty (D) Herbert Boyer

64. Which of the following is correct for serum?
(A) The fluid with RBC
(B) The fluid with WBC
(C) The fluid with both RBC and WBC
(D) The remaining fluid after the blood has coagulated or clotted.

65. Which of the following is true for Ig A ?
(A) Monomer (B) Dimer (C) Tetramer (D) Pentamer

66. Which of the following is true for CD4 cells?

(A) Work as receptors of HIV
(B) Keep away HIV
(C) Kill HIV
(D) None of the above.

67. What happens to antibody producing spleen cells and myeloma cells which result into hybridoma?

(A) First, they fuse
(B) They live together
(C) Myeloma cells degenerate
(D) Both of them degenerate

68. What does the Presence of HCG in urine of a woman indicate?

(A) Infection in lungs
(B) Pregnancy
(C) UTI
(D) Overweight

69. What is the organism which has developed a nutritional requirement through mutation, called?

(A) Diploid
(B) Mutagen
(C) Auxotroph
(D) Prototroph

70. Which of the following is true for carcinogenic screening ?

(A) Widal test
(B) Ames test
(C) Enterotube multi test
(D) ELISA test

71. Which of the following is true to an **episome**?

(A) Plasmid having to ability to integrate into host chromosomes
(B) An ineffective plasmid
(C) Outer part of the chromosome
(D) A cell without a plasmid.

72. Which of the following is correct for the 'flat sour spoilage'?

(A) The can becomes swollen by gas production.
(B) The spoilage occurs due to thermophilic microorganisms but the can is not swollen by gas production.
(C) The gas leaks outside.
(D) The can is swollen at the bottom.

73. Which of the following is correct for pasteurization (HTST)?

(A) Adding sorbic acid
(B) 72°C for 15 seconds
(C) 5°C
(D) 121°C for 15 minutes.

74. Which is true for functions of nitrites?

(A) Preserve red colour of meat and prevent germination of endospores of botulism-causing-pathogen.
(B) Provides green colour to meat.
(C) Gives red stain to the meat.
(D) Decolourises the meat and makes it tasty.

75. Which of the following is the correct function of geosmin?

(A) It makes the soil hard.
(B) It makes the soil soft.
(C) It provides characteristic musty odour to the soil.
(D) It changes the colour of the soil.

76. Which of the following is a **recalcitrant** ?

(A) Glucose (B) DDT (C) Cellulose (D) Starch

77. Which of the following is true for **Petroff-Hauser Chamber**?

(A) It is used to catch rats.
(B) It is used to trap insects.
(C) It is used to count total cell number.
(D) It is used to culture viruses.

78. A **diazotroph** is known for the organism :

(A) That release SO_2 (B) That fix up SO_2
(C) That fix up nitrogen (D) That fix up CO_2

79. What for the **biosensors** are used?

(A) To survey the number of bacteria.
(B) To locate biological active pollutants.
(C) To locate the presence of enemy in the war.
(D) To inactivate viruses.

80. Which of the following bacterial species is used as a **larvicide**?

(A) *Escherichia coli* (B) *Rhizobium*
(C) *Bacillus propillae* (D) *Thiobacillus*

81. Which of the following can be used in recovery of uranium and copper ores?

(A) *Azotobacter* (B) *Nostoc*
(C) *Thiobacillus ferroxidans* (D) *Nitrobacter*

82. Which of the following is correct regarding elements that depend on biogeochemical cycle?

(A) Uranium and Copper (B) Carbon and nitrogen
(C) Nitrogen and gold (D) Silver and phosphorus

83. Which of the following is a correct symbiont of cyanobacteria?

(A) *Marchantia* a bryophyte
(B) *Pinus* a gymnosperm
(C) *Anthoceros* a bryophyte and *Azolla* a floating-fern
(D) *Selaginella* a pteridophyte.

84. Which is the correct raw material for preparation of rum?

(A) Orange (B) Cane molasses (C) Grapes (D) Barley

85. Which of the following is correct for origin of Dairy microbiology in India?

(A) Karnal (B) Bangalore (C) Calcutta (D) Meerut

86. Which of the following is infected by Mycophages?

(A) Algae (B) Fungi (C) Bacteriophage (D) TMV

87. Which of the following viruses have reverse transcriptase?

(A) Cyanophage (B) Bacteriophage (C) T.M.V. (D) H.I.V.

88. Who was the first to describe causative agent of anthrax?

(A) Louis Pasteur (B) H.T. Ricketts (C) Robert Koch (D) P.A. Sharp

89. India born scientist who shared Nobel prize with Robert Holley and Marshall Nirenberg in 1968.

(A) S.M. Paul Khurana (B) Jagdish Chander Bose
(C) Hargobind Khurana (D) B.P. Pal.

90. Who reported transformation in bacteria in 1928?

(A) Fred Griffith (B) Kary B. Mulis
(C) Edmond H. Fisher (D) Paul Berg

91. Which colour-light gives best clarity under microscope out of the following?
(A) Red (B) Green (C) Blue (D) Yellow

92. Which of the following is used to see *Mycobacterium tuberculosis*?
(A) Dark-field Microscope (B) Ordinary Microscope
(C). Hand Lens (D) Fluorescent Microscope.

93. Which of the following is used in negative staining?
(A) Safranin (B) Fast green (C) Nigrosin (D) Acid-water

94. Which of the following is used as counter stain in Gram staining?
(A) Potassium Iodide (B) Mordant
(C) Crystal violet (D) Safranin

95. Which of the following is noncellular ?
(A) Bacterium (B) Archaeobacterium
(C) Fungus (D) Virus

96. Which of the following does not have a nucleus ?
(A) Bacterium (B) Fungus (C) Protozoa (D) Plant cell

97. Which of the following causes malaria ?
(A) *Entamoeba histolytica* (B) *Plasmodium vivax*
(C) *Balantidium coli* (D) *Trypanosoma gambiense*

98. What is the number of ascospores in a typical ascus?
(A) 6 (B) 12 (C) 8 (D) 18

99. Which of the following has multi-celled conidia?
(A) *Cystopus* (B) *Aspergillus* (C) *Alternaria* (D) *Rhizopus*

100. The grapes are crushed to bring out juice in production of wine. What is that called out of the following?
(A) Whey (B) Must (C) Wort (D) Malting

101. Which of the following has been isolated from the well known 'sourdough bread' which is 8 to 10 times more acidic ?
(A) *Salmonella* and *Rhizobium*
(B) *Penicillium* and *Escherichia coli*
(C) *Lactobacillus sanfrancisco* and *Saccharmomyces exigus*
(D) *Rhizopus* and *Mucor*

102. Who amongst the following discovered the cause of tuberculosis ?
(A) Robert Hooke (B) Alexander Fleming
(C) Robert Koch (C) Paul Ehrlich

103. Which of the following was the supporter of the wrong belief of **'spontaneous generation'** or theory of **biogenesis**?
(A) Robert Hooke (B) John Needham
(C) J.G. Cappuccino (D) Rene Dubos

104. Which of the following period is called **'golden age of microbiology'**?
(A) 1761 to 1811 (B) 1800 to 1850
(C) 1889 to 1939 (D) 1860 to 1910

105. What is the role of phase ring on the phase plate of **Zernike microscope**?
(A) Stops the passage of light rays.
(B) Stops the passage of green and red light only.
(C) Advances or retards the light rays 1/4 th of the wavelength
(D) Bears no effect.

106. Which of the following classified streptococci into serotypes in 1933?
(A) D. Bruce (B) Rebecca Lancefield
(C) A. Fraenkel (D) W. Reed

107. What is produced by *Bacillus subtitis* on fermentation of glucose, lactose and sucrose broths:
(A) Gas only (B) Acid only
(C) Both acid and gas (D) Neither acid nor gas

108. Which of the following is the new name for Common Wealth Mycological Institute (CMI) at Kew?
(A) Mycological Institute of USA (B) Mycological Institute of Sri Lanka
(C) Mycological Institute of Bangladesh (D) International Mycological Institute.

109. Which of the following bacterial species use hydrogen sulphide as source of energy?
(A) *Escherichia coli* (B) *Beggiatoa* and *Thiothrix*
(C) *Nitrosomonas* (D) *Nitrosococcus.*

110. Which of the following bacterial species form a pellicle of cellulose around them?
(A) *Salmonella typhi*
(B) *Lactococcus* and *Streptococcus*
(C) *Acetobacter xylinum* and *Sarcina ventriculi*
(D) *Escherichia coli.*

111. Which of the following is used as immersion oil for 100 X lens of light microscope?
(A) Mustard oil (B) Mineral oil
(C) Cedar wood oil (D) Refined oil

112. Whose results of first case of small pox were sent for publication in 'Transactions of Royal Society' and were rejected consequently, then he collected 23 cases and published his research in the form of a book.
(A) Louis Pasteur (B) K. Shiga
(C) Leeuwenhoek (D) Edward Jenner

113. What makes *Deinococcus* and *Deinobacter* radiation resistant bacteria?
(A) They have lipoproteins and lipopolysaccharides
(B) They have repair mechanism for damaged DNA and high concentration of carotenoids
(C) They have high concentration of sugars
(D) They are devoid of carotenoids

114. How does *Aquifex* differ from bacteria?
(A) They have scanty phospholipids
(B) They have glycerol diethers instead of phospholipids
(C) They have phospholipids instead of glycerol diethers
(D) They are endosymbionts.

115. Which of the following is used as a biofertilizer ?
(A) *Rhizopus* (B) *Aspergillus*
(C) Blue-green algae (BGA) (C) Red algae

116. Which of the following are archeobacteria and form "postage stam-like sheets of cells" in Red Sea?

(A) *Nostoc* (B) *Anabaena*
(C) *Planomonospora* (D) *Halobacterium*

117. How are **mesosomes** in bacteria artifacts and not true structures?

(A) They are not visible under light microscope
(B) They are not visible under phase contrast microscope
(C) They are not visible under dark field microscope
(D) They are not seen in electron micrographs requiring no dehydration.

118. What are moulds?

(A) Single celled algae and fungi (B) Single celled fungi
(C) Filamentous algae (D) Filamentous fungi

119. What is the name given to the plasmid which produces proteins that inhibits the growth of susceptible bacterial strains?

(A) R-plasmid (B) Col plasmid
(C) Ti-plasmid (D) PBR-322.

120. Which of the following is for procaryotic ribosomes?

(A) 16 S (B) 80 S (C) 70 S (D) 30 S

121. What is a fully developed virus particle called?

(A) Envelope (B) Capsid (C) Capsomere (D) Virion

122. Which is the correct number of groups in 'Bergey's Manual of Determinative Bacteriology' published in 1994?

(A) 19 (B) 33 (C) 35 (D) 40

123. Which of the following cause 'Mad cow disease'?

(A) Prions (B) Rickettsia (C) Chlamydia (D) None of the above

124. A positive strand DNA virus is :

(A) With DNA genome with minus configuration
(B) With a single stranded DNA of plus configuration
(C) DNA without any configuration
(D) With double stranded DNA with plus configuration.

125. Which of the following enzyme converts hydrogen peroxide into water and oxygen?

(A) Amylase (B) Lipase (C) Catalase (D) Hydrogenase

126. Extreme halophilic bacteria isolated from saline waters of dead sea require for normal growth:

(A) 05 percent salt (B) 15 percent salt
(C) 30 percent salt (D) 60 percent salt

127. At what temperature microorganisms are frozen rapidly in lyophilization?

(A) 0 to 5°C (B) 5 to 10°C
(C) – 54°C to – 72°C (D) 10 to 15°C

128. What is, the agar solidified in a vertical tube, called?

(A) Slant (B) Deep (C) Broth (D) None of the above

129. Which is the correct name given for bald circular patches on the head of children called?

(A) Tinea cruris (B) Tinea capitis
(C) Tinea imbricata (D) Tinea barbae

130. Which of the following form suitable substratum for the growth of dermatophytes?
(A) Sugars (B) Oil (C) Fat (D) Keratin

131. Which of the following grows rapidly on alkaline egg peptone?
(A) *Cyanobacteria* (B) *Azotobacter*
(C) *Vibrio cholerae* (D) *Rhizobium*

132. Which of the following is caused by the bacterium *Yersinia pestis*?
(A) Bubonic or pneumonic plague (B) Tetanus
(C) Yellow fever (D) Influenza

133. Which of the following stands for the function of mRNA produced in transcription?
(A) Carries the sequence which encode for amino acid sequence of one or more polypeptides specified by gene or set of genes in the chromosome.
(B) Works as an adapter which reads the transformation encoded in the mRNA and transfers the appropriate amino acid to growing polypeptide chain during protein synthesis.
(C) Associates with the proteins to form the intricate protein machine called ribosome
(D) Work as specialized RNAs with regulatory or catalytic functions

134. In which direction transcription bubble moves in transcription by RNA polymerase in *E.coli*.
(A) Transcription bubble moves from right to left.
(B) Transcription bubble moves from left to right keeping pace with RNA synthesis
(C) Transcription bubble does not move
(D) No positive supercoil is formed ahead of the transcription bubble and no negative super coils are formed behind the transcription bubble.

135. Which of the following is meant for **consensus sequence** of – 10 region.
(A) (5′) TTGACA (3′)
(B) Generally called Pribnow box after David Pribnow who first recognised it in 1975 is (5′) TATAAT (3′)
(C) – 10 region sequence reads from 3′ to 5′ right to left
(D) All sequences read 5′ ⟶ 3′ left to right.

136. RNA polymerase leaves its **foot print** on a promoter. What is foot printing in view of the following ?
(A) A technique derived from principles used in DNA sequencing.
(B) Spoting transcription bubble
(C) Biological inter conversion of DNA and RNA
(D) None of the above.

137. In the replication of *Escherichia coli* chromosome the regulatory apparatus for the synthesis of lagging strand is a travelling protein machine known as :
(A) Okazaki fragments (B) Fork movement
(C) Primosome (D) Origin

138. One of the key features of homologous genetic recombination is, one strand of each DNA is broken and joined to the other to form a crossover structure which is called?
(A) Alignment of homologous DNAs
(B) Holliday intermediate
(C) Heteroduplex DNA
(D) Error-prone repair

139. The DNA in the chromatin of all eucaryotic cells is very tightly associated with proteins called **histones** which package and order the DNA into structural units known as :

(A) Nucleoid
(B) Nucleosomes or histone complexes bound to DNA
(C) Plasmid
(D) Linker DNA

140. The DNA polymerases in *Escherichia coli* are

(A) DNA polymerase I
(B) DNA polymerase II
(C) DNA polymerase III
(D) DNA polymerase I, DNA polymerase II and DNA polymerase III.

141. Pertussis toxin proudced by *Bordetella pertussis* catalyzes ADP–ribosylation of GI blocking inhibition of adenylate cyclase by Gi which lead to :

(A) Blockage of acetylcholine receptor causing paralysis and death
(B) Block neurotransmission by preventing the opening of Na^+ channels
(C) Whooping cough, including hypersensitivity to histamines and lowered blood glucose.
(D) Adenocarcinomas (cancers) of the breast, stomach and ovary

142. The contour length of DNA of *Escherichia coli* has been estimated to be:

(A) 1.7 mm almost 850 times the length of the cell of *Escherichia coli*
(B) 17.5 μm
(C) 210 nm
(D) 25 nm

143. Adipose tissue whic consists of **adipocyter** (fat cells) releases fatty acids through the action of:

(A) Thyrotropin releasing hormone (TRH)
(B) Antidiuretic horomone (ADH)
(C) Prostaglandins
(D) Hormone epinephrine, as low levels of glucose in the blood triger the mobilization of triacylglycerols through the action of this hormone.

144. Cholera toxin secreted by *Vibrio cholerae* result in dehydration and electrolyte loss because:

(A) Continuous activation of the adenylate cyclase of intestinal epithelial cells and resultant high concentration of CAMP triggers continual secretion of Cl^-, HCO_3^- and water into intestinal lumen
(B) Activation of calcium binding protein calmodulin take place
(C) Ion channels are gated by ligands and resultant muscle contraction
(D) Nicotinic acetylcholine receptor responds to neurotransmitter acetylcholine

145. Who do bacteria in root nodules may fix hundreds of times more nitrogen than their free-living cousins under conditions generally encountered in soils.

(A) Predatory organisms cannot eat them
(B) They have assess to carbohydrates made by the plant and in the root nodules are bathed in a solution of an oxygen binding protein called **leghaemoglobin**
(C) The nitrogenase produced by the plant converts ammonium (NH_4^+) into atmopheric nitrogen (N_2)
(D) None of the above.

146. Nitrogen is fixed by nitrogen fixing cyanobacteria, free-soil dweller *Azotobacter* and symbiont *Rhizobium* due to :

(A) A complex protein called the nitrogenase complex
(B) The receptor mediated endocytosis
(C) A high-density lipoprotein (HDL)
(D) An enzyme lecithin-cholesterol acyl transferase (LCAT)

147. What differentiates cytochrome P – 450 usually present in the smooth endoplasmic reticulum from mitochondrial cytochrome oxidase?

(A) Cytochrome P - 450 reacts with O_2 and carbon monoxide but it can be differentiated from cytochrome oxidase because carbon monoxide complex of its reduced form absorbs light strongly at 450 nm, hence the name P - 450.
(B) Cytochrome P - 450 cannot catalyze hydroxilation reactions
(C) Cytochrome P - 450 cannot catalyze hydroxilation of steroids to yield the adrenocortical hormones.

148. Arachidonate is a parent to eicosanoids, a family of very potent biological signalling molecules which act as short-range messengers :

(A) Enzymes of smooth endoplasmic reticulum convert arachidonate into prostaglandins beginning with the formation of PGH_2 the intermediate precursor of many other prostaglandins and thromboxanes.
(B) Without involving prostaglandin endoperoxide synthase PGH_2 is formed
(C) Thromboxane synthase present in blood platelets cannot convert PGH_2 into thromboxane A_2
(D) Aspirin does not inhibit prostaglandin endoperoxide synthase by acetylating an essential Ser residue on the enzyme.

149. Which of the following is called **retrotransposon** from an yeast?

(A) Ty element
(B) Copia element
(C) Primordial exons
(D) LDL receptor and ribozymes

150. Which of the following mean for **termination codons** also called **stop codons** or **nonsense codons**

(A) AUG
(B) AUG, UAA, UAG and UGA
(C) UAA, UAG and UGA
(D) None of the above.

KEY

1. B	2. B	3. D	4. C
5. A	6. B	7. A	8. C
9. D	10. C	11. C	12. A
13. A	14. D	15. A	16. A
17. C	18. C	19. B	20. D
21. B	22. C	23. A	24. C
25. C	26. B	27. A	28. A
29. C	30. A	31. A	32. C
33. A	34. B	35. B	36. A

37. C	38. A	39. C	40. A
41. D	42. B	43. C	44. B
45. C	46. D	47. A	48. D
49. C	50. B	51. D	52. C
53. B	54. B	55. B	56. B
57. D	58. B	59. A	60. B
61. D	62. B	63. C	64. D
65. B	66. A	67. A	68. B
69. C	70. B	71. A	72. B
73. B	74. A	75. C	76. B
77. C	78. C	79. B	80. C
81. C	82. B	83. C	84. B
85. B	86. B	87. D	88. C
89. C	90. A	91. C	92. D
93. C	94. D	95. D	96. A
97. B	98. C	99. C	100. B
101. C	102. C	103. B	104. D
105. C	106. B	107. C	108. D
109. B	110. C	111. C	112. D
113. B	114. B	115. C	116. D
117. D	118. D	119. B	120. C
121. D	122. C	123. A	124. B
125. C	126. C	127. C	128. B
129. B	130. D	131. C	132. A
133. A	134. B	135. B	136. A
137. C	138. B	139. B	140. D
141. C	142. A	143. D	144. A
145. B	146. A	147. A	148. A
149. A	150. C		

SUGGESTED READINGS

Alberts, B., Bray, D., Lewis, J., Raff, M., Roberts, K. and Watson, J.D. 1994. Molecular Biology of the cell. 3rd edition. Garland Publishing, Inc, New York. PP XLIII+ 1– 1294 + Glossary and Index.

Anonymous, 1981. Standard methods for the Examination of Water and Wastewater. APHA, Washington. pp I-XLVII + 1–1134.

Anonymous, 2003. HIV testing Manual, Laboratory Diagnonosis, Biosafety and Quality Control. National AIDS Control Organization, Ministry of Health and Family Welfare, Govt. of India, New Delhi.

Arnold, H.L., Odom, R.B. and James, W.D. 1990. Andrew's diseases of the skin: Clinical dermatology. W.B. Saunders Company, Philadelphia. pp. 1–1062.

Atlas, R.M. 1986. Basic and practical microbiology, Macmillan Publishing Co., New York. pp. 741 + G-51 + 1–23.

Atlas, R.M. 1997. Principles of Microbiology. Wm C. Brown Publishers, Towa.

Atlas, R.M. and Bartha, R. 1998. (Indian Reprint). Microbial Ecology : Fundamentals and Applications. Benjamin/Cummings Science Publishing, California, pp. 1 – 694.

Austyn, J.M. and Wood, K.J. 1993. Principles of molecular immunology. Oxford University Press, Oxford. pp. 1–735.

Bakshi, R.K. 1974. Mycorrhiza and its role in Forestry. F.R.I., Dehra Dun, pp. 189 + Plates 12.

Black, Jacqulyn, G. 2008. Microbiology : Principles and Explorations. John Wiley & Sons (Asca) Pvt. Ltd., Haboken, N.J., pp. 1 – 846 + Appendices.

Behl, P.N. 1972. Practice of Dermatology. Oxford Printcraft India, Pvt. Ltd. In association with Arnold-Heinernann India, London. pp. 1–528

Brown, T.A. 2001. Gene Cloning and DNA Analysis : An Introduction. Blackwell Science. pp. 1 – 303.

Bruce, A.M. and Hawkes, H.A. 1983. Biological Filters. *In* A.L. Dawning. Used Water Treatment. Vol. 3. Academic Press.

Cappuccino, J. and Sherman, N. 1987. Microbiology: a laboratory manual. The Benjamin Cummings Publishing Company Inc., California.

Casida, L.E. (Jr) 1968. Industrial Microbiology. John Wiley and Sons, Inc. New York. pp. 1–460.

Collee, J.G., Fraser, A.G., Marmion, B.P. and Simmons, A (Editiors). 2014 (Reprint) 14th edition. Mackie & McCartney Practical Medical Microbiology. Churchil Livingstone, An Imprint of Elsevier.

Cruckshank, R., Duguid, J.P., Marmion, B.P. and Swain, R.H.A. 1975. (Ed.) Medical Microbiology. 12th edition. Vol. 2: The practice of medical microbiology. pp. 1–587

Dazzo, F.B. and Hubbel, D.H. 1975. Cross reactive antigens and lectins as determinants of symbiotic specificity in *Rhizobium*-clover association. Appl. Microbio. 30: 101.

Deb, W.C. 1982. Microbes and diseases of man: a text book of microbiology (including Helminthology). CBS Publishers and Distributors, Delhi. pp. 1514.

Deshmukh, S.K. Jain, P.C. and Agrawal, S.C. 186. A note on mycotoxicity of some essential oils. FITOTERAPIA 57 (4): 295-297.

Devlin, T.M. 1997. Textbook of Biochemistry with clinical Correlations. Wiley–Liss (A John Wiley & Sons Inc Publications), New York, pp. 1–1186.

Dey, T.K. 1974. A text book of bacteriology. Prof. Encee Dey-Publishing House, Calcutta. pp. 1-358.

Diener, T.O. 1987. The Viroids. Plenum Press, New York. pp XXI + 344.

DuBow, M.S. (Ed.) 1981. Bacteriophage Assembly. Alan R. Liss Inc., New York. pp 1-554.

DuQuense, T. 1987. Essential Drugs: A Common Formulatory. Heterodox, London. pp. 1-338.

Emmons, C.W., Bindord, C.H., Utz, J.P. and Know—Chung, K.J. 1977. Medical Mycology. Lea & Febriger, Philadelphia.

Finstein, M.S. 1972. Pollution Microbiology, a laboratory manual. Marcel Dekker, Inc., New York. pp 1-168.

Frazier, W.C. and Westhoff, D.C. 1992. Food Microbiology. Tata McGraw-Hill Publishing Company Limited New Delhi, pp 1-540. (A reprint of 1979 edition of McGraw-Hill Inc.)

Freifelder, D. 1990. Microbial Genetics. Narosa Publishing House, New Delhi, pp. 1-601. (A reprint of 1987 edition published by Jones and Barlett Publishers Inc. USA).

Frobisher, M., Hinsdill, R.D., Crabtree, K.T. and Goodheart, C.R. 1974. Fundamentals of Microbiology, 9th edition. W.B. Saunders Company, Philadelphia. pp. 1-850.

Goldsby, R.A., Kindt, T.J. and Osborne, B.A. 2001. Kuby Immunology. W.H. Freeman and Co., New York.

Goodwin, T.W. and Mercer, E.1. 1983. Introduction to Plant Biochemistry. Pergamon Press, Oxford. pp. 1–675.

Goyal, P., Chauhan, A. and Kaushik, P. 2015. Plants and Microbes : An Innovative Approach. SBW Publishers, Delhi, pp X + 1–452.

Hall, F.R. and Menn, J.J. (Ed.) 1999. Biopesticides : Use and Delivery. Humana Press Inc, Totowa, New Jersey. pp. 1–626.

Hammer, M.J. 1986. Water and Wasterwater Technology. S.I. Version. John Wiley & Sons, New York. pp. 1-536.

Harashima, K., Shiba, T. and Murata, N. 1989. Aerobic photosynthetic bacteria. Springerverlag, Berlin. pp. 169, 48 figures, 19 plates.

Harley, G.J.H. and Smith, S.E. 1983. Mycorrhizal symbiosis. Academic Press. London.

Hawker, L.E. and Linton, A.H. 1979. Microorganisms, form and environment. Edward Arnold (Publishers). Ltd., London. pp. 1-391.

Hayes, W. 1984. The Genetics of bacteria and their viruses. CBS Publishers & Distributors, Delhi. pp. 1-925.

Hoffbrand, A.V., Lewis, S.W. and Tuddenham, E.G.D. 1999. Postgraduate Haemotology. Butterworth-Heinemann, Linacre House, Jordan Hill, Oxford Ox2 8 DP, UK.

Hoiberg, D.H. 2001. Hundred Years with Nobel Laureates. Hundred special Articles written by Nobel Laureates exclusively for Encyclopaedia Britanica. Encyclopaedia Britanica (India) Pvt. Ltd., New Delhi and 1.K. International Pvt. Ltd., New Delhi, pp. 1 – 1098.

Hunter, G.W., Frye, W.W. and Swartzwelder, J.C. 1965. (Ed.) Manual of Tropical Medicine. W.B. Saunders Co. Philadelphia; and Charles E. Tuttle Co., Tokyo. pp. 1-892.

Ingraham, J.L. and Ingraham, C.A. 2000, Introduction to Microbiology. Brooks/ Cole Thomson Learning, 511 Foret Lodge Road, Pacific Groove, CA, USA pp xii + 1–804 + Appendices.

Jain, P.C. and Agrawal, S.C. 1978. Notes on the activity of some odoriferous organic compounds against some keratinophilic fungi. Trans. Mycol. Soc. Japan 19: 197-200.

Jensen, M.M. and Wright, D.N. 1989, Introduction to Microbiology for the Health Sciences. Prentice

- Hall International (UK) Ltd, London, pp XV + 1 – 704.

Joklik, W.K. 1976. Zinsser Microbiology. 16th edition. Appleton-Century Crofts. A Publication Division of Prentice-Hall Inc., New York.

Kasper, L.H. and Buzoni - Gatel, D. 1998. Some opportunistic Parasitic Infections in AIDS: Candidiasis, Pneumocystosis, Cryptosporidiosis and Toxoplasmosis. Parasitology Today, 14(4) : 150–156.

Kaul, T.N. 1983. Cultivated edible mushrooms. RRL, Jammu (Tawi) pp. 1-56.

Kaushik, P. 1983. Ecological and Anatomical Marvels of the Himalayan Orchids. Today and Tommorrow's Printers & Publishers, New Delhi. pp. viii + 123 + plates 71.

Kaushik, P. 1988. Indigenous Medicinal Plants, including microbes and fungi. Today and Tommorrow's Printers and Publishers, New Delhi. pp. 1-243.

Kaushik, P. 1988. Mycorrhizae in Forest Ecosystem. Arya Bhatt, 7(1): 39-47.

Kaushik, P. 1998. Mycorrhiza and Orchid Conservation, pp. 141-149. In I.S. Grover and A.K. Thukral (Eds.). Environment and Development. Scientific Publishers, Jodhpur.

Kaushik, P. 2000. Introductory Microbiology. 2nd edition. Emkay Publications, Post Box 9410, B-19 East Krishna Nagar, Swami Dayanand Marg, Delhi 51. pp. XII + 466.

Kaushik, P. 2003. Haridra (Turmeric) : Antibacterial Potential. Chowkhamba Sanskrit Series office. K37/99 Gopal Mandir Lane, P.O. Box 1008, Varanasi, pp. 1 – 124, Coloured Pl. 16.

Kaushik, P. 2006. Cytology, Genetics and Evolution. Faculty of Distance Education. Gurukul Kangri University, Hardwar. pp. 1 – 94.

Kaushik, P. 2009, 2nd Enlarged Edition. Indigenous Medicinal Plants including Microbes and Fungi. Today and Tommorrow's Printers & Publishers 4436/7 Ansari Road, Darya Ganj, New Delhi, pp. 1 – 664.

Kaushik, P. and Dhiman, A.K. 2000. Medicinal Plants and Raw Drugs of India. Bishen Singh Mahendra Pal Singh. 23-A, Connaught Place, Dehra Dun. pp. XII + 623 including colour plates.

Kaushik P. and Chauhan, A. 2009. Cyanobacteria : Antibacterial Activity, New India Publishing Agency, 101 Vikas Surya Plaza, CU Block, LSC Mkt. Pitam Pura, New Delhi.

Kaushik P. and Goyal, P. 2011. Medicinal Plants : Antibacterial Potential. Today and Tomorrow's Printers and Publishers, New Delhi.

Khurana, S.M. Paul. 1992. Potato viruses and viral diseases. Central Potato Research Institute (ICAR) Shimal (India) Technical Bulletin, 35:1-23.

Khurana, S.M. Paul, Singh R.A. and Kaley, D.M. 1988. Mycoplasma-associated diseases and their control in India. pp. 285-316. *In* K. Maramosch and S.P. Raychaudhuri (Eds). Mycoplasma disease of crops; basic and applied aspects. Springer-Verlag, New York.

Kumar, H.D. and Rai, L.C. 1976. Microbes and Microbial Processes, Affiliated East-West Press Pvt. Ltd., New Delhi.

Larone, D.H. 1993. Medically important fungi: a guide to identification. American Society for Microbiology, Washington D.C., pp. 1-230.

Lehninger, A.L., Nelson, D.L., and Cox, M.M. 1993. Principles of Biochemistry. CBS Publishers & Distributors. Delhi. pp. XLI + 1 - 1013. + Appendix.

Lemke, P.A. 1979. Viruses and Plasmids in Fungi. Marcel Dekker Inc., New York. pp. xi + 653.

Loutit, M.W. and Miles, J.A.R. 1978. Microbial Ecology. Springer Verlag, Berlin.

Madigan, M.T., Martinko, J.M. and Parker, J. 1997. Brock Biology of Microorganisms. Prentice Hall International, New Jersey, USA.

Mandahar, C.L. 1989. Plant Viruses, Vol. I. Structure and Replication. C.R.C. Press, USA.

Mandahar, C.L. 1990. Plant Viruses, Vol. II. Pathology. C.R.C. Press, USA.

Marshall, K.C. 1986. Advances in Microbial Ecology. Vol. 9. Plenum Press, New York. pp. 1-402.

Maulik, S. and Patel, S. 1997. Molecular Biochemistry : Therapeutic Applications and Strategies. Wiley – Liss (A John Wiley & Sons Inc Publications), New York. pp. 1–222.

Mehta, S.L. Lodha, M.L. and Sane, P.V. 1989. Recent advances in plant biochemistry. Publication and Information Division, ICAR Krishi Anusandhan Bhavan, Pusa, New Delhi, pp. 1-471.

Meyers, R.A. 1995. Molecular Biology and Biotechnology : A Comprehensive Desk Reference. Wiley – VCH, New York, pp. 1–1034.

Mishra, M.M. and Kapoor, K.K. 1986. (Ed.) Soil Biology. Deptt of Microbiology H.A.U. Hisar. pp. 1-275.

Mukherjee, K.L. 1988. Medical Laboratory Technology, A procedure Manual for Routine Diagnostic Tests. Vol. II. Tata McGraw Hill Publishing Company Ltd., 7 West Patel Nagar, New Delhi - 110 008. DP 1-891.

The book is in three volumes—

Vol. I Section I to III, Introduction Haematology, and Blood Banking

Vol. II Contains sections IV to VI, Microbiology, Serology and Clinical Pathology

Vol. III Sections VII to IX, Clinical Biochemistry, Histology, Cytology and Miscellaneous topics.

Oginsky, E.L. and Urnbreit, W.W. 1959. An introduction to bacterial physiology. W.H. Freeman and Company, San Francisco. pp. 1-442.

Onions, A.H.S., Allsopp. D. and Eggins, H.O.W. 1981. Smith's Introduction to Industrial Mycology. Edward Arnold, London. pp. 1-398.

Ormerod, J.C. 1983. the Phototrophic Bacteria: anaerobic life in light. Blackwell Scientific Publications, Oxford. pp. 1-246.

Osborne, D.J. and Jackson,M.B. 1989. Cell separation in plants : Physiology, Biochemistry and Molecular Biology. Springer verlag. pp 1-449.

Pelczar, M.J., Chan, E.C.S. and Krieg, N.R. 1993. Microbiology: Concepts and applications. McGraw Hill Inc., New York. pp. 1-896 + Appendix.

Pharis, R.P. and Rood (Eds) 1990. Plant Growth substances. Springer-Verlag. pp. 1-633.

Prescott, S.C. and Dunn, C.G. 1959. Industrial Microbiology. (III edition Revised by Cecil Gordon Dunn). McGraw-Hill Book Company, Inc., New York. pp. 1-945.

Prescott, L.M., Harley, J.P. and Klein, D.A. 2005. Microbiology. Mc Graw - Hill, Singapore, pp XXII + 1 – 992 + Appendices.

Rawn, J.D. 1989. Biochemistry. Harper & Row, Publishers, New York. pp. I1105.

Rao, C.V. 2002. An Introduction to Immunology. Narora Publishing House, 22 Daryaganj, Delhi Medical Association Road, New Delhi - 110 002.

Reed. G. (Ed.) 1982. Prescott & Dunn's Industrial Microbiology. 4th Edition. The AVI Publishing Company Inc. Westport, Connecticut. USA.

Rose, A.H. 1993. Advances in Microbial Physiology. Vol. 35. Academic Press, London. pp. 1-342.

Pirozynski, K.A. and Hawksworth, D.L. 1988. Co-evolution of fungi with plants and animals. Academic Press, Harcourt Brace Jovanovich, Publishers, London. pp. 1-285.

Ross, F.C. 1983. Introductory Microbiology. Charles E. Merrill Publishing Co. (A Bell & Howell Co.), Columbus, Ohio. pp. 1-615.

Salle, A.J. 1985. Fundamental Principles of Bacteriology. Tata McGraw-Hill Publishing Company Ltd., New Delhi. (Reprint of 1974 edition).

Samiy, A.H., Smith, L.H. and Wyngaasden, J.B. (Eds.)? Medical Microbiology and infectious diseases. Saunders, pp. 1-1935.

Sankaran, R. and Manja, K.S. (Eds.) 1995. Microbes for Better Living. Conference Secretariat, Defence Food Research Laboratory, Mysore. pp. 1-653.

Schlaman, H.R.M. 1992. Regulation of nodulation gene expression by Nod-D in *Rhizobium*. Journal of Bacteriology. August: 5177-5182.

Singh, R.S. 1984. Introduction to Principles of Plant Pathology (Third edition). Oxford & IBH Publishing Co. Pvt. Ltd., New Delhi.

Singleton, P and Sainsuburg, D. 1988. Dictionary of Microbiology and Molecular Biology. Second edition. John Wiley & Sons. Singapore. pp. 1-XII + 11019

Spencer, J.F.T. and Spencer, D.N. 1990. Yeast Technology, Springer-Verlag, Berlin. pp. 1-407.

Stanier, R.Y., Adelberg, E.A. and Ingraham, J.L. 1984. (Reprint) of 1977's fourth edition General Microbiology. The Mac Millan Ltd., London. pp. 1-871.

Mehrotra, M.D. 1991. Mycorrhizae of Indian Forest Trees. I.C.F.R.E. Dehra Dun. pp. 1-294.

Stanbury, P.F. and Whitaker, A. 1984. Principles of Fermentation Technology. Pergamon Press, Oxford. pp. 1-255.

Stryer, L. 1986. Biochemistry. CBS Publishers and Distributors, Delhi. pp. 1-949.

Subba Rao, N.S. 1989. (Reprint) Soil Microorganisms and Plant Growth. Oxford & IBH Publishing Co., Pvt, Ltd., New Delhi. pp. 1-314

Subba Rao, N.S. 1995. (A reprint of 1993 Edition). Biofertilizers in Agriculture and Forestry. Oxford and IBH Publishing Co., Pvt. Ltd., New Delhi.

Tortore, G.T., Funke, B.R. and Case, C.L. 1995. Microbiology, an introduction. 5th Edition. The Benjamin/Cummings Publishing Co. Inc., California. pp. 1-801.

Turner, W.B. 1971. Fungal Metabolities. Academic Press, London.

Ullrich, W.R., Aparicio, P.J., Syrett, P.J. and Castillo, F. 1987. Inorganic nitrogen metabolism. Springer-Verlag, Berlin. pp. 1-295.

Umbreit, W.W. (Ed.) 1965. Applied Microbiology. Vol. 7. Academic Press, New York. pp. 1-402.

Vially, R. (Ed.) 1992. Bacterial gas. Editions Technip Paris (France) pp. 1-242.

Watson, J.D., Baker, T.A., Bell, S.P., Gann, A., Levine, M. and Losick, R. 2004. Molecular Biology of the Gene. 5th edition. Pearson Education (Singapore) Pvt. Ltd., Indian Branch, 482 F. I. E. Patparganj, Delhi. pp. 1–XXVII + 1 – 732.

Watson, J.D., Hopkins, N.H.m Roberts, JW., Steitz, J.A. and Weiner, A.M. 1987. Molecular Biology of the Gene. 4th edition. The Bengamin/Cummings Publishing Co, Inc., California. pp. 1-1163 Index 1-26.

Weeks, B.S. and Alcamo, I.E. 206. AIDS: The Biological Basis. 4th edition, Jones and Bartlett Publishers, 40 Tall Pine Drive, Sudbury MA 01776, USA.

Weimer, D.B. (Ed.) 1992. Acquired immune deficiency syndrome. Karger, Basel (Switzerland). pp. 1-252.

West, E.S., Todd, W.R., Mason, H.S. and Bruggen, J.T.V. 1974. Textbook of Biochemistry. Oxford & IBH Publishing Co. Pvt. Ltd., New Delhi. pp. 1-1595.

Wilson, (Sir) G.S. and Miles, (Sir) A. 1975. Topley and Wilson's Principles of Bacteriology, Virology and Immunity. Vol. I, pp. 1-1247 + Index LVI, Vol. 2, pp. 1249-2706 + Index LVI. Adward Arnold Publishers Ltd., London.

Wistreich, G.A. and Lechtman, M.D. 1976. Microbiology and Human Disease, Glencose Press, California. pp. 1-905.

Yadav, J.S. Grover, S. and Batish, V.K. 1993. A Comprehensive Dairy Microbiology. Metropolitan Book Co Pvt. Ltd., New Delhi.

Zabel, R.A. 1992. Wood microbiology decay and its prevention. Academic Press Inc. Harcourt Brace Jovanovich Publishers, San Diego. pp. 1-476.

INDEX

B

C

D

E

F

G

H

I

P

Q

R

S

T

U

V

W

X

Y

Z